T5-AGX-448

Tectonic Evolution of the North Sea Rifts

Publication Number 181 of the
International Lithosphere Programme

Tectonic Evolution of the North Sea Rifts

Edited by

D. J. BLUNDELL and A. D. GIBBS

CLARENDON PRESS · OXFORD
1990

Oxford University Press, Walton Street, Oxford OX2 6DP
Oxford New York Toronto
Delhi Bombay Calcutta Madras Karachi
Petaling Jaya Singapore Hong Kong Tokyo
Nairobi Dar es Salaam Cape Town
Melbourne Auckland
and associated companies in
Berlin Ibadan

Oxford is a trade mark of Oxford University Press

Published in the United States
by Oxford University Press, New York

British Library Cataloguing in Publication Data
Tectonic evolution of the North Sea rifts.
1. North Sea. Natural gas deposits and petroleum deposits
I. Blundell, D. J. II. Gibbs, A. D.
553.280916336
ISBN 0–19–854595–9

Library of Congress Cataloging in Publication Data
Tectonic evolution of the North Sea rifts / edited by D. J. Blundell and A. D. Gibbs.
Includes bibliographical references.
1. Rift (Geology)—North Sea Region. 2. Plate tectonics—North Sea Region. 3. Geology—North Sea Region. 4. Earth—Crust.
I. Blundell, Derek J. (Derek John) II. Gibbs, A. D.
QE606.5.N75T43 1990 551.46′136—dc 20 90-7208
ISBN 0–19–854595–9

Typeset by Joshua Associates Ltd, Oxford
Printed in Great Britain by
Courier International Ltd, Tiptree, Essex

Preface

The idea for this book arose from the desire of Dr Peter Ziegler, as Chairman of the Inter-Union Commission on the Lithosphere, Working Group 3 that is responsible for intra-plate phenomena, to advance international understanding of rift systems. Mainly as a consequence of oil industry exploration, the present North Sea rift system is known in considerable detail; unravelling its past history, and understanding the mechanisms that controlled its evolution, is another matter. In the past decade there have been considerable advances in understanding. We believe these stem from a seminal paper published in 1978 by Professor Dan McKenzie on 'Some remarks on the development of sedimentary basins' which developed a simple model to give quantitative explanation to the evolution of a rift basin, through lithospheric stretching followed (as a consequence of the thermal disequilibrium caused by the resultant thinning of the lithosphere) by a superimposed sag basin as thermal equilibrium is regained, isostatic equilibrium having prevailed throughout. The central North Sea rift basin was one of the first tests of this model, and the study by Barton and Wood (1984) did much to confirm its applicability. The particular value of McKenzie's model is that, not only does it show how the whole lithosphere is involved in basin development and predict the sequence and rates of subsidence events, but it also provides a means of working out the subsidence and thermal histories of particular sedimentary formations within the basin, after due allowance for compaction (Sclater and Christie 1980). These provide a measure of the maturation history of source rocks that is so vital in the appraisal of hydrocarbon prospectivity.

However, in testing McKenzie's model, the amount of stretching—McKenzie's beta factor β—required to explain the observed subsidence and crustal thinning appeared to be too large when compared with the sum total of fault heave recorded on seismic sections across the North Sea rift. A number of geologists, Peter Ziegler foremost amongst them, began to question the model. Subsequently, a number of variations on McKenzie's theme have been produced to attempt to explain this and other discrepancies, with observations that have been found in various extensional basins around the world.

In a study of extensional basin development in the Basin and Range Province of western USA, the recognition of asymmetry as an essential feature—rather than the symmetry expected from McKenzie's model—led Wernicke (1985) to propose an alternative model for basins that develop from lithosphere extension along a major shear zone dipping at a low angle (perhaps 20° to 30°) through the whole lithosphere, and the consequent thermal and isostatic effects. Dubbed the pure-shear and simple-shear models respectively, they have been neatly compared by Buck *et al.* (1988), and effectively act as the end members to a range of possible models. However, the physical properties of the lithosphere from top to bottom are such that the mechanical behaviour of rocks at various levels of the crust and upper mantle in response to the horizontal tensile stress needed for stretching are likely to vary. Thus the nature of rifting is undoubtedly more complex than either the McKenzie or the Wernicke models would imply, since both treat the lithosphere as uniform.

A further consequence of the thermal disequilibrium arising from lithospheric stretching and thinning is the possibility of generating melts at or near its base which could rise to add magmatically to the crust, possibly reaching the surface as volcanics. McKenzie and Bickle (1988) introduced a new phase of understanding of this aspect of basin evolution quite recently by showing that, for given temperatures within the mantle, the amount of melt generated and its composition, relate directly to the amount of stretching, and that a critical amount of stretching could be necessary before any melt would be produced.

The North Sea rift system remains a key area to test the new ideas that are evolving from these various models. Since it is a major hydrocarbon province, it is also a key area in which to put the new ideas to profitable use. The structural development of the basins is a major factor in the delicate interplay of climate, sea level, and palaeogeography that influence sediment supply, transport, and depositional environments and then the burial, subsidence, and deformation that build the sedimentary architecture of the basin infill. The aim is to understand the processes well enough to give quantifiable predictions of hydrocarbon prospectivity.

This book attempts to put together a coherent set of research papers to review the present state of understanding and knowledge of the North Sea rift system in the light of the new ideas emerging from the theoretical models and the very detailed information with which to test them. It then reviews their relevance to understanding the history and hydrocarbon prospectivity of the basins. We invited the authors to prepare papers for this book, and to give a verbal presentation of their ideas at a symposium held during the European Union of Geosciences Assembly at Strasbourg in March 1989.

P. A. Ziegler sets the scene with a review of the tectonic and palaeogeographic development of the North Sea rift system, and poses the question of the apparent discrepancy in stretching between the upper crust and the lower crust, as observed in the sedimentary record within the basins, and from uplift of the Moho. Klemperer and Hurich then examine the accumulated evidence from deep seismic reflection profiling across the North Sea in order to determine crustal structure and argue the nature of the crustal extension. Additional evidence is provided from gravity data which, used in conjunction with the seismic, is particularly telling. Day and his collaborators present a new gravity map for the North Sea which gives a coherent view of the whole region for the first time.

Fichler and Hospers bring an advance in gravity interpretation by modelling the Viking Graben area in three dimensions (rather than two dimensions as used previously) to confirm the coincidence of the Moho determined from gravity observations as a density contrast boundary with the Moho determined from seismic data. Holliger and Klemperer take up this theme, focusing on the Central Graben to demonstrate the correspondence between gravity and seismic data in locating the Moho. They draw the conclusion that, if the Moho acts as an interface for both density and seismic p-wave velocity, it is reasonable to regard it as a compositional boundary between crust and mantle. Furthermore, the contrasts are such that in the Central Graben it is unlikely that there has been any significant addition of basic magma to the lower crust. The question of rift-related magmatism is considered by Latin, Dixon, and Fitton, who have tested the McKenzie and Bickle model in a detailed analysis of magmatism in the North Sea through the Jurassic to Cretaceous rifting stage.

The architectural framework of the rifts is discussed by Gibbs, who demonstrates the linkage between sets of faults. He takes note of the regional asymmetry of structures across the Viking and Central Grabens; and argues that the various models can be tested by appropriate cross-sectional balancing at various stages in the development of the basins. The architectural style, and the structural and thermal evolution of the Viking Graben is examined in more detail by Gabrielsen and his colleagues, in order to show how this influenced the nature and distribution of sediments within the basin. Roberts, Yielding, and Badley present a new kinematic model for the development of the central area of the North Sea as dominantly extensional rather than strike-slip (as has become fashionable as an explanation in recent years). This is followed by an account of the structural evolution of the Ringkøbing-Fyn high by Cartwright, examining the significance of this positive feature and its control on basin development both to the north and to the south.

Returning to the discrepancy problem, White addresses it directly, and argues that it is resolved if a detailed examination is made of an area which is known sufficiently well to constrain the controlling factors properly. He is able to demonstrate that the uniform stretching model of McKenzie is adequate for the East Shetland Basin, and that more complicated models are unnecessary. In contrast, Marsden and his colleagues present a model which takes account of the differing rheological properties of upper and lower crust, and incorporates a flexural response of the upper crust to the isostatic consequences of extension. They apply this model quantitatively to explain the Permo-Trias and Jurassic stages of rifting in the Viking Graben, and succeed in matching the structural and stratigraphic evolution to its present day architecture as observed on seismic sections. This paper advances understanding of the processes of rifting to a level where successful predictive interpretation can be undertaken; this gives a credible basis for examining the subsidence and thermal histories of sedimentary units of the basin fill and, in consequence, their hydrocarbon prospects. The book concludes with a short account by Spencer and Pegrum of the state-of-the-art in considering hydrocarbon plays.

The editors are greatly indebted to all the authors for agreeing to take part in this work. Editors and authors alike are truly grateful to the following for refereeing the various papers, and wish to record their thanks; C. Banks, C. Beaumont, A. Berthelsen, J. Brooks, D. Clark-Lowes, D. Fairhead, D. S. France, R. Hardman, M. Khan, D. McKenzie, R. McQuillin, R. Meissner, M. Menzies, J. Miles, J. Milsom, J. Platt, T. Reston, D. Smythe, A. Stein, M. Warner, and G. Williams.

1989

D.J.B.
A.D.G.

References

Barton, P. and Wood, R. (1984). Tectonic evolution of the North Sea basin: crustal stretching and subsidence. *Geophys. J. Roy. Astron. Soc.*, **79**, 987–1022.

Buck, W. R., Martinez, F., Steckler, M. S., and Cochran, J. R. (1988). Thermal consequences of lithospheric extension: pure and simple. *Tectonics*, **7**, 213–34.

McKenzie, D. P. (1978). Some remarks on the development of sedimentary basins. *Earth and Planetary Science Letters*, **40**, 25–32.

McKenzie, D. P. and Bickle, M. J. (1988). The volume and composition of melt generated by extension of the lithosphere. *J. Petrol.*, **29**, 625–79.

Sclater, J. G., and Christie, P. A. F. (1980). Continental stretching: an explanation of the post-mid-Cretaceous subsidence of the central North Sea Basin. *J. Geophys. Research*, **85 B**, 3711–39.

Wernicke, B. (1985). Uniform-sense normal simple shear of the continental lithosphere. *Can. J. Earth Science*, **22**, 108–25.

Contents

9 A kinematic model for the orthogonal opening of the late Jurassic North Sea rift system, Denmark–mid Norway

A. M. Roberts, G. Yielding, and M. E. Badley

10 The structural evolution of the Ringkøbing-Fyn High

J. Cartwright

11 Does the uniform stretching model work in the North Sea?

N. White

12 Application of a flexural cantilever simple-shear/pure shear model of continental lithosphere extension to the formation of the northern North Sea basin

G. Marsden, G. Yielding, A. M. Roberts, and N. J. Kusznir

Contributors

Anderson, O. B.
Danish Geodetic Institute, Gamlehave Alléz, DK-2920 Charlottenlund, Denmark.

Badley, M. E.
Badley Ashton and Associates, Winceby House, Winceby, Horncastle, Lincolnshire LN9 6RB, England.

Blundell, D. J.
Department of Geology, Royal Holloway and Bedford New College, University of London, Egham, Surrey TW20 0EX, England.

Cartwright, J.
Department of Geology, Imperial College, University of London, South Kensington, London SW7 2BP, England.

Day, G. A.
British Geological Survey, Murchison House, West Mains Road, Edinburgh EH9 3LA, Scotland.

Dixon, J. E.
Grant Institute of Geology and Geophysics, University of Edinburgh, West Mains Road, Edinburgh EH9 3JW, Scotland.

Engsager, K.
Danish Geodetic Institute, Gamlehave Alléz, DK-2920 Charlottenlund, Denmark.

Faerseth, R. B.
Norsk Hydro, PO Box 200, N-1321 Stabekk, Norway.

Fichler, C.
Division of Petroleum Engineering and Applied Geophysics, Norwegian Institute of Technology, University of Trondheim, N-7034 Trondheim-NTH, Norway.

Finnstrom, E. G.
Division of Petroleum Engineering and Applied Geophysics, Norwegian Institute of Technology, University of Trondheim, N-7034 Trondheim-NTH, Norway.

Fitton, J. G.
Grant Institute of Geology and Geophysics, University of Edinburgh, West Mains Road, Edinburgh EH9 3JW, Scotland.

Gabrielsen, R. H.
Geological Institute, University of Bergen, Allegaten 41, N-5007 Bergen, Norway.

Gibbs, A. D.
Midland Valley Exploration, 14 Park Circus, Glasgow G3 6AX, Scotland

Holliger, K.
Geophysical Institute, ETH Hönggerberg, CH-8093 Zurich, Switzerland.

Hospers, J.
Division of Petroleum Engineering and Applied Geophysics, Norwegian Institute of Technology, University of Trondheim, N-7034 Trondheim-NTH, Norway.

Hurich, C. A.
Seismological Observatory, University of Bergen, Allegaten 41, N-5007 Bergen, Norway.

Idil, S.
Norsk Hydro, PO Box 200, N-1321 Stabekk, Norway.

Klemperer, S. L.
Department of Earth Sciences, University of Cambridge, Bullard Laboratories, Madingley Road, Cambridge CB3 0EZ, England.

Klovjan, O. S.
Norsk Hydro, PO Box 200, N-1321 Stabekk, Norway.

Kusznir, N. J.
Department of Earth Sciences, University of Liverpool, PO Box 147, Liverpool L69 3BX, England.

Latin, D. M.
Grant Institute of Geology and Geophysics, University of Edinburgh, West Mains Road, Edinburgh EH9 3JW, Scotland.

Liebe, T.

Institute of Geophysics, University of Hamburg, Bundesstrasse 55, D-200 Hamburg 13, Federal Republic of Germany.

Makris, J.

Institute of Geophysics, University of Hamburg, Bundesstrasse 55, D-200 Hamburg 13, Federal Republic of Germany.

Marsden, G.

Department of Earth Sciences, University of Liverpool, PO Box 147, Liverpool L69 3BX, England.

Pegrum, R. M.

Statoil, PO Box 300, Forus, N-4001 Stavanger, Norway.

Plaumann, S.

Niedersächsisches Landesamt für Bodenforschung, Alfred-Bentz-Haus, Postfach 51 01 53, 3000 Hannover 51, Federal Republic of Germany.

Roberts, A. M.

Badley Ashton and Associates, Winceby House, Winceby, Horncastle, Lincolnshire LN9 6PB, England.

Spencer, A. M.

Statoil, PO Box 300, Forus, N-4001 Stavanger, Norway.

Steel, R. J.

Norsk Hydro, PO Box 200, N-1321 Stabekk, Norway.

Strang van Hees, G.

Technische Hogeschool Delft, Postbus 5030, 2600 GA Delft, Netherlands.

Walter, S. A.

British Geological Survey, Murchison House, West Mains Road, Edinburgh EH9 3LA, Scotland.

White, N.

Department of Earth Sciences, University of Cambridge, Bullard Laboratories, Madingley Road, Cambridge CB3 0EZ, England.

Yielding, G.

Badley Ashton and Associates, Winceby House, Winceby, Horncastle, Lincolnshire LN9 6PB, England.

Ziegler, P. A.

Geologisch-Paläontologisches Institut Universität Basel, Bernoullistrasse 32, CH-4056 Basel, Switzerland.

1 Tectonic and palaeogeographic development of the North Sea rift system

P. A. Ziegler

Abstract

The Mesozoic North Sea rift forms an integral part of the Arctic–North Atlantic rift system. Rifting activity in the North Sea commenced during the earliest Triassic, peaked during the Late Jurassic–earliest Cretaceous, and terminated during the Palaeocene. During the early Middle Jurassic a major rift dome developed in the central North Sea. Geophysical data indicate an important discrepancy between upper and lower crustal attenuation across the North Sea rift system. This raises doubts about the concept that during rifting the crust is thinned by mechanical stretching only. It is argued that during the rifting process the Moho-discontinuity can be seriously destabilized.

1.1. Introduction

The Mesozoic North Sea rift system consists of the Viking, Central and Moray Firth–Witch Ground grabens and the Horda-Egersund half graben. The north–south trending Viking–Central Graben has a length of some 1000 km and cross-cuts the structural grain of the Caledonian basement. Devonian, Carboniferous, and Permian sediments form part of the pre-rift sequence, the distribution of which is only partly known due to its deep burial beneath Mesozoic syn–rift deposits and thick Cainozoic post-rift sediments.

The Mesozoic North Sea rift system forms an integral part of the Arctic–North Atlantic mega-rift system, the evolution of which culminated in earliest Eocene crustal separation between Greenland and northern Europe. In the North Sea area rifting commenced during the earliest Triassic, intensified during the Middle Jurassic to earliest Cretaceous, and subsequently gradually abated. Its Cainozoic post-rift evolution is characterized by the development of a broad thermal sag basin (Ziegler 1982, 1988). The North Sea is the most important hydrocarbon province that is associated with the Arctic–North Atlantic mega-rift. Its productive fairway is closely associated with the Viking Graben and the Central Graben. Ultimate recoverable reserves in established accumulations are of the order of 32 billion barrels of oil and 120 trillion cubic feet of gas (Fig. 1.1; Ziegler 1980).

1.1.1. *Basement complex*

Much of the crystalline basement underlying the North Sea area became consolidated during the Caledonian orogenic cycle. The eastern deformation front of the Caledonian fold belt can be traced from the Oslo Graben through southernmost Norway and the North Sea into the border area of Denmark and Germany. The easternmost parts of the North Sea are underlain by a late Precambrian basement complex that forms part of the Fenno-Scandian Shield. The western deformation front of the Caledonides is recognized in the Hebrides, from where it extends north-westwards into central East Greenland. The West–Hebrides platform and the outer parts of the West–Shetland shelf are underlain by Precambrian crystalline basement (Ziegler 1986, 1988).

The Scottish–Norwegian Caledonides, crossing the northern North Sea, are characterized by a

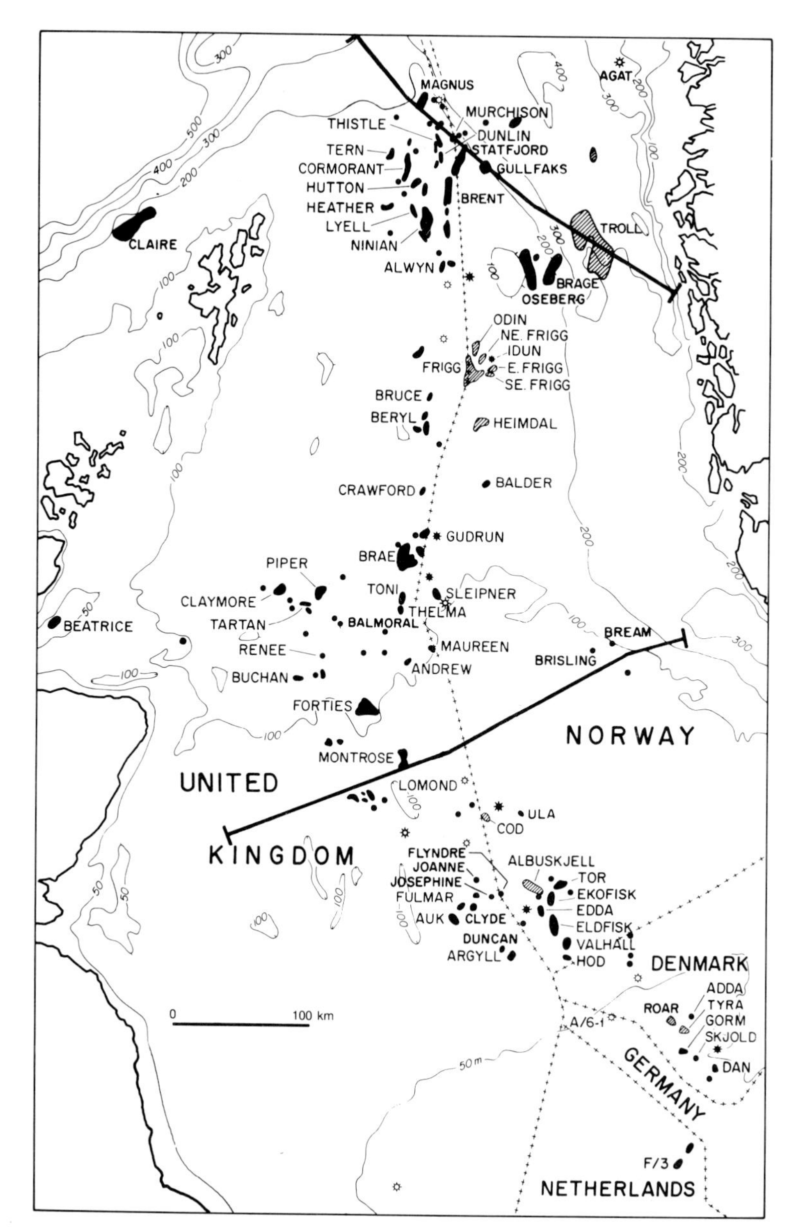

Fig. 1.1. Location map showing oil and gas fields, and location of cross-sections for Fig. 1.18.

north-east–south-west striking structural grain. In contrast, the North German–Polish Caledonian fold belt that branches off from the Scottish–Norwegian Caledonides in the central North Sea is probably characterized in the central North Sea by a north-south trending structural grain and in the south-central and southern North Sea by a south-easterly to easterly trending one. Late Palaeozoic and in part Mesozoic fault trends suggest at least a partial re-activation of these basement trends. Stratigraphic evidence from the Oslo Graben area, the Ardennes, and southern England indicates that the late Caledonian orogenic cycle came to a close during the early Gedinnian (Ziegler 1982, 1986, 1988).

1.1.2. *Devonian and Carboniferous basins*

During the Devonian the post-orogenic collapse of the Arctic–North Atlantic Caledonides (of which the Scottish-Norwegian Caledonides of the northern North Sea form part) was accompanied by a major sinistral translation between Laurentia–Greenland and the Fenno-Scandian shield. In the northern North Sea area these movements gave rise to the rapid subsidence of wrench-induced basins (Midland Valley, Orcadian basin, Hornelen–Solund basins) in which thousands of metres of Devonian Old Red Sandstone clastics accumulated (Ziegler 1988, 1989).

The Carboniferous evolution of the British Isles was governed by tensional stresses. Several north-east-trending grabens (Northumberland and Midland Valley) as defined onshore appear to strike into the central North Sea, where they are poorly defined. A further Carboniferous basin is recognized in the Moray Forth area (Ziegler 1988).

The area now occupied by the southern North Sea formed during the Devonian and early Carboniferous part of the tensional Rhenohercynian basin that developed in the area of the Mid-European and North German–Polish Caledonides. With the late Viséan onset of the Variscan (Hercynian) orogenic cycle, the tensional setting of this basin terminated and the Variscan foreland basin developed in its place. During the late Carboniferous, the area of the central and southern North Sea was occupied by the distal (cratonward) parts of the Variscan foreland basin. In this basin a southward-expanding thick wedge of paralic coal measures accumulated. During the late Westphalian terminal phases of the Variscan orogeny, the Carboniferous grabens of the British Isles became inverted in response to intraplate compressional stresses (Ziegler 1982, 1986, 1988, 1989).

During the latest Carboniferous-early Permian, north-west Europe became transected by the late Hercynian, post-orogenic system of conjugate shear faults (Fig. 1.2). In the North Sea area wrench faulting was associated with widespread intrusive and extrusive magmatism, the subsidence of the highly volcanic Oslo Graben, and deformation of the sedimentary fill of the Variscan foreland basin. Palaeomagnetic data also indicate that the Great Glen fault transecting the Scottish Highlands became reactivated during this time (Storetvedt 1987). Regional (thermal?) uplift of the North Sea area during the Stephanian and Autunian was accompanied by the deep truncation of Carboniferous and Devonian sediments.

Seismic and well data from the North Sea area allow only a tentative reconstruction of the subcrop patterns of Palaeozoic series beneath the regional early Permian unconformity. Particularly in the central and northern North Sea the configuration of the Devonian and Carboniferous basins cannot be resolved on the basis of the available seismic data. However, locally there is evidence for the presence of thick Devonian and/or Carboniferous sediments beneath the base Triassic or base Permian unconformity. The presence of such basins requires a commensurate amount of crustal thinning that obviously can neither be quantified nor areally defined. Despite this, it must be assumed that by early Permian time the North Sea area was characterized by a mature continental crust with an average thickness of some 35 km, at least in areas devoid of thick Devonian–Carboniferous series. However, the widespread occurrence of Permo-Carboniferous volcanic and intrusives suggests that large parts of the lithosphere of the North Sea area were thermally destabilized. This is borne out by its late-early to late Permian evolution.

1.1.3. *Permian basins*

At the end of the Autunian, wrench fault and volcanic activity abated in north-west Europe; the northern and southern Permian basins began to subside during the Saxonian, presumably in response to the decay of thermal anomalies that were induced during the Stephanian–Autunian phase of wrench faulting (Fig. 1.3; Ziegler 1982, 1988; Sørensen 1985).

The ill-defined northern Permian basins extends in an east–west direction across the central North Sea. It is separated from the southern Permian basin by the mid-North Sea–Ringkøbing–Fyn-Møns trend of highs, that came into existence during the Stephanian-Autunian. The southern Permian basin extends from the coast of England across the southern North Sea and northern Germany into Poland; its configuration is well constrained by numerous boreholes and seismic data. The basin is, broadly, saucer shaped; its subsidence was accompanied by minor faulting only, except in its easternmost parts, where the north-west striking Polish Trough came into evidence by the Saxonian. Lower

Permian Rotliegend continental redbeds and basinal shales and evaporites reach a maximum thickness of some 1500 m in the southern Permian basin. Rotliegend clastics in the northern Permian basin attain thicknesses of the order of 600 m (Ziegler 1982; Glennie 1984, 1986). At the transition to the late Permian, the Zechstein Sea advanced from the Norwegian–Greenland Sea rift southward and flooded the Rotliegend basins of north-west Europe (Fig. 1.4). In both basins the basal Zechstein

Keys to Figs 1.2–1.17

LITHOLOGICAL SYMBOLS

Sand and conglomerate
Sand
Sand and shale
Carbonate and sand
Carbonate
Carbonate and shale
Shale, some carbonate
Shale
Organic shale
Halite
Sulphate
Oolitic iron ore
Coal
Batholiths
Volcanics, local
Major extrusives

SPECIAL SYMBOLS ON PALAEOGEOGRAPHIC MAPS

Direction of clastic influx
Direction of intra-basinal clastic transport
Erosional edge of map interval
Direction of marine incursion
300 Thickness of map interval in metres

Abbreviations used in Figs 1.2–1.18

BF	Broad Fourteens Basin
CB	Channel Basin
CG	Central Graben
CR	Cardigan Bay
EA	Eichsfeld-Altmark Swell
EB	Egersund Basin
EL	Emsland Low
ET	Emsland Trough
FH	Friesland High
FSB	Fennoscandian Border Zone
GT	Glückstadt Trough
GTL	Gifhorn Trough Lineament
HB	Horda Basin
HG	Horn Graben
HR	Hunsrück Mountains
HS	Hunte Swell
HSB	Hampshire Basin
HZ	Harz Mountains
KB	Kish Bank Basin
LS	Lower Saxony Basin
MA	Mendip Axis
MF	Moray Firth Basin
MM	Moreton-in-the-Marsh Axis
MNH	Mid-North Sea High
MW	Market-Weighton Axis
MX	Manx-Furness Basin
ND	North Danish Basin
NF	Normandy Fault
NL	West Netherland Low
NN	North Netherland Swell
NO	Nordfjord High
NR	Neurupiner Lineament
OZ	Osning Zone
PO	Portsdown Swell
PYF	Pays-de-Bray Fault
RGR	Rostock-Gramzower Lineament
RFH	Ringkøbing-Fyn High
SH	Sub-Hercynian Basin
SHL	Sunn-Hordland Area
SP	Sole Pit Basin
SPF	Sticklepath Fault
SWB	Solway Basin
TB	Trier Embayment
TF	Thuringian Forest
TW	Thuringian-West Brandenburg Depression
T-IJ-H	Texel-IJsselmeer High
US	Unst Basin
VB	Vosges-Black Forest Highs
VG	Viking Graben
VL	Vlieland Basin
WA	Western Approaches Trough
WB	Wessex Basin
WD	Weser Depression
WG	Worcester Graben
WN	West Netherland Basin
ZR	Zandvoort Ridge

Fig. 1.2. Stephanian-Autunian palaeogeography. Legend and explanation of abbreviations used in this and subsequent figures.

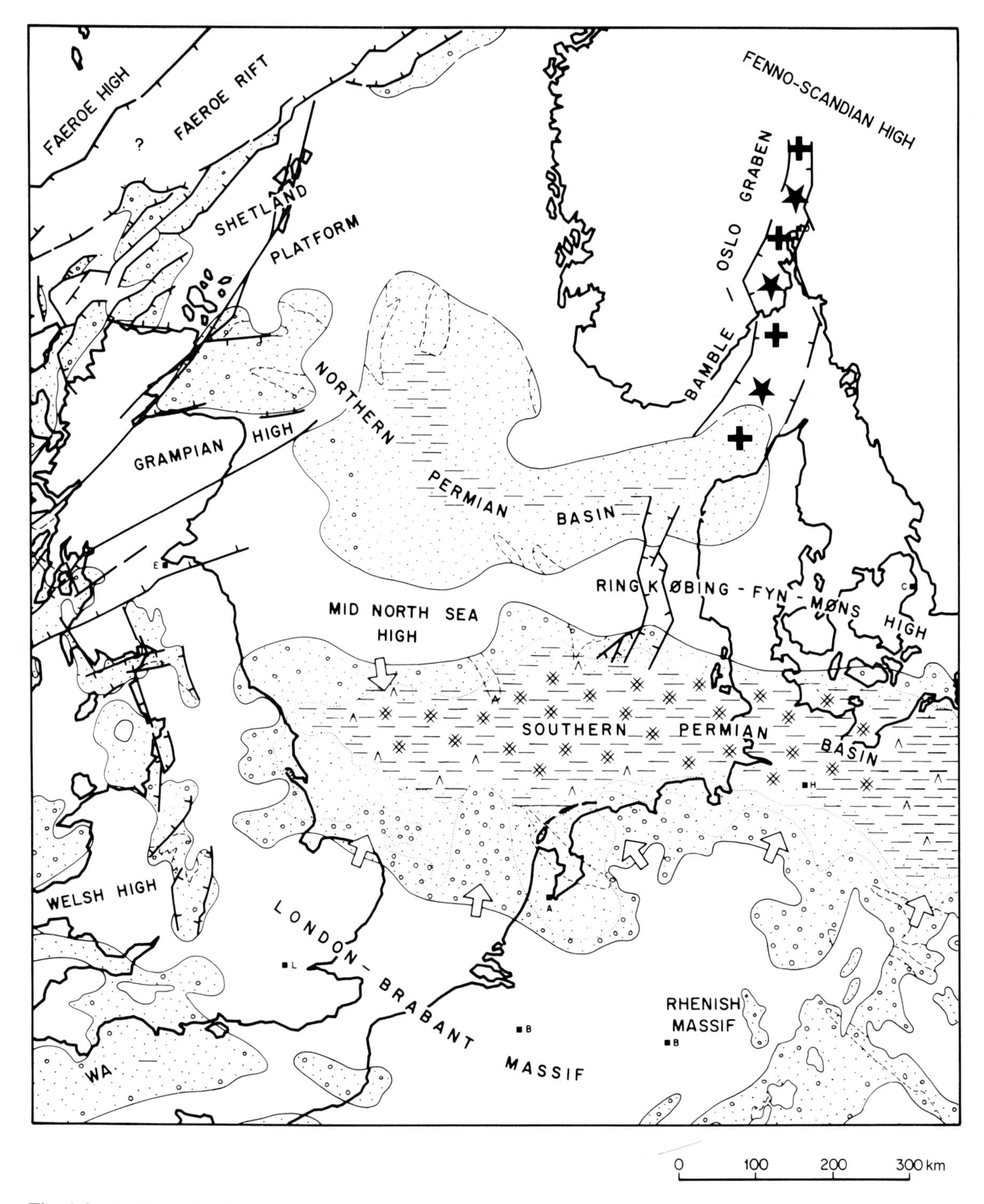

Fig. 1.3. Rotliegend palaeogeography.

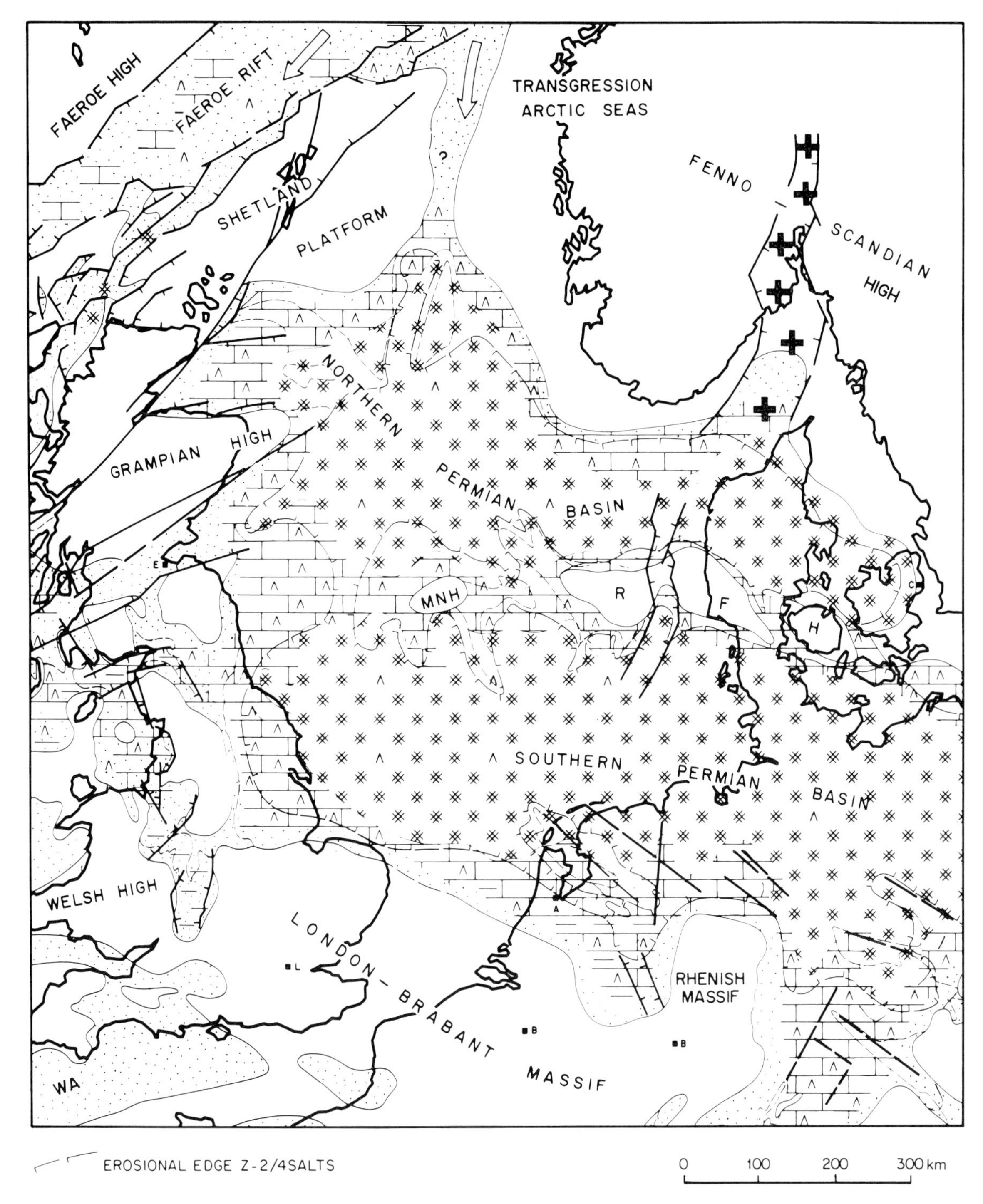

Fig. 1.4. Zechstein palaeogeography.

transgressive Kupferschiefer corresponds to a chronostratigraphic marker. This thin, highly organic shale was deposited below wave base and thus indicates that the northern and southern Permian basins subsided below mean sea level prior to the apparently catastrophic Zechstein transgression.

From the Norwegian-Greenland Sea rift, the late Permian sea advanced southwards into the Faeroe-Rockall rift and entered the southern and northern Permian basins, presumably via the Irish Sea area and the English Midlands. As there is no evidence for the occurrence of Zechstein series in the northern Viking Graben, it is uncertain whether it had already started to subside during the late Permian, thus providing a second, important avenue for the Zechstein transgression.

Repeated glacio-eustatic sea level fluctuation governed the accumulation of the highly cyclical Zechstein carbonate, sulphate, and halite series. The distribution of the carbonate/sulphate banks that developed along the basin margins during Zechstein cycles 1 and 2, and of the thick basinal Zechstein–2 salts, indicate that both the northern and southern Permian basins were elongated in an east-west direction (Ziegler 1982; Taylor 1984). The mid-North Sea–Ringkøbing-Fyn trend of highs corresponded to a string of shoals. Although Zechstein facies patterns suggests that the Horn Graben (transecting the Ringkøbing-Fyn High in a north-south direction) in the prolongation of the Oslo Graben did subside during the Late Permian, comparable evidence is lacking for the Danish segment of the North Sea Central Graben. Furthermore, seismic data suggest that the Zechstein salts preserved in the southern Viking Graben form part of the downfaulted pre-rift sequence that is eroded over the adjacent rift flanks as a consequence of their mid-Mesozoic uplift.

Overall, it appears that during the Late Permian rifting propagated from the Norwegian-Greenland Sea area southward into the Faeroe-Rockall Trough and that the northern and southern Permian basins continued to subside in response to thermal contraction of the lithosphere and its loading by water and sediments (Ziegler 1988). Along the southern margin of the southern Permian basin minor faulting controlled Zechstein facies patterns in Germany and the Netherlands (M. A. Ziegler 1989). In the eastern parts of the southern Permian basin the Polish Trough continued to subside differentially.

In the central North Sea the Zechstein series attained primary thicknesses of the order of 1500 to 2000 m (Olsen 1987). Their present thickness distribution is, however, highly variable due to salt diapirism and erosion at the base of the mid-Jurassic unconformity. As there is little evidence for Late Permian rifting activity governing Zechstein facies patterns in the North Sea, the Zechstein series is considered to form part of the pre-rift sequence.

1.1.4. *Triassic–early Jurassic rifting stage*

At the transition from the Permian to the Triassic, rifting activity accelerated in the Norwegian–Greenland Sea area and in the Tethys domain. By Early Triassic time, north-west and central Europe as a whole were subjected to regional tensional stresses causing the differential subsidence of a complex set of multidirectional grabens and troughs (Fig. 1.5). Tensional reactivation of Permo-Carboniferous fracture systems was probably responsible for the localization of some of the Triassic grabens (e.g. North Danish Basin) whereas others, such as the Viking and Central Grabens were entirely new features (Ziegler 1988).

Rifting activity continued through Triassic times into the Jurassic with little evidence for discrete rifting pulses.

Stratigraphic evidence indicates that during the earliest Triassic the Norwegian-Greenland Sea rift propagated rapidly into the North Sea area causing the differential subsidence of the Viking and Central Grabens, the Horda–Egersund half graben and the Moray Firth–Witch Ground graben system. At the same time the Horn Graben became reactivated. The Viking and Central Grabens clearly crosscut Caledonian basement trends. However, individual fault segments, particularly in the northern Viking Graben, suggest a reactivation of Caledonian basement discontinuities (Johnson and Dingwall 1981).

The Triassic graben system of the North Sea is almost perpendicular to the axis of the Northern Permian basin and the mid-North Sea-Ringkøbing-Fyn trend of highs. Particularly in the Danish part of the Central Graben there is clear evidence for the Early Triassic onset of its subsidence.

Triassic sediments attain maximum thicknesses of some 2000 m in the Central Graben, 3000 m in the

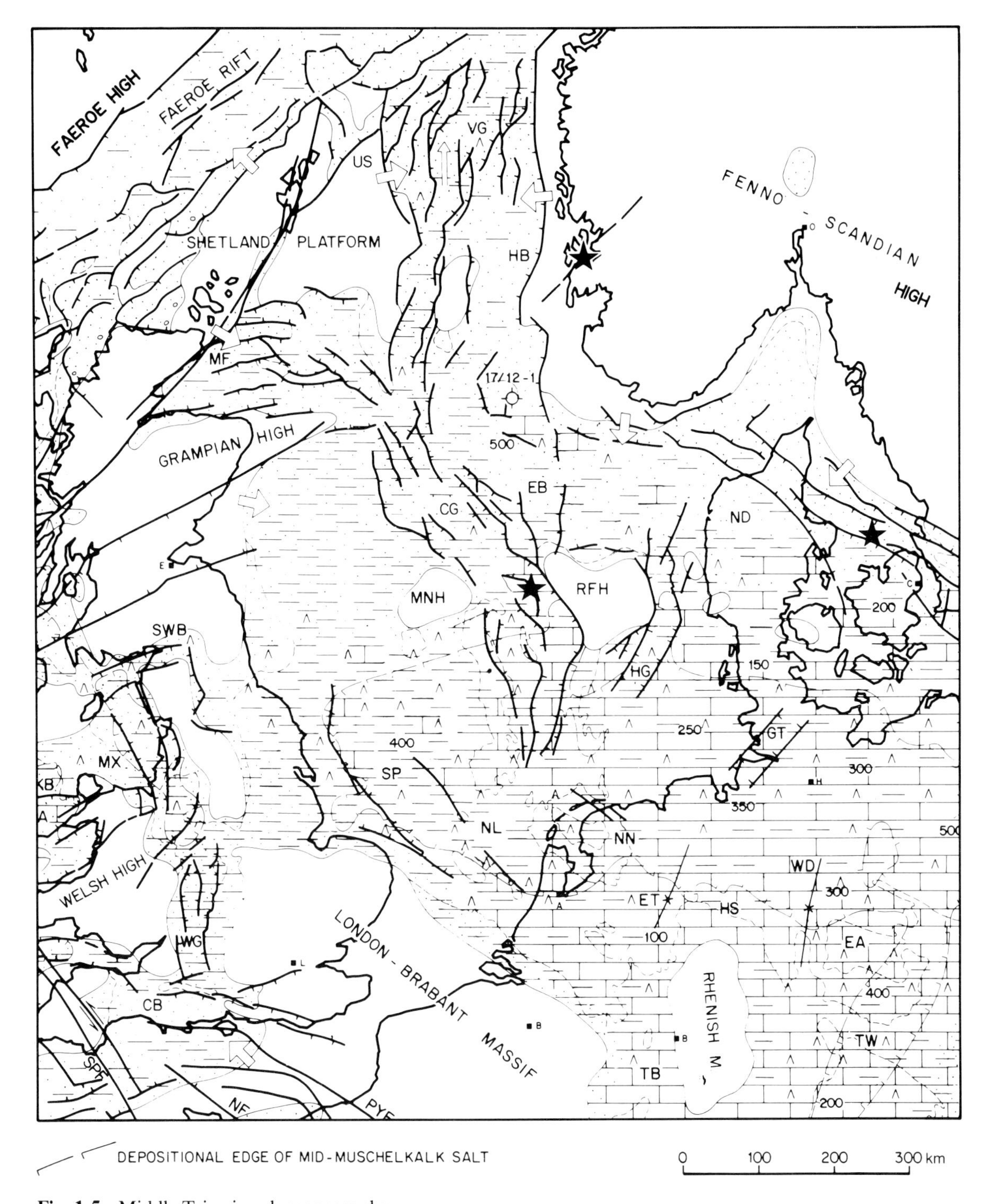

Fig. 1.5. Middle Triassic palaeogeography.

northern Viking Graben, 4000 m in the Horn Graben, and 6000 m in the North Danish half graben. Restored isopachs of the Triassic series indicate, however, that the northern and southern Permian basins continued to subside during the Triassic (Ziegler 1982). This suggests that the onset of Triassic rifting was not associated with major lithospheric thermal perturbances. In fact, the Triassic rifts of the North Sea display a very low level of volcanic activity, and there is only limited evidence for simultaneous uplift of the rift flanks (e.g. southern Viking Graben).

Overall, Permian basin edges became progressively overstepped during the Triassic. During the latest Permian, the Zechstein seas withdrew from the North Sea area possibly in response to its stress-induced broad, low relief, doming. In the northern and central North Sea, Triassic series are developed in a red bed facies. Lower and Middle Triassic shales and sandstones (Smith Bank Formation) are overlain by the more sandy Upper Triassic Skagerrak Formation which was mainly derived from the Fenno-Scandian Shield. Cyclical Middle and Late Triassic marine incursions, originating from the Tethys shelves, did not advance beyond the Danish part of the Central Graben and touched only the south-eastern parts of the Egersund basin (Fig. 1.5; Ziegler 1982; Fisher 1986).

In the central North Sea, accumulation of the Triassic series and syndepositional faulting triggered diapiric deformation of the Zechstein salts during the Middle and Late Triassic. This, in combination with the Middle Jurassic deep truncation of the Triassic series in the central North Sea area, permits only a tentative reconstruction of the Triassic subsidence patterns; it also impedes the assessment of the amount of crustal extension that occurred during the Triassic. However, in the northern Viking Graben, where the Triassic series rests on basement and is conformably overlain by Lower Jurassic sediments, the interface between the basement and the Triassic red beds is largely non-reflective. Thus the amount of fault-controlled crustal extension which occurred during the Triassic cannot be quantified within an acceptable degree of certainty; the net amount of 18 km of intra-Triassic crustal stretching across the northern Viking Graben, as indicated in Fig. 1.18, should be regarded as a highly tentative value.

During the Rhaetian and Hettangian, open marine conditions were established in the southern and south-central North Sea, whereas continental, coal-bearing sandstones (Statfjord Sands, Fig. 1.6) accumulated in the subsiding Viking Graben and on the Horda Platform. These sands were derived from the Fenno-Scandian Shield, the Scottish Highlands, and the Shetland Platform; they form an important reservoir for major oil accumulations in the northern Viking Graben (e.g. deeper pools in Brent and Statfjord fields; Bowen 1975; Karlsson 1986; Spencer *et al.* 1987).

During the late Sinemurian, the Arctic and Tethys seas linked up via the Viking Graben and Horda Basin (Fig. 1.7) as evidenced by the deposition of the open marine shales of the Dunlin group. Lower Jurassic series reach maximum thicknesses of some 500 m in the northern Viking Graben, 750 m in the North Danish basin, and 250–500 m in the southern parts of the Central Graben.

Although lower Jurassic series have been largely removed from the central North Sea area by early mid-Jurassic erosion, it is assumed that the Central Graben and the Horda-Egersund half graben, similar to the Viking Graben and the Dutch part of the Central Graben, continued to subside differentially during the Early Jurassic. The amount of crustal dilation occurring during the Early Jurassic cannot, however, be quantified. In the North Sea area volcanic activity was at a very low level during the Early Jurassic; moreover, there is no evidence for rift-induced thermal doming, nor for significant rift shoulder uplift.

1.1.5. *Mid-Jurassic thermal doming stage*

At the transition from the Early–Middle Jurassic the central North Sea area became uplifted and formed a broad arch that was transected by the Central Graben. The dimensions of this arch can be reconstructed by mapping areas of continued Lower and Middle Jurassic sedimentation, and by tracing the subcrop of Lower Jurassic and Triassic series against the base mid-Jurassic unconformity (mid-Kimmerian unconformity: Fig. 1.8). Uplift of this rift dome was coupled with the interruption of connections between the Arctic and Tethys Seas (Ziegler 1982, 1988).

The central North Sea dome extended in a north–south direction from the southern Viking Graben into the southern North Sea over a distance of some 700 km and in a west–east direction, from

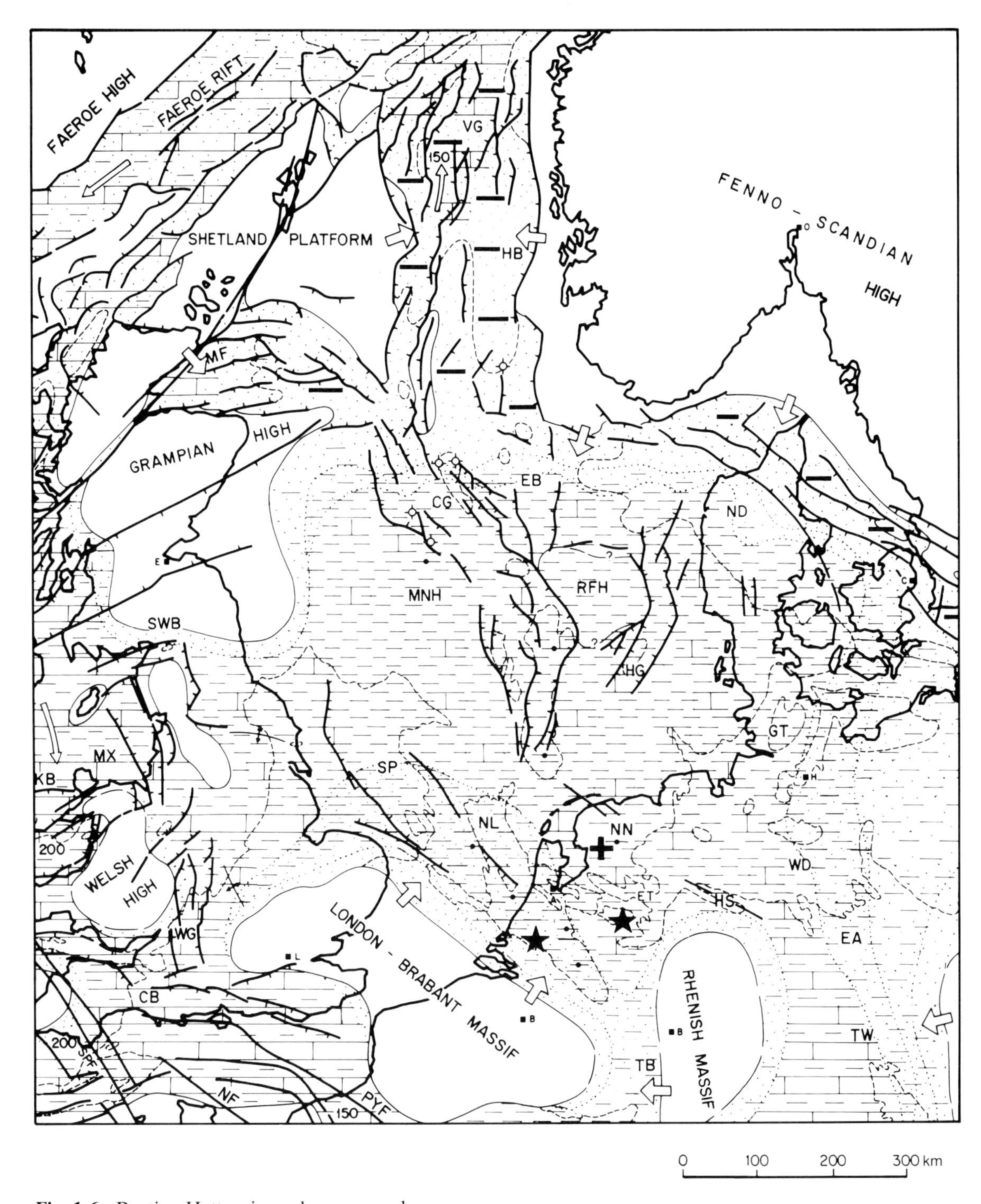

Fig. 1.6. Raetian-Hettangian palaeogeography.

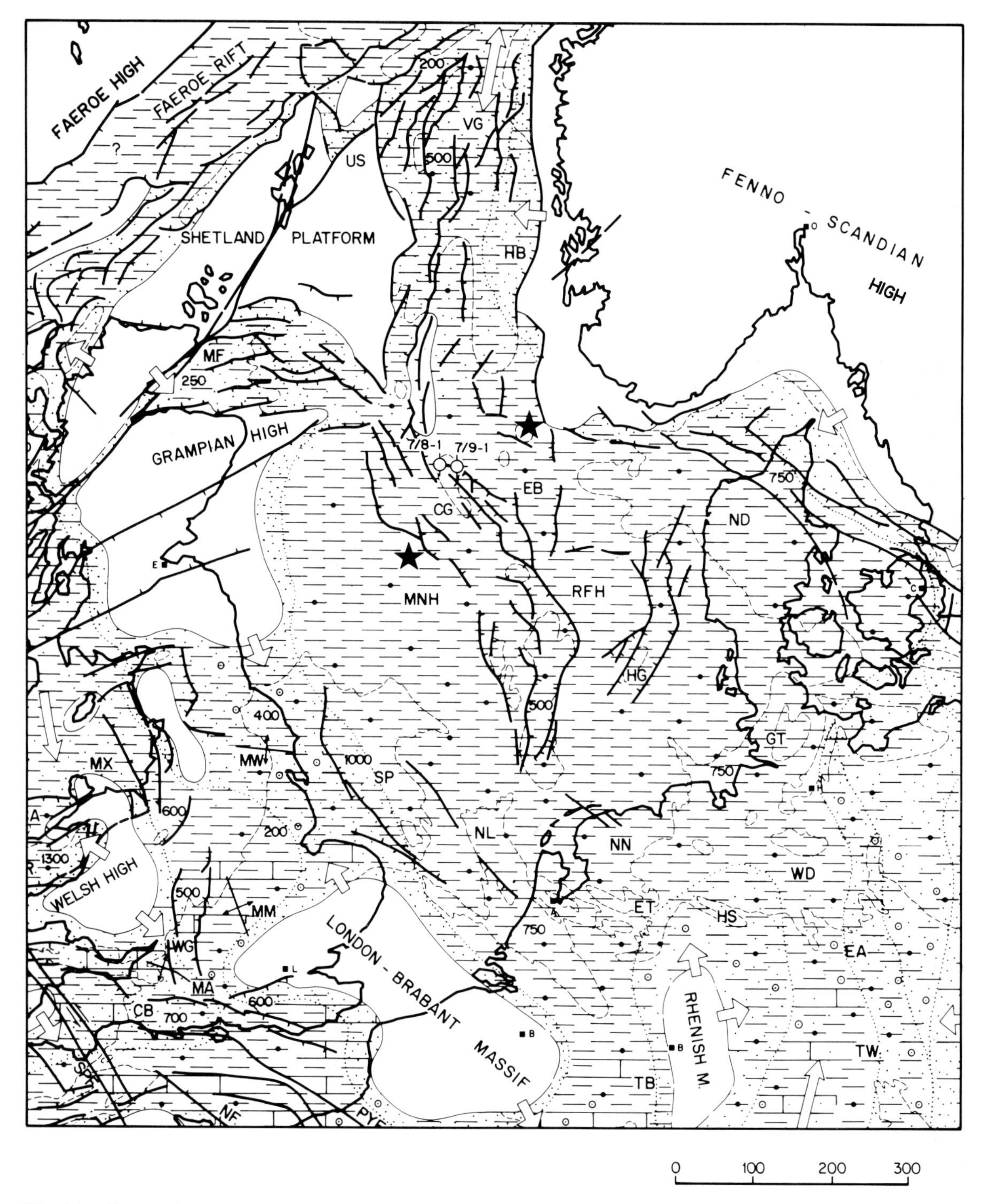

Fig. 1.7. Sinemurian-Toarcian palaeogeography.

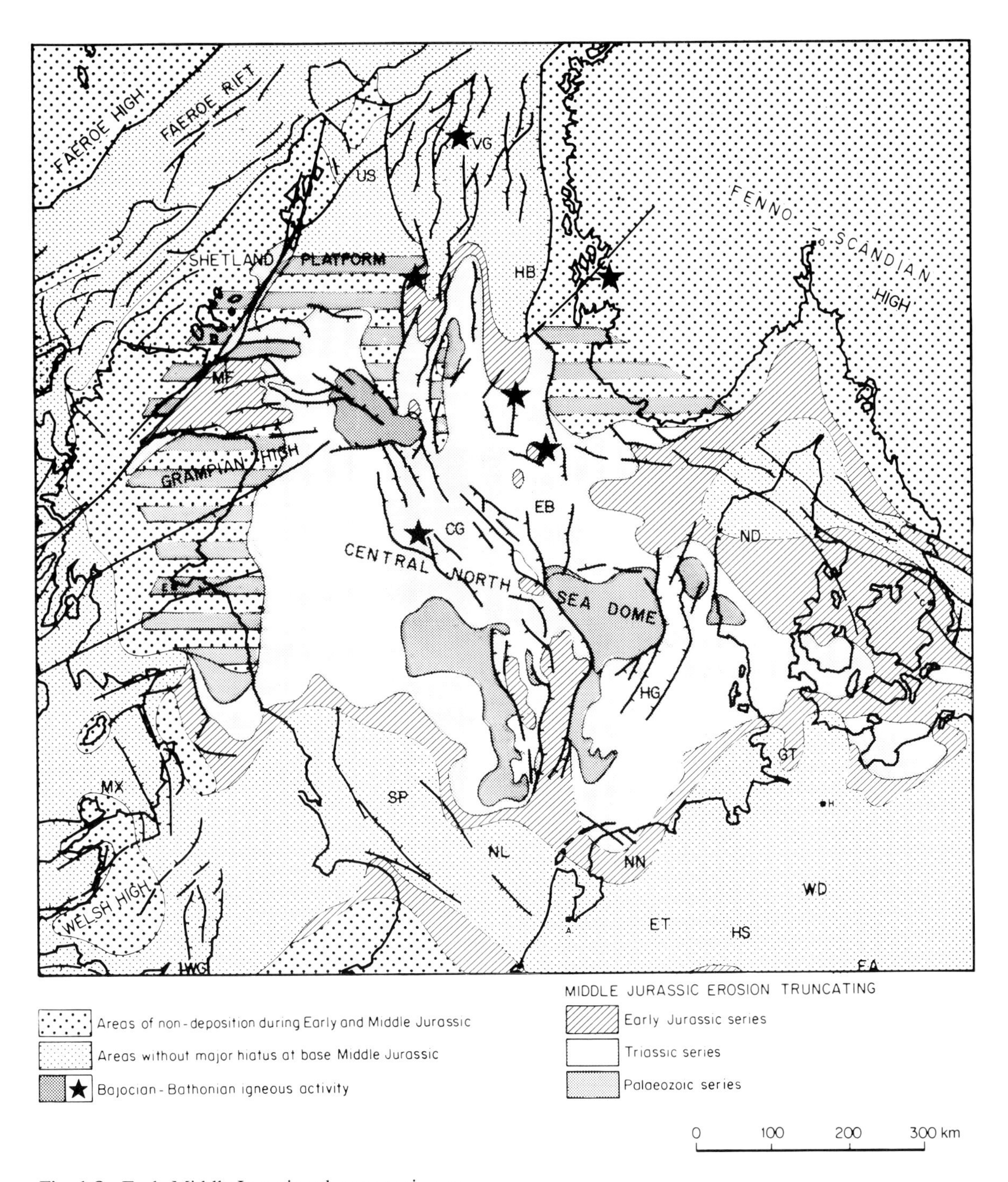

Fig. 1.8. Early Middle Jurassic palaeotectonic map.

England to Denmark over a distance of about 1000 km. Uplift of this dome was accompanied by the development of a large volcanic complex at the triple junction between the Viking, Central, and the Moray Firth–Witch Ground grabens. Subsidiary volcanic centres occur in the southern Viking Graben, in the Egersund basin, in coastal Norway, and in the Central Graben. Volcanics display the bimodal mafic–felsic alkaline chemistry that is typical for intracratonic rifts (Dixon and Fitton 1981). Stratigraphic data indicate that the bulk of these volcanics was extruded during the Bajocian; on the other hand radiometric age determination suggest that extrusive and intrusive activity persisted into the Early Cretaceous (Ritchie *et al*. 1988 Latin *et al.*, this volume).

Whereas the northern Viking Graben continued to subside differentially during the Aalenian to Bathonian, the axial parts of the Central Graben became uplifted and subjected to erosion, particularly during the late Aalenian and early Bajocian. On its even more drastically elevated rift flanks, erosion cut deeply into Triassic series, in places into Permian strata and locally even to basement levels (Fig. 1.8). The structural relief of the central North Sea dome was probably of the order of 1500–2500 m.

Erosion products, consisting mainly of recycled Triassic sandstone, were deposited in adjacent subsiding basins in major deltaic complexes, such as the regressive-transgressive Brent Group in the Viking Graben (Graue *et al*. 1987), the sandstones of the Ravenscar Group in Yorkshire (Hemmingway 1974), the Haldager Sandstone of the North Danish basin (Hamar *et al*. 1983), and the Sonninia and Coronata Sandstones of northern Germany (Hoffman 1949; Fig. 1.9). The sandstones of the Brent Group attain thicknesses of 300–400 m and form an outstanding hydrocarbon reservoir in the northern Viking Graben (e.g. Brent, Ninian, Statfjord, Gullfaks fields, all in the billion barrel class: Ziegler 1980; Brown 1984; Karlsson 1986; Spencer *et al*. 1987).

Within the Central Graben, sedimentation resumed during the middle Bajocian and Bathonian under continental to lacustrine conditions, initially in tectonically silled basins and later on a broader scale (Pentland/Bryne Formation, Fig. 1.10). During the Bathonian, increased subsidence of the Viking and Central Grabens combined with a relative rise in sea level caused the back-stepping of the Brent–Pentland delta system into the southern Viking Graben (Harris and Fowler 1987) and first marine transgressions into the Central Graben. This may reflect the onset of collapse of the North Sea rift dome, as suggested by the widespread accumulation of the continental upper Bathonian and Bajocian Haldager Sandstone in the Egersund–Horda basin (Hamar *et al*. 1983). During the Callovian and Oxfordian the central North Sea dome area continued to subside, as illustrated by a regional transgression in the southern Viking Graben, the Central Graben, and the Horda–Egersund–North Danish basins (Fig. 1.11).

At the same time clastic influx into the basins of the southern North Sea and northern Germany abated. By Oxfordian and Kimmeridgian time the rift shoulders of the Central Graben became gradually inundated. On the other hand, areas flanking the Dutch part of the Central Graben became apparently uplifted during the Callovian (Cleaver Bank and Broad Fourteens highs). Late Middle and early Late Jurassic foundering of the Central North Sea rift dome was accompanied by significant amounts of crustal extensions. This is evident by the development of deep half-grabens, particularly in the southern Viking Graben and the Central Graben, in which Middle and Late Jurassic sediments attain thickness of 1000 m and more (Cayley 1987).

Transgressive Bajocian to lower Kimmeridgian sandstones (Pentland and Hugin formations) contain important hydrocarbon accumulations in the Central Graben and the Outer Moray Firth basin (e.g. Fulmar, Ula, Piper, Claymore: Maher 1981; Spencer *et al*. 1986). A Callovian–Oxfordian deltaic system, prograding from coastal Norway onto the northern Horda Platform, forms the reservoir of the giant Troll gas field (Hellem *et al*. 1986). By Oxfordian–Kimmeridgian times, deeper water conditions were established in the Viking and Central Grabens with paralic conditions persisting only in its southernmost, Dutch parts (Figs 1.10, 1.11, and 1.12).

In summary, the mid-Jurassic uplift of the central North Sea area has substantially contributed to the development of major sandstone reservoirs that contain the bulk of the hydrocarbon reserves in the Viking Graben and that are also of importance for the hydrocarbon potential of the Central Graben. Furthermore, sands shed southward from the North Sea rift dome to northern Germany contain a number of oil accumulations.

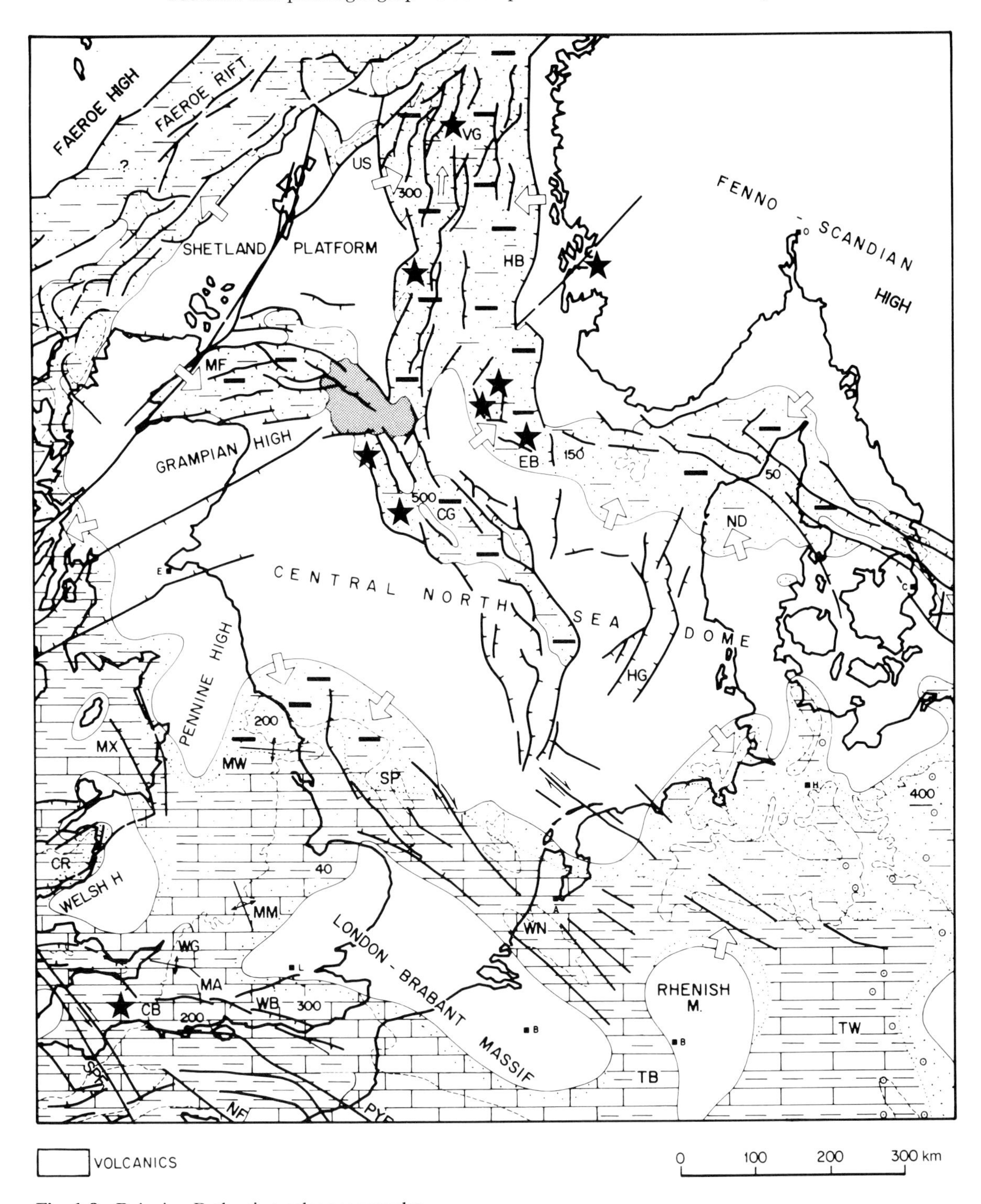

Fig. 1.9. Bajocian-Bathonian palaeogeography.

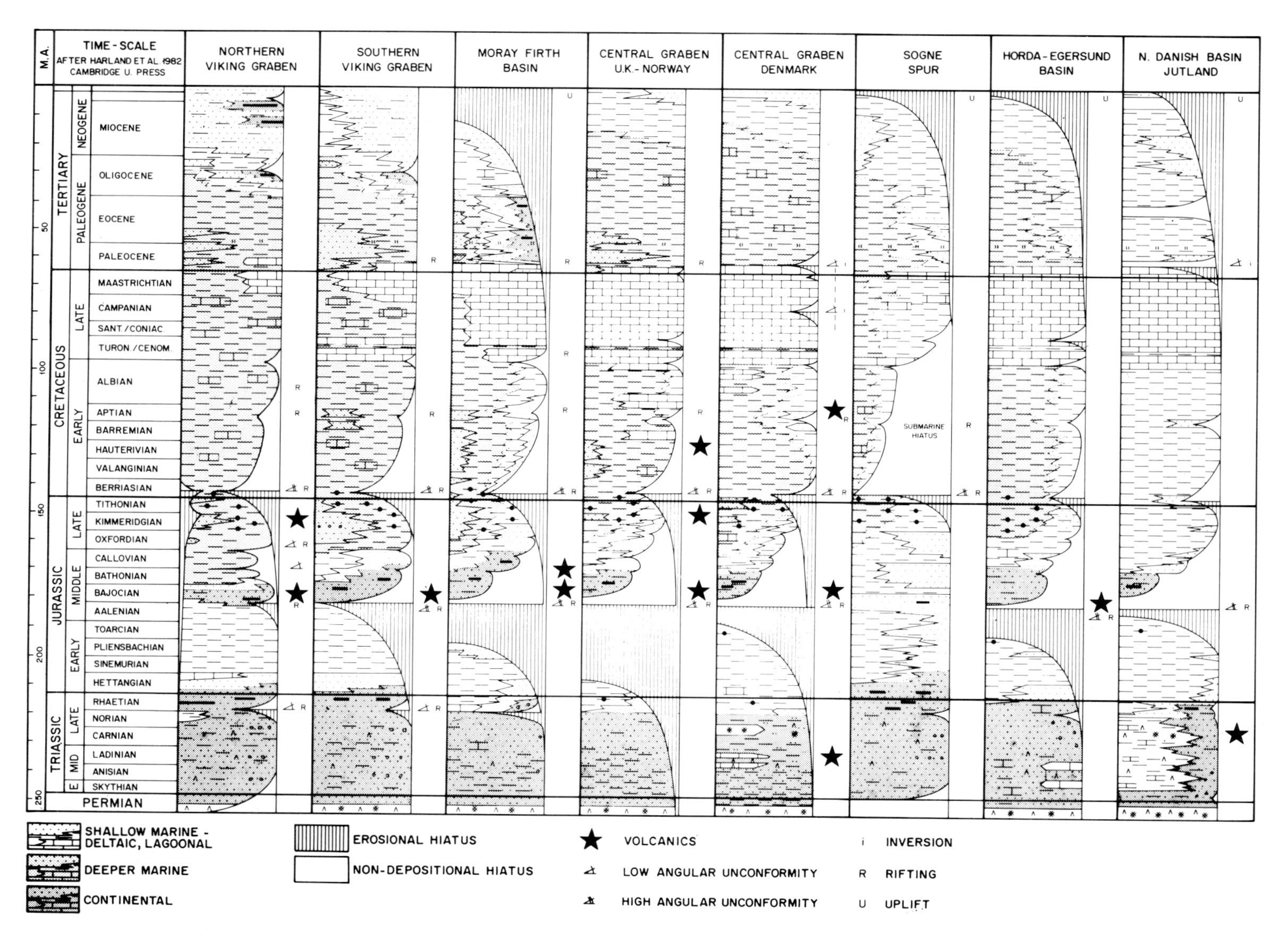

Fig. 1.10. Chronostratigraphic-lithostratigraphic correlation chart. Time scale after Harland *et al.* (1982).

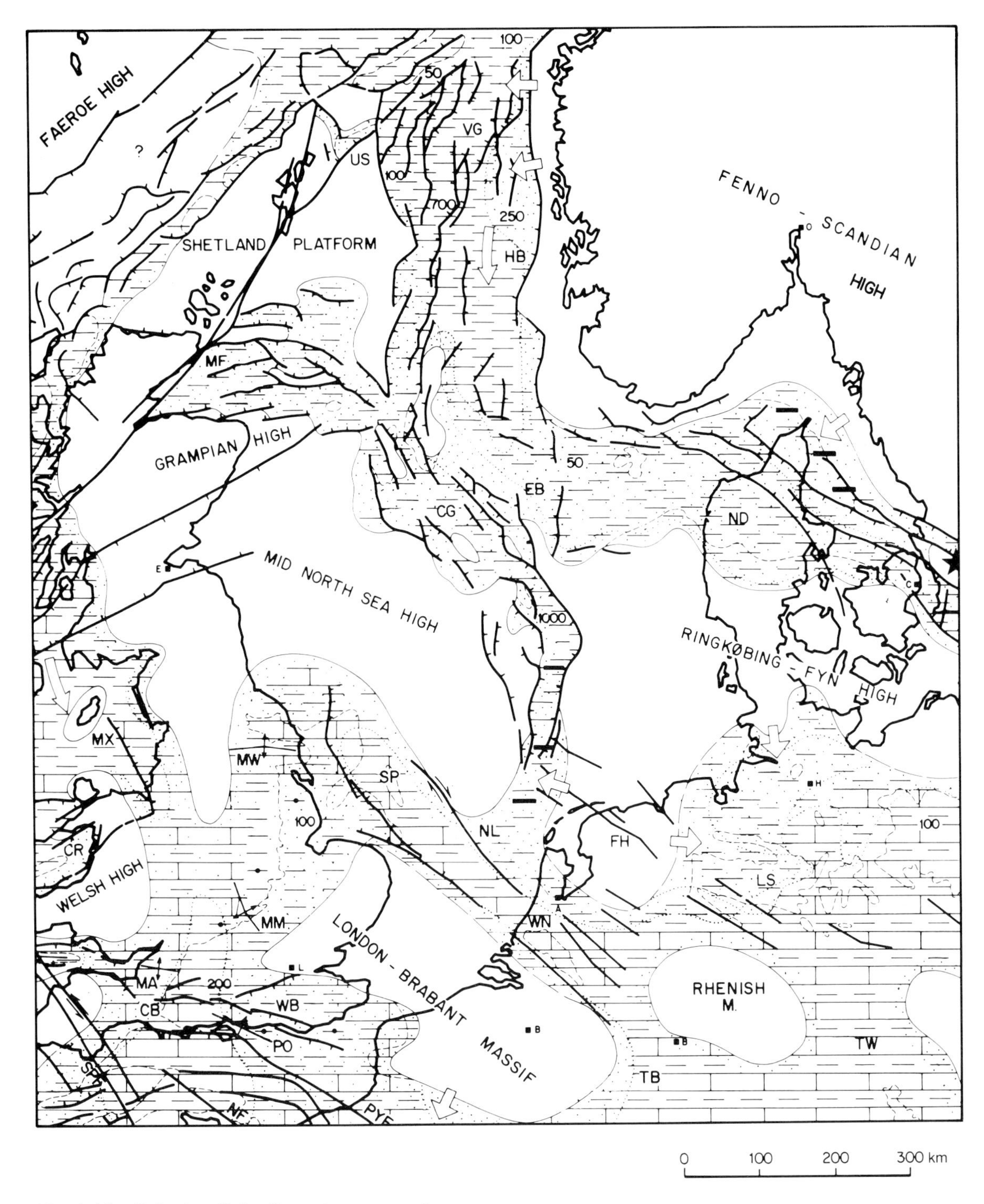

Fig. 1.11. Callovian-Oxfordian palaeogeography.

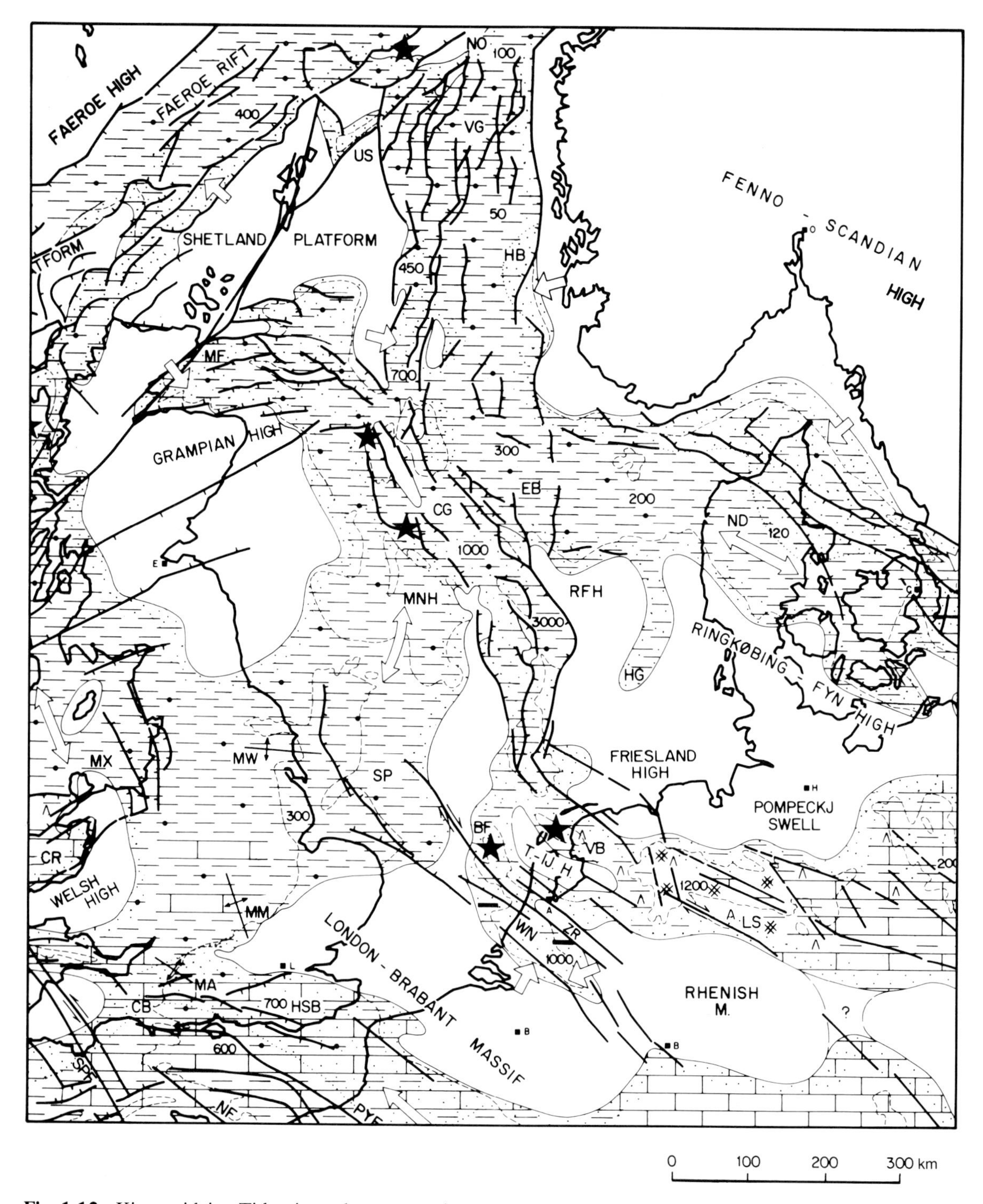

Fig. 1.12. Kimmeridgian-Tithonian palaeogeography.

1.1.6. *Late Jurassic–early Cretaceous rifting stage*

During the Kimmeridgian to Berriasian/Valanginian(?), the rate of crustal extension across the North Sea rift system apparently accelerated, and tectonic activity now concentrated on the Viking and Central Grabens and the Moray Firth–Witch Ground graben system. The occurrence of local dextral convergent wrench deformations (for instance in the Central Graben) indicates that crustal stretching in the North Sea was not purely orthogonal during the late Jurassic and early Cretaceous. Crustal dilation across the Central Graben was compensated at its southern termination by a system of dextral wrench faults that controlled the subsidence of the Sole Pit, Broad Fourteens, West Netherland, and Lower Saxony basins (Ziegler 1982, 1988).

Whereas the fault system delimiting the Horda Platform and the Egersund Basin to the east became largely inactive during the early Cretaceous, tectonic activity continued along the Fenno-Scandian border zone forming the north-eastern boundary of the North Danish basin. The Horn Graben was apparently inactive during the late Jurassic.

During the Oxfordian and Kimmeridgian, deeper water conditions were established in the Viking and Central Grabens, and also in the Moray Firth–Witch Ground graben system (Figs 1.11 and 1.12). By late Kimmeridgian time the Mid-North Sea High had subsided to the degree that an open marine connection was established between the central North Sea and southern England.

Throughout the central and northern North Sea Kimmeridgian to Berriasian shales are developed in a kerogenous source rock facies; their thickness and richness is very variable. These deeper water shales are the principal source for hydrocarbon accumulations in the central and northern North Sea (Barnard and Cooper 1981; Cornford 1984; Baird 1986).

During the Callovian and Oxfordian the Shetland Platform and the Hebrides Shelf became relatively uplifted, possibly in conjunction with thermal doming of the Faeroe-northern Rockall Rift zone (Fig. 1.11). Particularly during the Oxfordian and Kimmeridgian, clastics were shed eastward from these uplifted areas into the southern Viking Graben and the Moray Firth-Witch Ground troughs where they accumulated in submarine fan complexes. In the southern Viking Graben these deposits provide the reservoirs for a number of important, largely stratigraphically-trapped oil and gas/condensate accumulations (Brae field area: Stow *et al.* 1982; Brown, 1984; Spencer *et al.* 1987; Turner *et al.* 1987).

The Faeroe-northern Rockall dome foundered rapidly during the Kimmeridgian. At the same time the area occupying the triple junction between the Faeroe Trough, Viking Graben and mid-Norway rift zone became uplifted (Fig. 1.12). Subaereal exposure of the Nordfjord Ridge and the northern parts of the Tampen Spur (north of Gullfaks field) gave rise to the erosion of the Brent and Statfjord Sandstone and of Triassic series (Karlsson 1986; Nelson and Lamy 1987). Erosion products were shed southward into the Viking Graben area where they were deposited as submarine fans, forming the reservoir of the Magnus fields (Magnus Sandstone: De'Ath and Schuyleman 1981).

During the late Kimmeridgian, the supply of clastics from Norwegian coastal areas to the northern parts of the Horda basin abated, and remained at a relatively low level during the Early Cretaceous.

At the transition from the Jurassic to the Cretaceous, rifting activity accelerated in the entire Arctic-North Atlantic rift system. In the North Sea, this so-called 'late Kimmerian' rifting pulse affected mainly the Viking and Cental Grabens and the Moray Firth-Witch Ground graben system. The late-Kimmerian tectonic activity in the general North Sea area was accompanied by a relative drop of sea level, reflecting perhaps less a truly eustatic lowering of sea level than stress-induced broadscale positive lithospheric deflections (Cloetingh *et al.* 1987).

Latest Jurassic-Early Cretaceous intensified rifting activity in the Faeroe-West Shetland Trough and the mid-Norway basin was paralleled by the rapid subsidence of the Nordfjord dome along a set of listric normal faults (Duindam and van Hoorn 1987; Nelson and Lamy 1987).

In the Viking Graben, rapid differential subsidence of rotational fault blocks in response to accelerated crustal extension resulted in the development of a submarine relief of some 1000 m. The leading edges of some of the major fault blocks became subaereally exposed, as evident by the development of wave-cut platforms on, for instance, the Gullfaks structure (Karlsson 1986). The flanks of these highs (as well as the crests of submarine highs) were swept clean by contour currents, whereas pelagic shales accumulated in intervening lows in which sedimentation was more or less continuous across the Jurassic-Cretaceous boundary. In the Central Graben, contemporaneous rift tectonics are less obvious as much of the tensional deformation at pre-Permian levels was taken up by plastic deformation in the Zechstein salts and was thus dissipated to shallower levels.

In the North Sea area, the largely submarine (and, in many places, composite) late-Kimmerian unconformity is regionally correlative (Rawson and Riley 1982). Its origin is probably related to drastic changes in the current regime caused by the rapid subsidence of the Viking and Central Grabens, and an important drop in relative sea level. This change in current regime was associated with a stronger oxygenation of the bottom waters, as reflected by the regional termination of kerogenous shale deposition.

Following the late-Kimmerian rifting pulse, the rate of crustal extension across the North Sea Graben system diminished gradually. Within the Viking and Central Grabens, the throw on many faults decreases and fades out upward in Lower Cretaceous shales (Figs 1.18). However, master faults delineating the Viking and Central Grabens, remained active throughout early and late Cretaceous times. During the Early Cretaceous, continued differential subsidence of this graben system was accompanied by the gradual infilling of its relief with deep-water shales and minor pelagic carbonates, ranging in age from Berriasian to Albian (Fig. 1.13 and 1.14). Lower Cretaceous shales attain a thickness of some 1000 m in the Viking Graben and 250–500 m in the Central Graben. In the inner Moray Firth basin, Lower Cretaceous, partly turbiditic sandstone series reach a thickness of up to 1000 m.

During the Early Cretaceous, a general rise in sea level resulted in the progressive overstepping of basin edges and a gradual decrease in clastic influx into the North Sea area. In the Viking and Central Grabens, only minor hydrocarbon accumulations have been established in Lower Cretaceous series which, on a regional scale, lack good reservoir development.

Palinspastic reconstructions of regional structural cross-sections indicate that reflection seismically mappable fault-offsets across the northern Viking Graben account for 19 km of crustal stretching during the Jurassic and Early Cretaceous. In comparison crustal dilation across the central North Sea by Triassic to Early Cretaceous faulting amounts to some 25 km, of which about 8.5 km can be assigned to the Late Jurassic and Early Cretaceous (Fig. 1.18).

1.1.7. *Late Cretaceous and Palaeocene late rifting stage*

During the Late Cretaceous, rifting activity abated further in the North Sea area whereas the Norwegian–Greenland Sea and the Faeroe–West Shetland rift system remained active (Bukovics and Ziegler 1985; Duindam and van Hoorn 1987). In the Norwegian–Greenland Sea, crustal separation was achieved at the transition from the Palaeocene to the Eocene.

In the graben systems of the North Sea, many of the faults that control the structural relief of the late-Kimmerian unconformity die out upwards within the Upper Cretaceous strata; only a few master-faults show continuous displacement growth during the Late Cretaceous and Palaeocene. Intra-Senonian block-faulting is evident locally. Palinspastic reconstructions of regional cross sections through the northernmost part of the Viking Graben indicate that crustal extension by faulting amounted to some 2.5 km during the late Cretaceous (Fig. 1.18); the bulk of these displacements was taken up in the northwestern parts of this cross section along the boundary fault of the Magnus Block which subparallels the strike of the Faeroe-West Shetland and the mid-Norway basin fault systems. For the Central Graben, fault offsets indicate an extension value of about 1.5 km during the Late Cretaceous and Palaeocene.

Isopach maps and facies patterns of the Upper Cretaceous and Palaeocene series of the North Sea basin suggest that its late-rifting stage evolution was mainly governed by regional thermal subsidence (Ziegler 1982). The global Late Cretaceous rise in sea level played an important role in the progressive overstepping of the basin margins and a rapid decrease in clastic influx. This trend was reversed during the middle Palaeocene.

In the North Sea basin, clear water conditions prevailed during the Cenomanian, as evidenced by the onset of chalk deposition. This sedimentary regime persisted until Danian times (Fig. 1.15). Upper Cretaceous to Danian chalks attain thicknesses of 1000–2000 m in the Central Graben, thin over its

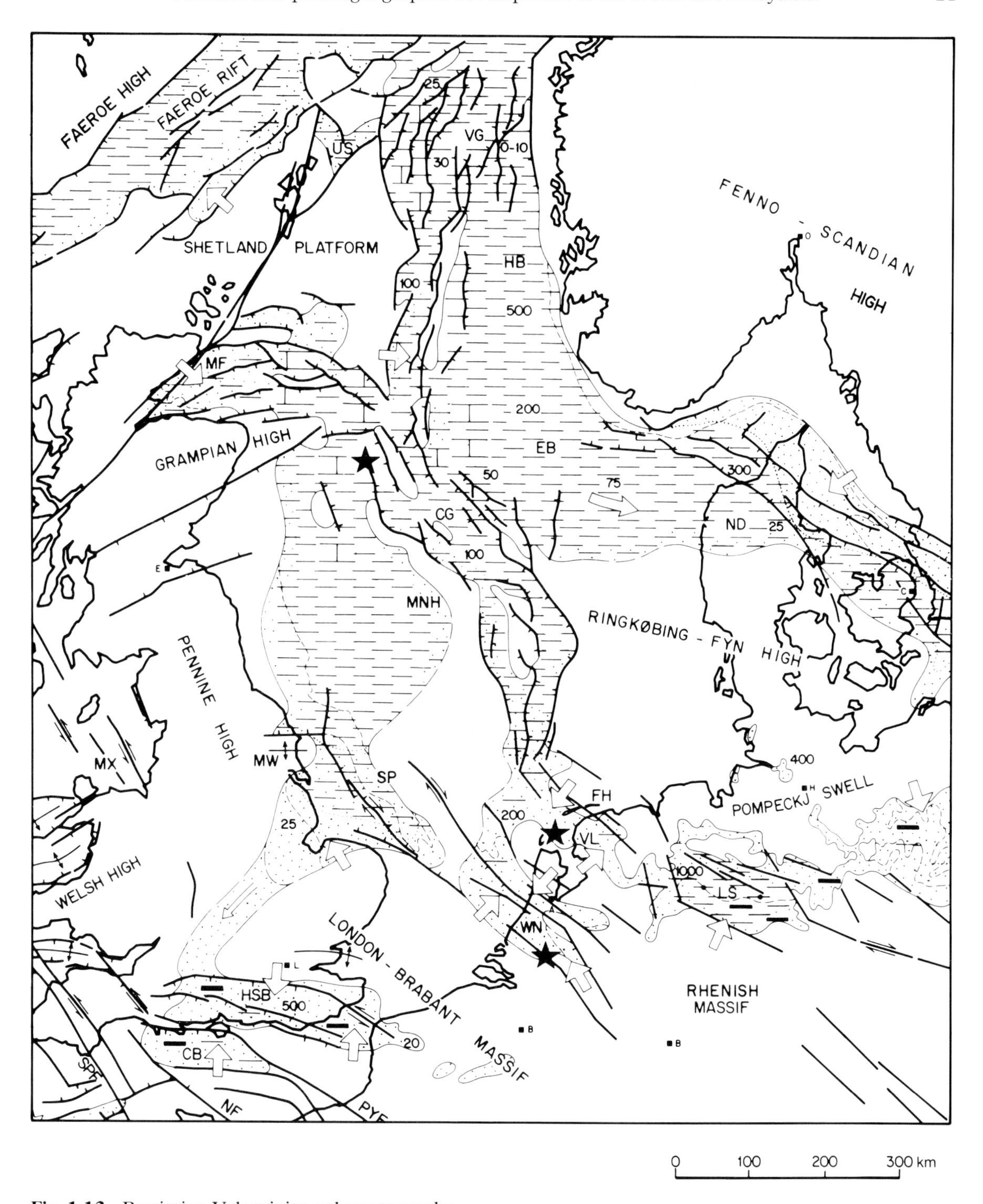

Fig. 1.13. Berriasian-Valanginian palaeogeography.

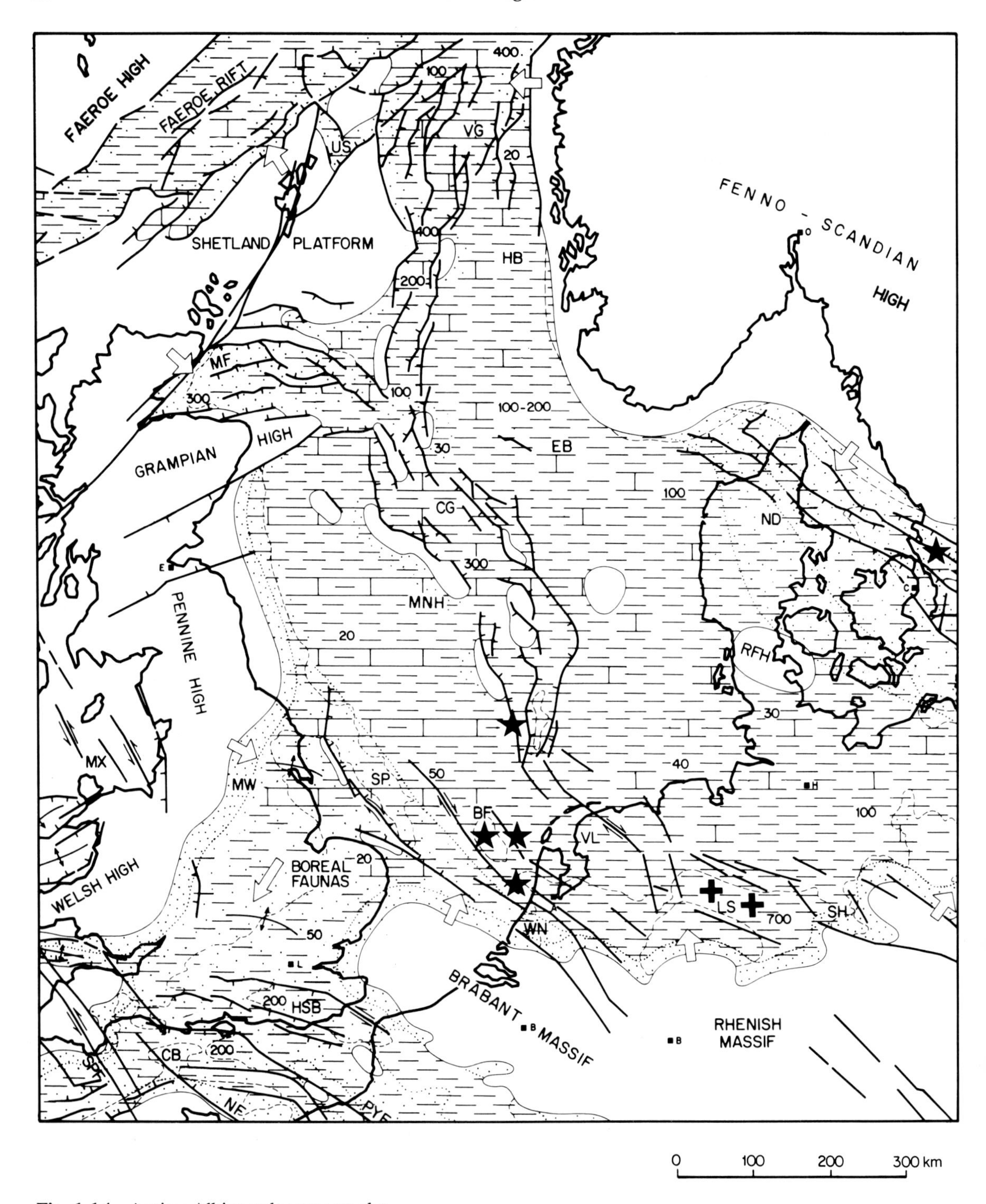

Fig. 1.14. Aptian-Albian palaeogeography.

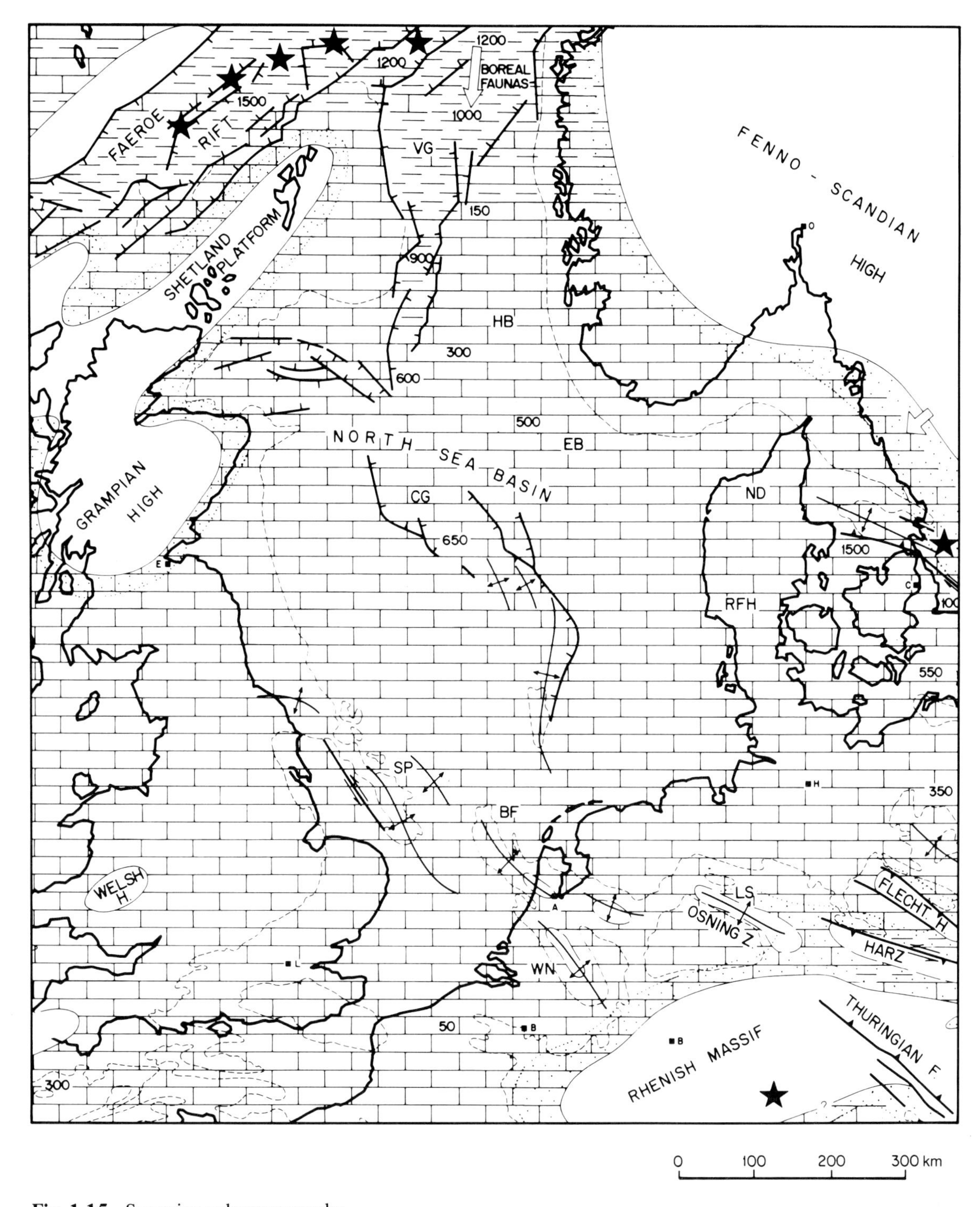

Fig. 1.15. Senonian palaeogeography.

shoulders to some 250 m, and thicken toward their erosional edges near the present day coastlines of Norway and the United Kingdom. In the Viking Graben, chalks give way northwards to pelagic marls and clays that range in thickness from 1000 to 2500 m. It is likely that Late Cretaceous basin edges had substantially overstepped the present-day coastlines of Norway and of the British Isles.

Upper Cretaceous chalks and marls progressively infilled the sea-floor topography of the Central and Viking Grabens, as suggested by their onlap against intra-basinal highs and the graben flanks (Fig. 1.18). Although Upper Cretaceous sedimentation rates exceeded subsidence rates and gradually rising sea levels, shallow-water conditions were never established in the axial parts of the North Sea graben system (Hancock and Scholle 1975; Watts *et al.* 1980; Hatton 1986; Kennedy 1987).

Maastrichtian and Danian chalks form the reservoir of major hydrocarbon accumulations trapped in salt-induced anticlinal structures, particularly in the Norwegian and Danish parts of the Central Graben (e.g. Ekofisk, Eldfisk, Albuskjell, Dan., Gorm fields: Ziegler 1980; D'Heur 1986; Damtoft *et al.* 1987; Spencer *et al.* 1987). These high-porosity, low-permeability reservoirs are characterized by very low hydrocarbon recovery factors (less than 10 per cent; Sørensen *et al.* 1986).

During the Senonian, igneous activity increased along the Faeroe–West Shetland and mid-Norway rift zone, culminating during the Palaeocene in a major volcanic event (Thulean volcanism) that affected much of the Arctic–North Atlantic borderlands. Following earliest Eocene crustal separation between Greenland, the Rockall–Hatton Bank, and Norway, this volcanic activity ceased abruptly (Ziegler 1988). The regional earliest Eocene ash-marker of the North Sea basin (Jacqué and Thouvenin 1975) is thought to be related to a short-lived pulse of explosive subaerial volcanism that accompanied the actual crustal separation phase in the northern North Atlantic and the Norwegian–Greenland Sea (Roberts *et al.* 1984).

During the Senonian and Palaeocene, north-west Europe as a whole became affected by intra-plate compressional stresses that can be related to the Alpine collision of Africa and Europe (Ziegler 1987, 1988). These stresses induced the inversion of Mesozoic grabens and troughs at distances up to 1300 km to the north of the Alpine collision front. Such inversion movements are evident in the North Danish and Egersund basins, and in the British, Danish, and Dutch parts of the Central Graben (Fig. 1.16). Particularly in the Danish sector of the central North Sea, these deformations are responsible for structural trap formation (e.g. Tyra gas field). Moreover, earthquakes that accompanied these inversion movements triggered mass flows of chalk during the Danian, as evident in the reservoirs of the Norwegian and Danish chalk fields (Hatton 1986). At the transition from the Palaeocene to the Eocene, these compressional intra-plate stresses apparently relaxed.

The timing of inversion movements in the central and southern North Sea, and the build-up of the Thulean volcanic event suggests that these Alpine intraplate compressional stresses counteracted tensional forces governing the evolution of the northern North Atlantic and the Norwegian–Greenland Sea (Ziegler 1988). Furthermore, these stresses may have contributed towards the Senonian and Palaeocene subsidence of the North Sea basin, the mid-Palaeocene uplift of the northern British Isles, and regional relative sea-level fluctuations by inducing broad scale lithospheric deflections (Cloetingh *et al.* 1987; Kooi and Cloetingh 1989).

In conjunction with the Thulean thermal event, areas bordering the Rockall and Faeroe–West Shetland trough became progressively domed during the latest Cretaceous and Palaeocene. From the uplifted northern British Isles, clastics were initially shed during the Danian into the Viking Graben. Clastic influx into the North Sea basin accelerated sharply at the end of the Danian. This coincides with the regional termination of chalk deposition. During the mid-Palaeocene to early Eocene, deltaic complexes prograded from the Shetland Platform and the Scottish Highlands towards the margins of the partly fault-controlled Viking and Central Grabens (Figs 1.10 and 1.16). Repeated slope-failure triggered density currents that transported sands into the axial parts of the Viking and Central Grabens, where they accumulated in complex submarine fan systems in water depths of the order of 500 m (Heritier *et al.* 1979; Morton 1982; Conort 1986). These sandstones contain important gas accumulations in the Viking Graben (e.g. Frigg and Heimdal fields: Spencer *et al.* 1987) and oil and gas/condensate fields mainly in the British parts of the Central Graben (e.g. Forties, Montrose: Fowler 1975; Carman and Young 1981). All these fields are charged with

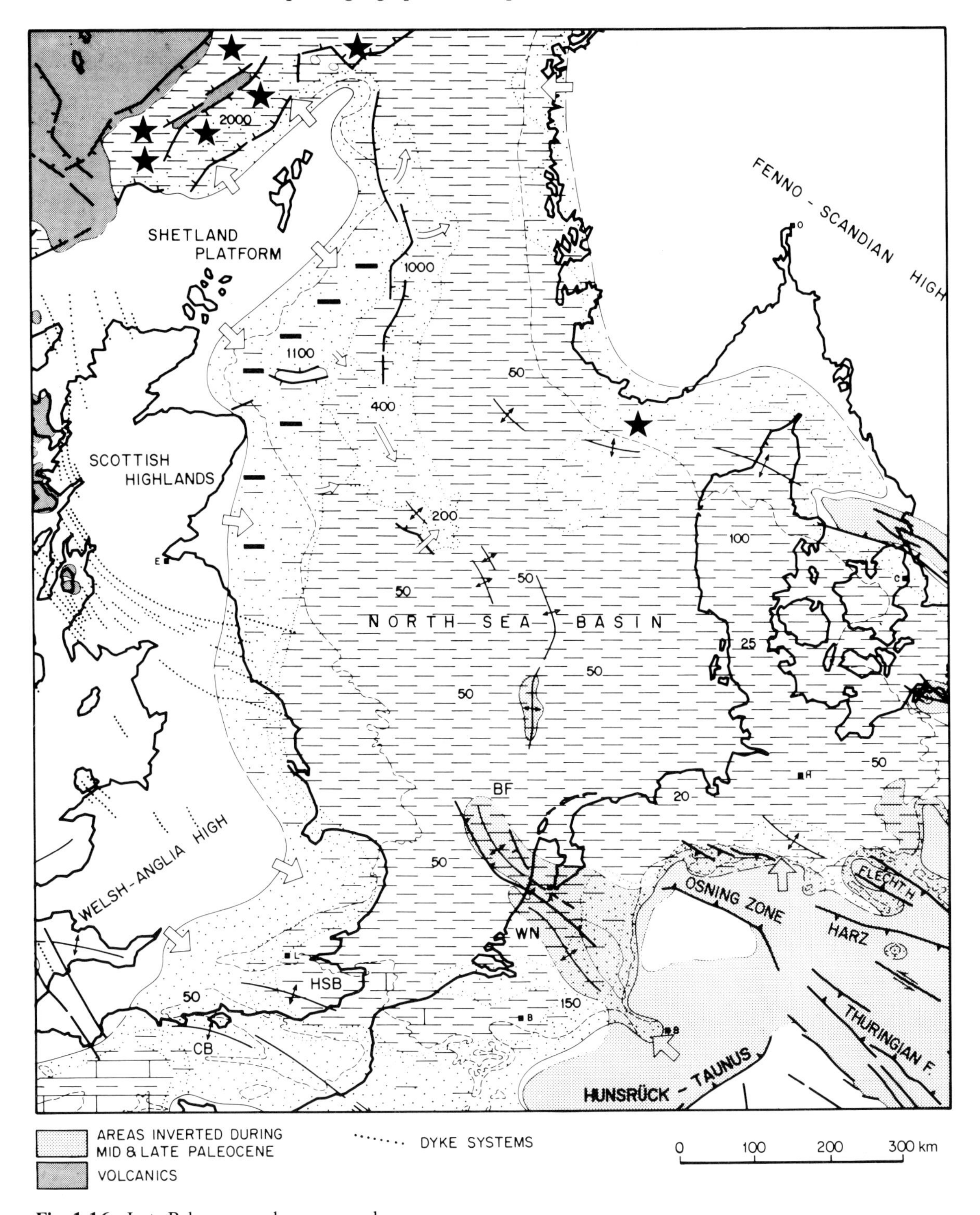

Fig. 1.16. Late Paleocene palaeogeography.

hydrocarbons generated from Kimmeridgian source rocks.

1.1.8. *Eocene to recent post-rifting stage*

Following early Eocene crustal separation in the northern North Atlantic and the Norwegian–Greenland Sea, the North Sea basin became tactonically quiescent; its subsequent evolution was essentially governed by thermal relaxation of the lithosphere and its loading by sediments.

The regional structure map of the top Chalk series shown in Fig. 1.17 describes the geometry of the Cainozoic North Sea basin, and indicates that it developed by broad crustal downwarping. In the central parts of this saucer-shaped basin, Cainozoic series reach a thickness of 3.5 km; the basin axis is aligned with the trace of the Viking and Central Grabens.

There is no evidence for post-Palaeocene reactivation of the Mesozoic North Sea graben system, and the Eocene and younger Rhine rift clearly dies out northwards in the coastal areas of the Netherlands. In the central and southern North Sea, local subsidence anomalies and faulting are caused by the diapirism of the Permian salts that played an important role as a trap-forming mechanism at Mesozoic and lower Tertiary reservoir levels (e.g. Ekofisk province in southern Norway). Additional traps containing major hydrocarbon accumulations developed during the Tertiary by differential compaction over deep-seated horst blocks (e.g. Forties and Frigg fields: Ziegler 1980).

The thickness of Eocene and younger sediments generally expands from the margins of the North Sea basin toward its axis. Notable exceptions are the Oligocene deltaic complexes that prograded from the Shetland Platform into the deeper waters of the Viking Graben (Ziegler 1982).

With the Neogene development of the north European river system, deltas began to prograde during the Miocene and Pliocene into the southern North Sea, and also from the Fenno-Scandian border zone into the central North Sea; however, clastic influx from the Norwegian coast remained at a generally low level. In the central North Sea, Neogene and Quaternary open marine clays are some 2000 m thick. In this area shallow marine conditions were established during the Miocene, suggesting that sedimentation rates outpaced subsidence rates. During the Miocene and Pliocene, repeated sea-level fluctuations strongly influenced sedimentation patterns in the North Sea, particularly in its marginal parts, where regionally correlative disconformities are evident, for instance at the base and top of the Miocene. In the central North Sea, Quaternary series range in thickness between 500 and 1000 m (Caston 1979; Nilsen *et al.* 1986), and thus reflect a sharp acceleration of subsidence rates. This may be related to a renewed build-up of NNW–SSE oriented compressional intra-plate stresses as evident from borehole break-out data and the analysis of earthquake focal mechanisms (Klein and Barr 1986; Kooi and Cloetingh 1989).

Along the eastern margin of the North Sea basin Cainozoic and Mesozoic series became deeply truncated during the Pleistocene, probably in conjunction with the isostatic adjustments of the Fenno-scandian Shield to Pleistocene glaciation and deglaciation. Furthermore, it is possible that compressional intra-plate stresses have contributed to the uplift of the Fenno-scandian Shield (Fig. 1.18).

The great thickness of the Cainozoic series has contributed much towards the maturation of the Upper Jurassic source rocks in the Viking and Central Grabens. However, in the Egersund-Horda basin (paralleling the coast of southern Norway) the thickness of Cainozoic deposits is generally insufficient for the regional maturation of the Kimmeridgian source rock that is well developed in these basins. In the Egersund basin only local hydrocarbon kitchens developed during the Neogene in salt-induced Cretaceous subsidence centres.

In the North Sea, Oligocene and younger series contain only minor biogenic gas accumulations.

1.2. Crustal configuration and amount of upper crustal extension

The present day crustal configuration of the North Sea basin (as derived from gravity, reflection, and refraction seismic data) shows that the zone of maximum crustal thinning coincides closely with the trace of the Viking and Central Graben (Fig. 1.18: Donato and Tully 1981; Barton and Wood 1984; Hospers *et al.* 1985; Beach *et al.* 1987; Zervos 1987; Ziegler 1988). Refraction and deep reflection seismic lines indicate that the Moho rises from a depth of 33–34 km under the coast of Norway to 22–24 km and 20 km under the Viking and Central Grabens, respectively, and descends again to a depth

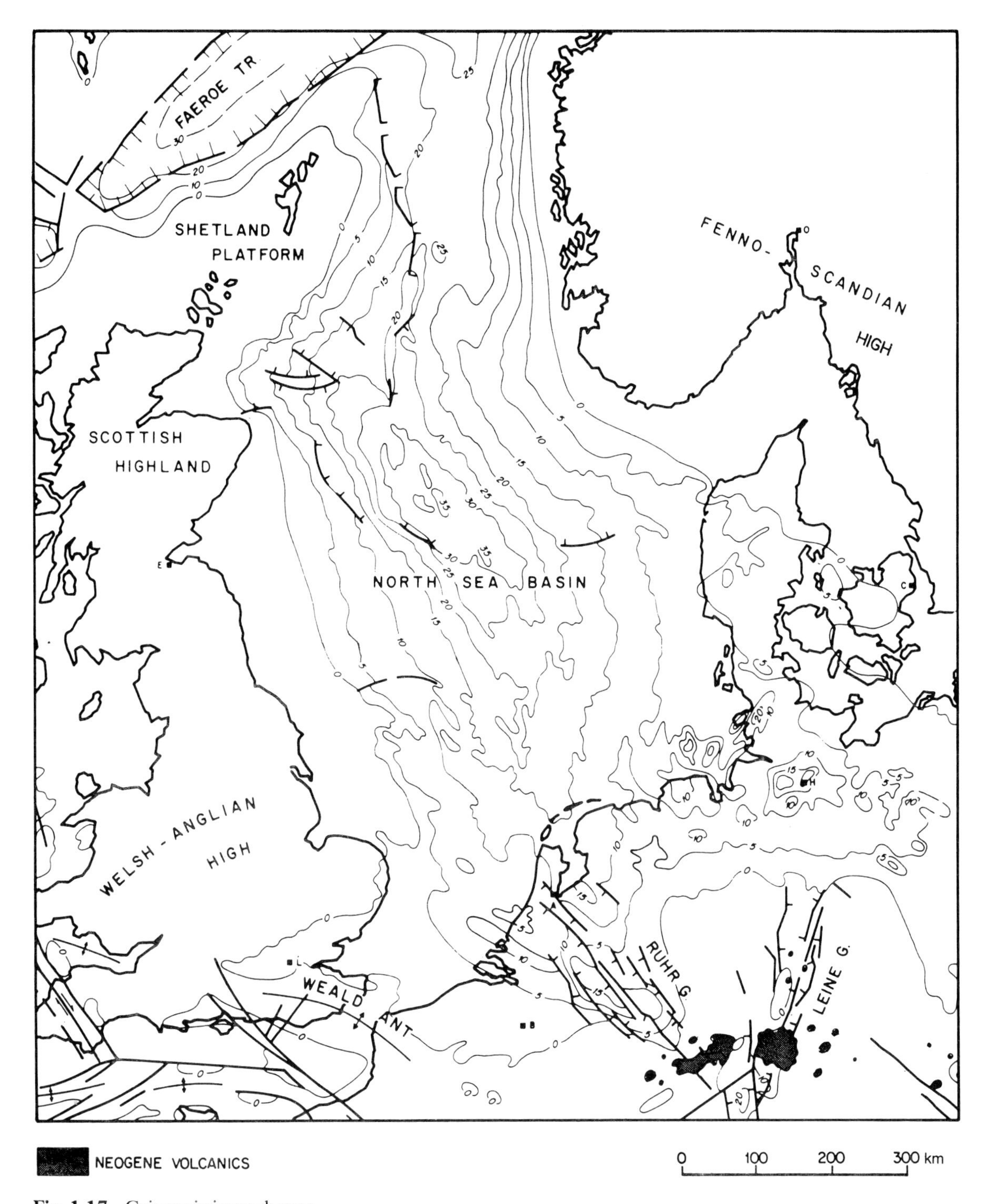

Fig. 1.17. Cainozoic isopach map.

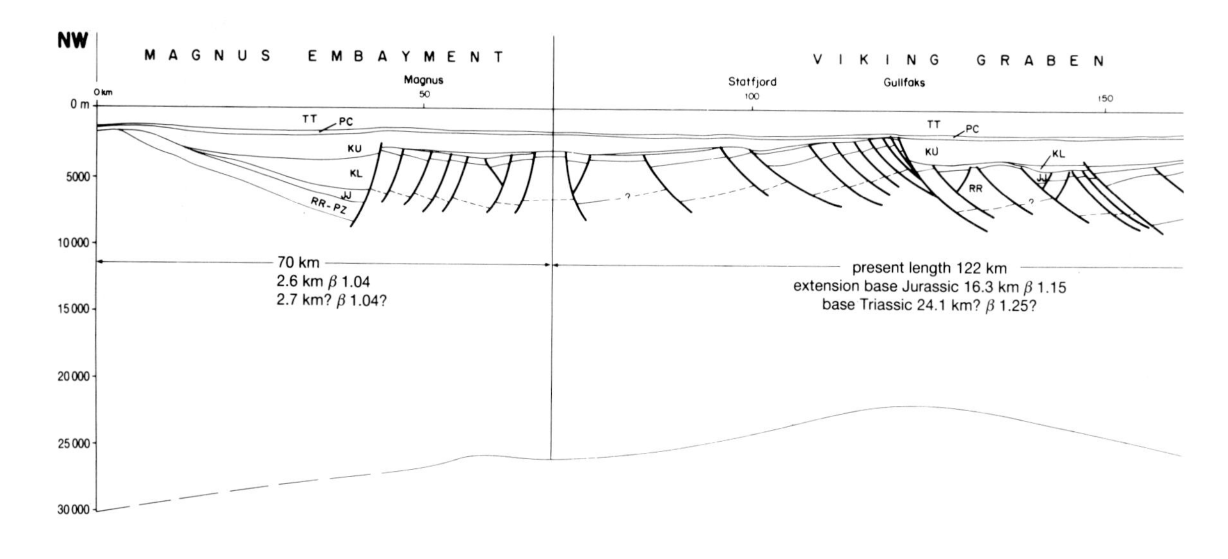

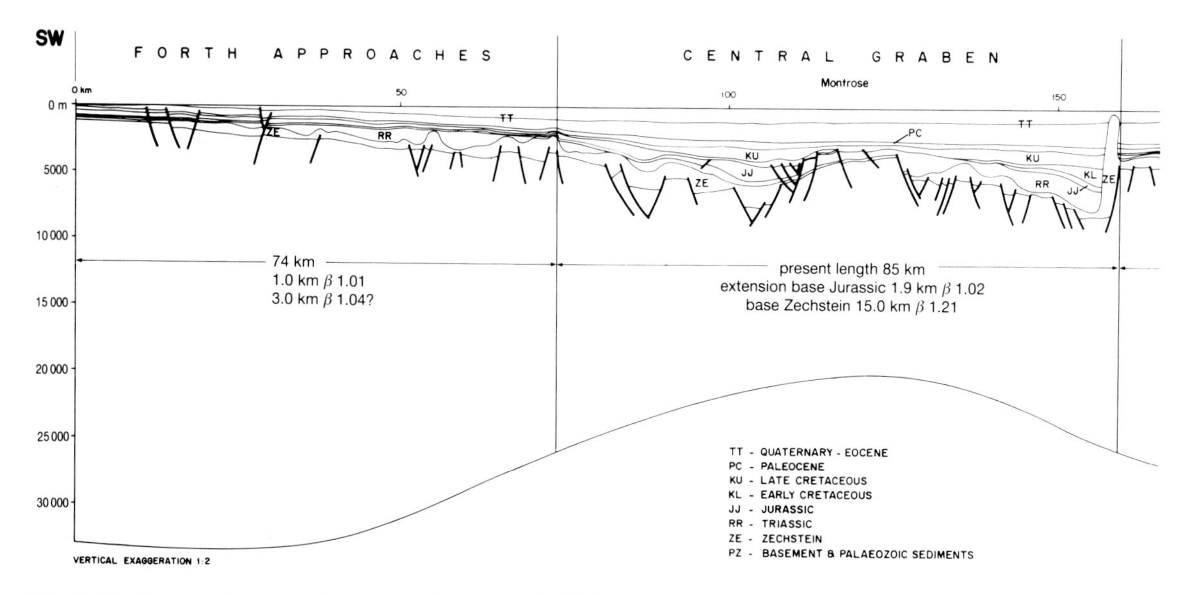

Fig. 1.18. Regional structural cross-sections through Viking and Central Grabens.

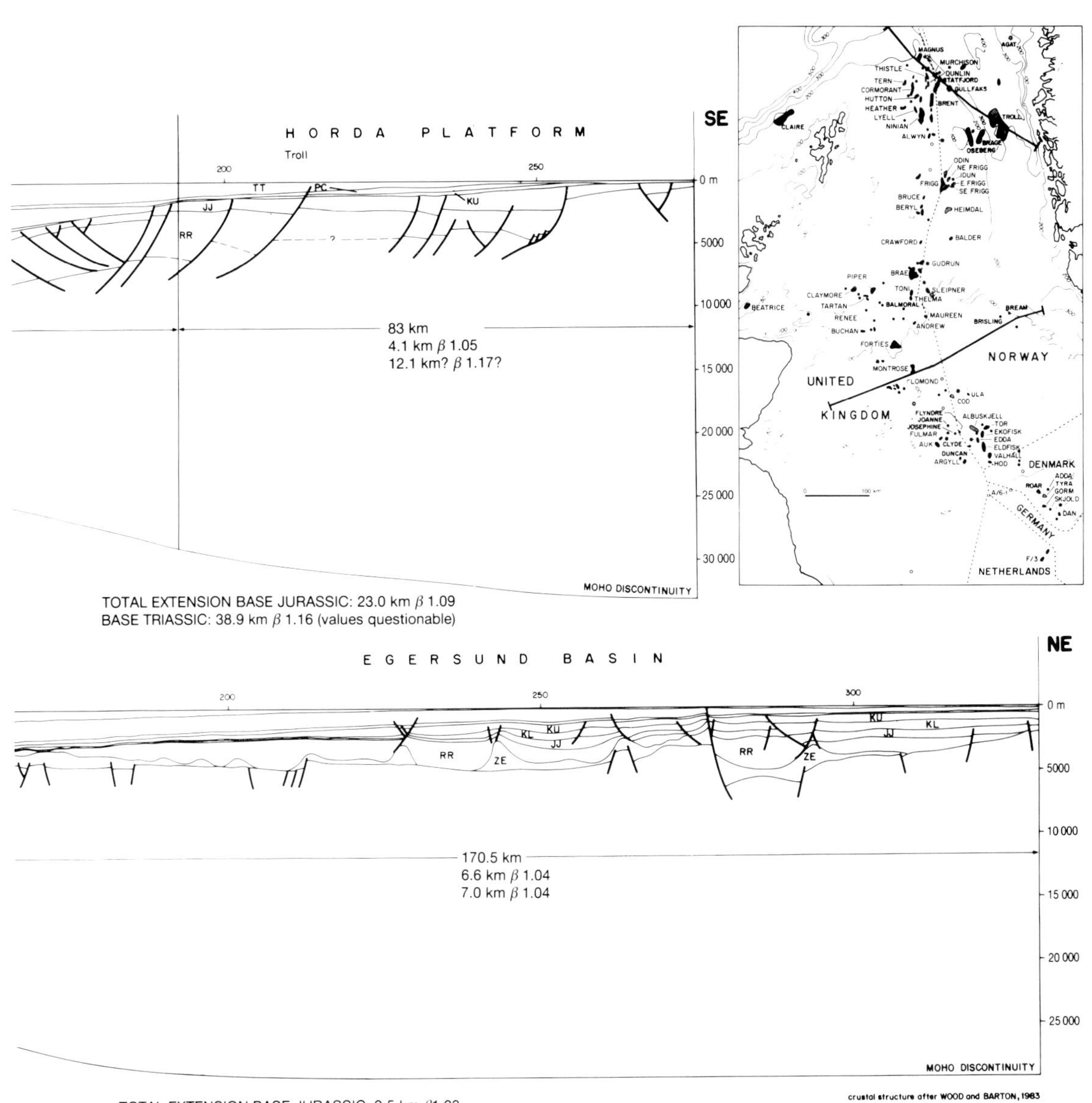

TOTAL EXTENSION BASE JURASSIC: 23.0 km β 1.09
BASE TRIASSIC: 38.9 km β 1.16 (values questionable)

TOTAL EXTENSION BASE JURASSIC: 9.5 km β1.03
BASE ZECHSTEIN: 25.0 km β 1.08

of 30 and 32 km under Britain and the Shetland platform, respectively. Under the axial parts of the Viking and Central Grabens, in which sediments reach a thickness of 8–10 km, the continental crust is 60 per cent thinner than beneath Britain, the Shetland Platform, and the Fenno-scandian Shield. The available refraction data indicate that the upper mantle of the North Sea area is characterized by a velocity of 8.1–8.2 km s^{-1} (Barton and Wood 1984). Deep reflection profiles show that the lower crust is highly reflective and diffractive under the Viking and Central Grabens, and that its reflectivity decreases toward the Norwegian coast and the Shetland platform, where a fairly discrete Moho reflection is evident (Beach 1985; Beach *et al.* 1987; Gibbs 1987a, b; Ziegler 1988).

Assuming that the crust of the North Sea was thinned during its Mesozoic rifting phase by mechanical stretching only (McKenzie 1978), then the amount of extension across Central Graben would be about 100–105 km (Wood and Barton 1983; Barton and Wood 1984; Sclater *et al.* 1986) and for the Viking Graben of the order 100–130 km. In the latter case the large range of uncertainty is due to the poor definition of the base of the syn-rift sediments; the high value corresponds to the hypothetical base Triassic as shown in Fig. 1.18. As discussed earlier, it is uncertain and probably unlikely that at the end of the Palaeozoic, the crust of the North Sea had a uniform thickness of 32 to 34 km.

Multichannel reflection seismic data across the Viking Graben and the Horda Platform show that extension of syn-rift sediments by faulting is, at base Jurassic levels, of the order of only 20 km (Fig. 1.18). Although the base of the early rift-syn Triassic sediments cannot be mapped on a regional scale, it is unlikely that the total amount of Mesozoic crustal extension across the northern Viking Graben exceeds 30 km. This is in keeping with the geometry of intra-Triassic reflections, that show no major divergence from the hangingwall to the footwall portion of individual fault blocks. Moreover, there is no evidence for low-angle normal faults being cut by a second generation of steeper fault systems. Correspondingly, upper crustal stretching factors for the North Viking Graben transect by major Mesozoic faulting are of the order of 1.16. This compares with a stretching factor of about 1.5 for the southern parts of the West Shetland platform where Permo-Triassic redbeds, contained in half grabens, are characterized by strongly diverging reflection patterns. For the West Shetland Trough, Late Jurassic to Palaeocene stretching factors by major faulting are of the order of 1.3 as indicated by reflection seismic data (Duindam and van Hoorn 1987; Kirton and Hitchen 1987). This shows that reflection seismic data are indeed able to resolve the structural configuration of basins characterized by higher extension factors than those observed in the North Sea, an aspect that has sometimes been contested.

In the central North Sea, extension values by major faulting (as obtained at the level of the base Zechstein pre-rift sequence) are in the range of 25 km (Fig. 1.18). Although it is possible that this amount can be increased by some 10 km on account of numerous small faults that are subject to interpretational differences or that cannot be detected by the reflection seismic tool, it is unlikely that displacements along such faults could account for a doubling or even a quadrupling of the amount of net extension as required by the stretching model (McKenzie 1978; Wood and Barton 1983; Ziegler 1983; Barton and Wood 1984; Shorey and Sclater 1988).

In the central North Sea, as in the Viking Graben, there is no seismic evidence for flat-laying extensional faults being cut by a second generation of steeper faults. Moreover, fault blocks as defined at the base Zechstein salt level are generally characterized by gentle dipslopes. This is not compatible with the arguments advanced by Shorey and Sclater (1988) that claim that 'hidden' faults could account for a doubling of the measured extension value to some 60 km.

Similarly, the amount of upper crustal extension postulated by the stretching model for the Witch Ground and the outer Moray Firth area is at variance with the amount of extension observed on reflection seismic lines (Christie and Sclater 1980; Smythe *et al.* 1980).

For the Danish part of the Central Graben and the Horn Graben, reflection seismic data suggest a combined extension value of no more than 15 km.

Palinspastic reconstructions of the two regional seismic lines crossing the Viking and the Central Grabens suggest that the bulk of the crustal extension across the North Sea rift system occurred during Triassic to earliest Cretaceous times. There is no evidence to support separation of a Triassic–Early Jurassic rifting phase from a late Jurassic–Early Cretaceous phase, other than the rift-induced mid-Jurassic thermal doming of the central North Sea. Although not readily quantifiable, there is evidence that the rate of crustal stretching accelerated during the Late Jurassic and earliest Cretaceous, and thereafter abated rapidly.

Quantitative subsidence analyses for the central North Sea suggest basin-wide extension of 50–80 km (Christie and Sclater 1980; Wood and Barton 1983; Barton and Wood 1984; Hellinger *et al.* 1988). The validity of this type of approach to the quantification of crustal extension is questionable, particularly because in long-lived rifts the remnant thermal anomaly at the end of the rifting stage cannot be related directly to a crustal stretching factor (Ziegler 1988). Moreover, the realization that basin subsidence can be strongly influenced by intraplate stresses of a compressional as well as a tensional nature (Cloetingh 1988; Kooi and Cloetingh 1989) raises further doubts about the validity of the concept that subsidence curves for passive margins, and in particular for abandoned rifts, can be unreservedly translated into crustal stretching factors. In the case of the North Sea basin, such analyses have to account for the latest Cretaceous-Palaeocene compressional palaeostress field, and also the Recent stress field (Klein and Barr 1986).

The discrepancies observed between 'stretching' factors derived from crustal configuration, upper crustal extension based on reflection seismic data, and quantitative subsidence analyses cannot be reconciled with each other in terms of the original stretching model (McKenzie 1978). On the other hand, clay and sand model experiments suggest that extension by major faulting may amount to only 30–50 per cent of the total known extension (Krantz and Sclater 1988). Whether the results of these model experiments can be applied to the 'real world' remains an open question that needs to be resolved in order to understand the geodynamics of rifting.

Although upper crustal extension values (as determined from reflection seismic data) do not appear to vary substantially across the northern and the central North Sea, Cainozoic sediments in the central North Sea attain thicknesses up to 3.5 km, whereas in the northern North Sea they range between 2 and 2.5 km. The subsidence centre of the Cainozoic North Sea basin coincides with the largest gravity anomaly (Hospers *et al.* 1985; Zervos 1987) and, perhaps not coincidentally, also with the culmination of the mid-Jurassic North Sea Rift dome.

The available data point towards an important discrepancy between upper crustal mechanical stretching and lower crustal 'attenuation' factors, and the possibility that the mass of the lower crust is not being conserved during rifting. Similar evidence comes from other rifted basins (Pinet *et al.* 1987), suggesting that during rifting processes the geophysically defined Moho may be displaced upwards by physico-chemical processes, possibly involving the permeation of the lower crust by mantle-derived melts. This process may be associated with the development of the reflection-seismically defined lamination of the lower crust. However, it is realized that apart from basaltic sills, metamorphic layering and shear zones can also contribute to lower crustal reflectivity (Beach *et al.* 1987; Ziegler 1988).

If most of the lower crustal reflections can, indeed, be related to injection of mantle-derived material (Matthews 1986; Meissner 1986; Bois *et al.* 1988), it seems plausible that this process may ultimately cause a change of the physical properties of the lower crust and a gradual upward displacement of the geophysically-defined crust/mantle boundary, akin to the process discussed by Beloussov (1960, 1962). The crustal configuration of the central North Sea and its post-Middle Jurassic evolution suggest that such processes may become particularly effective in areas where, during a rift-doming stage, major amounts of asthenospheric melts have ascended to the crust/mantle boundary.

1.3. Conclusions

The Mesozoic North Sea rift forms an integral part of the Arctic–North Atlantic mega-rift system.

At the transition from the Permian to the Triassic the Norwegian–Greenland Sea rift propagated southwards into the North Sea area. In the North Sea, crustal extension causing the differential subsidence of the Viking, Central, and Moray Firth–Witch Ground Grabens and the Horda–Egersund half graben reached a peak during the Jurassic and earliest Cretaceous. Rifting activity abated during the Cretaceous and terminated with the earliest Eocene crustal separation in the Norwegian–Greenland Sea and the northern North Atlantic. The total rifting period in the North Sea spans some 175 Ma.

An important feature in the evolution of the North Sea rift system is the early middle Jurassic uplift of a major rift dome in the central North Sea, causing a reversal in the subsidence pattern of its Central Graben. Uplift of this dome was responsible for the development of important hydrocarbon reservoirs, both in the central and northern North Sea. Although less well documented, it appears that the areas of the Faeroe-West Shetland rift and of the triple junction between the Viking Graben and the Norwegian–Greenland Sea rift became thermally domed during the Callovian-Oxfordian and the Kimmeridgian respectively. Uplift of these domes is responsible for the accumulation of the Upper Jurassic sandstone reservoirs in the southern and northern Viking Graben respectively. All of these rift domes foundered shortly after their uplift, long before rifting activity in the respective grabens had ceased.

The latest Cretaceous to Palaeocene Thulean volcanic event, preceding crustal separation in the northern North Atlantic and the Norwegian–Greenland Sea, was paralleled by the build-up of intra-plate compressional stresses in north-west Europe. These can be related to the Alpine collision of Africa and Europe. The earliest Eocene crustal separation between Eurasia, Laurentia, and Greenland coincides with the relaxation of these intra-plate stresses.

Palaeocene–early Eocene thermal uplift of areas bordering the Rockall–Faeroe–West Shetland trough gave rise to the accumulation of deepwater fan deposits in the Viking and Central Grabens.

The axis of the Cainozoic North Sea thermal sag basin coincides with the Viking and Central Grabens. This illustrates that the zone of maximum crustal thinning corresponds also to the zone of maximum lithospheric attenuation, which supports the pure shear, depth-dependent stretching model (McKenzie 1978; Rowley and Sahagian 1986).

Geophysical data suggest, however, that a major discrepancy exists between upper and lower crustal thinning across the North Sea rift system, which raises the question whether the geophysically defined crust/mantle boundary remains stable during rifting processes. Crustal attenuation during rifting may be achieved by a combination of crustal stretching and physico-chemical processes affecting the crust/mantle boundary. This hypothesis needs to be substantiated by additional geophysical data, from North Sea and other rifts; however, if applicable, this concept would require a basic revision of the currently-favoured stretching models that have been advanced for the development of extensional basins (Badley *et al.* 1988). Furthermore, none of these models take into account the temporary uplift of such rift domes as the early Middle Jurassic central North Sea arch. The development of such domes can be related to the diapiric intrusion of asthenospheric melts to the crust-mantle boundary where they spread out laterally upon finding their density equilibrium (asthenolith; tensional failure model, Turcotte 1981; Ziegler 1982, 1988; Turcotte and Emerman 1983: magmatic underplating, Keen and de Voogd 1988).

In the case of the central North Sea dome, the early Middle Jurassic thermal uplift of the Central Graben area exceeded its contemporaneous isostatic subsidence (caused by continued crustal stretching). Evidence for continued Middle Jurassic crustal stretching across the North Sea rift system is provided by the subsidence pattern of individual fault blocks in the northern Viking Graben (Brown *et al.* 1987) and, in the Central Graben by the subcrop and onlap pattern of Triassic and Jurassic strata, respectively, against the regional mid-Kimmerian unconformity. The amount of Middle Jurassic extension that occurred in the Viking and Central Grabens is, however, difficult to quantify on a regional scale.

The foundering of temporary rift domes prior to termination of rifting activity in the respective grabens can be related to a number of processes, such as cooling and contraction of the asthenolith, thermomechanical subcrustal erosional processes involving the injection of mantle-derived material into the lower crust, causing an increase of its density, and possible phase changes (Moretti and Pinet 1987; Pinet *et al.* 1987).

These processes may be associated with a gradual ascent of the crustal isotherms and a commensurate upward displacement of the ductile-brittle deformation transition zone. This could be partly responsible for the increasingly strong rotation of individual fault blocks during the latest Jurassic-earliest Cretaceous phases of accelerated crustal extension which affected the formerly domed Faeroe-West Shetland trough, the North Sea Graben system and also the area of the Nordfjord ridge.

From a geodynamic point of view it is interesting to note that, particularly in the North Sea, the late-Kimmerian rifting pulse, spanning perhaps less than ten million years, was not associated with significant volcanic activity, nor with renewed doming of the

Central Graben area. It is speculated that tectonic subsidence (caused by crustal extension) exceeded thermal uplift (induced by lithospheric thinning) and that, as a consequence of the mid-Jurassic thermal pulse, the density of the upper mantle had decreased to the degree that renewed intrusion of large volumes of asthenospheric melts to the crust/mantle boundary was impeded.

Acknowledgements

The author is indebted to his colleagues in Shell's North Sea Exploration teams for their contributions to the synthesis presented in this paper. Special thanks are extended to Dr Berend van Hoorn for his contribution to the construction of the regional cross-sections contained in Fig. 1.18, to Mrs Ineke Hilberding who assisted in constructing the palaeogeographic maps, to Mrs Josje Kriest for preparing the regional structural cross-sections, to Mr Ruud van Aarle for drafting, and to Mrs Monique Molenaar and Mrs Wilma Ruggenberg for typing the manuscript. Permission to publish this paper was granted by Shell Internationale Petroleum Mij. B.V., and is gratefully acknowledged.

References

Badley, M. E., Price, J. D., Rambech Dahl, C., and Agdestein, T. (1988). The structural evolution of the northern Viking Graben and its bearing upon extensional mode of basic formation. *J. Geol. Soc., London*, **145**, 455–72.

Baird, R. A. (1986). Maturation and source rock evaluation of Kimmeridge Clay, Norwegian North Sea. *Am. Assoc. Petrol. Geol. Bulletin*, **70**, 1–11.

Barnard, P. C. and Cooper, B. S. (1981). Oils and source rocks of the North Sea area. In *Petroleum geology of the continental shelf of north-west Europe* (ed. L. V. Illings and G. D. Hobson), 169–75, London, Heyden and Son.

Barton, P. and Wood, R. (1984). Tectonic evolution of the North Sea basin: crustal stretching and subsidence. *Geophys. J. Roy. Astron. Soc.*, **79**, 987–1022.

Beach, A. (1985). Some comments on sedimentary basin development in the northern North Sea. *Scot. J. Geol.*, **21**, 493–512.

Beach, A., Bird, T., and Gibbs, A. (1987). Extensional tectonics and crustal structure: deep seismic reflection data from the northern North Sea Viking Graben. In *Continental extensional tectonics* (Ed. M. P. Coward, J. F. Dewey and P. L. Hancock), 467–76. Geol. Soc. Special Publication 28. Oxford, Blackwell Scientific Publications.

Beloussov, V. V. (1960). Development of the earth and tectogenesis. *J. Geophys. Res.*, **65**, 4127–46.

Beloussov, V. V. (1962). *Basic problems in geotectonics*. New York, McGraw Hill.

Bois, C., Cazes, M., Hirn, A., Mascle, A., Matte, P., Montadert, L., and Pinet, B. (1988). Crustal laminations in deep seismic profiles in France and neighbouring areas. In *Deep seismic reflection profiling of the continental lithosphere* (ed. D. H. Matthews and BIRPS group). *Geophys. J. Roy. Astron. Soc.*, **89** (in press).

Bowen, J. M. (1975). The Brent oil field. In *Petroleum and the continental shelf of north west Europe. Vol. I. Geology* (ed. A. W. Woodland), 353–77. Barking, Applied Science Publishers Ltd.

Brown, S. (1984). Jurassic. In *Introduction to the petroleum geology of the North Sea* (ed. K. W. Glennie), 103–31. Oxford, Blackwell Scientific Publications.

Brown, S., Richards, P. C., and Thomson, A. R. (1987). Patterns in the deposition of the Brent Group (Middle Jurassic) UK, North Sea. In *Petroleum Geology of north west Europe* (ed. J. Brooks and K. Glennie), 899–915.

Bukovics, C. and Ziegler, P. A. (1985). Tectonic development of the mid-Norway continental margin. *Marine Petrol. Geol.*, **2**, 2–22.

Carman, G. J. and Young, R. (1981). Reservoir geology of the Forties oilfield. In *Petroleum geology of the continental shelf of north west Europe* (ed. L. V. Illings and G. D. Hobson), 371–9. London, Heyden and Son.

Caston, V. N. D. (1979). The Quarternary history of the North Sea. In *Acta University Uppsala Symposium University Uppsala Annum Quingentesimum Celebrantis*, **2**, 23–8.

Cayley, G. T. (1987). Hydrocarbon migration in the Central North Sea. In *Petroleum geology of north west Europe* (ed. J. Brooks and K. Glennie), 549–55. London, Graham and Trotman.

Christie, P. A. and Sclater, J. G. (1980). An extension origin of the Buchan and Witchground Graben in the North Sea. *Nature*, **283**, 729–32.

Cloetingh, S. (in press). Intraplate stress: a new element in basin analysis. In *New perspectives in basin analysis* (ed. K. Kleinsphen and C. Pada). New York, Springer Verlag.

Cloetingh, S., Lambeck, K., and McQueen, H. (1987). Apparent sea-level fluctuations and a palaeostress field for the North Sea region. In *Petroleum geology of north west Europe* (ed. J. Brooks and K. Glennie), 49–57. London, Graham and Trotman.

Conort, A. (1986). Habitat of Tertiary hydrocarbons, South Viking Graben. In *Habitat of hydrocarbons on the Norwegian continental shelf* (ed. A. M. Spencer *et al.*), 159–70. Norwegian Petroleum Society; London, Graham and Trotman.

Cornford, C. (1984). Source rocks and hydrocarbons of the North Sea. In *Introduction to the petroleum geology of the North Sea* (ed. K. Glennie), 171–204. Oxford, Blackwell Scientific Publications.

Damtoft, K., Andersen, C., and Thomsen, E. (1987). Prospectivity and hydrocarbon plays of the Danish Central Trough. In *Petroleum geology of north west Europe* (ed. J. Brooks and K. Glennie), 403–17. London, Graham and Trotman.

De'Ath, N. G. and Schuyleman, S. F. (1981). The geology of the Magnus Field. In *Petroleum geology of the continental shelf of north-west Europe, Vol. 1, Geology* (ed. L. V. Illings and G. D. Hobson), 342–51. London, Heyden and Son.

D'Heur, M. (1986). The Norwegian chalk fields. In *Habitat of hydrocarbons on the Norwegian continental shelf* (ed. A. M. Spencer *et al.*), 77–89. London, Graham and Trotman.

Dixon, J. E. and Fitton, J. G. (1981). The tectonic significance of post-Carboniferous igneous activity in the North Sea basin. In *Petroleum geology of the continental shelf of north-west Europe* (ed. L. V. Illings and G. D. Hobson), 121–37. London, Heyden and Son.

Donato, J. A. and Tully, M. C. (1981). A regional interpretation of North Sea gravity data. In *Petroleum geology of the continental shelf of north-west Europe* (ed. L. V. Illings and G. D. Hobson), 65–75. London, Heyden and Son.

Duindam, P. and van Hoorn, B. (1987). Structural evolution of the West Shetland continental margin. In *Petroleum geology of north west Europe* (ed. J. Brooks and K. Glennie), 765–73. London, Graham and Trotman.

Fisher, M. J. (1986). Triassic. In *Introduction to the petroleum geology of the North Sea* (ed. K. W. Glennie), 113–33. Oxford, Blackwell Scientific Publications.

Fowler, C. (1975). The geology of the Montrose Field. In *Petroleum and the continental shelf of north west Europe, Vol. 1: Geology* (ed. A. W. Woodland), 467–77. Barking, Applied Science Publishers Ltd.

Gibbs, A. D. (1987a). Linked tectonics of the northern North Sea basins. In *Sedimentary basins and basin-forming mechanisms* (ed. C. Beaumont and A. J. Tankard). *Canadian Soc. Petrol. Geol. Mem.*, **12**, 163–71.

Gibbs, A. D. (1987b). Deep seismic profiles in the northern North Sea. In *Petroleum geology of north west Europe* (ed. J. Brooks and K. Glennie), 1025–8. London, Graham and Trotman.

Glennie, K. W. (1984). Early Permian-Rotliegend. In *Introduction to the petroleum geology of the North Sea* (ed. K. Glennie), 41–60. Oxford, Blackwell Scientific Publications.

Glennie, K. W. (1986). Development of N. W. Europe's southern Permian gas basin. In *Habitat of Palaeozoic gas in N. W. Europe* (ed. J. Brooks, J. C. Goff and B. van Hoorn). *Geol. Soc., Special Publication* **23**, 3–22.

Graue, E., Helland-Hansen, W., Johnsen, J., Lomo, L., Nottvedt, A., Ronning, K., Ryseth, A., and Steel, R. (1987). Advance and retreat of Brent Delta System, Norwegian North Sea. In *Petroleum geology of north west Europe* (ed. J. Brooks and K. Glennie), 915–37. London, Graham and Trotman.

Hamar, G. P., Fjaeran, T., and Hesjedal, A. (1983). Jurassic stratigraphy and tectonics of the south-southeastern Norwegian offshore. In *Petroleum geology of the southeastern North Sea and the adjacent onshore areas* (ed. J. P. H. Kaasschieter and T. J. A. Reijers), 103–15. Amsterdam, De Bussy Ellermans Harms B. V.

Hancock, J. M. and Scholle, P. A. (1975). Chalk of the North Sea. In *Petroleum and the continental shelf of north west Europe: Vol. I, Geology*, (ed. D. W. Woodland), 413–27. Barking, Applied Science Publishers Ltd.

Harland, W. B., Cox, A. V., Llewellyn, P. G., Pickton, C. A. G., Smith, A. G., and Walters, R. (1982). *A geological time scale*. Cambridge University Press.

Harris, J. P. and Fowler, R. M. (1987). Enhanced prospectivity of the mid-late Jurassic sediments of the South Viking Graben, northern North Sea. In *Petroleum geology of north west Europe* (ed. J. Brooks and K. Glennie), 879–98. London, Graham and Trotman.

Hatton, I. R. (1986). Geometry of allochthonous Chalk Group *Marine Petrol. Geol.*, **3**, 79–98.

Hellem, T., Kjemperud, A., and Ourebo, O. K. (1986). The Troll Field: a geological/geophysical model established by the PL085 Group. In *Geology of the Norwegian oil and gas fields* (ed. A. M. Spencer *et al.*), 217–41. London, Graham and Trotman.

Hellinger, S. J., Sclater, J. G., and Giltner, J. (1988). Mid-Jurassic through mid-Cretaceous extension in the Central Graben of the North Sea: Part 1, Estimates from subsidence. *Basin Research*, **1**, 191–200.

Hemmingway, J. E. (1974). Jurassic. In *The geology and mineral resources of Yorkshire* (ed. D. H. Raynes and J. E. Hemmingway). *Yorkshire Geol. Soc.*, 161–223.

Heritier, F. E., Lossel, P., and Wathne, E. (1979). Friggs Field—Large submarine fan trap in Lower Eocene rocks of North Sea Viking Graben. *Am. Assoc. Petrol. Geol. Bull.*, **63**, 1999–2020.

Hoffmann, K. (1949). Zur Palaeogeographie des Nordwestdeutschen Lias und Dogger. In *Erdoel un*

Tektonic in Nordwest Deutschland (ed. A. Benz), 97–113. Hannover-Celle.

Hospers, J., Finnstrøm, E. G., and Rathore, J. S. (1985). A regional gravity study of the North Sea (56–62N), *Geophys. Prospecting*, **33**, 543–66.

Jacqué, M. and Thouvenin, J. (1975). Lower Tertiary tuffs and volcanic activity in the North Sea. In *Petroleum geology of the continental shelf of north west Europe: Vol. I, Geology* (ed. A. W. Woodland), 455–65. Barking, Applied Science Publishers Ltd.

Johnson, R. J. and Dingwall, R. G. (1981). The Caledonides: their influence on the stratigraphy of the northwest European continental shelf. In *Petroleum geology of the continental shelf of north west Europe* (ed. L. V. Illings and G. D. Hobson), 85–98. London, Heyden and Son Ltd.

Karlsson, W. (1986). The Snorre, Statfjord and Gullfaks oilfields and the habitat of hydrocarbons on the Tampen Spur, offshore Norway. In *Habitat of hydrocarbons on the Norwegian continental shelf* (ed. A. M. Spencer *et al.*), 181–97. London, Graham and Trotman.

Keen, C. E. and de Voogd, B. (1988). The continent-ocean boundary at the rifted margin off eastern Canada: new results from deep seismic reflection studies. *Tectonics*, **7**, 107–24.

Kennedy, W. J. (1987). Sedimentology of Late Cretaceous-Palaeocene Chalk reservoirs, North Sea Central Graben. In *Petroleum geology of north west Europe* (ed. J. Brooks and K. Glennie), 469–81. London, Graham and Trotman.

Kirton, S. R. and Hitchen, K. (1987). Timing and style of crustal extension north of the Scottish mainland. In *Continental extensional tectonics* (ed. M. P. Coward, J. F. Dewey, and P. L. Hancock). *Geol. Soc. London, Special Publication*, **28**, 501–10. Oxford, Blackwell Scientific Publications.

Klein, R. J. and Barr, M. V. (1986). Regional state of stress in western Europe. In *Rock stress and rock stress measurements* (ed. O. Stephensson), 33–44. Lund, Centek Publishers.

Kooi, H. and Cloetingh, S. 1989. Intraplate stress and the tectonic-stratigraphic evolution of the central North Sea. In *Extensional tectonics and stratigraphy of the North Atlantic margins* (ed. A. J. Tankard and H. R. Balkwill). *Am. Assoc. Petrol. Geol. Me*, **46**, 541–58.

Krantz, S. A. and Sclater, J. G. (1988). Internal deformation in clay models of extension by block faulting. *Tectonics*, **7(4)**, 823–32.

McKenzie, D. P. (1978). Some remarks on the development of sedimentary basins. *Earth and Planetary Science Letters*, **40**, 25–32.

Maher, C. E. (1981). The Piper oilfield. In *Petroleum geology of the continental shelf of north west Europe* (ed. L. V. Illings and G. D. Hobson), 358–71. London, Heyden and Son.

Matthews, D. H. (1986). Seismic reflections from the lower crust around Britain. In *The nature of the lower continental crust* (ed. J. B. Dawson, D. A. Carswell, J. Hall, and K. H. Wedepohl). *Geol. Soc. Special Publication* **24**, 11–22. Oxford, Blackwell Scientific Publications.

Meissner, R. (1986). The continental crust, a geophysical approach. In *International geophysical series*, vol. 34 (ed. W. L. Donn). London, Academic Press.

Moretti, I. and Pinet, B. (1987). Discrepancy between lower and upper crustal thinning. In *Sedimentary basins and basin-forming mechanisms* (ed. C. Beaumont and A. J. Tankard), *Can. Soc. Petrol. Geol., Mem.*, **12**, 233–9.

Morton, A. C. (1982). Lower Tertiary sand development in the Viking Graben, North Sea. *Amer. Assoc. Petrol. Geol.*, *Bulletin* **66**, 1542–59.

Nelson, P. H. H. and Lamy, J. (1987). The More/West Shetlands area: a review. In *Petroleum geology of north west Europe* (ed. J. Brooks and K. Glennie), 775–84. London, Graham and Trotman.

Nilsen, O. B., Sørensen, S., Thiede, J., and Skarbo, O. (1986). Cenozoic differential subsidence of North Sea. *Am. Assoc. Petrol. Geol.*, *Bulletin* **70**, 276–98.

Olsen, J. C. (1987). Tectonic evolution of the North Sea region. In *Petroleum geology of north west Europe* (ed. J. Brooks and K. Glennie), 389–403. London, Graham and Trotman.

Pinet, B., Montadert, L., Mascle, A., Cazes, M., and Bois, C. (1987). New insights on the structure and the formations of sedimentary basins from deep seismic profiling in western Europe. In *Petroleum Geology of north west Europe* (ed. J. Brooks and K. Glennie), 11–31. London, Graham and Trotman.

Rawson, P. F. and Riley, L. A. (1982). Latest Jurassic-early Cretaceous events and the 'Late Cimmerian unconformity' in the North Sea area. *Am. Assoc. Petrol. Geol. Geol., Bulletin* **66**, 2628–48.

Ritchie, J. D., Swallow, J. L., Mitchell, J. G., and Morton, A. L. (1988). Jurassic age from intrusives and extrusives within the Forties Igneous Province. *Scot. J. Geol.*, **23(1)**, 81–8.

Roberts, D. G., Backman, J., Morton, A. C., Murray, J. W., and Keene, J. B. (1984). Evolution of volcanic rifted margins: synthesis of leg 81 results on the western margin of the Rockall Plateau. In *Initial reports of the Deep Sea drilling Project, v. LXXXI* (ed. D. G. Roberts, D. Schnitker, *et al.*), 883–911. Washington US Government Printing Office.

Rowley, D. B. and Sahagian, D. (1986). Depth-dependent stretching: a different approach, *Geology*, **14**, 32–5.

Sclater, J. G., Hellinger, M., and Shore, M. (1986). An analysis of the importance of extension in accounting for post-Carboniferous subsidence of the North Sea basin. *University of Texas, Inst. for Geophys.*, Technical Report No. 44.

Shorey, M. D. and Sclater, J. G. (1988). Mid-Jurassic through Mid-Cretaceous extension in the Central Graben of the North Sea, part 2: Estimates from faulting observed on a seismic reflection lines. *Basin Research*, **1**, 201–15.

Smythe, D. K., Skuce, A. G., and Donato, J. A. (1980). Geological objections to an extensional origin of the Buchan and Witchground Graben in the North Sea. *Nature*, **287**, 467–8.

Sørensen, K. (1985). Danish basin subsidence by Triassic rifting on a lithosphere cooling background. *Nature*, **319**, 660–3.

Sørensen, S, Jones, M., Hardman, R. F. P., Leutz, W. K., and Schwarz, P. H. (1986). Reservoir characteristics of high- and low-productivity chalks from the Central North Sea. In *Habitat of hydrocarbons on the Norwegian Continental Shelf* (ed. A. M. Spencer *et al.*), 91–110. London, Graham and Trotman.

Spencer, A. M., Home, P. C., and Wilk, V. (1986). Habitat of hydrocarbons in the Jurassic Ula trend, Central Graben, Norway. In *Habitat of hydrocarbons on the Norwegian continental shelf* (ed. A. M. Spencer *et al.*), 111–27. London, Graham and Trotman.

Spencer, A. M., Campbell, C. J., Hanslien, S. H., Nelson, P. H. H., Nijsaether, E., Ormaasen, E. G., and Holter, E. (ed.; 1987). *Geology of the Norwegian oil and gas fields*. London, Graham and Trotman.

Storetvedt, K. M. (1987). Major late Caledonian and Hercynian shear movements on the Great Glen fault. *Tectonophysics*, **143**, 253–67.

Stow, D. A. V., Bishop, C. D., and Mills, S. J. (1982). Sedimentology of the Brae oil field, North Sea: fan models and controls. *J. Petrol. Geol.*, **5**, 129–48.

Taylor, J. C. M. (1984). Late Permian-Zechstein. In *Introduction to the petroleum geology of the North Sea* (ed. K. Glennie), 61–83. Oxford, Blackwell Scientific Publications.

Turcotte, D. L. (1981). Rifts—Tensional failures of the lithosphere. In *Papers presented to the Conference of Processes of Planetary Rifting, Christian Brother's retreat house, Napa Valley, California. 3–5 Dec. 1981*, 5–8. Lunar and Planetary Institute contribution 451.

Turcotte, D. L. and Emerman, S. H. (1983) Mechanisms of active and passive rifting. *Tectonophysics*, **94**, 39–50.

Turner, C. C., Cohen, J. M., Connell, E. R., and Cooper, D. M. (1987). A depositional model for the South Brae oil field. In *Petroleum geology of north west Europe* (ed. J. Brooks and K. Glennie), 853–64. London, Graham and Trotman.

Watts, N. L., Lapré, L. F., van Schijndel, F. S., and Ford, A. (1980). Upper Cretaceous and Lower Tertiary chalks of the Albuskjell area, North Sea: deposition in a slope and base of slope environment. *Geology*, **8**, 217–21.

Wood, R. and Barton, P. (1983). Crustal thinning and subsidence in the North Sea. *Nature*, **302**, 134–6.

Zervos, F. (1987). A compilation and regional interpretation of the northern North Sea gravity map. In *Continental extensional tectonics* (ed. M. P. Coward, J. F. Dewey, and P. L. Hancock). *Geol. Soc. London Special Publication* **28**, 477–93. Oxford, Blackwell Scientific Publications.

Ziegler, M. A. (1989). North German Zechstein facies patterns with regard to their older substrate. *Geologische Rundschau* **78(1)**, 105–127.

Ziegler, P. A. (1980). Northwestern European basin: geology and hydrocarbon provinces. In *Facts and principles of world petroleum occurrence* (ed. A. D. Miall). *Can. Soc. Petrol. Geol., Mem.* **6**, 653–706.

Ziegler, P. A. (1982). *Geological Atlas of Western and Central Europe*. Amsterdam, Elsevier Scientific Publishing Co.

Ziegler, P. A. (1983). Crustal thinning and subsidence in the North Sea: matters arising. *Nature*, **304**, 561.

Ziegler, P. A. (1986). Geodynamic model for the Palaeozoic crustal consolidation of western and central Europe. *Tectonophysics*, **126**, 303–28.

Ziegler, P. A. (1987). Late Cretaceous and Cenozoic intra-plate compressional deformations in the Alpine foreland—geodynamic model. *Tectonophysics*, **137**, 389–420.

Ziegler, P. A. (1988). Evolution of the Arctic-North Atlantic rift system. *Am. Assoc. Petrol. Geol., Mem.*, **43**, 198.

Ziegler, P. A. (1989). Laurussia—the Old Red continent. In *Devonian of the world* (ed. N. J. McMillan, A. F. Embry, and D. J. Glass). *Can. Soc. Petrol. Geol., Mem.*, **14**, **1**, 15–48.

2 Lithospheric structure of the North Sea from deep seismic reflection profiling

S. L. Klemperer and C. A. Hurich

Abstract

In the last six years nearly 10 000 km of deep reflection profiling has been carried out in the North Sea, providing an opportunity to study the tectonic evolution of the North Sea on a crustal scale. This paper reviews the use of deep reflection data to interpret the amount of crustal extension, the type of extension on a lithospheric scale, and the interplay between the early Palaeozoic accretion and formation of the continental crust of the North Sea and its subsequent reactivation during extension. In the deep Mesozoic grabens the crust has been thinned by a factor of two, leading to separation of Britain from Norway by about 100 km. The Mesozoic extension mechanism was approximately symmetrical, coaxial stretching, not lithospheric simple shear. Many extensional structures must have been newly formed during the Mesozoic, and evidence of reactivation of Palaeozoic compressional faults is limited. Some models of North Sea development which place heavy reliance on tectonic control by Palaeozoic compressional structures may be ill-founded.

2.1. Introduction

North-western Europe, a site of oceanic subduction and continental collision during the early Palaeozoic, underwent late-orogenic collapse to form the 'Devonian' basins, then was subjected to intracontinental rifting during the Permo-Triassic and Jurassic, producing the North Sea triple rift system (e.g. Ziegler 1982; Glennie 1986). When studying the North Sea at the lithospheric scale, a challenging problem is to understand the relationships between these different geological episodes both in a tectonic and in a structural sense. This paper is a review and synthesis of the results of deep seismic reflection profiling in the North Sea as they pertain to the crystalline crust and uppermost mantle, to a depth of about 50 km. We do not discuss the stratigraphy or intra-basinal structures of the North Sea even though deep seismic reflection profiles provide good regional coverage of these features. Nor do we discuss the other available geophysical data for the North Sea (*gravity*, Day *et al*. this volume; Holliger and Klemperer 1989, and this volume; Fichler and Hospers this volume; *seismic refraction data*, Christie and Sclater 1980; Christie 1982; Barton and Wood 1984; Barton 1986).

In this paper, we describe the available deep seismic reflection data base, then discuss the reflectivity patterns observed in the North Sea. Reflection profiles allow the categorization in three dimensions of the 'reflectivity' of the lithosphere—number, size, shape and distribution of reflections. Vertical reflectivity changes allow the distinction between crust and mantle (Barton 1986; Holliger and Klemperer 1989, and this volume), and hence enable the mapping out of crustal thickness variations and crustal stretching factors. Lateral reflectivity variations may mark changes in crustal type, of lithology or tectonic history (Freeman *et al*. 1988; Klemperer *et al*. 1990). The high resolution of reflection profiling also allows the mapping of tectonic structures—faults or shear zones—throughout the crust and uppermost mantle.

We can use the North Sea seismic reflection data

to identify crustal-scale structures, and where possible, to determine the origin and development of these structures during successive tectonic episodes. In particular we are interested in the evidence for the reactivation or non-reactivation during extension of pre-existing Caledonian structures. Deep seismic profiles help to locate some Caledonian crustal-scale structures (for example, the Iapetus suture, the Great Glen fault, and the Hardanger-Karmøy-Stavanger shear zones) and imply that these structures have helped control subsequent basin development. Other major faults important in the early Palaeozoic (for example, the Tornquist zone and the Highland Boundary fault) have no clear expression on deep seismic profiles across the North Sea, and have probably not been important at a crustal scale during the most recent, Mesozoic, episode of North Sea extension.

Finally in this paper we review evidence from deep profiles for the nature of the Mesozoic extension, and show how the seismic data support symmetric stretching mechanisms for the Mesozoic basin formation.

2.2. The deep reflection data base

Figure 2.1a illustrates the location of nearly 10 000 km of deep reflection profiles (recorded to 15 s or greater) that have been recorded in the North Sea. Approximately one-third of the data are GECO speculative surveys, one-third are surveys generously recorded by Mobil Norway and Mobil North Sea on behalf of academic groups in the UK and Norway, and the remaining surveys are other academic profiles (BIRPS) and commercial profiles (Britoil, Murphy, NOPEC, Shell, Western Geophysical). We are grateful to these companies which have kindly made their seismic data available for

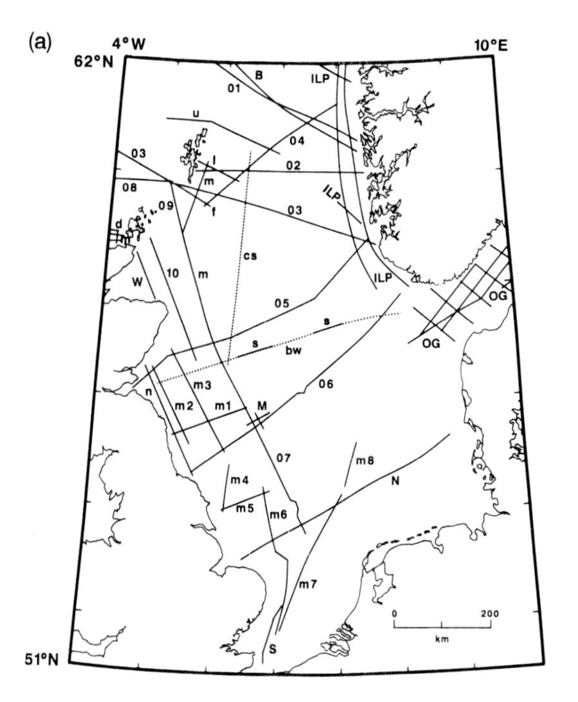

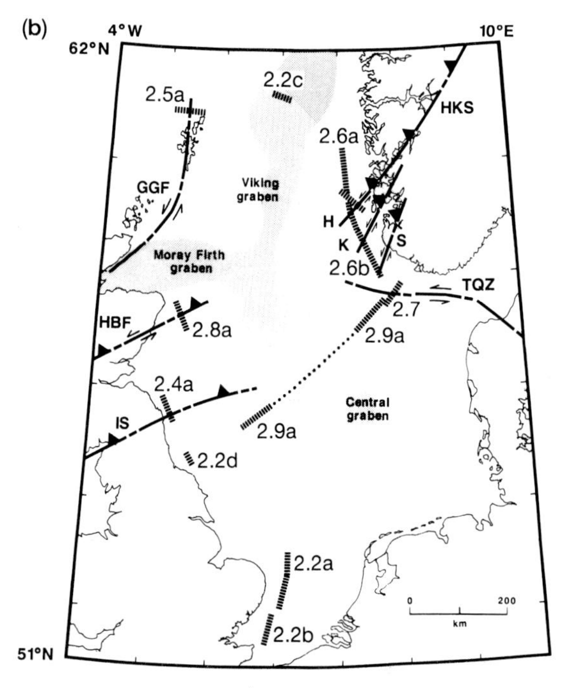

Fig. 2.1.a. Map of North Sea area showing location of marine deep seismic reflection profiles recorded to at least 15 s two-way time (solid lines) consulted during this study. For the full list of acronyms, see Table 2.1; numerals correspond to GECO NSDP profiles; lower-case letters are BIRPS profiles; proprietary data are indicated with capital letters. Dotted lines are refraction profiles: cs: Christie and Sclater (1980); bw: Barton and Wood (1984). **b.** Location map as Fig. 2.1a, with major Caledonian structures shown as dashed lines (GGF: Great Glen fault; HBF: Highland Boundary fault; IS: Iapetus suture; H, K, and S: Hardanger, Karmøy, and Stavanger shear zones; TQZ: Tornquist zone) and the Mesozoic triple graben system shown by stipple. Thick dashed lines indicate the parts of reflection profiles used in this paper, identified by their figure number.

study. All the profiles shown were recorded from 1982 onwards, and many have only recently become available, so that much primary scientific work remains to be done. Figure 2.1a locates and names the profiles, while Fig. 2.1b is an index map to the illustrations of seismic data shown elsewhere in this paper, and to tectonic structures discussed in this paper. Table 2.1 lists the surveys shown, their availability, and published descriptions of the profiles.

2.2.1. *Reflectivity structure of the North Sea: lateral variability*

It has recently become fashionable to relate the reflective character of the continents and its lateral variations to regional tectonic history, either with models that attempt to be globally applicable (Allmendinger *et al.* 1987*a*; Wever *et al.* 1987) or with models valid only for one region (Freeman *et al.* 1988; Marillier *et al.* 1989). A necessary part of such attempts is the description of reflective character. A distinction is often made between reflection profiles showing mainly diffractions, indicating structures in the crust that are smaller than a Fresnel zone (i.e. less than two or three kilometres), and profiles dominated by reflections, indicating more laterally continuous structures, sometimes reaching 10 or 20 km in length.

A commonly held generalization about lithospheric reflectivity in north-west Europe is that, beneath sedimentary basins, the upper part of the crust is fairly transparent, the lower part of the crust is highly reflective (the 'reflective lower crust'), and the uppermost mantle is transparent. The characteristic reflective lower crust has been widely reported around the British Isles (BIRPS data, e.g. Matthews 1986), in France (ECORS data, e.g. Bois *et al.* 1988), and in Germany (DEKORP data, e.g. DEKORP Research Group, 1985). Allmendinger *et al.* (1987*b*) and Cheadle *et al.* (1987) first suggested that the reflective lower crust has an association with regions of extension. Currently popular hypotheses for the formation of reflective patterns during extension are that individual reflectors within the reflective lower crust may represent shear zones (Reston 1988) or mafic intrusions (Warner, 1990). Other data suggests that the reflective lower crust has a more general assocation with Phanerozoic orogenic belts in which deformation—extensional or compressional—has affected the whole crust, whereas the absence of a reflective lower crust, or the presence of a diffuse and often diffractive seismic fabric, is thought to have an association with little-deformed cratons or continental interiors (Allmendinger *et al.* 1987*a*).

It must be stressed that this analysis is qualitative and subjective; that the problem of distinguishing different patterns of reflectivity in a quantitative manner has not yet been successfully tackled; and that there is a continuum of patterns of reflectivity between the examples described. However, the results of profiling in the North Sea give qualified support to the proposal that reflectivity patterns are modified during extension. Seismic sections crossing from the North Sea basin to adjacent platformal areas typically (but not ubiquitously) show a variation from highly reflective to less reflective lower crust (cf. McGeary 1987; Blundell *et al.*, in press Fig. 2.2a). Examples of 'diffractive' crust from the London platform, and of 'reflective' crust from the Shetland Terrace are shown in Figs 2.2b and 2.2c respectively. But it is also possible to select counter-examples. Figure 2.2d shows reflective lower crust beneath the English Caledonides in a coastal area with less than 5 per cent Mesozoic extension (e.g. Fig. 2.3) and only thin Carboniferous cover. The reflective lower crust also extends at least to the Norwegian coast, beneath areas of very thin Mesozoic rocks (e.g. Fig. 2.6a). The existence of these counter examples shows that models linking reflective character with tectonic history must be treated with caution.

Though no ubiquitous relationship exists between reflectivity and Mesozoic extension, the distinction in general between the lower crust of undeformed basement blocks and that of rift basins suggests that the lower crustal reflective fabric is deformed, or can even be created, during extension. The lateral reflectivity changes over the region of the North Sea seem to demonstrate that the whole thickness of the reflective lower crust of the basinal areas of the North Sea was modified during the Mesozoic extension. This conclusion is important because it implies that extensional processes affect the whole of the lower crust in a distributed manner, and that it cannot be correct to think of extension occurring only on a single lithospheric shear zone separating two internally undeformed crustal blocks, as in the model of Wernicke (1985) which has been applied to the North Sea principally by Beach (1985, 1986), by Beach and colleagues (1987) and by Gibbs (1987*a*, *b*; 1989*a*, *b*). However, without a sound

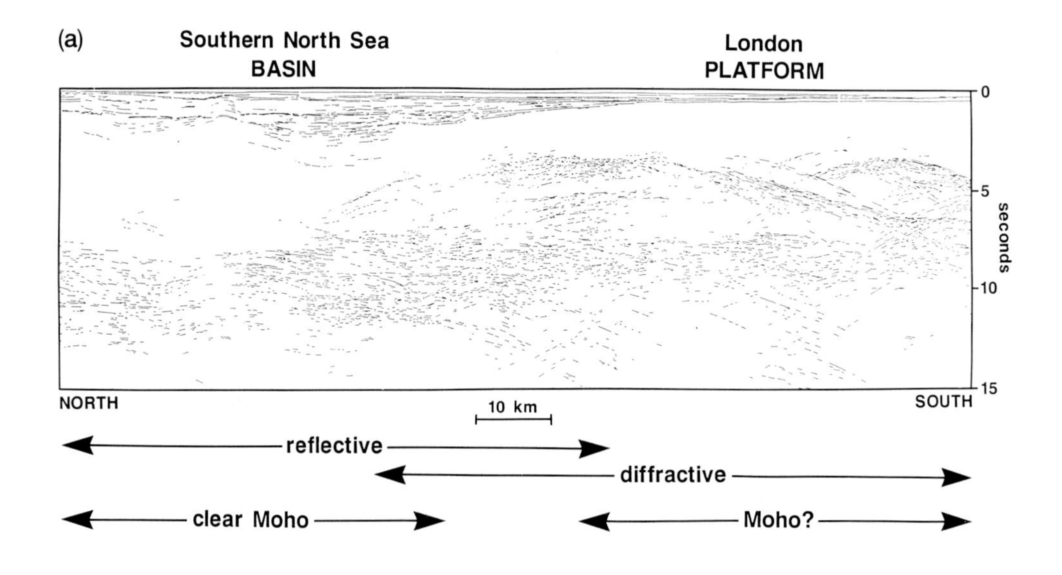

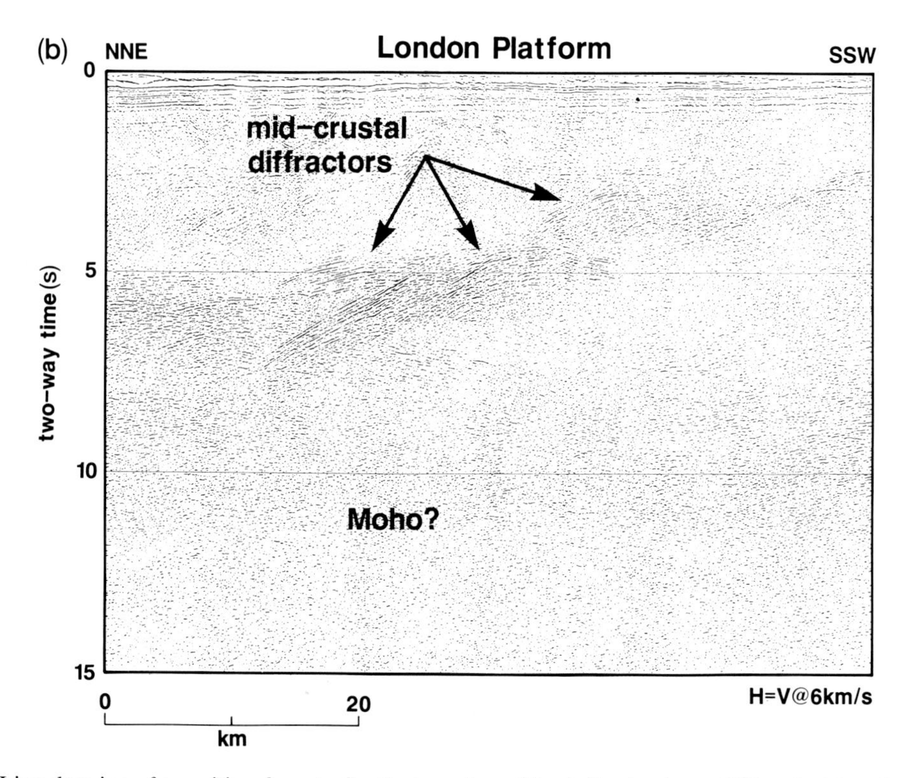

Fig. 2.2.a. Line drawing of transition from 'reflective' southern North Sea basin to 'diffractive' London Platform (i.e. an objective representation of primary seismic reflections and diffractions, excluding multiples, sideswipes and noise, used to allow small-scale reproduction of seismic data). Data from BIRPS profile MOBIL 6, located in Fig. 2.1b (after Blundell *et al.*, in press). **b.** Seismic data from profile Shell UK 82-101 to illustrate prominent diffractions in the middle crust of the London Platform (see Fig. 2.1b). **c.** Example of reflective lower crust from the Shetland Terrace on western margin of Viking Graben. From GECO profile NSDP 84-01 (see Fig. 2.1b). **d.** Example of reflective lower crust from an area of negligible Mesozoic extension, on the northeast coast of England within the Caledonides. Data from BIRPS profile NEC. (See Fig. 2.1b.)

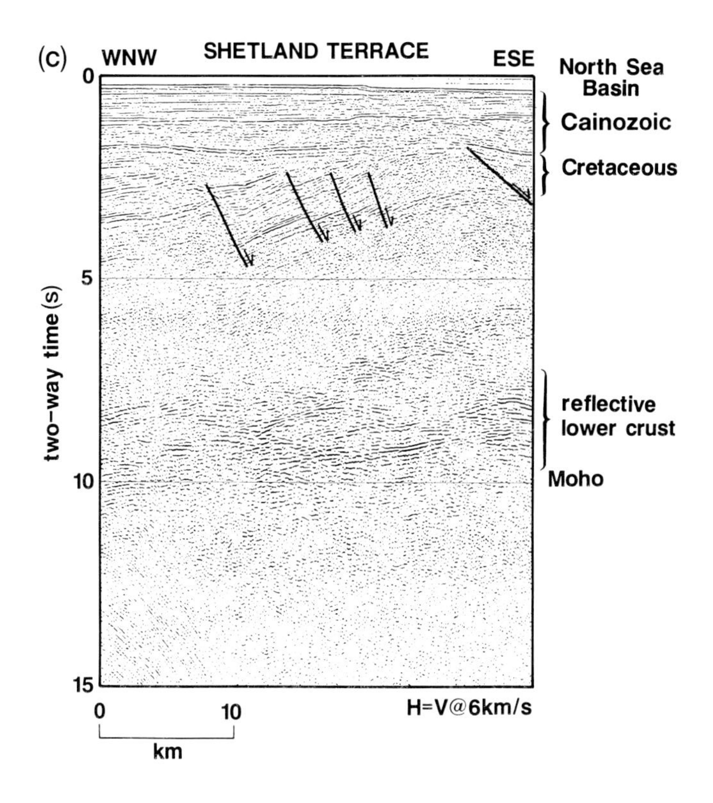
(c)
WNW
SHETLAND TERRACE
ESE
North Sea Basin
Cainozoic
Cretaceous
reflective lower crust
Moho
two-way time (s)
0
5
10
15
0
10
km
H=V@6km/s

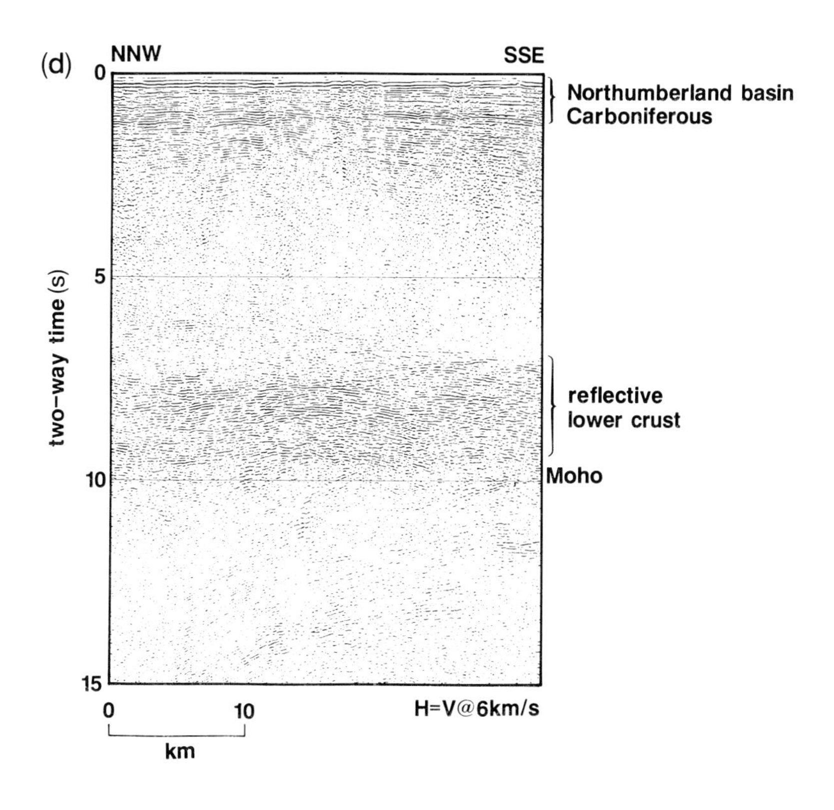
(d)
NNW
SSE
Northumberland basin Carboniferous
reflective lower crust
Moho
two-way time (s)
0
5
10
15
0
10
km
H=V@6km/s

Table 2.1 Deep seismic reflection database: North Sea profiles recorded to at least 15 seconds two-way time

Year	Acronym	Profile	Location	References	Source
1981	m	*BIRPS MOIST	West Orkney Basin	1	2036
1982	S	Shell UK 82-101	English Channel	Fig. 2.2b	2228
1983	B	Britoil NNS 83-22	Viking Graben	2–7	6276
	N	Nopec/GECO SNST 83-07	S Central Graben, Horn Graben	8	3165
	s	*BIRPS SALT	Central Graben	8, 9, 10	3409
1984	u, l, f, m	*BIRPS SHET (Unst, Lerwick, FairIsle) (Moray)	Shetland Platform, Outer Moray Firth	6, 11, 12, Fig. 2.5	6276
	d	*BIRPS DRUM	West Orkey Basin	13–16	8536
	01, 02, 03, 04	†GECO NSDP 84-01, 02, 03, 04	Viking Graben, West Shetland area	6, 7, 17–22, Fig. 2.2c	6276
	M	Murphy M37-84-10, 61	Mid-North Sea High	None	4680
1985	05, 06, 07	†GECO NSDP 85-05, 06, 07	Central Graben, Mid-North Sea High	7, 8, 13, 20, 23, Figs 2.7, 2.9a	6276
	08, 09, 10	GECO NSDP 85-08, 09, 10	West Shetland area, Moray Firth	Fig. 2.8	6276
	n	*BIRPS NEC	Mid-North Sea High	23, 24, 25, Figs 2.2d, 2.4a	7276
	W	Western Geophysical NMF-30	Moray Firth	26	2858
1987	m1 to m8	*BIRPS MOBIL	UK East Coast and S. North Sea	25, 27, Fig. 2.2a	7650
	ILP-10, 11, 12	Norwegian Lithosphere Program (Bergen) MOBIL lines ILP-10, 11, 12	Horda Platform	28, 29, 30, Figs 2.6a, 2.6b	7650
	OG-01 to 13	Norwegian Lithosphere Program (Oslo) MOBIL, lines OG-1 to 13	Skagerrak, Oslo, Gra-ben	31–34	7650

Year: year of acquisition.
Acr: profile acronym in Figure 2.1a. Numbers = GECO NSDP profiles; small letters = BIRPS profiles; capital letters = proprietary commercial profiles, and unreleased Norwegian Lithosphere Program profiles.
References: 1: Brewer and Smythe 1984; 2: Beach 1985; 3: Beach 1986; 4: Beach *et al.* 1987; 5: Frost 1987; 6: Klemperer 1988; 7: Klemperer and White 1989; 8: Holliger and Klemperer this volume; 9: Barton *et al.* 1984; 10: Barton 1986; 11: McGeary 1987; McGeary 1989; 13: Flack *et al.* 1990; 14: McGeary and Warner 1985; 15: Warner and McGeary 1987; 16: Flack and Warner 1990; 17: Gibbs 1987*a*; 18: Gibbs 1987*b*; 19: Holliger and Klemperer 1989; 20: Gibbs 1989 *a*; 21: Pinet 1989; 22: Fichler and Hospers this volume; 23: Klemperer and Matthews 1987; 24: Freeman *et al.*, 1988; 25: Klemperer *et al.*, 1990; 26: Coward *et al.* 1989; 27: Blundell *et al.*, in press; 28: Hurich *et al.* 1986; 29: Hurich *et al.* 1987; 30: Hurich and Kristoffersen 1988; 31: Husebye *et al.* 1988; 32: Kinck *et al.* 1988; 33: Larsson and Husebye 1988; 34: Larsson and Husebye, submitted.
References to figures are to figure numbers in this paper.
Source: units are airgun volume/cubic inch, multiplied by source pressure/2000 p.s.i. This number provides a crude measure of signal penetration.
*: BIRPS data available at cost of reproduction from the Marine Geophysics Programme Manager, British Geological Survey, Murchison House, West Mains Road, Edinburgh EH9 3LA, UK.
†: GECO/BIRPS data available at cost of reproduction from Merlin Geophysical Ltd, Merlin House, Boundary Road, Woking GU21 5BX, UK.
N.B. Some additional commercial marine 15 second profiles exist with the area of Fig. 2.1, but were not made available during this study. Several short land profiles also exist.

physical understanding of the nature of the lower crustal reflectors, it is not possible to use the presence or absence of reflectors to constrain the amount of extension undergone in the North Sea.

Though major regional changes in crustal reflectivity may relate to Mesozoic tectonics (i.e. the most recent major tectonic episode in this region), on a much more local scale (<100 km) it is proving possible to recognize changes in lower crustal reflective fabrics that have early Palaeozoic (Caledonian) trends (Freeman *et al.* 1988; Klemperer *et al.* 1990). Mapping out these older reflective patterns may yet help to constrain the variable nature of the basement on which the North Sea developed.

2.2.2. *Reflectivity structure: the reflection Moho*

Though lateral variability of reflection data is often rather poorly defined and of uncertain significance, vertical variations in reflectivity are frequently well-defined. In particular, the transition from reflective or diffractive lower crust to nearly reflection-free upper mantle is often rather abrupt and is sometimes marked by particularly bright and continuous reflectors (Fig. 2.2c, 2.2d, 2.4a, 2.5, 2.7, 2.8, and 2.9a). It has become usual to call the base of the reflective lower crust the reflection Moho (Klemperer *et al.* 1986) since much evidence both from the North Sea (Barton 1986) and from around the world (Mooney and Brocher 1987) supports the correspondence of the reflection Moho with the Moho *sensu stricto* defined on the basis of refraction velocities.

While only two modern refraction profiles capable of detecting the Moho exist in the North Sea (Fig. 2.1a), the reflection profiles provide a database, ten times larger, on which the reflection Moho can be mapped. It is possible to use the present-day thickness of the pre-rift crustal basement (from the pre-Permian surface to the Moho) as a direct measure of the degree of extension in different parts of the North Sea. This procedure, and the assumptions and errors, are given in detail and applied to the northern North Sea by Klemperer (1988). The main assumptions are:

(1) that the reflection Moho is the base of the crust, as suggested by the equivalence of the reflection Moho with the refraction Moho (Barton 1986) and with the gravity Moho in the North Sea (Holliger and Klemperer 1989; this volume);
(2) that crustal volume is conserved during extension, as is suggested by the small amount of volcanics in the North Sea (Latin *et al.* this volume);
(3) that crustal thickness was uniform before rifting began, and that the Moho acted as a passive marker during extension; and
(4) that the average seismic velocity of the basement is constant.

A preliminary map of crustal stretching values for the North Sea (β-values, here, the ratio of a 32 km-thick reference crust to the present-day thickness of pre-Zechstein crustal basement) is given in Fig. 2.3. A complete integration of all the available reflection data (Fig. 2.1a) with the excellent gravity coverage (Day *et al.*, this volume) and the sparse refraction data (most notably Solli 1976; Christie and Sclater 1980; and Barton and Wood 1984), to produce complete Moho and β-value maps, is beyond the scope of this paper. In particular, the assumption of a laterally constant basement seismic velocity of 6.2 km s^{-1} that was used to construct Fig. 2.3 is known to be wrong, and is probably the reason that the (unshaded) region of $1.2 < \beta < 1.5$ appears far broader on the Scandinavian side than on the British side of the Viking and Central Grabens. Refraction data suggests an increase in crustal velocity across the eastern margin of the Central Graben, to about 6.5 km s^{-1} offshore Norway (Barton and Wood 1984), so that the actual crustal thickness is more symmetric about the axis of the Viking and Central Grabens than appears in Fig. 2.3. None the less, even the simplified map of Fig. 2.3 contains some important results. The main graben structures, and even a number of the intra-basin highs, are well defined. Maximum crustal stretching values (integrated over all stretching episodes from the Permian onwards) in both the Viking Graben and the Central Graben exceed 2.0. Average values across the whole of the North Sea are in the range of 1.2 to 1.4, so that the British Isles must have moved about 100 km away from Scandinavia during the Mesozoic extension. The pattern of crustal thinning is approximately symmetric, and centred beneath the main graben structures, which in turn are centred beneath the thickest parts of the post-rift thermal-subsidence basins (Day *et al.* 1981; Klemperer and White, 1989). This coaxial alignment of rifting and subsidence is characteristic of basins produced by stretching models (e.g. McKenzie 1978) and also of

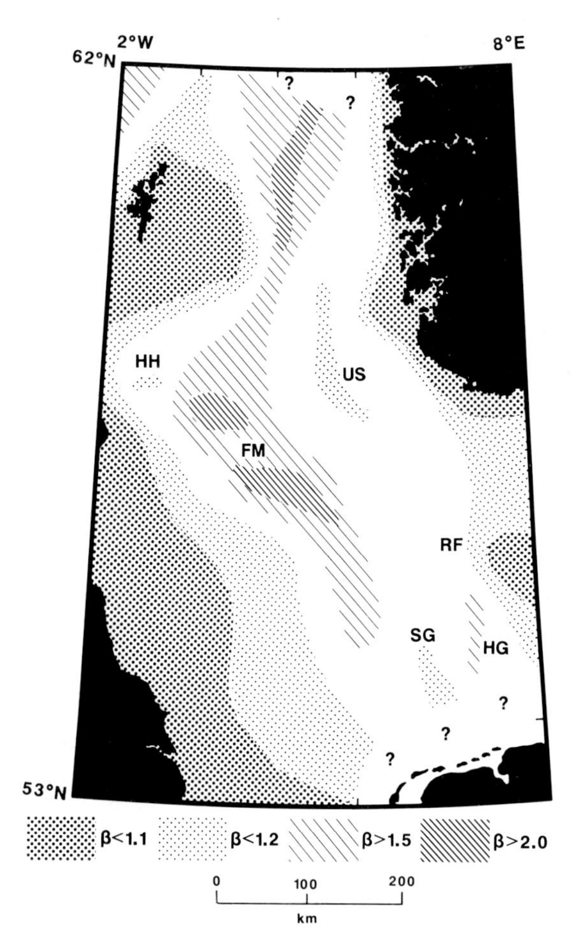

Fig. 2.3. Simplified map of crustal β-values (ratio of 32 km reference crustal thickness to present depth of Moho beneath base Zechstein, assuming a crustal velocity of 6.2 km s^{-1}, i.e. $\beta < 1.1$ corresponds to more than 9.4 s from base Zechstein to reflection Moho, $\beta < 1.2$ to more than 8.6 s; $\beta > 1.5$ to less than 6.9 s; and $\beta > 2.0$ to less than 5.2 s). Data sources are migrated profiles NSDP 84-01 to -04 (Klemperer 1988), NSDP 85-05 and -06 and NOPEC profile (Holliger and Klemperer, this volume), other unmigrated reflection profiles shown in Fig. 2.1a, and refraction profiles of Christie (1982), Barton and Wood (1984), and Eugeno-S Working Group (1988). HH: Halibut horst; US: Utsira-Sele high; FM: Forties-Montrose high; RF: Ringkøbing-Fyn high; SG: Schill Ground high; HG: Horn Graben.

those basins formed by whole-lithosphere normal simple shear along low-angle detachments (e.g. Wernicke 1985). However, more detailed study of the geometry of the thermal-subsidence basin demonstrates that stretching models are most appropriate for the North Sea basin (White, this volume).

2.3. Caledonian structures of the North Sea and their reactivation during late Palaeozoic and Mesozoic basin formation

The North Sea lies above the orogenic belt formed where three continental plates, Laurentia (Scotland and the allochthons of western Norway), Baltica (the Scandinavian shield), and Avalonia or Cadomia and associated Gondwanan terranes (England and Wales and western Europe), collided in early- to mid-Palaeozoic time (e.g. Soper and Hutton 1984; Ziegler 1982). It has frequently been speculated that major elements of the Mesozoic triple rift system (the Central, Viking, and Moray Firth Grabens) were controlled by the Palaeozoic collisional structures. For example, Watson (1985) suggests that the Viking Graben overlies a Palaeozoic transform plate boundary, and Frost (1987) that the Viking Graben overlies a Palaeozoic collisional suture. Neither hypothesis has been proved, largely because basement penetrations by drilling across the North Sea are too sparse for accurate determination of the separate continental blocks (e.g. Frost *et al*. 1981).

This illustrates the difficulty with such interpretations, that it is only rarely possible to be confident about the location of early Palaeozoic structures within the areas of deeply buried basement. Only in the marginal regions of the North Sea is it possible to test the hypothesis of reactivation of Caledonian thrusts or strike-slip faults as younger normal faults or transfer faults. Even in marginal areas of the North Sea, arguments for reactivation are generally based on little more than the subparallel strike of Caledonian and Mesozoic structural trends, e.g. in the Viking Graben area and west of the Shetlands (Johnson and Dingwall 1981), or e.g. in the Magnus basin (Gibbs 1987*a*).

None the less, claims have sometimes been made for a major role for Palaeozoic faults as controlling structures during Mesozoic extension. Pegrum (1984*a*, *b*) suggests that the Tornquist zone extends from northern Denmark to Fair Isle (between the Shetlands and Scotland) and was the locus of over 300 km of post-Caledonian strike-slip motion. Graversen (1988, and submitted) claims that the development of the Central Graben was controlled by NE–SW intra-continental transform faults. Doré and Gage (1987) identify as 'fundamental crustal dislocations' both a NE-trending Highland Boundary fault alignment and the NW-trending Tornquist alignment which meet at the Mesozoic rift triple junction to divide the North Sea into northern and southern 'tectono-sedimentary domains'. Other such models have been published, and are referenced in the discussion below. Clearly, from the diversity of fundamental trends that are claimed, not all these models can be correct; nor need any of them be.

Seismic reflection profiling has an important contribution to make in demonstrating whether or not Caledonian faults have controlled Mesozoic structure, because seismic profiles can be used to trace Caledonian faults from unreactivated onshore outcrop to offshore areas where they can potentially be related to Mesozoic structures. In this section we show three examples of major early Palaeozoic structures—the Iapetus suture, the Great Glen fault, and the Hardanger-Karmøy-Stavanger shear zones—that can be imaged, and hence precisely located, by deep seismic profiles, and which seem to have been utilized during post-collisional extension. We then show two examples of major early Palaeozoic structures—the Highland Boundary fault, and the Tornquist zone—for which the deep seismic profiles argue against a continuing role as major crustal structures during Mesozoic extension.

2.3.1. *Iapetus suture and the Mid-North Sea High*

The most fundamental tectonic boundary in the British Isles is the Iapetus suture which separates North American (Laurentian) from European (Avalonian) continental blocks. These terranes, formerly on opposite sides of the Iapetus ocean, were joined by collision at the end of the Caledonian orogeny (Wilson 1966). Though the existence of the suture is required by palaeontological evidence (Cocks and Fortey 1982), it is never exposed in Britain, being everywhere buried by late Palaeozoic sedimentary cover.

Deep seismic reflection profiles have been recorded across the suture, and have traced it 900 km along strike from west of Ireland (Klemperer 1989), and between Ireland and Britain (Brewer *et al.* 1983), and then to the east of Britain into the North Sea on a further four profiles located in Fig. 2.1 (NEC: Freeman *et al.* 1988. MOBIL-2 and -3: Klemperer *et al.* 1990. NSDP 85-7: Freeman *et al.* 1988). Arguments for the interpretation of north-dipping reflections, and/or a north-dipping boundary between crustal blocks with different reflective fabrics, as the Iapetus suture on each of the six profiles are given in detail in those papers. An example of these north-dipping reflections from BIRPS profile NEC, in this instance separating a very reflective lower crust north of and above the suture from a more transparent zone beneath the suture, is shown in Fig. 2.4a. The trend from SW Ireland to NE England of the north-dipping reflectors identified in Fig. 2.4a clearly marks them as a Caledonian crustal structure, not a Mesozoic extensional feature, whether or not they correspond precisely with the Iapetus suture. The north-dipping reflectors can be traced out 130 km towards the Central Graben (Fig. 2.4b). Many previous extrapolations of the Iapetus suture, lacking any real control, extended due NE from England (e.g. Watson 1985) or even swung around towards a more northerly trend, projecting towards the Mesozoic extensional triple junction (e.g. McKerrow 1982). In contrast, the new reflection data suggest that the trend of the Iapetus suture swings from NE across the Irish to ENE across the western North Sea (Klemperer *et al.*, 1990; Fig. 2.4b). Because the new reflection data precisely locate this mid-Palaeozoic suture, it is now possible to relate it to younger tectonic features more meaningfully than has previously been possible.

Though the Iapetus suture trends nearly orthogonally to the Mesozoic Central Graben, the earlier episode of basin formation in the Carboniferous and Permian was dominated by east to north-east trends. In particular, the Mid-North Sea High (MNSH) was a long-lived positive element which is recognizable in palaeogeographic maps for the late Carboniferous, Permian, and early Triassic (Ziegler 1982). On seismic data the MNSH cannot be precisely defined, for example as a fault-bounded block might be, so its limits are somewhat arbitrary. In Fig. 2.4b the MNSH is shown as the area where Zechstein salt is not present, after Ziegler (1982, his plate 3). West of the Central Graben there is a spatial association

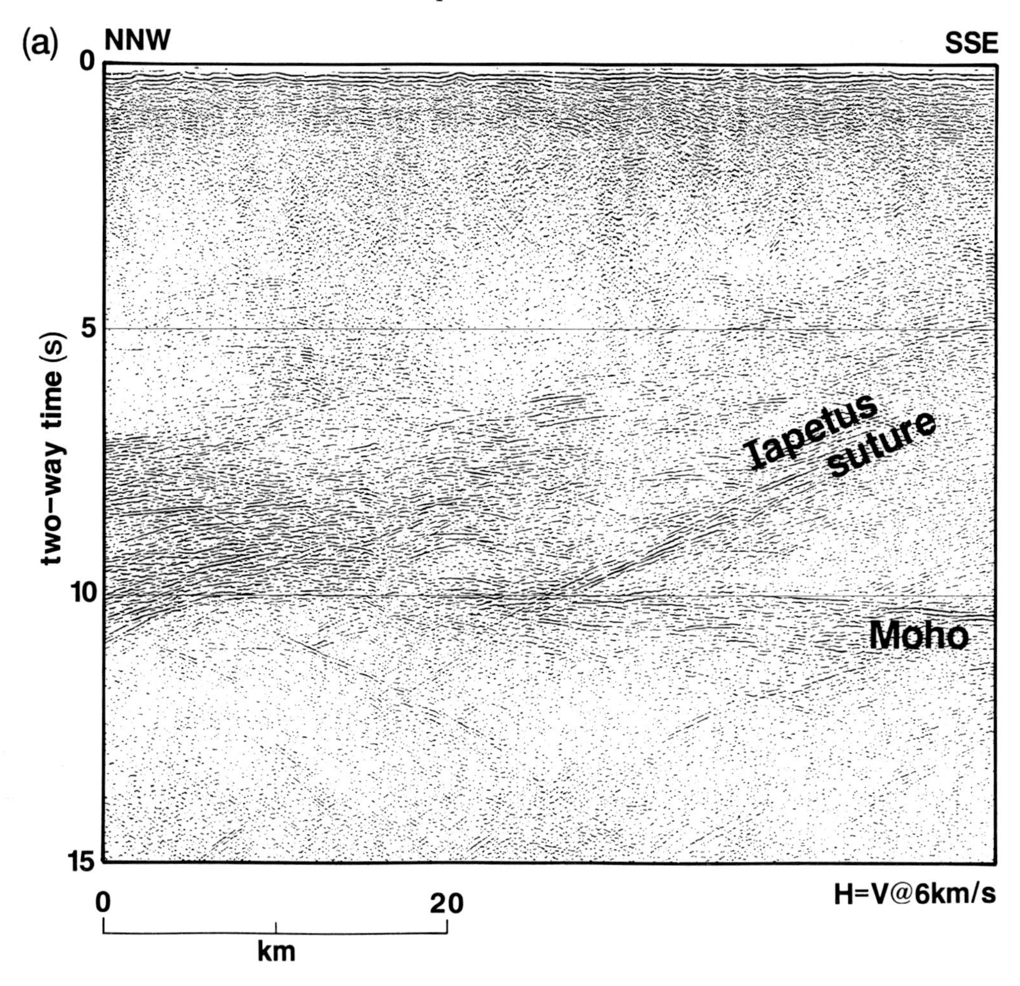

Fig. 2.4.a. Seismic data illustrating north-dipping reflectors identified as Iapetus suture zone (Freeman *et al.* 1988; Klemperer *et al.* 1990). The north-dipping reflectors sole into the reflection Moho at 10 s after appropriate migration. From BIRPS profile NEC. (See Fig. 2.1b.)

between the Iapetus suture defined on reflection data and the MNSH. It seems plausible that the crust of the suture zone, thickened by early Palaeozoic thrusting, should have intermittently been a slight topographic high for 150 Ma following the continental collision. Indeed, there is still present today a zone of slightly thickened crust, that may represent the northern limit of the suture zone (Freeman *et al.* 1988), extending with Caledonian trend into the western North Sea (Klemperer *et al.*, 1990; located in Fig. 2.4b by opposed arrowheads indicating dip on the Moho). Additional local effects, such as the buoyancy forces due to the presence of low-density granites, seem insufficient to explain the regional extent of the MNSH and the magnitude of relief across the High (Donato *et al.* 1983).

Relevant deep seismic data are only available west of the Central Graben, and, although the MNSH continues across the graben to Denmark, there is no evidence that the Iapetus suture continues so far, nor indeed, any consensus as to how the Iapetus suture meets and truncates (or is truncated by) the Tornquist convergent zone, the Palaeozoic suture between Baltica and Gondwanan terranes (Soper 1988). On Fig. 2.4b are shown the Caledonian deformation fronts located by Ziegler (1982, his

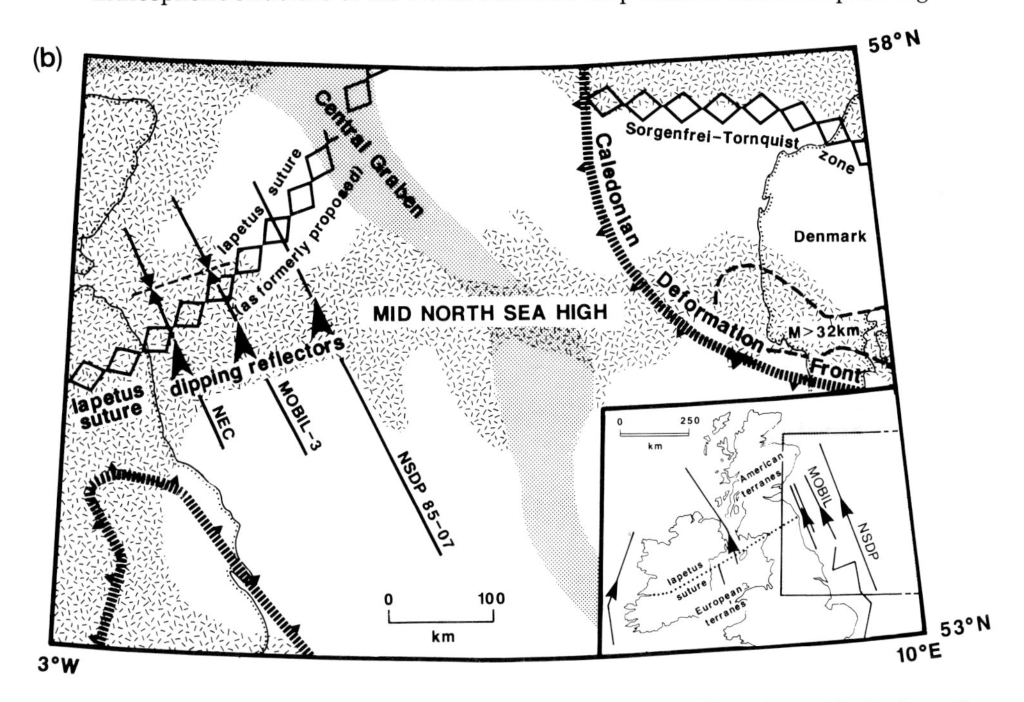

Fig. 2.4.b. Map illustrating relationship between the Iapetus suture mapped from the north-dipping reflectors (the tip of each arrowhead is at the position where the suture reflectors meet the Moho) and the Mid-North Sea high (drawn as the area where Zechstein salt is not present, after Ziegler (1982, his plate 3). Synform symbol (opposed arrowheads) on profiles NEC and MOBIL-3 locate the 40 km-wide zone of crust still thickened at the present day (after Klemperer *et al.* 1990). Zone of thickened crust beneath Ringkøbing-Fyn high is shown by a dashed line and marked as M > 32 km, after Kinck *et al.* (1988) and Eugeno-S Working Group (1988). Profile MOBIL-2 omitted for clarity. Caledonian deformation fronts after Ziegler (1982); Sorgenfrei-Tornquist zone after Eugeno-S Working Group (1988). Inset shows the location of the other BIRPS profiles (west of the main map area) on which the Iapetus suture can be located by north-dipping reflectors shown by arrowheads (after Klemperer, 1989).

plate 1) as the cratonward limits of Silurian thrusting. It is not claimed here that the location of the MNSH east of the Central Graben is controlled by the Caledonian deformation front, but rather that, if the MNSH in the western North Sea is controlled by the Caledonian-age Iapetus suture, then it is possible that structures of similar age localized the MNSH where it continues to the east. Though there are no good constraints on the crustal thickness of the MNSH between the Central Graben and the Danish coast, beneath Denmark the crust of the Rynkøbing-Fyn High, the eastward extension of the MNSH, is 2 to 4 km thicker than adjacent crust (Eugeno-S Working Group 1988).

At the end of the Palaeozoic the MNSH ceased to be a major sedimentological divide and it was transected by Mesozoic rifting in the Central Graben (Fig. 2.4b), though local thinning of Triassic and Jurassic strata is seen above the presumed buried granites of the MNSH (Donato *et al.* 1983). Thus the deep seismic data suggest that the mid-Palaeozoic Iapetus suture provided tectonic control to late Palaeozoic basin formation. Faults within the suture zone may have been reactivated, but because we cannot trace the suture to the pre-Devonian surface on our deep seismic data we cannot prove this possibility. In contrast the Mesozoic rifting in the central North Sea was orthogonal to, and so presumably was not controlled by, nor reactivated, the Caledonian suture.

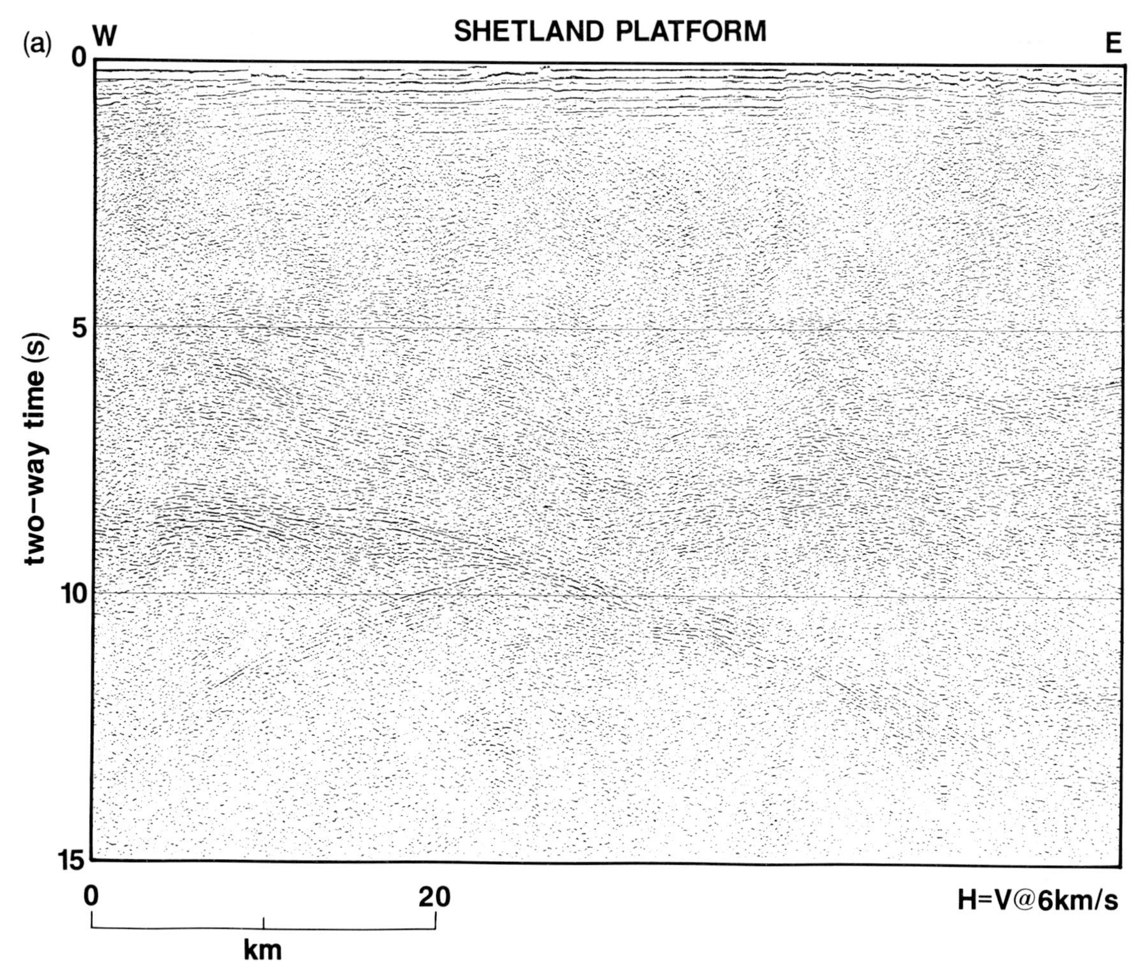

Fig. 2.5.a. Seismic data from BIRPS profile SHET-Unst, showing prominent diffractions with their apices vertically beneath the surface trace of the Walls Boundary/Great Glen fault. (See Fig. 2.1b.) **b.** Interpretation of Fig. 2.5a, after McGeary (1989).

2.3.2. *Great Glen fault and Inner Moray Firth, Fair Isle, and Sandwick basins*

The Great Glen fault is the most important of the late Caledonian strike-slip faults which trend NE across Scotland (e.g. Watson 1984; Soper and Hutton 1984). The timing and magnitude of displacement along the Great Glen fault are disputed, but there may have been 100 to 600 km of sinistral displacements in Devonian time, followed by 50 to 300 km of post-Devonian dextral movement (e.g. Smith and Watson 1983; Storetvedt 1987). The fault is well-known across Scotland, and though offshore its trace is complex, an important splay (if not the major trace of the Great Glen fault itself) runs northwards through the Shetlands as the Walls Boundary fault (McGeary 1989). Reactivation of the Great Glen fault has previously been demonstrated during Devonian or Permo-Triassic formation of the Sandwick basin (Hitchen and Ritchie 1987), during Permo-Triassic formation of the Fair Isle basin (Bott and Browitt 1975), and during Mesozoic formation of the Inner Moray Firth basin (McQuillin *et al.* 1982).

What is important in the context of this paper is that a deep seismic profile across the Great Glen/Walls Boundary fault shows that this strike-slip fault exists as a crustal-penetrating feature even today

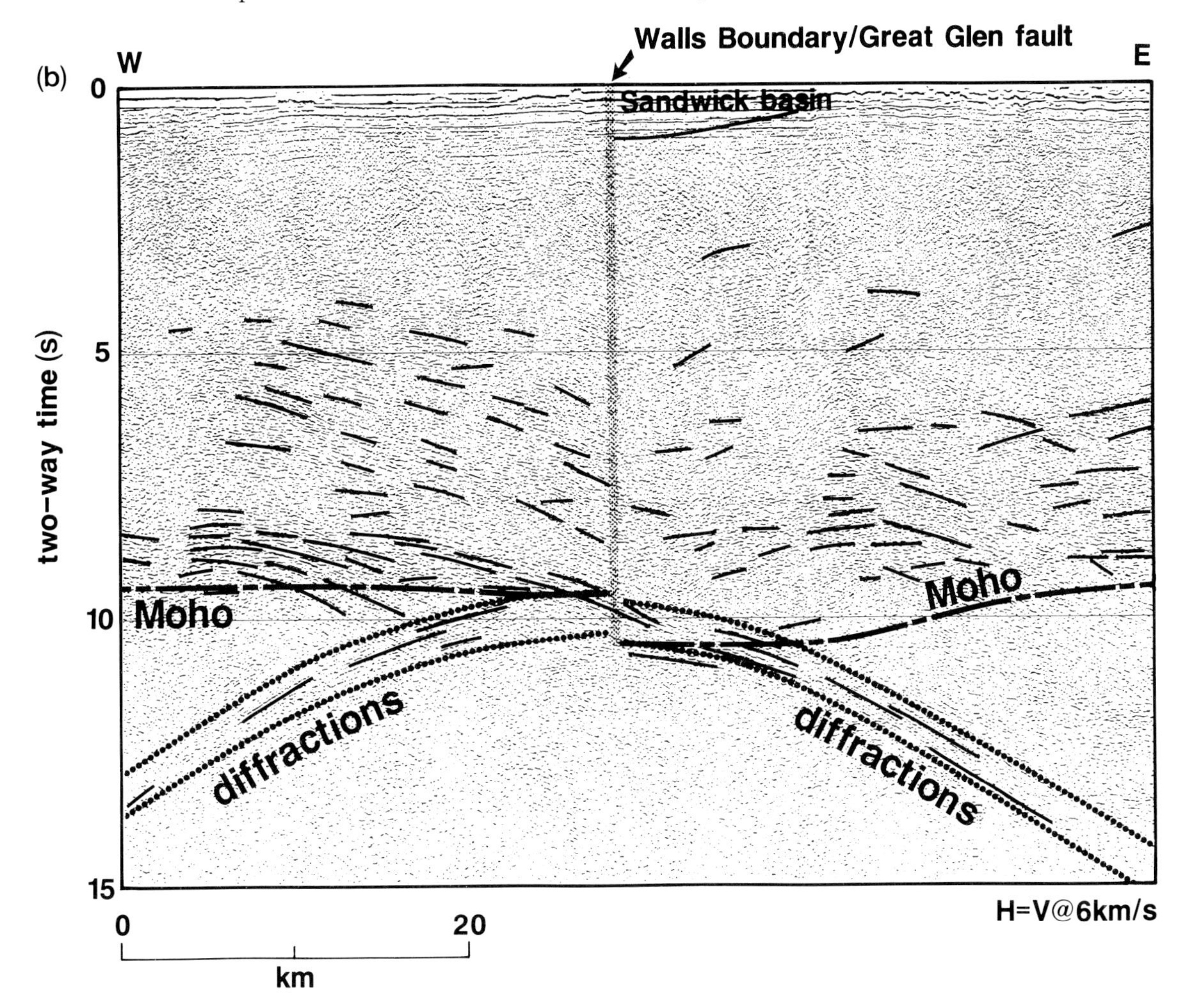

(McGeary 1989). It is not normally possible to directly image vertical strike-slip faults with reflection profiles, but these faults can be recognized by a vertical zone of disturbance across which it is impossible to map continuous reflectors, and by diffractions from reflectors truncated by the vertical fault. BIRPS profile Unst (Fig. 2.1) shows prominent diffractions from the Moho, vertically beneath the surface trace of the Walls Boundary fault (Fig. 2.5a). These diffractions appear to be caused by a vertical offset of the Moho (Fig. 2.5b).

The existence of the sedimentary basins bounded by the Great Glen/Walls Boundary fault implies that this Caledonian feature underwent reactivation from the Devonian into the Mesozoic. The existence of an offset Moho at the present day suggests that either the strike-slip fault plane was utilized throughout the whole crust during Upper Palaeozoic and Mesozoic reactivation, or at least that the degree of post-Caledonian deformation of the lower crust in this area was insufficient to obliterate the evidence for the Caledonian structure. The indirect imaging of the Walls Boundary fault in this manner shows that deep seismic reflection profiling is an appropriate technique to try to locate such major strike-slip faults.

2.3.3. *Hardanger-Karmøy-Stavanger shear zones*

Three zones of north-west-dipping reflections occur in deep seismic data along the south-west coast of Norway (ILP-10 and 12; Fig. 2.6). Onshore

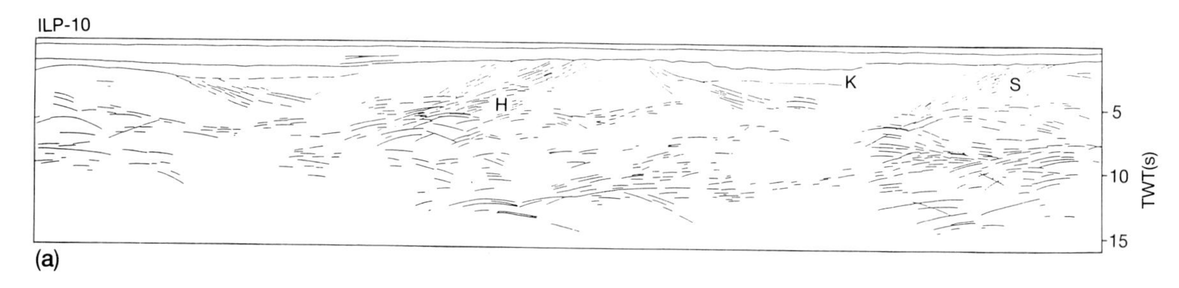

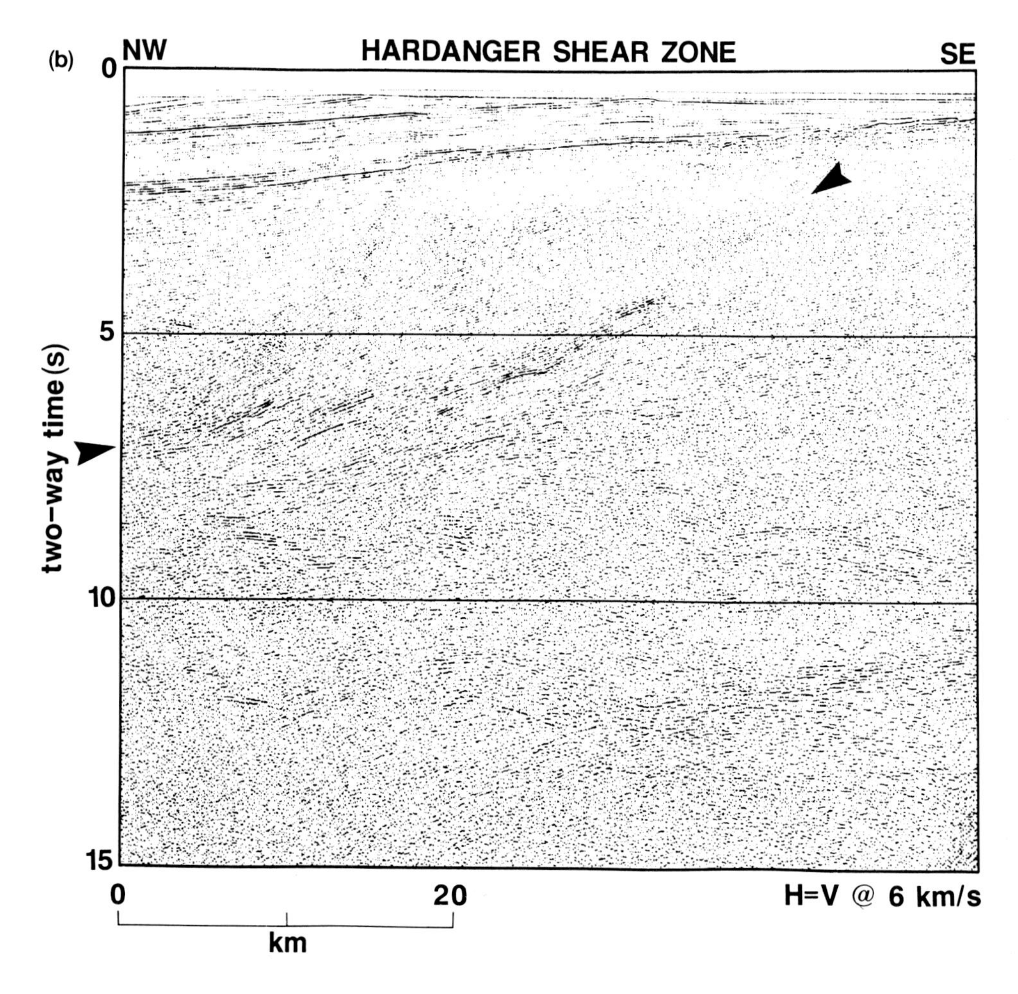

Fig. 2.6.a. Line drawing of Norwegian Lithosphere Program MOBIL profile ILP-10, after Hurich and Kristoffersen (1988). (See Fig. 2.1b.) **b.** Seismic data from profile ILP-12 showing north-dipping reflections from the Hardanger shear zone. (See Fig. 2.1b.) **c.** Caledonian shear zones, extensional faults and Bouguer gravity along the south-west coast of Norway, after Hurich and Kristoffersen (1988).

(c)

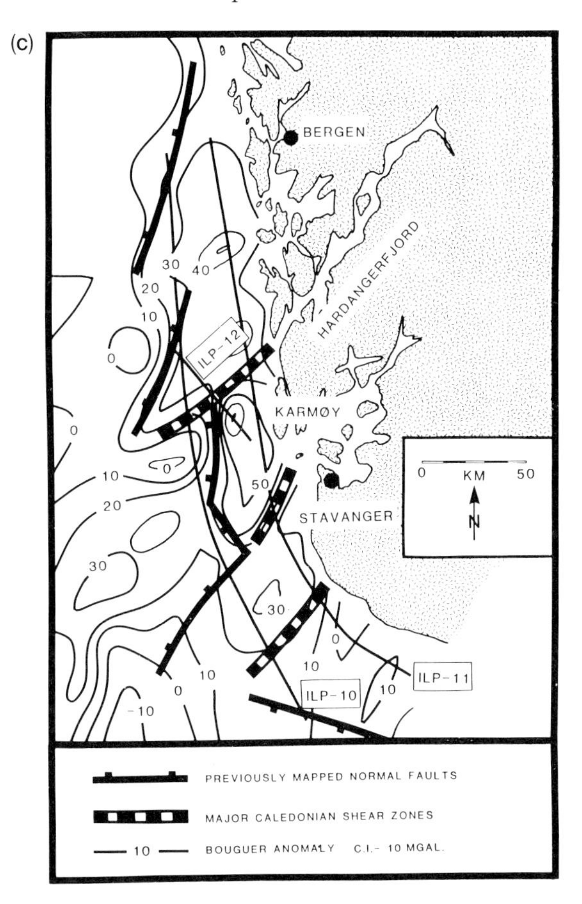

projection of the reflections at Hardangerfjord suggests a correlation between that reflective zone and a several kilometre-thick shear zone dipping 35–40° to the north-west. Reflection zones at Karmøy and Stavanger (Fig. 2.6a and c) project onshore into a broad area of penetratively deformed basement rocks. The three reflective zones can be traced about 50 km offshore onto the Horda Platform along NE–SW strike, parallel to the dominant Caledonian trend in southern Norway. Hurich and Kristoffersen (1988) interpret the reflective zones as Caledonian thrust faults that penetrate to at least lower crustal levels and bound an area of thickened crust. The seismic data and regional geology indicate that portions of these thrust faults underwent several periods of late-Caledonian and post-Caledonian reactivation as normal faults. It is this reactivation that is of interest in the present discussion.

Increasingly geological evidence (McClay *et al.* 1986; Norton 1986; Seranne and Seguret 1987) indicates that late deformation in the Caledonian hinterland was dominated by extensional collapse of the orogen. In most cases the extension occurred on older thrust faults, but in some cases may have occurred on newly-formed low-angle normal faults (Norton 1987). Development of the extensional faults localized deposition of the Old Red Sandstone facies in the so-called Devonian basins. The Hardangerfjord shear zone (Fig. 2.6b) is an example of a crustal-scale Caledonian thrust that was probably reactivated in extension during late Caledonian deformation. The Caledonian hinterland, which acted as basement during post-Caledonian extension in the North Sea, was penetrated by zones of weakness in the form of kilometre-scale shear zones.

The majority of post-Caledonian normal faults on the Horda Platform have a dominant north-south trend, about 60° from the Caledonian strike (Fig. 2.6c), and so must be newly-formed faults and not reactivated Caledonian faults. However, deep seismic data and shallow commercial seismic data demonstrate that some of the post-Caledonian normal faults strike along the Caledonian trend, or are bounded by Caledonian structures. For example, the direct offshore extension of the Karmøy shear zone is a normal fault bounding a half-graben filled with sediments of unknown but probable Permo-Triassic age (Fig. 2.6a and c). Presumably, the normal fault is at least in part a reactivation of the Caledonian Karmøy shear zone. Thickening of the lower Cretaceous section across the fault indicates that, like other late Palaeozoic faults on the Horda platform, this fault continued to be active in the Mesozoic. Another example of Caledonian influence on the younger structures is a north-south trending normal fault bounded on the north and south by the Hardangerfjord and Karmøy shear zones. In this case, the Hardangerfjord shear zone may have acted as a transfer fault, shifting normal movement on the coast-parallel Øygarden fault system, about 30 km to the east towards the coast (Fig. 2.6c).

These examples from the area of southern Norway demonstrate that, although major zones of weakness were present in the post-Caledonian basement, post-Caledonian reactivation of these zones was limited to a few specific areas. However, when the old fault zones were reactivated in the late Palaeozoic, they continued to be active throughout the Mesozoic, with many newly-formed late-Palaeozoic faults.

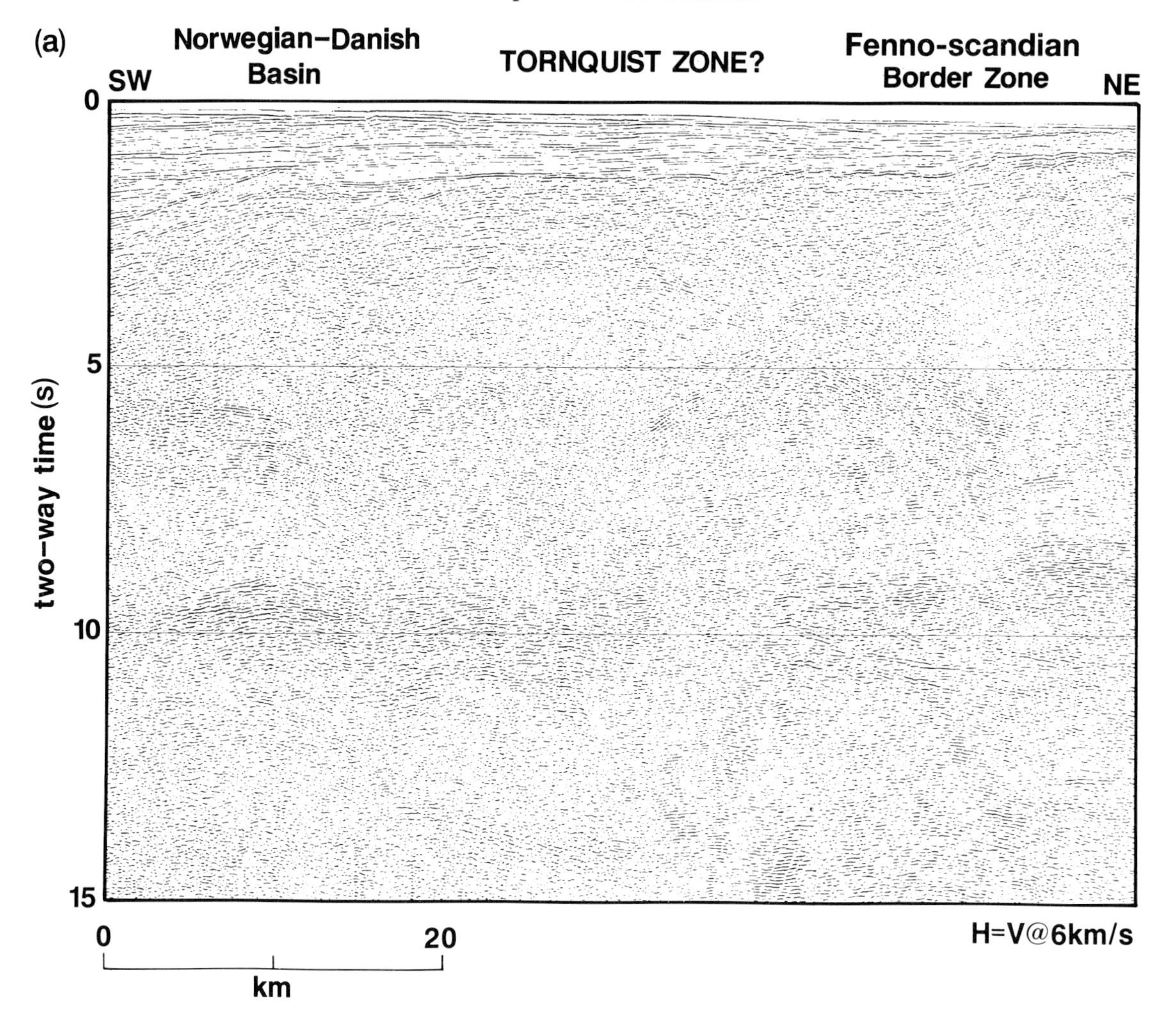

Fig. 2.7.a. Seismic data from GECO profile NSDP 85-06 across the Sorgenfrei-Tornquist zone. (See Fig. 2.1b.) **b.** Interpretation of Fig. 2.7a.

2.3.4. *Tornquist zone*

The Tornquist zone (TQZ) has a well-defined NW trend from the Black Sea to Denmark. In Poland and further south-east it is called the Tornquist–Teisseyre zone, and forms the SW edge of the Precambrian East European platform. In southernmost Sweden and northern Denmark it lies within the Precambrian platform, and is marked by an alignment of faults including the Fjerritslev fault, the Fenno-Scandian Border zone, and Mesozoic inversion structures (Pegrum 1984*a*, *b*; Eugeno-S Working Group 1988). Some authors believe that the Tornquist zone extends into the North Sea, or even via the Iceland–Faeroes Ridge to Greenland (Pegrum 1984*a*, *b*; Doré and Gage 1987), while others have suggested that it terminates at the Oslo Graben (Ziegler 1982). If the Tornquist zone terminates at the Permian Oslo Graben, it is presumably only a Palaeozoic structure, whereas if it extends further west it might have a Mesozoic–Cainozoic history as well.

The Eugeno-S Working Group (1988) states that the Tornquist zone from Denmark to the Caledonian Deformation front (Fig. 2.4b) is characterized by Late Cretaceous-Early Tertiary inversion tectonics. They present no evidence for any Palaeozoic history for this part of the Tornquist zone, and so

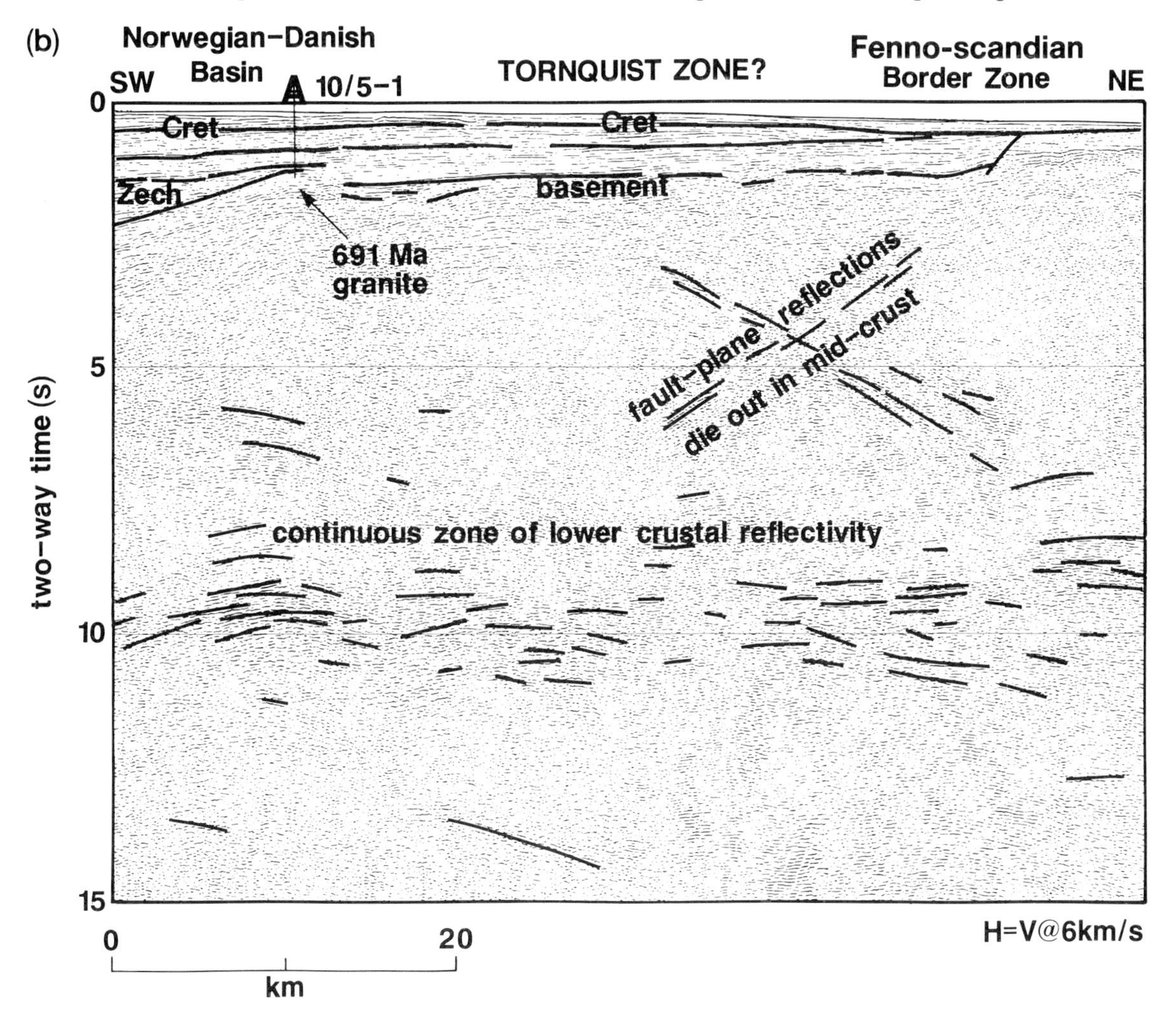

suggest that this segment of the TQZ should be separately named the Sorgenfrei–Tornquist zone to distinguish it from the Tornquist–Teisseyre zone. At the other extreme, Pegrum (1984*b*) suggests 350 km of late Palaeozoic dextral displacement on the TQZ from Scotland to Poland, followed by continual though smaller transtensional and transpressional reactivation of the TQZ throughout the Mesozoic. Pegrum (1984*a*) suggests that the location of the Jurassic Forties volcanic field in the rift triple-junction area was controlled by the TQZ. Doré and Gage (1987) also believe that the TQZ extended to the triple junction and exerted major structural and sedimentary control on the development of Mesozoic rift basins. Clearly it is important, in the context of the geological development of the North Sea, to understand which of these views is correct, and whether the TQZ is the most important structural control in the eastern North Sea, or whether the TQZ does not exist beneath the North Sea.

Deep seismic studies have the potential to locate the Tornquist zone where it exists, and to identify reactivation features if they exist. In eastern Europe the TQZ is clearly located by a 5–10 km change in crustal thickness within and across the zone (Meissner *et al.* 1987), and crustal thickening also occurs across the TQZ in southern Sweden and northern Denmark (Meissner *et al.* 1987; Eugeno-S Working Group 1988). Detailed results of 1987 deep reflection profiling in the Skagerrak west of Denmark are not yet released, but the deep profiles

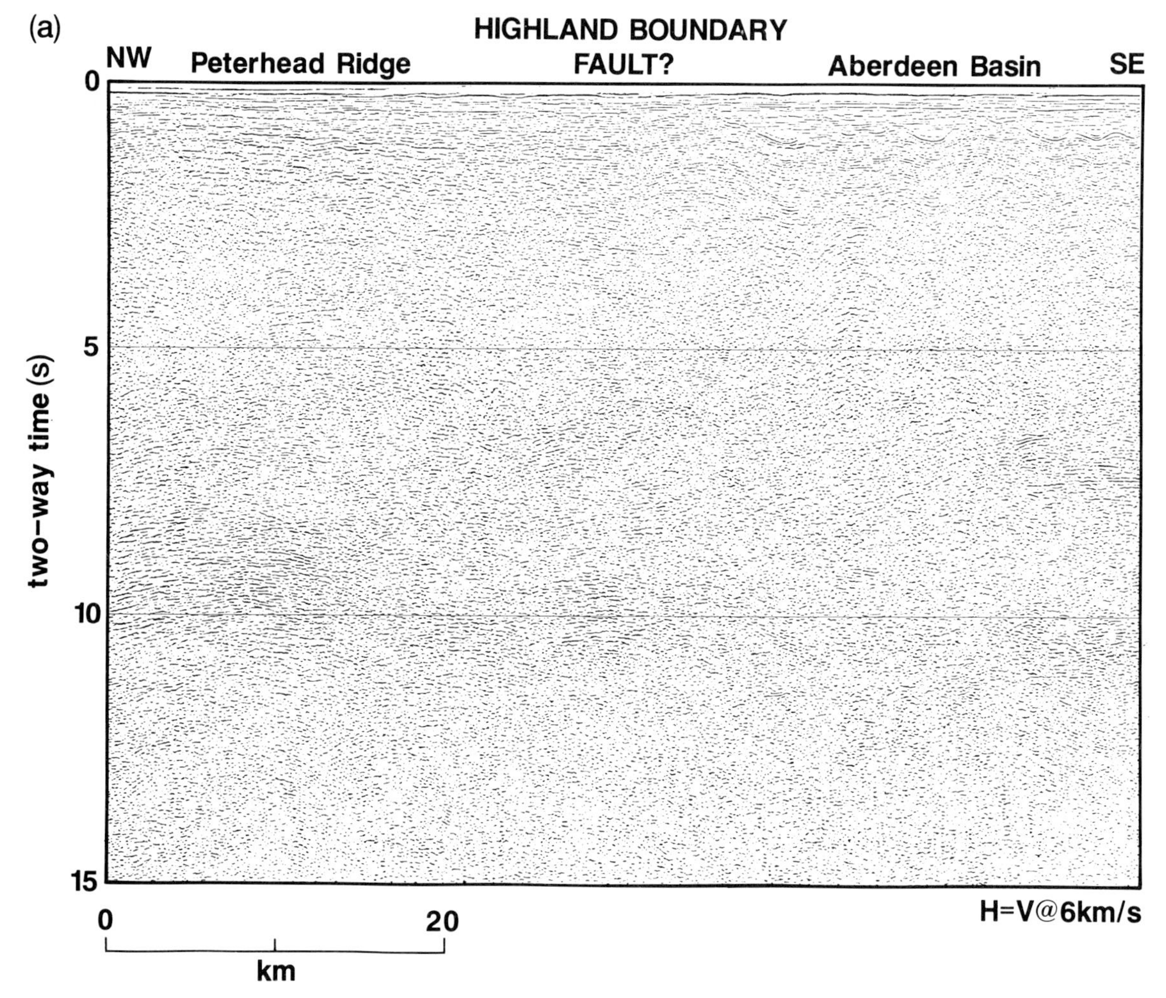

Fig. 2.8.a. Seismic data from GECO profile NSDP 85-10 across the north-eastward projection of the Highland Boundary fault. (See Fig. 2.1b.) **b.** Interpretation of Fig. 2.8a.

appear not to show evidence for changes in crustal thickness across the Fjerritslev fault zone and the Fenno-Scandian border zone (Husebye *et al*. 1988).

Where the GECO deep profile NSDP 85-06 (Fig. 2.1) crosses the line of the Tornquist zone suggested by Pegrum (1985*a*, *b*), the reflection Moho appears flat (Fig. 2.7) and a thin band of reflectivity (about 3 km thick) at the base of the crust is continuous across the region. The crossing reflections within the upper crustal basement, which uncross and dip at 50° to 60° after migration (with appropriate allowance for the 30° deviation of the reflection profile from a dip line), probably represent steep faults of the Fenno-Scandian Border zone. The SW-dipping reflectors project up to a Mesozoic normal fault with only a small throw (1 to 1.5 km) while the NE-dipping reflections do not project towards any clear Mesozoic fault. Neither set of dipping reflectors extends deeper than about 20 km, and certainly not to the Moho. Whether these reflectors represent Palaeozoic or Mesozoic faults, there is no support from this reflection profile for the existence of a crustal-penetrating strike-slip fault that has been active during or since the dominantly Mesozoic North Sea rifting.

The TQZ certainly has less of an effect on the crustal structure of the northeast margin of the North Sea than does the Great Glen fault along the

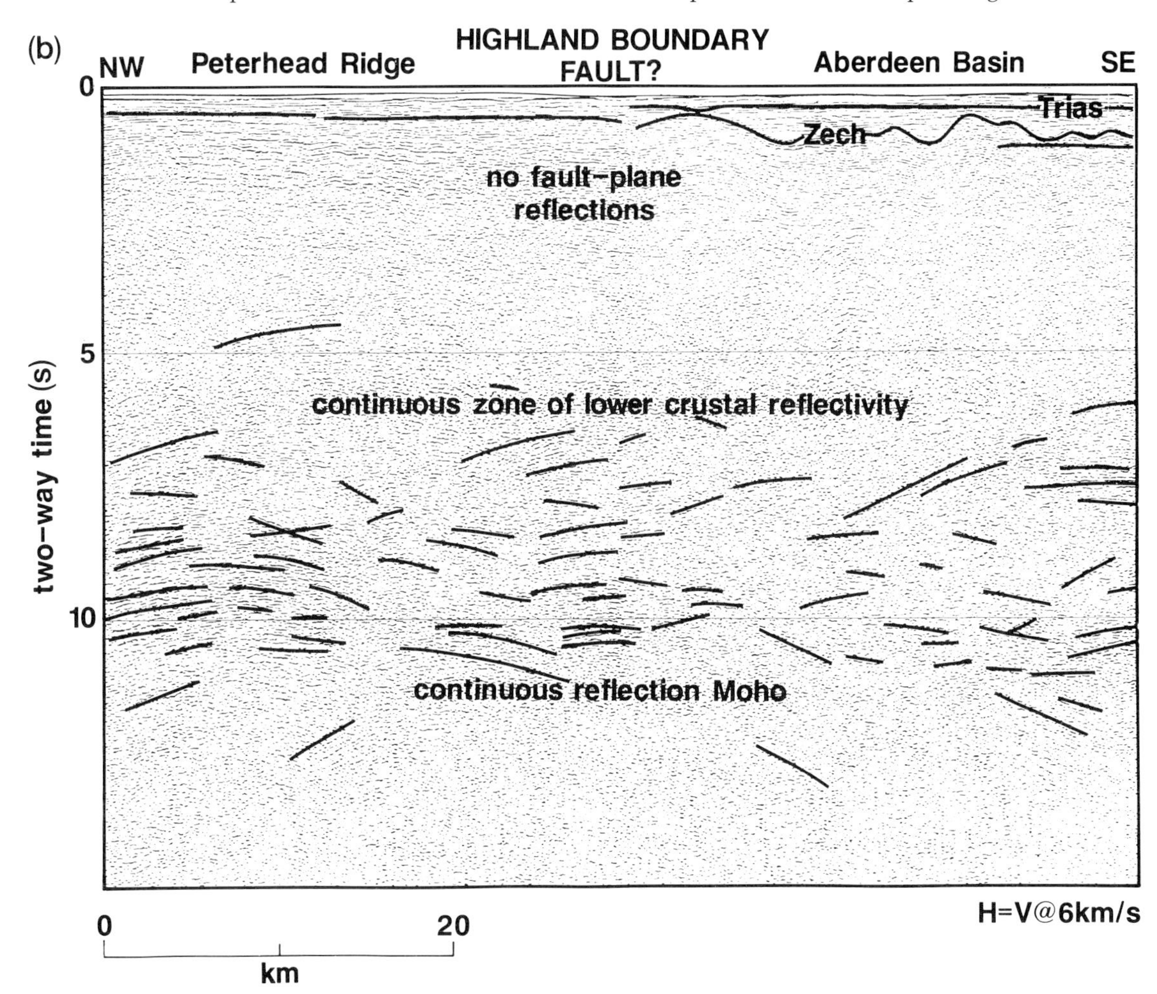

northwest margin. From these reflection data it is entirely possible that the Tornquist-Teisseyre zone never extended along the Sorgenfrei-Tornquist zone into what is now the North Sea. The zone of undoubted faulting along the Sorgenfrei-Tornquist zone (Eugeno-S Working Group 1988), the Fjerritslev fault which clearly cuts Permian and Mesozoic strata (Husebye *et al*. 1988), and the complex pattern of faults along this line towards the rift triple junction (Pegrum 1984*a*), may be simply a chance alignment of structures along the projection of the Palaeozoic Tornquist–Teisseyre zone from eastern Europe into the North Sea.

2.3.5. *Highland Boundary fault*

Just as the Tornquist zone in eastern Europe has a well-accepted lower Palaeozoic history but an uncertain extension into the North Sea, so also does the Highland Boundary fault (HBF). The HBF acted as a terrane boundary in the Ordovician, and in Scotland separates the metamorphic Caledonides of the Grampian terrane to the north from the ophiolitic Highland Bounder terrane to the south (Hutton 1987). A late Silurian ignimbrite oversteps the HBF in Scotland limiting post-Silurian strike-slip movement there to less than a few tens of kilometres (Thirlwall 1989), but the HBF did act as a normal fault during the Devonian, down-faulting the Midland Valley basin to the south (Bluck 1984). The HBF was identified by Dikkers (1977) with a 'Forties' lineament extending from Scotland to the Horda fault zone west of Norway, that he interpreted as having influenced the change in trend of the rift system from the Viking Graben to the Central

Graben. The few basement well penetrations within the North Sea suggest that the HBF projects offshore in a general way as the 'Dundee-Stavanger line' (Frost *et al*. 1981), separating a high-grade 'Grampian' region to the north from younger, lower-grade metamorphics to the south.

No direct structural continuation of the HBF into the North Sea has been reported, but there is a visual alignment of the HBF with such features as the Peterhead fault zone (south margin of the Buchan Graben) (Institute of Geological Sciences 1982), the junction of the Viking Graben and Central Graben, the Ling Graben in the Norwegian sector, and even the Caledonian Faltungsgraben structures—the Hardangerfjord shear zone—on the Norwegian mainland (Doré and Gage 1987). Doré and Gage postulate that 'the Highland Boundary fault–Ling Graben Alignment represents a major crustal discontinuity that continued to act as a palaeogeographic control in post-Caledonian times'.

Such a major crustal discontinuity is an obvious target for deep seismic studies. Certainly, the Hardangerfjord shear zone appears to be a major crustal structure on deep seismic reflection data recorded close to the Norwegian coast (Fig. 2.6a; Hurich and Kristoffersen 1988), but this feature cannot be equated with the HBF because autochthonous Precambrian 'Baltic' basement exists for hundreds of kilometres both north-west and south-east of Stavanger (e.g. Hossack and Cooper 1986). Thus, traditional understanding of the Caledonide orogen clearly precludes the extension of the HBF directly across the North Sea into Norway at Stavanger as proposed by Doré and Gage (1987).

Deep seismic reflection profiles also cross the trend of the HBF close to Scotland (NSDP 85-10 and BIRPS profile Moray, Fig. 2.1). Figure 2.8 illustrates NSDP 85-10 where it crosses the offshore projection of the HBF. There is no evidence for any significant late Palaeozoic or Mesozoic fault at the appropriate position on the profile, though the projection of the HBF does separate the Peterhead Ridge from the Aberdeen Basin. Lower crustal reflectivity occurs from about 6–10 s travel-time (about 18–30 km depth) across the whole of Fig. 2.8a and there is no evidence for any offsets of these reflectors, nor indeed for any blank zone through which a vertical fault plane or zone might run. There is no change in crustal reflectivity pattern across the position of the HBF, suggesting that crustal type is now not very different on either side of the HBF. There is little doubt that the HBF was a terrane boundary during lower Palaeozoic time, and little doubt therefore that it extended at least 70 km east of the present coast of Scotland, to the location of Fig. 2.8, but there is no deep seismic evidence for the HBF there today. This suggests that, rather than remaining a prominent crustal discontinuity throughout late Palaeozoic and Mesozoic basin formation, the HBF was largely inactive and that subsequent tectonic episodes—including perhaps formation of the present reflectivity patterns—have obliterated the trace of the fault.

Thus deep reflection profiling is able to locate some of the fundamental Palaeozoic structures and show their relation to late Palaeozoic and Mesozoic basin formation. It is noteworthy that the reactivation or tectonic control discussed in this paper (Mid-North Sea High, basins along the Great Glen fault and Hardangerfjord shear zone) was most significant during the late Palaeozoic, with less demonstrable reactivation during Mesozoic time (Inner Moray Firth basin and basins along the Karmøy shear zone). Of the Caledonian structures reactivated during Devonian and Carboniferous time, some were reactivated again in the Triassic and Jurassic; but we know of no Caledonian structures which were dormant during the late Palaeozoic and which were reactivated only in the Mesozoic.

In contrast to the reactivated structures, some proposed lineaments or alignments along projections of equally significant Caledonian structures have no seismic image at a crustal scale today. Though these features—including the Tornquist zone and the Highland Boundary fault within the North Sea—may have been locally reactivated in the upper crust, it seems unwise to attribute to them great significance as crustal discontinuities after the ending of the Caledonian orogeny.

2.4. Lithospheric structures active during Mesozoic rifting

In this section we turn attention to the main structures that controlled Mesozoic rifting on a lithospheric scale. In the Central Graben, which is almost orthogonal to Caledonian strike, it is probable that the major faults were newly formed in the Mesozoic rather than being reactivated Palaeozoic structures. In the Viking Graben, which is parallel to the local

Caledonide strike, faults active in the Mesozoic may have had a Palaeozoic history as well (though this is generally unsubstantiated) and, as in the Central Graben, many of the faults may have been newly formed in the Mesozoic.

There has been considerable controversy in the North Sea as to whether Mesozoic extension at the crustal and lithospheric scale is controlled by localized asymmetric faults, as in the Wernicke model of whole-lithosphere normal simple shear (1985), or whether on this large scale it is more appropriate to consider extension as a pure shear phenomenon, as in the McKenzie stretching model (1978; see also review by White, this volume). There is now a considerable literature supporting the application of each model to the North Sea. Klemperer (1988), White (1989), and Klemperer and White (1989) advocate stretching models, and develop interpretations of the deep seismic data to support this view, whereas papers by Beach (1985, 1986), Beach and colleagues (1987), and Gibbs (1987*a*, *b*; 1989 *a*, *b*) all argue for lithospheric simple shear, and develop contrary interpretations of the deep seismic data. It is not necessary to review the arguments of those papers here, but some brief points may be made.

Interpretations of deep seismic data for the lower crust, far beyond the reach of the drill, are inevitably speculative. However, no seismic section yet recorded across the North Sea shows a single reflector that can be interpreted as a shear zone extending from a fault within or bounding a sedimentary basin beyond, or even down to, the Moho (e.g. Gibbs, 1989 *b*; Klemperer and White, 1989). Therefore interpretations of crustal or lithospheric scale simple shear zones in the North Sea are based largely on perceived asymmetries in the fault patterns within the sedimentary basins and perceived asymmetries of reflectivity distribution in the lower crust (e.g. Beach 1986; Klemperer and White, 1989). The alternative interpretation of the same reflectivity pattern is possible, in terms of pure shear in the lower crust and extension by stretching (e.g. Klemperer 1988). Since interpretations of the deep seismic data are speculative and so inevitably contentious, other tests between the different extensional models must be sought. The most convincing test is provided by the geometries of syn-rift and post-rift basins, which for the North Sea are well explained by symmetric stretching models, and do not fit the predictions of the asymmetric simple-shear models (White 1989; Klemperer and White, 1989; White, this volume).

Rather than speculate further on the structural significance of the complex patterns of crustal reflectivity, we turn to the much simpler pattern of reflectivity in the upper mantle. Beach (1986) was the first to draw attention to the existence of clear sub-Moho reflectors in the North Sea, describing an east-dipping zone of reflections in the mantle beneath the eastern side of the Viking Graben. Other similar mantle reflections in the North Sea area have since been described by Gibbs (1987*a*, *b*) as a regional east-dipping structure, and part of a hypothesized lithospheric zone of simple shear. Klemperer (1988) and Klemperer and White (1989) confirm the existence of these east-dipping reflectors beneath the east margin of the Viking Graben and the Central Graben, but illustrate similar but west-dipping structures in the mantle beneath the west margin of the Viking Graben and Central Graben. Thus the overall pattern in the uppermost mantle is of symmetric outward-dipping structures. An example of these reflectors is shown in Fig. 2.9a; this example has been chosen because profile NSDP 85-06 has not previously been published and other, perhaps clearer, examples of mantle reflectors beneath the North Sea are given in Gibbs (1987a; NSDP 84-01), Klemperer (1988; NSDP 84-03, 84-04, and BIRPS UNST), Klemperer and White (1989; NSDP 84-02, 84-04, and 85-05), and Flack *et al.* (1990; NSDP 85-05 and 85-07). A map showing the distribution of all the known intra-mantle reflectors in the North Sea is shown as Fig. 2.9b.

A considerable number of dipping reflections within the upper mantle is now known around the British Isles (Flack *et al*., 1990), the best studied being the Flannan reflector which extends to at least 80 km depth and for more than 100 km along strike (Flack and Warner, 1990). The location of many of these intra-mantle reflectors vertically beneath major crustal faults (Flack *et al*., 1990), their dip (typically around 10° to 30°), and their very high reflection coefficients (up to 0.1) have all been used to suggest that these reflectors are mantle fault zones or shear zones (Warner and McGeary 1987). This interpretation is accepted for the intra-mantle reflectors in the North Sea area because of their location beneath the east and west margins of the graben system. Further, the strike of the intra-mantle reflectors, where this can be determined by intersecting profiles, is clearly that of the Mesozoic grabens; in

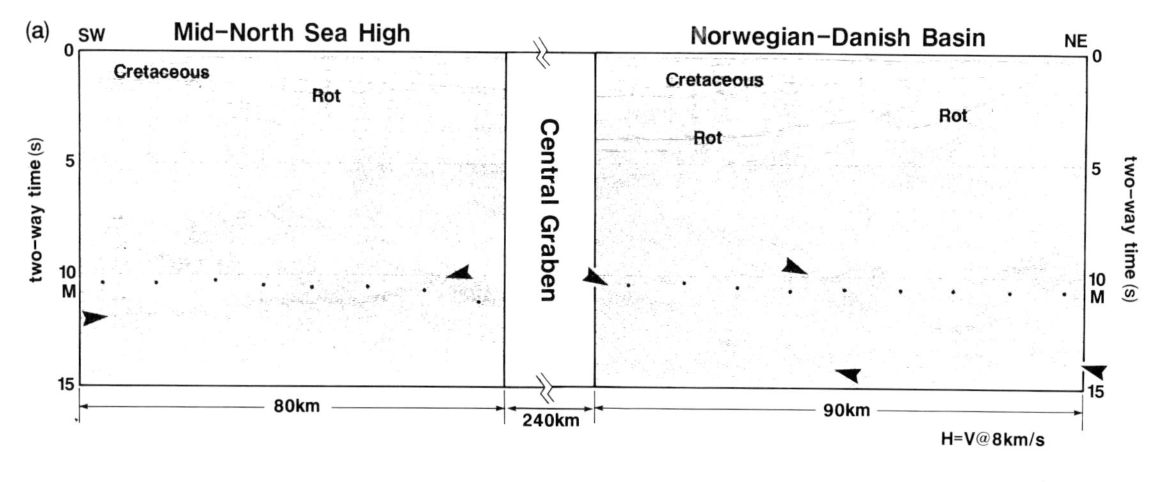

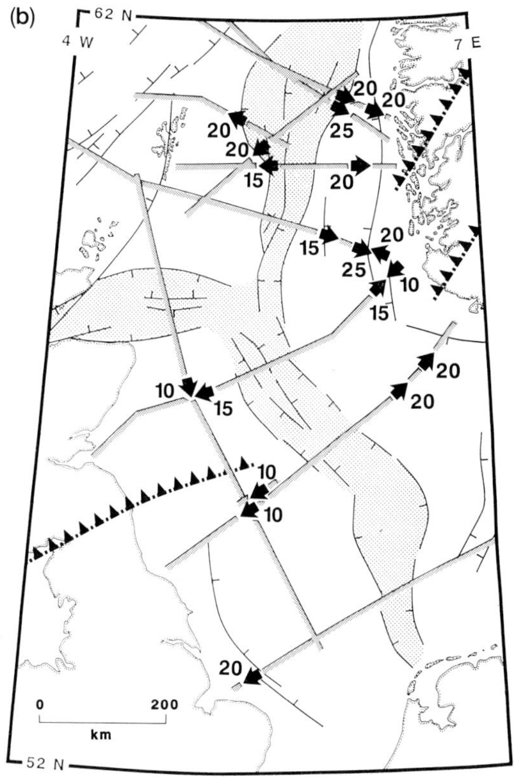

the central North Sea, it is almost orthogonal to the Caledonian structures mappable on deep seismic reflection data (Fig. 2.9b). Thus these intra-mantle reflectors may be shear zones formed during Mesozoic rifting, though difficulties with this interpretation remain (principally the failure to observe clear offsets of the Moho where the intra-mantle reflectors project up to the Moho; Klemperer 1988). None the less the seismic data can be used to support a symmetric stretching model, in which macroscopic pure shear in the (reflective) lower crust decouples asymmetric half-graben formation in the upper crust

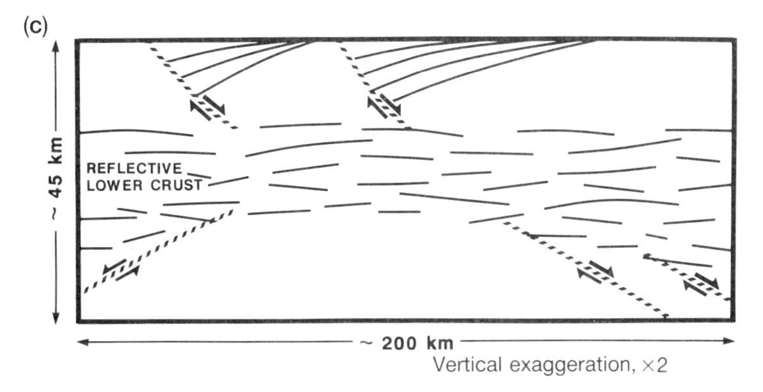

Fig. 2.9.a. Segments of GECO profile NSDP 85-06 showing intra-mantle reflectors dipping south-west on the south-west side of the Central Graben, and dipping north-east on the north-east side of the graben. Data shown are 2-D f-k migrated at 6 km s^{-1} (the average velocity to the prominent reflectors in the mantle) and are displayed true scale at 8 km s^{-1} to show the sub-Moho reflectors in their true geometry. The Moho is marked by dots and M. Rot = top of the Rotliegendes. (See Fig. 2b.) **b.** Map of locations in the North Sea at which mantle reflectors are observed. Arrows and numbers give apparent dip after 2-D depth migration, in the line of section; note the general symmetry of outward dips on either side of the graben system. Stippled areas are main rift structures, and thin lines give the schematic Mesozoic fault pattern. Thrust symbols mark major Caledonian structures, after Freeman *et al.* (1988) and Hossack and Cooper (1986). Note the orthogonal strike of the Caledonian structures and the mantle reflectors which are parallel to the strike of the Mesozoic rifts. **c.** Decoupled symmetric stretching model (Klemperer 1988). Note the asymmetric rift basins (linked half-grabens), the decoupling layer of effective pure shear in the ductile lower crust (perhaps equivalent to the reflective lower crust?), and the symmetric outward-dipping shear zones in the upper mantle which define a mantle 'horst' vertically beneath the crustal graben.

from a pattern of symmetric, outward-dipping shear zones in the upper mantle (Fig. 2.9c. Klemperer 1988).

2.5. Summary

This paper has attempted to show the utility of deep seismic reflection profiling in understanding the regional tectonic development of the North Sea. On the broadest scale, laterally changing patterns of reflectivity in the lower crust may be useful as a guide to the extensional deformation undergone by the crust. The distinct vertical change in reflectivity patterns above and below the reflection Moho allows the mapping out of crustal thickness, and hence the regional quantification of crustal thinning by extension. Individual structural features can be projected from the Palaeozoic structural highs surrounding the North Sea into the basin areas; in some cases these features have a clear seismic signature through the whole crust, and an apparent relationship with late Palaeozoic and sometimes Mesozoic reactivation. Other major Palaeozoic structural boundaries cannot be seismically imaged beneath the North Sea, and may have been overprinted rather than utilized by Mesozoic extension. Extensional faults of a crustal scale are not imaged, leading to the conclusion that deformation in the lower crust at least may be distributed rather than localized. In contrast, isolated reflectors in the upper mantle may be discrete shear zones.

Acknowledgements

The NSDP programme was shot by GECO as a speculative survey with participation by Arco, BP, Britoil, Elf, Esso, BIRPS, Shell, Statoil, and Norsk Hydro. BIRPS is indebted to GECO for allowing participation in their survey and publication of their data. Britoil, GECO, Murphy, NOPEC, Shell, and Western Geophysical kindly allowed access to their proprietary deep seismic profiles. This project was assisted by other members of the BIRPS group, and by reviews from Nicky White, Dave Smythe, and one anonymous reviewer. BIRPS is funded by the Natural Environment Research Council. SLK is funded by a Royal Society University Research Fellowship. Cambridge Earth Sciences Contribution 1452.

References

Allmendinger, R. W., Nelson, K. D., Potter, C. J., Barazangi, M., Brown, L. D., and Oliver, J. E. (1987*a*). Deep seismic reflection characteristics of the continental crust. *Geology*, **15**, 304–10.

Allmendinger, R. W., Haugeh, T. A., Hauser, E. C., Potter, C. J., Klemperer, S. L., Nelson, K. D., Knuepfer, P., and Oliver, J. E. (1987*b*). Overview of the COCORP 40°N Transect, western United States: the fabric of an orogenic belt. *Geol. Soc. of America Bulletin*, **98**, 308–19.

Barton, P. J. (1986). Comparison of deep reflection and refraction structures in the North Sea. In *Reflection seismology: a global perspective* (ed. M. Barazangi and L. Brown). Washington, *American Geophysical Union Geodynamics Series*, **13**, 297–300.

Barton, P. and Wood, R. (1984). Tectonic evolution of the North Sea basin: crustal stretching and subsidence. *Geophys. J. of the Roy. Astron. Soc.*, **79**, 987–1022.

Barton, P., Matthews, D. H., Hall, J., and Warner, M. (1984). Moho beneath the North Sea compared on normal-incidence and wide-angle seismic records. *Nature*, **308**, 55–6.

Beach, A. (1985). Some comments on sedimentary basin development in the northern North Sea. *Scot. J. Geol.*, **21**, 493–512.

Beach, A. (1986). A deep seismic reflection profile across the northern North Sea. *Nature*, **323**, 53–5.

Beach, A., Bird, T., and Gibbs, A. (1987). Extensional tectonics and crustal structure: deep seismic reflection data from the northern North Sea Viking Graben. In *Continental extensional tectonics* (ed. M. P. Coward, J. F. Dewey, and P. L. Hancock), *Geol. Soc. Special Publication*, **28**, 467–76.

Bluck, B. J. (1984). Pre-Carboniferous history of the Midland Valley of Scotland. *Trans. Roy. Soc., Edinburgh: Earth Sciences*, **75**, 275–95.

Blundell, D. J., Hobbs, R. W., Klemperer, S. L., Scott-Robinson, R., Long, R., West, T., and Duin, E. (in press). Crustal structure beneath the central and southern North Sea from BIRPS deep seismic reflection profiling. J. Geol. Soc.

Bois, C., Cazes, M., Hirn, A., Mascle, A., Matte, P., Montadert, L., and Pinet, B. (1988). Contributions of deep seismic profiling to the knowledge of the lower crust in France and adjacent areas. *Tectonophysics*, **145**, 253–75.

Bott, M. H. P. and Browitt, C. W. A. (1975). Interpretation of geophysical observations between the Orkney and Shetland islands. *J. Geol. Soc.*, **131**, 353–71.

Brewer, J. A. and Smythe, D. K. (1984). MOIST and the continuity of crustal reflector geometry along the Caledonian-Appalachian orogen. *J. Geol. Soc.*, **141**, 105–20.

Brewer, J. A., Matthews, D. H., Warner, M. R., Hall, J., Smyth, D. K., and Whittington, R. J. (1983). BIRPS deep seismic reflection studies of the British Caledonides. *Nature*, **305**, 206–10.

Cheadle, M., McGeary, S., Warner, M. R., and Matthews, D. H. (1987). Extensional structures on the UK continental shelf: a review of evidence from deep seismic profiling. In *Continental extensional tectonics* (ed. M. P. Coward, J. F. Dewey, and P. L. Hancock). *Geol. Soc., Special Publication*, **28**, 445–65.

Christie, P. A. F. (1982). Interpretation of refraction experiments in the North Sea. *Phil. Trans. Roy. Soc., of London*, **A305**, 101–12.

Christie, P. A. F. and Sclater, J. G. (1980). An extensional origin for the Buchan and Witchground Graben in the North Sea. *Nature*, **283**, 729–32.

Cocks, L. R. M. and Fortey, R. A. (1982). Faunal evidence for oceanic separations in the Palaeozoic of Britain. *J. Geol. Soc.*, **139**, 465–78.

Coward, M. P., Enfield, M. A., and Fischer, M. W. (1989). Devonian basins of northern Scotland: extension and inversion related to the late Caledonian-Variscan tectonics. In *Inversion tectonics* (ed. M. A. Cooper and G. D. Williams). *Geol. Soc. Special Publication*, **44**, 275–308.

Day, G., Cooper, B. A., Anderson, W. F. J., Burgers, H. C., Ronnevik, H. C., and Schoneich, H. (1981). Regional seismic structure maps of the North Sea. In *Petroleum geology of the continental shelf of north-west Europe* (ed. L. V. Illing, and G. D. Hobson), 76–84. London, Institute of Petroleum, Heyden and Son.

Day, G. A., Anderson, O. B., Ensager, K., Finnstrom, E. G., Hospers, J., Liebe, T., Makris, J., Plaumann, S., Strang van Hees, G., and Walter, S. A. (this volume). North Sea gravity map.

DEKORP Research Group (1985). First results and preliminary interpretation of deep-reflection seismic recordings along profile DEKORP2-South. *J. Geophys.*, **57**, 137–63.

Dikkers, A. J. (1977). Sketch of a possible lineament pattern in northwest Europe. *Geol. en Mijnbouw*, **56**, 275–85.

Donato, J. A., Martindale, W., and Tully, M. C. (1983). Buried granites within the Mid-North Sea High. *J. Geol. Soc.*, **140**, 825–37.

Doré, A. G. and Gage, M. S. (1987). Crustal alignments

and sedimentary domains in the evolution of the North Sea, northeast Atlantic margin and Barents Shelf. In *Petroleum geology of north-west Europe* (ed. J. Brooks and K. Glennie), 1131–48. London, Graham and Trotman.

Eugeno-S Working Group (1988). Crustal structure and tectonic evolution of the transition between the Baltic Shield and the North German Caledonides (the Eugeno-S Project). *Tectonophysics*, **150**, 253–348.

Fichler, C. and Hospers, J. (this volume). Gravity modelling in the Viking Graben area, North Sea.

Flack, C. A. and Warner, M. R. (1990). Three-dimensional mapping of seismic reflectors from the crust and upper mantle, north-west of Scotland. *Tectonophysics*, **173**, 469–82.

Flack, C. A., Klemperer, S. L., McGeary, S., Snyder, D. B., and Warner, M. R. (1990). The reflective upper mantle of the U.K. *Geology*, **18**(6) 528–32.

Freeman, B., Klemperer, S. L., and Hobbs, R. W. (1988). The deep structure of northern England and the Iapetus Suture zone from BIRPS deep seismic reflection profiles. *J. Geol. Soc.*, **145**, 727–40.

Frost, R. E. (1987). The evolution of the Viking Graben tilted fault-block structures: a compressional origin. In *Petroleum geology of north-west Europe* (ed. J. Brooks and K. Glennie), 1009–24. London, Graham and Trotman.

Frost, R. T. C., Fitch, F. J., and Miller, J. A. (1981). The age and nature of the crystalline basement of the North Sea basin. In *Petroleum geology of the continental shelf of north-west Europe* (ed. L. V. Illing and G. D. Hobson), 43–57. London, Institute of Petroleum, Heyden and Son.

Gibbs, A. D. (1987*a*). Linked tectonics of the northern North Sea basins. In *Sedimentary basins and basin-forming mechanisms* (ed. C. Beaumont and A. J. Tankard), **12**, 163–71.

Gibbs, A. D. (1987*b*). Deep seismic profiles in the northern North Sea. In *Petroleum geology of north-west Europe* (ed. J. Brooks and K. Glennie), 1025–8. London, Graham and Trotman.

Gibbs, A. D. (1989*a*). A model for linked basin development around the British Isles. In *Extensional tectonics and stratigraphy of the North Atlantic Margins* (ed. A. J. Tankard and H. R. Balkwill). *Am. Assoc. Petrol. Geol. Memoir*, **46**, 501–9.

Gibbs, A. D. (1989*b*). Structural styles in basin formation. In *Extensional tectonics and stratigraphy of the North Atlantic Margins* (ed. A. J. Tankard and H. R. Balkwill). *Am. Assoc. Petrol. Geol. Memoir*, **46**, 81–93.

Glennie, K. W. (1986). *Introduction to the petroleum geology of the North Sea*. Oxford, Blackwell Scientific Publications, 2nd edn.

Graversen, Ø. (1988). Continental transform faults and evolution of graben systems in central and northwest Europe. *Geol. Assoc. Can. Program with Abstracts* (St. John's Newfoundland) **13**, A48.

Graversen, Ø. (submitted). Continental transform faults—a structural barrier in geological correlation. *Norsk Geologiske Tiddskrift*.

Hitchen, K. and Ritchie, J. D. (1987). Geological review of the west Shetland area. In *Petroleum geology of north west Europe* (ed. J. Brooks and K. Glennie), 737–49. London, Graham and Trotman.

Holliger, K. and Klemperer, S. L. (1989). A comparison of the Moho interpreted from gravity data and from deep seismic reflection data in the northern North Sea. *Geophys. J.*, **97**, 247–58.

Holliger, K. and Klemperer, S. L. (this volume). Gravity and deep seismic reflection profiles across the North Sea rifts.

Hossack, J. R. and Cooper, M. A. (1986). Collision tectonics in the Scandinavian Caledonides. In *Collision tectonics* (ed. M. P. Coward and A. C. Ries). *Geol. Soc. Special Publication*, **19**, 287–304.

Hurich, C. A. and Kristoffersen, Y. (1988). Deep structure of the Caledonide orogen in southern Norway: new evidence from marine seismic reflection profiling. *Norges geologiske undersøkelse Special Publication*, **3**, 96–101.

Hurich, C. A., Egilson, T. and Kristoffersen, Y. (1986). *Norwegian Lithosphere Project, Annual Report 1986*. Bergen, Jordskjelvstasjonen.

Hurich, C. A., Egilson, T. and Kristoffersen, Y. (1987). *Norwegian Lithosphere Project, Annual Report 1987*. Bergen, Jordskjelvstasjonen.

Husebye, E. S., Ro, H. E., Kinck, J. J., and Larsson, F. R. (1988). Tectonic studies in the Skagerrak province: the 'Mobil Search' cruise. *Norges geologiske undersøkelse Special Publication*, **3**, 14–20.

Hutton, D. H. W. (1987). Strike-slip terranes and a model for the evolution of the British and Irish Caledonides. *Geol. Mag.*, **124**, 405–25.

Institute of Geological Sciences (1982). *Peterhead, sheet 57° N-02° W 1:250,000 series, solid geology*.

Johnson, R. J. and Dingwall, R. G. (1981). The Caledonides: their influence on the stratigraphy of the northwest European continental shelf. In *Petroleum geology of the continental shelf of northwest Europe* (ed. L. V. Illing and G. D. Hobson), 85–97. London, Institute of Petroleum, Heyden and Son.

Kinck, J. J., Husebye, E. S., and Lund, C.-E. (1988). The S. Scandinavia crust—structural complexities from seismic reflection and refraction profiling. In *Nordic symposium in Earth Sciences on imaging and understanding the lithosphere*, 19–22 Oct. 1988 (ed. S. Gregersen, H. Korhonen, S. Bjørnsson, E. S. Husebye, C.-E. Lund, and L. B. Loughran). Sweden, Tanomstrand.

Klemperer, S. L. (1988). Crustal thinning and nature of

extension in the northern North Sea from deep seismic reflection profiling. *Tectonics*, **7**, 803–21.

Klemperer, S. L. (1989). Seismic reflection evidence for the location of the Iapetus suture west of Ireland. *J. Geol. Soc.*, **146**, 409–12.

Klemperer, S. L. and Matthews, D. H. (1987). Iapetus suture located beneath the North Sea by BIRPS deep seismic reflection profiling. *Geology*, **15**, 195–8.

Klemperer, S. L. and White, N. J. (1989). Coaxial stretching or lithospheric simple shear in the North Sea? Evidence from deep seismic profiling and subsidence. In *Extensional tectonics and stratigraphy of the North Atlantic margins* (ed. A. J. Tankard and H. R. Balkwill) *Am. Assoc. Petrol. Geol. Memoir*, **46**, 511–22.

Klemperer, S. L., Hauge, T. A., Hauser, E. C., Oliver, J. E., and Potter, C. J. (1986). The Moho in the northern Basin and Range Province, Nevada, along the COCORP 40°N seismic reflection transect. *Geol. Soc. Amer. Bulletin*, **97**, 603–18.

Klemperer, S. L., Hobbs, R. W., and Freeman, B. (1990). Dating the source of lower crustal reflectivity using BIRPS deep seismic profiles across the Iapetus suture. *Tectonophysics*, **173**, 445–54.

Larsson, F. R. and Husebye, E. S. (1988). Crustal lamination—Skagerrak tectonic province. In *Nordic symposium in Earth Sciences on imaging and understanding the lithosphere*, 19–22 Oct. 1988 (ed. S. Gregersen, H. Korhonen, S. Bjørnsson, E. S. Husebye, C.-E. Lund and L. B. Loughran. Sweden Tanomstrand.

Larsson, F. R. and Husebye, E. S. (submitted). Crustal reflectivity in the Skagerrak area. *Tectonophysics*.

Latin, D. M., Dixon, J. E., and Fitton, J. G. (this volume). Rift-related magmatism in the North Sea basin.

McClay, K. R., Norton, M. G., Coney, P., and Davies, G. H. (1986). Collapse of the Caledonian orogen and the Old Red Sandstone. *Nature*, **323**, 147–9.

McGeary, S. (1987). Nontypical BIRPS on the margin of the northern North Sea: the SHET survey. *Geophys. J. Roy. Astron. Soc.*, **89**, 231–8.

McGeary, S. (1989). Reflection seismic evidence for a Moho offset beneath the Walls Boundary strike-slip fault. *J. Geol. Soc.*, **146**, 261–9.

McGeary, S. and Warner, M. R. (1985). Seismic profiling of the continental lithosphere. *Nature*, **317**, 795–7.

McKenzie, D. P. (1978). Some remarks on the development of sedimentary basins. *Earth and Planetary Science Letters*, **40**, 25–32.

McKerrow, W. S. (1982). The northwest margin of the Iapetus ocean during the early Palaeozoic. *Am. Assoc. Petrol. Geol. Memoir*, **34**, 521–33.

McQuillin, R., Donato, J. A., and Tulstrup, J. (1982). Development of basins in the Inner Moray Firth and the North Sea by crustal extension and dextral displacement of the Great Glen fault. *Earth and Planetary Science Letters*, **60**, 127–39.

Marillier, F., Keen, C. E., Stockmal, G. S., Quinlan, G., Williams, H., Colman-Sadd, S. P., and O'Brien, S. J. (1989). Crustal structure and surface zonation of the Canadian Appalachians: implications of deep seismic reflection data. *Can. J. Earth Sciences*, **26**, 305–21.

Matthews, D. H. (1986). Seismic reflections from the lower crust around Britain. In *The nature of the lower continental crust* (ed. J. B. Dawson, D. A. Carswell, J. Hall, and K. H. Wedepohl). *Geol. Soc. Special Publication*, **24**, 11–22.

Meissner, R., Wever, T., and Flüh, E. R. (1987). The Moho in Europe—implications for crustal development. *Annales Geophysicae*, **5B**, 357–64.

Mooney, W. D. and Brocher, T. M. (1987). Coincident seismic reflection/refraction studies of the continental lithosphere: a global review. *Reviews of Geophysics*, **25**, 723–42.

Norton, M. G. (1986). Late Caledonian extension in western Norway: a response to extreme crustal thickening. *Tectonics*, **5**, 195–204.

Norton, M. G. (1987). The Nordfjord-Sogn detachment, W. Norway. *Norsk Geologiske Tiddskrift*, **67**, 93–106.

Pegrum, R. M. (1984*a*). The extension of the Tornquist Zone in the Norwegian North Sea. *Norsk Geologiske Tiddskrift*, **64**, 39–68.

Pegrum, R. M. (1984*b*). Structural development of the Russian-Fennoscandian platform. In *Petroleum geology of the North European margin* (ed. A. M. Spencer), 359–69. London, Graham and Trotman.

Pinet, B. (1989). Deep seismic profiling and sedimentary basins. *Bulletin de la Societé Géologique de France* **(8)**, **4**, 749–66.

Reston, T. J. (1988). Evidence for shear zones in the lower crust offshore Britain. *Tectonics*, **7**, 929–45.

Seranne, M. and Seguret, M. (1987). The Devonian basins of western Norway: tectonics and kinematics of an extending crust. In *Continental extensional tectonics* (ed. M. P. Coward, J. F. Dewey, and P. L. Hancock). *Geol. Soc. Special Publication*, **28**, 537–48.

Smith, D. I. and Watson, J. (1983). Scale and timing of movements on the Great Glen fault, Scotland. *Geology*, **11**, 523–6.

Soper, N. J. (1988). Timing and geometry of collision, terrane accretion and sinistral strike-slip events in the British Isles. In *The Caledonian-Appalachian orogen* (ed. A. L. Harris and D. J. Fettes). *Geol. Soc. Special Publication*, **38**, 481–92.

Soper, N. J. and Hutton, D. W. (1984). Late Caledonian sinistral displacements in Britain: implications for a three-plate collision model. *Tectonics*, **3**, 781–94.

Storetvedt, K. M. (1987). Major late Caledonian and Hercynian shear movements on the Great Glen fault. *Tectonophysics*, **143**, 253–67.

Thirlwall, M. F. (1989). Movement on proposed terrane boundaries in northern Britain: constraints from Ordovician-Devonian igneous rocks. *J. Geol. Soc.*, **146**, 373–6.

Warner, M. R. (1990). Basalts, water and shear zones in the lower continental crust? *Tectonophysics*, **173**. 163–74.

Warner, M. R. and McGeary, S. (1987). Seismic reflection coefficients from mantle fault zones. *Geophys. J. Roy. Astronom. Soc.*, **89**, 223–30.

Watson, J. (1984). The ending of the Caledonian orogeny in Scotland. *J. Geol. Soc.*, **141**, 193–214.

Watson, J. (1985). Northern Scotland as an Atlantic-North Sea divide. *J. Geol. Soc.*, **142**, 221–43.

Wernicke, B. (1985). Uniform sense simple shear of the continental lithosphere. *Can. J. Earth Sciences*, **22**, 108–25.

Wever, T., Trappe, H., and Meissner, R. (1987). Possible relations between crustal reflectivity, crustal age, heat flow, and viscosity of the continents. *Annales Geophysicae*, **5B**, 255–66.

White, N. J. (1989). Nature of lithospheric extension in the North Sea. *Geology*, **17**, 111–14.

White, N. J. (this volume). Does the uniform stretching model work in the North Sea?

Wilson, J. T. (1966). Did the Atlantic close and then re-open? *Nature*, **211**, 676–81.

Ziegler, P. A. (1982). *Geological atlas of western and central Europe*. The Hague, Shell International.

3 North Sea Gravity Map

G. A. Day, O. B. Andersen, K. Engsager, E. G. Finnstrom, J. Hospers, T. Liebe, J. Makris, S. Plaumann, G. Strang van Hees, and S. A. Walter

Abstract

Marine gravity data sets from eight organizations have been assembled; together they form a virtually complete cover of the North Sea between 51°N and 62°N. The whole data set has been adjusted and merged with adjacent land data to form the first detailed North Sea gravity compilation. A Bouguer anomaly gravity map is presented at a scale of 1:4 000 000.

3.1. Introduction

A great deal of gravity data has been acquired in the North Sea over the last 20 years or so. A number of publications, and maps at various scales, have been produced based on these surveys (Sunderland 1972; Zervos 1986; and references in next section). Adequate cover now exists to map the entire North Sea from 51°N to 62°N (except for a small area off southern Norway) at a scale of 1:1 000 000 or larger. The data available from a number of agencies in countries around the North Sea have now been merged and adjusted to create a levelled data base for the North Sea and surrounding land areas.

3.2. Sources of data

1. British Geological Survey
BGS data comprise the greater part of the data set, and cover approximately the western half of the North Sea (Tully and Donato 1985). Most data were acquired in the years 1970 to 1983 as a series of individual surveys conducted on behalf of the UK Department of Energy. For the greater part of the data set the line spacing is 8 km or less, which is increased to a maximum of 14 km in a few small areas.

2. Danish Geodetic Institute
These data were acquired in the Skagerrak and west of Denmark (Andersen and Engsager 1977). The line spacing is variable but generally less than 10 km.

3. Delft University of Technology
Data acquired by Delft, and by Delft in association with the University of Hamburg, occupy the south-east portion of the North Sea (Haagmans *et al.* 1988). The line spacing of the main survey is approximately 18 km.

4. Federal Institute of Geosciences and Natural Resources, Hanover
This survey consists of a number of east-west lines crossing the northern part of the North Sea (Plaumann 1979). The line spacing is approximately 18 km.

5. Hamburg University Institute of Geophysics
The data used are from *Valdivia*, cruise No. 13, covering most of the German sector of the North Sea with a line spacing in the north of approximately 19 km and in the south of approximately 10 km. Some data from a *Meteor* cruise linking this area with the Skagerrak are also included.

6. Norwegian Geographical Survey
The data, which occupy a band extending approximately 100 km west of the Norwegian coast, were acquired on a survey conducted jointly with the US Army Topographic Command in 1970 (Hospers and Finnstrom 1984; Hospers *et al.* 1985). Most of the survey comprises east-west lines with a spacing of about 10 km, although larger data gaps do occur.

There are ten tie lines, orientated either north-west–south-east or north-east–south-west.

7. Western Atlas International
Data from four commercial surveys have been used in this compilation. Norway North Sea data, with a line spacing generally better than 20 km, largely fill the gap between the Norwegian Geographical Survey data and BGS data, and overlap the Norwegian Survey data. Stord Basin data, with a line spacing of about 15 km, lies immediately offshore Norway, mainly between 59° and 60°N. The Norway South survey extends south of this to the southern limit of Norwegian waters where it overlaps the Danish Geodetic Institute data. A small amount of the Denmark Continental Shelf survey data, with a line spacing of approximately 6 km, has been used in the western part of Danish waters.

8. UK Hydrographic Department data
Data acquired by the Hydrographic Department of the UK Ministry of Defence north of 62°N, on behalf of the UK Department of Energy, were used in the adjustment process, but these data do not contribute to the map itself.

9. Data for the land area of UK were supplied by the British Geological Survey (Hipkin and Hussain 1983); data for mainland Europe were supplied by Hamburg University.

3.2.1. *Datums*

All BGS data are reduced using the International Gravity Formula 1967 (Hipkin and Hussain 1983). Base stations were tied to the National Gravity Reference Net 1973, which itself is linked to the International Gravity Standardization Net (IGSN 1971: Masson-Smith *et al.* 1974). Data sets made available to BGS by the other contributors were also based on the 1967 formula and referred to IGSN71.

3.3. Merging of data sets

Marine gravity data may contain errors due to any of the following causes.

1. The gravity meter itself may be malfunctioning, and small errors are difficult to detect while the ship is at sea. Some instruments suffer from tares.
2. Drift measurements must be averaged between port calls and depend on the accuracy of base ties and, where there is a large change in gravity value between harbour base stations, the calibration of the instrument.
3. Meter calibration is not always known with great accuracy.
4. Very large accelerations caused by ship motion must be damped by the gravity meter, making the operation weather-dependent. The threshold of satisfactory operation is determined by the type of gravity meter, its position in the ship, the size of the ship and the way it responds to wave motion, and the relationship between the ship's course and the swell direction, which is relevant for instruments that are prone to cross-coupling errors.
5. Imprecise position data cause location errors, but more importantly, produce errors in calculating the Eötvos effect, probably the greatest contributor of high frequency noise in the data.

With the exception of the Western Atlas International data, all the data have been examined and checked by the appropriate research groups. This initial processing and error analysis of the data is detailed in the reports referenced in the previous section. Western Atlas have performed some filtering and network adjustment so that all data sets are

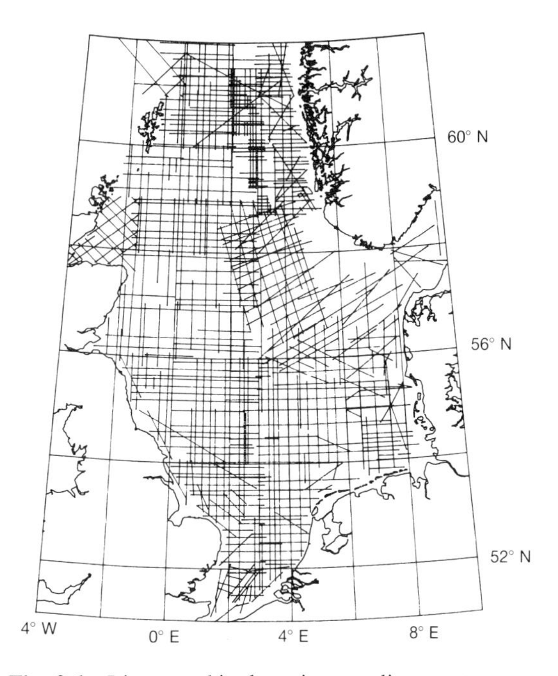

Fig. 3.1. Lines used in the primary adjustment.

consistent within themselves, corrections having been made for errors such as instrument drift and tares.

Given this starting point, the free air gravity data have been merged using the ARK-GMP software package developed by ARK Geophysics Ltd. The data from the various sources were subsampled where necessary to give a mean data spacing of 1500 to 2000 m. The Hamburg University data were also filtered to remove high frequency noise. The processing then consisted of constant (DC) shifts and progressive bending of the data to reduce intersection errors to close to zero, thus integrating the various data sets and removing any systematic errors due, for example, to poor base station ties or poor meter calibration within individual survey projects. In addition, the increase in survey line intersections created by merging the data sets provides better control of experimental error and processing uncertainties such as the Eötvos correction.

Initially a primary net of 474 lines was chosen. These lines were generally spaced about 20 km apart to give regular coverage over the entire area. Figure 3.1 shows the lines used in this primary net. The lines selected were generally long traverse lines with a large number of intersections so that good control of the data was possible. The first processing stage was to apply DC shifts to these lines to minimize the intersection errors. This was an iterative procedure using 15 iterations. Having thus removed any systematic errors between surveys, bending of the survey lines was carried out. This was also an iterative process, and was performed in several stages, first with maximum bending of

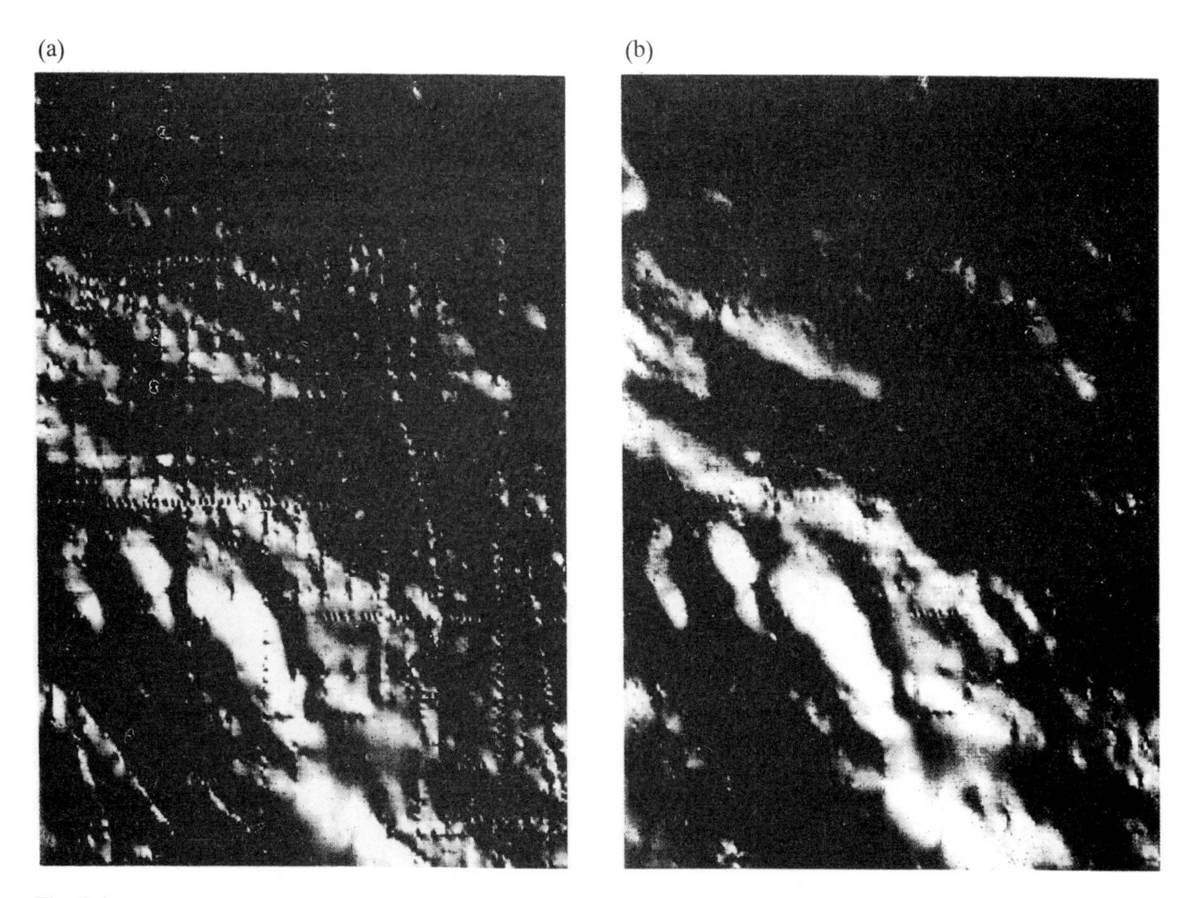

Fig. 3.2. Shaded relief image of part of the southern North Sea data (a) before network adjustment (b) after adjustment. Illumination from NE with a sun elevation of 45°.
Grid spacing 800 metres.

0.25 mGal/km, then subsequently with bending up to 0.5 mGal/km, 1.0 mGal/km, and finally up to 1.5 mGal/km (15 iterations at each stage).

At each stage the mis-ties were examined and lines with large residual mis-ties investigated. It was sometimes possible to attribute these errors to gaps in a line or a possible data spike. Such errors were corrected manually, whilst in other cases a line or part of a line with large errors was deleted from the data set. At the end of this processing sequence there were seven non-zero mis-ties, the maximum being 1.50 mGal.

The remaining lines used in the compilation (Figs 3.3 to 3.6) were then merged with these adjusted primary lines, and adjusted progressively in a similar sequence to that used for the primary lines. First, the primary lines were frozen and the secondary lines allowed to DC shift. The primary lines were then allowed to DC shift, with the secondary lines allowed to bend up to a maximum of 0.25, 0.5, and finally 1.0 mGal/km. All lines were then allowed to bend up to a maximum of 0.5 mGal/km. (Each of the above stages was carried out through 15 iterations.) Again, at each stage the mis-ties were examined and locally poor data edited or suspect lines deleted from the data set.

After adjusting in this way, pseudo-relief imaging of the data was used to help identify lines which had introduced steep local gradients in the contour surface. Figure 3.2 shows such a display for the Anglo-Dutch Basin area before and after adjusting the data (the positive anomalies in the southern half of the figure are caused by the Sole Pit Inversion); the effects of network adjustment can be clearly seen. Some linear artefacts of the processing can be seen as high frequency noise. This was attributed to close near-parallel lines, some of which were deleted, as were any other lines or part lines which appear suspect on the pseudo-relief image. The adjustment of the whole data set as detailed previously was then

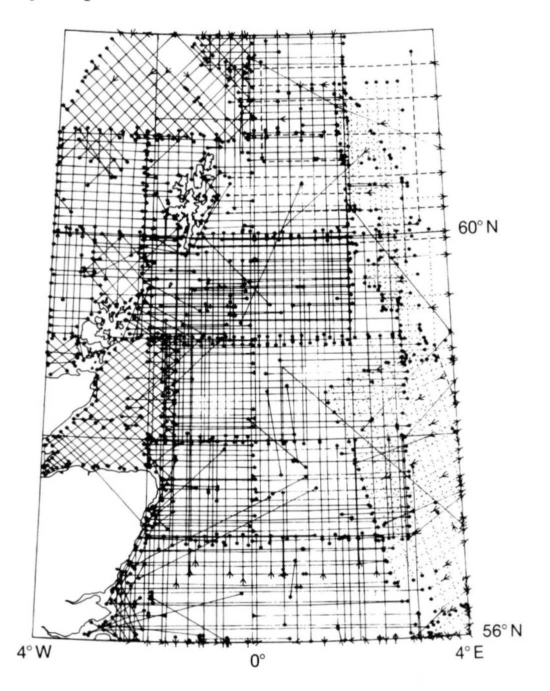

Fig. 3.4. Lines used in north-west quadrant. Solid = British Geological Survey, broken = FIGNAR Hanover, dotted = Western Atlas International.

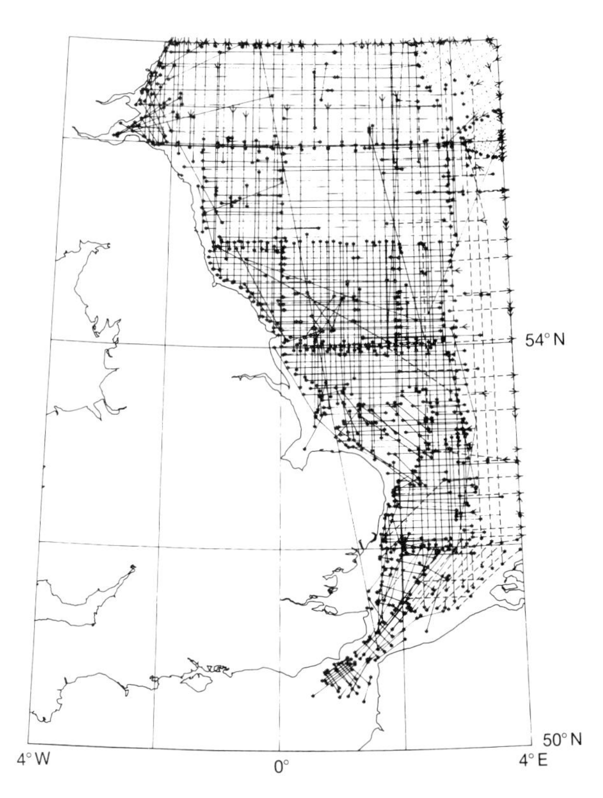

Fig. 3.3. Lines used in south-west quadrant. Solid = British Geological Survey, broken = Delft Technological University, dotted = Western Atlas International.

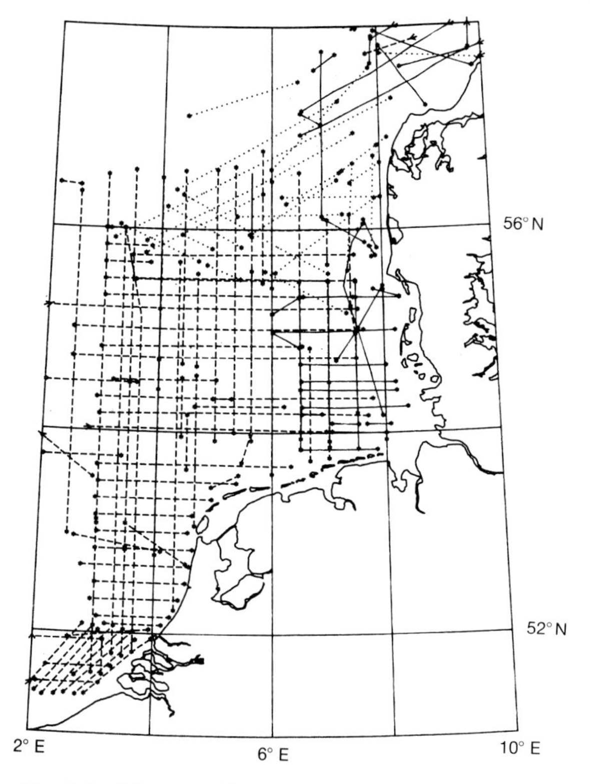

Fig. 3.5. Lines used in south-east quadrant. Solid = Hamburg University, broken = Delft Technological University, dotted = Danish Geodetic Institute.

repeated. Table 3.1 shows statistics for free air gravity mis-ties. Before network adjustment 91.9 per cent of mis-ties were less than 5.0 mGal and 98.9 per cent less than 10 mGal. The maximum mis-tie was 22.1 mGal. After network adjustment there remain 6 mis-ties above 2.0 mGal (maximum 3.3 mGal) out of a total of 10 654 intersections.

It should be noted that much of the high frequency noise seen on the pseudo-relief image is of very low amplitude, 1 to 2 mGal, which is within the noise envelope of the data. This could be removed by application of a low pass filter but real features of similar frequency and amplitude seen on the image would also be removed in this process. This paper concerns production of a Bouguer anomaly grid over the entire North Sea and, as such, the minimum amount of filtering has been carried out so that future workers using the data may process the grid according to their own requirements.

The water depths were adjusted in a similar way to the free air data, but this time all the data were adjusted together. Initially all lines were DC shifted and then bending was applied, first up to 10 m/km and then finally 20 m/km (15 iterations at each stage). Table 3.2 shows statistics for depth mis-ties.

The Bouguer gravity anomaly was then calculated from the network-adjusted free air and depth values according to the relationship:

$$\text{Bouguer anomaly} = \text{free air gravity} + 0.068765 \times \text{depth}$$

This assumes a Bouguer correction density of 1.64 g/cm^3, which represents the correction for a rock density of 2.67 g/cm^3 and 1.03 g/cm^3 for the water column. A value of 2.67 g/cm^3 was used for the Bouguer reductions for the surrounding land area.

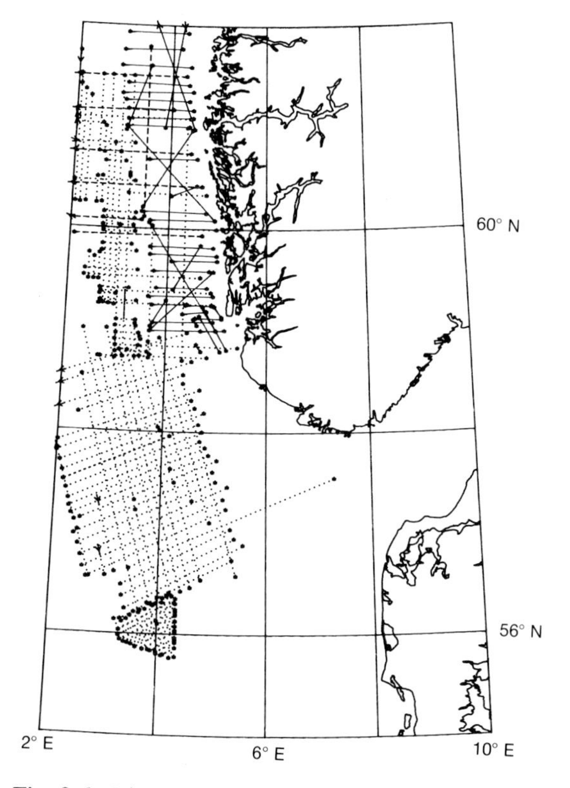

Fig. 3.6. Lines used in north-east quadrant. Solid = Norwegian Geographical Survey, broken = FIGNAR Hanover, dotted = Western Atlas International.

Table 3.1 Free air gravity mis-ties

Mis-tie value (mGal)		No. of intersections	
Lower bound	<Upper bound	Before adjustment	After adjustment and editing
0.0	1.0	4626	10514
1.0	2.0	2783	54
2.0	3.0	1315	5
3.0	4.0	640	1
4.0	5.0	424	0
5.0	6.0	246	0
6.0	7.0	206	0
7.0	8.0	142	0
8.0	9.0	85	0
9.0	10.0	69	0
10.0	11.0	39	0
11.0	12.0	18	0
12.0	13.0	18	0
13.0	14.0	12	0
14.0	15.0	7	0
15.0	16.0	4	0
16.0	17.0	9	0
17.0	18.0	3	0
18.0	19.0	6	0
19.0	20.0	1	0
20.0	21.0	1	0
21.0	22.0	0	0
22.0	23.0	1	0

Table 3.2 Depth mis-ties

Mis-tie value (m)		No. of intersections	
Lower bound	<Upper bound	Before adjustment	After adjustment and editing
0.0	10.0	10416	10793
10.0	20.0	292	0
20.0	30.0	64	1
30.0	40.0	8	0
40.0	50.0	3	0
50.0	60.0	6	0
60.0	70.0	3	0
70.0	80.0	0	0
80.0	90.0	0	0
90.0	100.0	1	0
100.0	110.0	1	0
110.0	120.0	0	0
120.0	130.0	1	0

No attempt was made to reduce mis-ties by adjusting the position values of the data. By far the greater part of the data set has been collected using modern navigation methods which (given that there are no large gravity gradients in the North Sea) provide an accuracy such that the uncertainty in the gravity value due to position uncertainty is smaller than the other errors of observation. (With a gravity gradient of 4 mGal per kilometre a position error of 100 m would contribute an error of less than 0.5 mGal.) Except in two or three discrete places, gravity gradients in the North Sea are very much less than 4 mGal per kilometre. For the few data sets where position errors may be greater than 100 m position-related errors are unlikely to exceed 1 mGal and such errors will in any case be reduced by the adjustment process.

The adjusted marine data set was merged with data from the adjacent UK land area and gridded with a cell size of 4 km × 4 km. This grid was then merged with the grid data provided by the University of Hamburg for mainland Europe and the enlarged grid contoured using the Dynamic Graphics Inc. Interactive Surface Modelling (ISM) Program with grid smoothing (factor 1). The contour map was plotted using a BGS mapping program ALLPLOT (Dowswell 1988), which incorporates the contouring routine CFILL, a subset of COLMAP (Green 1989).

Original data sets used in this compilation are held by the bodies listed in the section on sources of data. Requests for such data should be directed to these sources. Adjusted data and gridded datasets from this compilation are available on sale from the British Geological Survey, Edinburgh.

Acknowledgements

Acknowledgement is made to the numerous people, including ships' staff and marine scientists, data base, data processing, and computer staff, whose efforts made preparation of this map possible. The work also required a considerable investment, and the contribution of various funding agencies is also fully acknowledged. We are grateful to Western Atlas International for permission to use their data. This paper is published by permission of the Director of the British Geological Survey (NERC).

References

Andersen O. B. and Engsager, K. (1977). Surface ship gravity measurements in Danish waters 1970–1975. *Memoir de l'Institute Geodesique de Danemarque*, Series 3, Tome 43.

Dowswell, H. J. (1988). ALLPLOT map plotting package user guide. *BGS Internal Marine Report* 88/11.

Green, C. A. (1989). COLMAP: A colour mapping package for 2-D geophysical data. *BGS Technical Report* WK/89/19.

Haagmans, R. H. N., Husti, G. J., Plugers, P., Smit, J. H. M., and Strang van Hees, G. L. (1988). NAVGRAV Navigation and gravimetric experiment at the North Sea. *Netherlands Geodetic Commission* publication on Geodesy, New Series, **32**.

Hipkin, R. G. and Hussain, A. (1983). Regional gravity anomalies. 1. Northern Britain. *Institute of Geological Sciences*, Report No. 82/10.

Hospers, J. and Finnstrom, E. G. (1984). The gravity field of the Norwegian sector of the North Sea. *Norwegian Geol. Survey. Bull.*, **396**, 25–34.

Hospers, J., Finnstrom, E. G., and Rathore, J. S. (1985). A regional gravity study of the Northern North Sea (56–62°N). *Geophys. Prospecting*, **33**, 543–66.

Masson-Smith, D. J., Howell, P. M., and Abernethy-Clerk, A. B. D. E. (1974). The National Gravity Reference Net 1973 (NGRN73). Professional papers of the *Ordnance Survey* new series, **26**.

Plaumann, S. (1979). Eine Schwerekarte der Nordsee für den Bereich östlich der Shetland Inseln. *Geologisches Jaarbuch Teihe* E14, 11–23, Hannover.

Sunderland, J. (1972). Deep sedimentary basin in the Moray Firth. *Nature*, **236**, 24–5.

Tully, M. C. and Donato, J. A. (1985). 1:1 000 000 northern North Sea Bouguer anomaly gravity map. *Brit. Geol. Survey*, *Report*, **16**, No. 6.

Zervos, F. A. (1986). Geophysical investigation of sedimentary basin development: Viking Graben, North Sea. PhD thesis, University of Edinburgh.

4 Gravity modelling in the Viking Graben area

C. Fichler and J. Hospers

Abstract

Three-dimensional gravity modelling along the NSDP 84 deep seismic profiles has resulted in a map of the Moho covering the Viking Graben area. The gravity model displays the major sedimentary boundaries, the top of the crystalline basement and the Moho. The sedimentary boundaries were provided by the interpreted deep seismic sections and available structure maps. The modelling was mainly concerned with the position of the Moho. The initial Moho model was deduced from the deep seismic sections which image the Moho beneath the platforms, but only sporadically beneath the graben. The resulting 'gravity' Moho coincides in general with the 'seismic' Moho where observed. The topography of the Moho reflects, in a highly smoothed way, the top of the crystalline basement. The Moho rises from a depth of approximately 31 km under the platforms to 19 km beneath the deepest parts of the graben. The results are thought to favour a stretching model of graben development.

4.1. Introduction

The Viking Graben is the northern arm of the North Sea graben system, situated between southern Norway and the Shetland Islands. The study area is shown in Fig. 4.1.

The main tectonic evolution of this area is outlined as follows (Ziegler 1982). In early Palaeozoic times the area was formed by collision of the continents of Laurentia and Fennosarmatia (Glennie 1984). The location of the Iapetus Suture in the North Sea is only approximately known. In Permian and early Mesozoic times the evolution of the North Sea Basin was initiated. Two major active rift phases occurred in the Viking Graben area; the first one dated between Permian and Early Triassic times and the second one between Mid-Jurassic and Early Cretaceous times (Badley *et al.* 1988). Between the rift phases a compressional pulse may have occurred (Frost 1987). During and after the tectonically active phases the graben development continued with further subsidence and deposition of sediments.

Diverse opinions exist at present concerning the mechanism of graben formation. Models related to the simple-shear model of Wernicke (1985) and the stretching model of McKenzie (1978) will be discussed later in this paper.

The Bouguer gravity field of the area has been described by Hospers *et al.* (1985). A regional gravity interpretation has been carried out by Donato and Tully (1981), concerned with the sedimentary structure and history of subsidence. More recently, Zervos (1987) has interpreted the gravity field with regard to the Moho, based on results from conventional seismic sections. More information about the deeper crust has become available through deep seismic profiling (Beach 1986; Beach *et al.* 1987; Gibbs 1987; Klemperer 1988). Holliger and Klemperer (1989) have carried out two-dimensional (2-D) gravity modelling along the deep seismic profiles.

This study presents three-dimensional (3-D) gravity modelling where the attention is focused on the topography of the Moho. The model is based on results obtained from the NSDP 84 deep seismic profiles, maps of the major sedimentary boundaries and the most recently available maps of the top of the crystalline basement (Hospers and Ediriweera 1988, 1990).

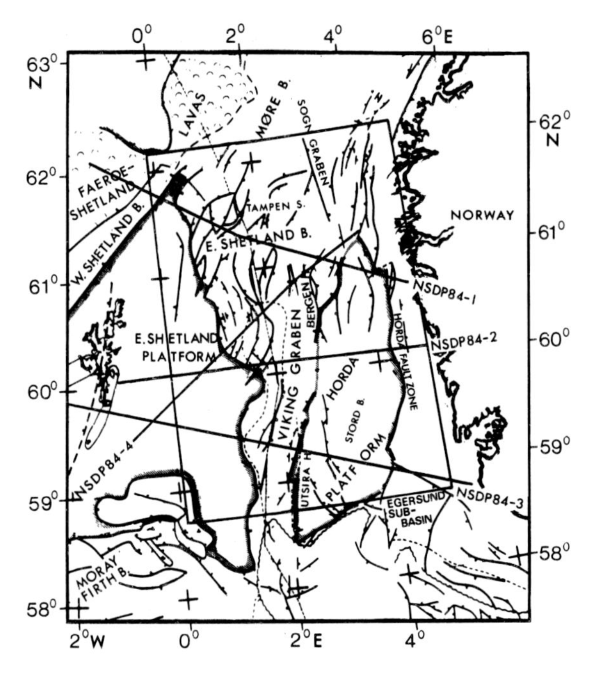

Fig. 4.1. Tectonic map of the northern part of the North Sea with the study area framed (simplified after Hamar, 1980, by permission). The locations of the deep reflection seismic profiles NSDP 84 are shown (B = basin, E = east, H = high, S = spur, W = west).

4.2. Gravity modelling

The gravity interpretation presented here is based on the free-air anomaly map displayed in Fig. 4.2 (Ediriweera 1988). The gravity anomalies show only small values, ranging from +15 to −15 mGal in the graben area and over most of the map. Hospers and his colleagues (1985) showed that the gravity anomalies on a local scale are primarily associated with basement highs, large bodies of heavy basic or ultrabasic rock in the crystalline basement, igneous intrusions, and major basement faults. Stronger anomalies are found near the Norwegian coast and over the batholith described by Donato and Tully (1982) on the western shoulder of the graben.

On a regional scale, the gravity field is dominated by relatively weak gravity anomalies over the graben structures, indicative of near-isostatic equilibrium. A simple estimation of the gravity effect caused by a sedimentary thickening from 1 km to up to 12 km in the graben area would result in anomalies of at least −100 mGal. These high values are not observed which, in itself, indicates the presence of a Moho uplift under the graben.

The present interpretation uses free-air gravity values along the NSDP 84 (North Sea Deep Seismic Profiling 1984) lines, where the underlying structure is best known. The locations of the profiles are given in Fig. 4.1.

Recently 2-D modelling has been applied here by Holliger and Klemperer (1989). This is justified as, to a first approximation, the graben can be regarded as a 2-D feature. However, a simple 2-D model may not be sufficient to give a detailed description of the sedimentary structure and anticipated Moho topography. This is of increasing importance at greater depths because of the potential field character of the gravity data. 3-D modelling was therefore carried out using the method of Talwani and Ewing (1960) which describes the body being modelled by means of horizontal lamellae of polygonal shape. The minimization of the difference between the gravity calculated from the model and the observed gravity was carried out manually. Since gravity modelling does not lead to a unique model, criteria for the acceptance of the parameter values had to be chosen. The conditions selected were that the Moho depths should be near the seismically defined depths, and that the topography of the Moho should be as smooth as possible.

The sediments down to the top of the crystalline basement were divided into layers of different density (water: 1.03 g/cm^3; Cainozoic: 2.1 g/cm^3; Cretaceous: 2.35 g/cm^3; Jurassic + Triassic + Permian? + Devonian?: 2.7 g/cm^3). These density values were taken from borehole measurements in the study area (Zervos 1987). A standard deviation of less than 0.1 g/cm^3 was estimated from the density distribution of well samples of different stratigraphic units (Donato and Tully 1981; Zervos 1987). The crystalline crust (2.8 g/cm^3) and the mantle (3.33 g/cm^3) were treated as having uniform densities. These density values agree in general with the densities used for the Viking Graben by Donato and Tully (1981, 1982), Zervos (1987) and Holliger and Klemperer (1989). A granite batholith in the East Shetland Platform (Donato and Tully 1982) was incorporated into the crystalline crust (vertical cylinder, top at 2 km, base at 10 km, radius = 20 km, position of cylinder axis: 1.24°E, 59.93°N, density: 2.68 g/cm^3).

The simplest possible model, with uniform density, has been chosen for the crystalline crust.

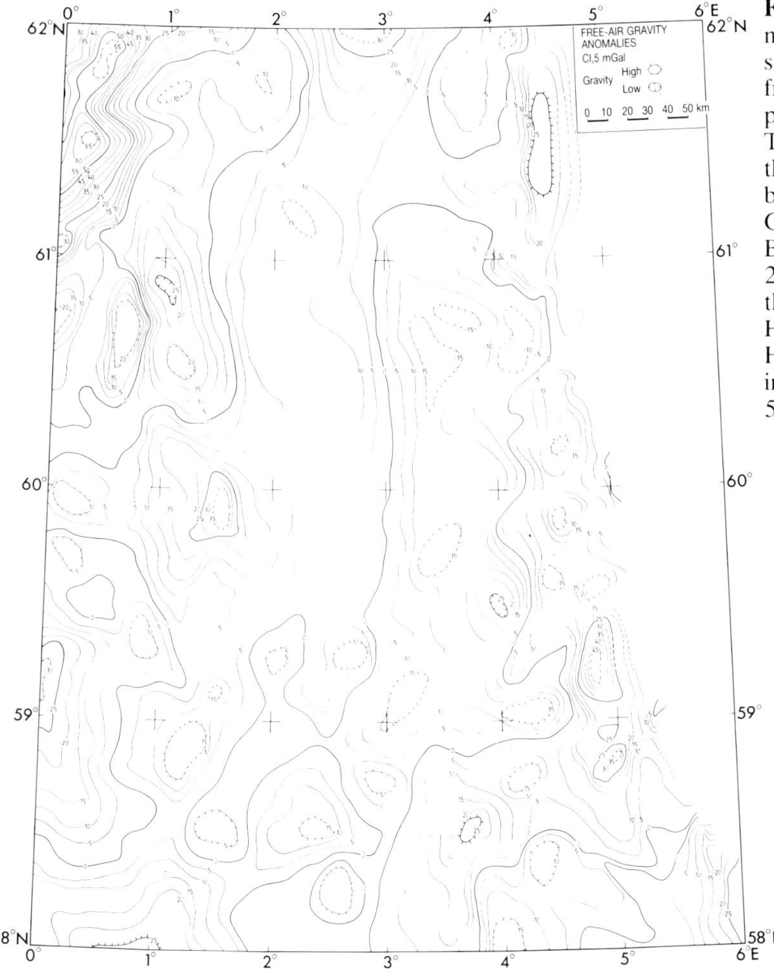

Fig. 4.2. Free-air gravity anomaly map (Ediriweera 1988), obtained by subtracting the Bouguer correction from the Bouguer anomalies published by Hospers *et al.* (1985). The gravity data shown are tied to the IGSN 71 reference net, and have been reduced using the International Gravity Formula of 1967. The Bouguer correction density is 2.67 g/cm^3. Further details about these gravity data are given by Hospers and Finnstrøm (1984) and Hospers *et al.* (1985). The contour interval on the map shown is 5 mGal.

This density value must be regarded as an average value of the crustal densities; it will represent a useful approximation, even in the case of density changes with depth. A deviation from the real average density will thus lead to an upward or a downward shift of the whole Moho topography, whereas the structural characteristics will be preserved. More serious effects are expected in the case of lateral density variations. One might expect lateral density changes near the graben centre as a result of the strong extension, for example by intruded basic rocks (Keen 1987), as discussed below. However, lateral heterogeneities which may create considerable gravity effects have not been found by refraction seismic studies across the Viking Graben (Solli 1976) and adjacent areas (Christie 1982; Barton and Wood 1984). It is therefore assumed that the simple density model used in this study will provide a useful approximation, until more density information becomes available.

The geometry of the initial gravity model was derived from the deep seismic results (Fichler and Hospers 1990) summarized below, and from structure maps (Hamar 1980; Day *et al.* 1981; Hospers and Ediriweera 1988, 1990).

The major sedimentary boundaries, indicated by the density contrasts described above, were displayed as depth maps with isolines of polygonal shape for the gravity modelling. The construction of these maps was based on the sedimentary horizons derived from the deep seismic interpretation (Fig. 4.3). The lateral extent of the horizons defined on the depth sections was determined for the sea bed by means of a bathymetric map, for the sedimentary boundaries by structure maps (Hamar 1980; Day *et al.* 1981) and for the top basement by the maps of Ediriweera (1988), and Hospers and Ediriweera (1988, 1990). The geometry of the crust-mantle boundary was incorporated by means of a provisional Moho map based on the seismic data.

The need for accuracy on the one hand and the necessity to limit the amount of data on the other needed to be taken into account. Unequal contour intervals were chosen in order to reduce the number of lamellae. Detailed modelling is required only near to the profiles and near to the surface (Talwani 1973). The area surrounding the profiles that needs to be accurately modelled increases with depth. Consequently, the maps for layers near the surface (water bottom, Fig. 4.4a, and base Cainozoic, Fig. 4.4b) are detailed near the profiles and only roughly outlined in between. The base Cretaceous map (Fig. 4.4c) and the top basement map (Fig. 4.4d) present detailed information on a broader band along the profiles. In order to avoid cut-off effects at the ends of the sections the structures have been extended appropriately outside of the frame. The thickening of the continental crystalline crust underneath the Caledonides (Sellevoll 1973) was included.

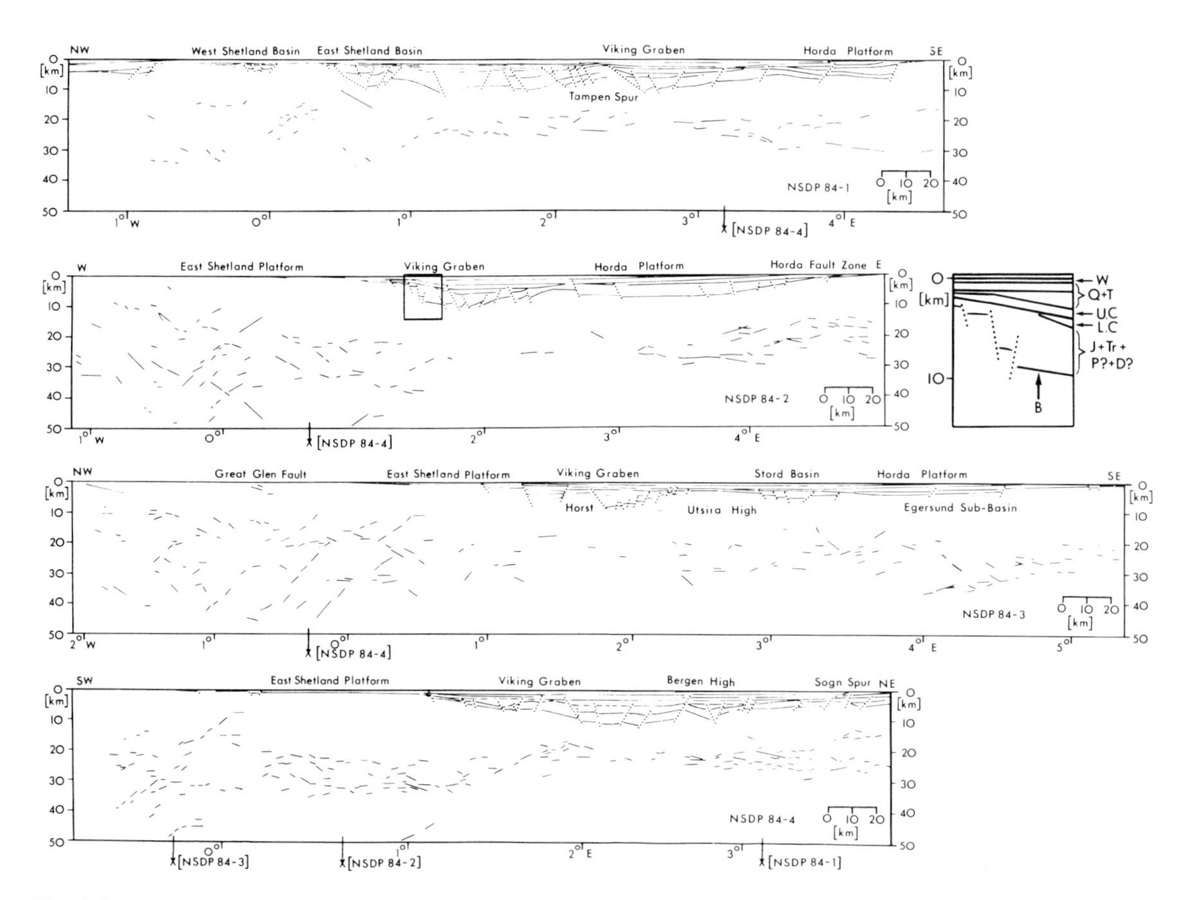

Fig. 4.3. Results from the depth conversion of the NSDP 84 sections. The reflections are shown in their unmigrated positions. Line crossings are marked by an X. The position of all stratigraphic boundaries is only approximate. Key for the enlarged part of the NSDP 84 line 2: W = water, Q = Quaternary, T = Tertiary, U.C = Upper Cretaceous, L.C = Lower Cretaceous, J = Jurassic, Tr = Triassic, P = Permian and D = Devonian. B marks the top of the crystalline basement.

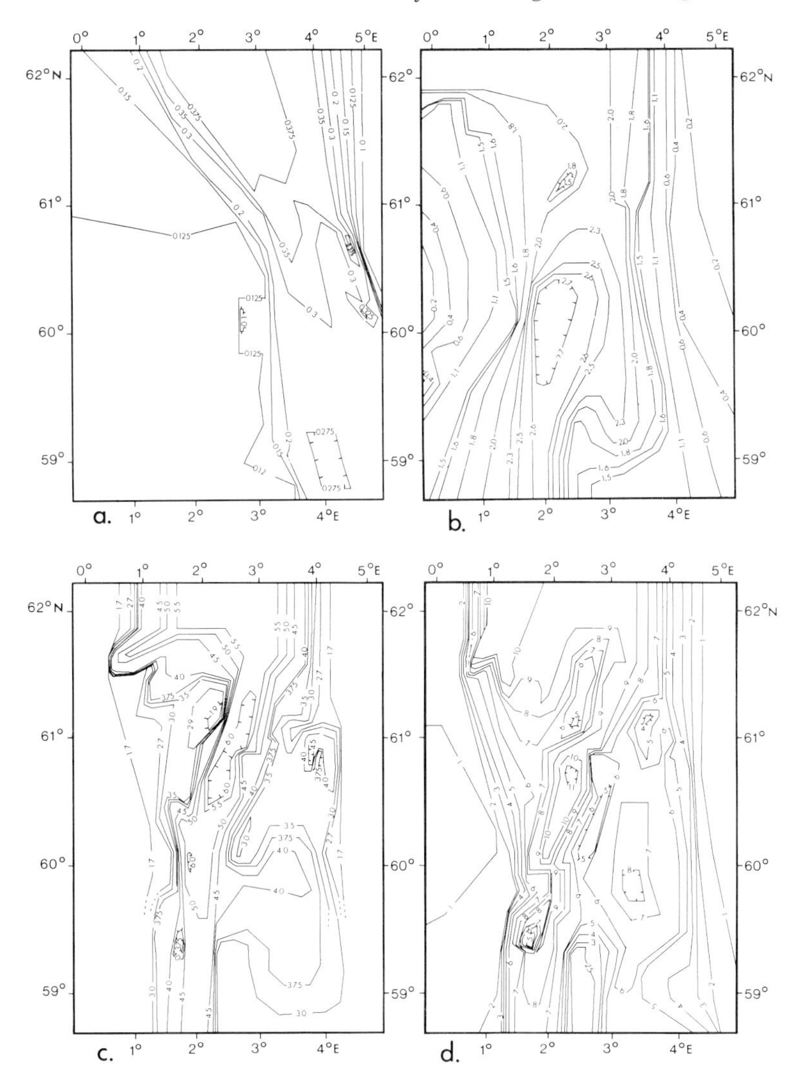

Fig. 4.4. Maps of the sedimentary boundaries used in the gravity modelling: **a** water bottom, **b** base Cainozoic, **c** base Cretaceous, **d** top of the crystalline basement. The depth contours are given in the polygonal shape as used for the modelling and are not equidistant. Depths are expressed in km below mean sea level.

The actual 3-D gravity modelling is carried out as follows. Water and sedimentary layers, with their appropriate densities, are imagined to be replaced by crustal material (density 2.8 g/cm^3). The corresponding gravity effect is calculated and used as a correction to the observed free-air gravity anomalies. This yields the 'sediment-reduced free-air anomaly' of Fig. 4.5a–d. The remaining gravity anomalies are assumed to be due to the density contrast at the Moho.

4.3. Input from deep reflection seismic profiling

The interpretation of the NSDP 84 lines 1 to 4 (acquired and processed by GECO A/S) summarized below, is described in more detail by Fichler and Hospers (1990). The depth-converted sections are shown in Fig. 4.3. The reflection evidence shown agrees in general with that presented by Gibbs (1987) and Klemperer (1988), who interpreted the same data.

For the depth conversion the sedimentary cover down to the top of the crystalline basement is

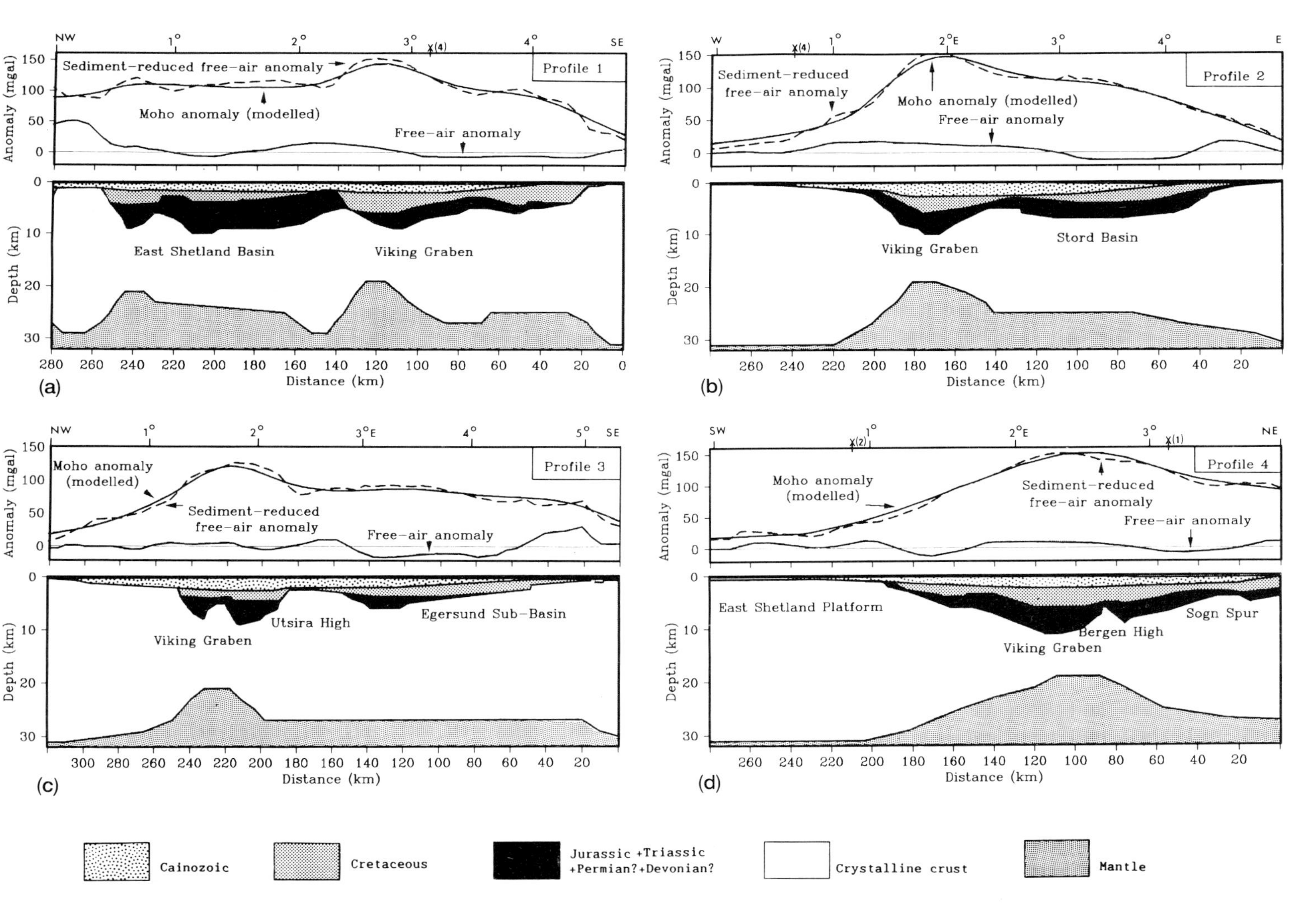

Fig. 4.5. Results from gravity modelling. Profiles 1–4 are situated along the NSDP 84 lines 1–4, respectively. The 'free-air anomaly' is the observed anomaly.

From this, the 'sediment-reduced free-air anomaly' has been calculated by correcting for the gravity effect of water and sediments. The 'Moho anomaly' corresponds to the Moho model shown.

divided into layers of different velocities (water 1.5 km/s; Quaternary: 2.1 km/s; Tertiary: 2.2 km/s; Upper Cretaceous: 2.9 km/s; Lower Cretaceous + Jurassic + Triassic: 3.8 km/s; Permian? + Devonian?: 4.4 km/s). The crystalline crust is divided into an upper (6.2 km/s) and a lower crust (6.8 km/s) in cases where a distinct change in the reflectivity pattern occurs. The top of the lower crust is defined as the transition from the transparent to the reflective part of the crust. In other cases, where no changes occur, the whole crust is treated as having one velocity (6.4 km/s). The position of the Moho is deduced either from a change in the reflection pattern or from pronounced reflections.

The sedimentary cover is imaged in a relatively detailed way (Fig. 4.3) and provides the major input for the gravity modelling. The sedimentary cover varies from a thin layer less than 1 km thick on the platform areas to 8 km in the Stord Basin, and up to 11 km in the deepest parts of the Viking Graben. Even with the limited resolution of the sections due to the compressed vertical scale, the major characteristics of the history of graben formation are visible. The layers down to the Cretaceous do not show major faulting, but there is considerable thickening over the graben. This is indicative of subsidence of the crust in the tectonically quiet final phase of the graben formation. The earlier rifting is documented through the faults in the underlying sediments, which can be partly followed down into the uppermost part of the basement.

The half-graben structure of the Viking Graben, with a large eastward dipping boundary fault on the western side, is clearly visible. The other faults of the graben system dip mainly towards the east, although faults which dip towards the west are also common. The imaging of faults on the sections is limited to the sediments and the uppermost part of the crystalline crust.

Depth conversion of the reflections from the crystalline crust and the upper mantle results in characteristic reflectivity patterns. The upper crust is transparent on all sections. The lower crust shows a variety of reflection patterns. Transparent, layered, and more complicated patterns are common, and appear to be associated with different tectonic regions.

The Moho apparently rises from a depth of 31 km under the platform areas to approximately 20 km under the graben. Uncertainties remain, especially beneath the graben area, regarding the position of the Moho.

4.4. Results from gravity modelling

The results of the gravity modelling are shown in Fig. 4.5 as cross sections along the seismic profiles, and in Fig. 4.6 as a Moho map.

Figures 4.5a–d show the crustal model with the resulting Moho in the lower part of each figure. The observed gravity anomaly and the gravity anomalies associated with the model are shown in the upper part of the figure. The modelled anomaly curves fit the observed gravity quite well, differences being less than 10 mGal. The so-called 'edge effect' is clearly visible at places of strongly increasing sedimentary thickness, e.g. on the eastern shoulder of the Stord Basin on profile 2 (Fig. 4.5b). The edge effect results from the summation of the opposite gravity effects caused by sedimentary thickening and uplifted mantle.

Both Moho types agree well beneath the platform areas on all profiles as well as beneath the graben on profiles 1 and 3 (Figs. 4.5a and c). The coincidence implies that the Moho is the petrological boundary between crust and mantle. The seismic Moho appears to be somewhat lower than the gravity Moho beneath the graben on profiles 2 and 4 (Fig. 4.5 b and d). Under the platform areas, where they are covered by only a thin layer of sediments and water, the Moho reaches a maximum depth of 31 km. With increasing sedimentary thickness the Moho depth decreases and reaches a minimum of 19 km beneath the thickest sedimentary layers. The Moho topography reflects, in a highly smoothed way, the top of the crystalline basement.

The characteristics of the Moho topography discussed here agree in general with the results of 2-D modelling carried out by Holliger and Klemperer (1989).

On the Moho map (Fig. 4.6) a ridge trending approximately NNE–SSW between 60° and 61.5°N and reaching up to 19 km in depth can be traced along the deepest basement trough. South of 60°N the ridge changes its trend to N–S. North of 61°N another less distinct ridge, situated farther west under the East Shetland Basin, has the same trend.

Recently published Moho maps (Zervos 1987; Holliger and Klemperer 1989) agree in general with the Moho map presented here. The regional

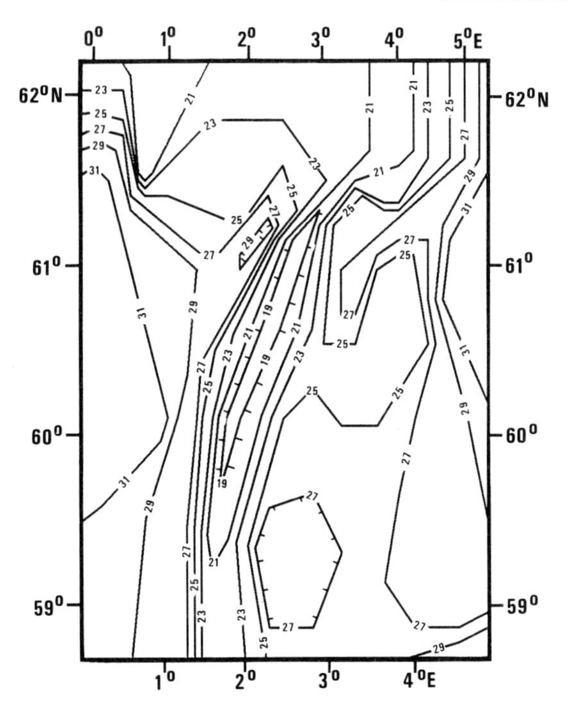

Fig. 4.6. Map of the Moho depth in km below mean sea level (contour interval is 2 km).

characteristics of the uplifted Moho beneath the graben are common. However, the Moho map presented here shows more detail. This is demonstrated by minor lows in the Moho topography in the areas of the Utsira High, the Tampen Spur and the Bergen High (see location map, Fig. 4.1).

The Moho map of Zervos (1987) resulted from 3-D gravity modelling combined with the interpretation of conventional seismic data. The minimum depth of the Moho found there, is, at 20 km, slightly greater than that shown here. The Moho ridge here (Fig. 4.6) is also longer and has steeper slopes. (This discrepancy could be easily explained by the non-uniqueness of the gravity modelling, which is reduced here through the additional use of deep seismic data.) Another Moho high at 2°E, 59°N, as proposed by Zervos (1987), is not observed in the present data set. This proposed high is crossed by the NSDP 84 line 3, but no indications are seen there.

The Moho map of Holliger and Klemperer (1989) resulted from the interpretation of the NSDP 84 lines. The major differences with the Moho map presented here are apparently caused by the different choice of the contour interval. They used an interval of 5 km whereas an interval of 2 km has been chosen here.

4.5. Discussion

Gravity modelling combined with deep seismic interpretation has proved to be a useful method in the investigation of the structure of the Moho. Ambiguity in interpretation is considerably reduced compared with the use of only one method. For the deeper crust, especially beneath the graben, the seismic data leave uncertainties; here the gravity modelling provides further information. In addition, 3-D modelling makes use of the potential field character of the gravity data and allows a modelling of the Moho topography not only in the profile-depth plane but also between lines.

The amount of crustal stretching has been investigated. It should be noted that the crustal thinning calculated below represents the integrated effect of the whole graben history with its different tectonic phases. Crustal stretching can be described in terms of the extension factor β (McKenzie 1978). If a crustal slab of constant width is considered, the stretched slab will have had its original length multiplied by a factor β and its original thickness divided by the same factor. A thickness of 31 km has been used here for the unstretched crust. The extension factor β as deduced from the sections (Fig. 4.5) reaches a maximum value of about 3 near the graben centre on profiles 2 and 4. This value is close to the β-value of 3.2, where Le Pichon and Sibuet (1981) found that oceanic crust may begin to form.

However, no indications have been found of a proto-ocean with an oceanic-type crust having been formed under the Viking Graben. The free-air gravity anomaly map (Fig. 4.2) does not show any abnormalities in these areas either. In addition, aeromagnetic data (Gulf 1977, 1980) demonstrate the presence of roundish continental-type anomalies and the absence of oceanic-type linear magnetic anomaly patterns. We therefore conclude that within the study area extreme local stretching of the continental crust has occurred without this causing a rupture leading to the development of a proto-ocean. The discrepancy between the seismic and the gravity Moho on lines 2 and 4 may indicate a lateral change in the crustal densities. In the most severely stretched areas, basic rocks may have intruded the deeper crust and/or underplating may have occurred (Keen 1987, Latin *et al.* 1989). Holliger and Klemperer (1989) suggest that this could be explained by a model relating lithospheric extension and basaltic melt production (McKenzie and Bickle 1988).

There are differing opinions about the mechanism of formation of the Viking Graben. Beach (1986) and Gibbs (1987) follow the idea of a lithospheric simple-shear model as described by Wernicke (1985), and propose a system of detachment faults cutting through the whole crust. Giltner (1987), Badley *et al.* (1988) and Klemperer (1988) suggest a stretching model related to the uniform stretching model of McKenzie (1978). Heterogeneous stretching is suggested by Coward (1986).

Stretching as the physical reaction to tension has proved to be an adequate explanation for the other graben areas of the North Sea graben system (Christie and Sclater 1980; Barton and Wood 1984; Holliger and Klemperer, this volume). The Viking Graben, as a member of the same graben system in the same geological province, could have developed similarly. Subsidence curves according to the stretching model have been calculated and explain the observed sedimentary thicknesses, most recently by Giltner (1987) and Badley *et al.* (1988).

The results presented here are thought to favour a stretching model. From the gravity modelling it is seen that the topography of the Moho reflects the structure of the top of the crystalline basement. This indicates an asymmetrical thinning not only in the upper brittle crust through faulting but also in the lower crust. This would appear to contradict a graben development according to the Wernicke model, where the crustal geometry is expected to show a shifting of the position of the high of the Moho relative to the low of the top of the basement, or requires quite complicated processes to achieve the observed geometry (Gibbs 1987). A stretching origin is further supported by the lack of evidence of deep reaching faults on the seismic sections (Fig. 4.3).

An interesting observation in this context is a distinct difference in the reflection pattern of the lower crust east and west of the Viking Graben with its suggested origin older than the graben formation. These different types of lower crust may have reacted with a different degree of resistance to the applied tension, which might have caused variable stretching.

It appears likely that at least some initial differences in the deep crustal structure existed before the graben development started. Earlier tectonic processes connected with the Caledonian orogeny and the possible nearness of the Iapetus Suture could have affected the lower crust. It appears quite possible that the location of the Viking Graben has been affected by these older tectonic structures.

4.6. Conclusions

1. 3-D gravity modelling combined with deep seismic interpretation has proved to be a useful method to investigate the deep crustal structure of the Viking Graben area.
2. A map of the topography of the Moho has been constructed. The Moho depth varies from 19 km near the centre of the graben to 31 km under the adjacent platforms. The topography of the Moho reflects, in a highly smoothed way, the top of the crystalline basement.
3. The gravity Moho (defined by a major density contrast) and the seismically defined Moho agree in general, which implies that this boundary is the petrological boundary between crust and mantle. Differences in depth, however, occur in the most extensively stretched parts of the graben, which may indicate intrusion of basic rocks and/or underplating.
4. The results are thought to favour a stretching model of graben development.

Acknowledgements

One of us (C.F.) has been supported by a NATO science postdoctoral fellowship (via Deutscher Akademischer Austauschdienst, DAAD) in 1987 and a postdoctoral fellowship from the Royal Norwegian Council for Scientific and Industrial Research (NTNF) in 1988 and 1989. She also acknowledges the help of colleagues at Geophysikalisches Institut, Universität Karlsruhe, West Germany.

The work reported here is part of a project initiated and directed by the other author (J.H.) and supported by BP Petroleum Development Ltd, Chevron Exploration North Sea, Conoco Norway,

Inc., Elf Aquitaine, Esso Exploration and Production, Mobil Exploration Norway, Svenska Petroleum, Tenneco Oil Co., and Total Marine. This support is gratefully acknowledged.

Norwegian International Lithosphere Programme Contribution No. 85.

References

Badley, M. E., Price, J. D., Rambech Dahl, C., and Agdestein, T. (1988). The structural evolution of the northern Viking Graben and its bearing upon extensional modes of basin formation. *J. Geol. Soc.*, **145**, 455–72.

Barton, P. and Wood, R. (1984). Tectonic evolution of the North Sea Basin: crustal stretching and subsidence. *Geophys. J. Roy. Astronom. Soc.*, **79**, 987–1022.

Beach, A. (1986). A deep seismic reflection profile across the northern North Sea. *Nature*, **323**, 4, 53–55.

Beach, A., Bird, T., and Gibbs, A. (1987). Extensional tectonics and crustal structure: deep seismic reflection data from the northern North Sea Viking Graben. In *Continental extensional tectonics* (ed. M. P. Coward, J. F. Dewey and P. L. Hancock). *Geol. Soc. Special Publication*, **28**, 467–76. Oxford, Blackwell Scientific Publications.

Christie, P. A. F. (1982). Interpretation of refraction experiments in the North Sea. *Phil. Trans. Roy. Soc.*, London, **A 305**, 101–12.

Christie, P. A. F. and Sclater, J. G. (1980). An extensional origin for the Buchan and Witchground Graben in the North Sea. *Nature*, **283**, 729–32.

Coward, M. P. (1986). Heterogeneous stretching, simple shear and basin development. *Earth and Planetary Science Letters*,**80**, 325–36.

Day, G., Cooper, B. A., Andersen, C., Burgers, W. F. J., Rønnevik, H. C., and Schöneich, H. (1981). Regional seismic structure maps of the North Sea. In *Petroleum geology of the continental shelf of north-west Europe* (ed. L. V. Illing and G. D. Hobson), 76–84. London, Heyden.

Donato, J. A. and Tully, M. C. (1981). A regional interpretation of the North Sea gravity data. In *Petroleum geology of the continental shelf of north-west Europe* (ed. L. V. Illing and G. D. Hobson), 65–75. London, Heyden.

Donato, J. A. and Tully, M. C. (1982). A proposed granite batholith along the western flank of the North Sea Viking Graben. *Geophys. J. Roy. Astronom. Soc.*, **69**, 187–95.

Ediriweera, K. K. (1988). A geological and geophysical investigation of the Viking Graben area, North Sea. Thesis, Norwegian Institute of Technology, Trondheim, Norway, 255 pp.

Fichler, C. and Hospers, J. (1990). Deep crustal structure of the northern North Sea Viking Graben: results from deep reflection seismic and gravity data. *Tectonophysics*, **178**, 241–54.

Frost, R. E. (1987). The evolution of the Viking Graben tilted fault block structures: a compressional origin. In *Petroleum geology of north west Europe* (ed. J. Brooks and K. Glennie), 1009–24. London, Graham and Trotman.

Gibbs, A. D. (1987). Deep seismic profiles in the northern North Sea. In *Petroleum geology of north west Europe* (ed. J. Brooks and K. Glennie), 1025–8. London, Graham and Trotman.

Giltner, J. P. (1987). Application of extensional models to the Northern Viking Graben. *Norsk Geologisk Tidsskrift*, **67**. Oslo, 339–52.

Glennie, K. W. (ed., 1984). *Introduction to the petroleum geology of the North Sea*. Oxford, Blackwell Scientific Publications.

Gulf (1977). *Bergen quadrangle, total magnetic intensity*. Map on scale 1:500,000. Gulf Science and Technology Co., Pittsburgh.

Gulf (1980). *Total magnetic intensity, North Sea*. Map on scale 1:2,000,000. Gulf Science and Technology Co., Pittsburgh.

Hamar, G. P. (1980). *Tectonic map of the North Sea*, scale 1:1,000,000. Statoil, Stavanger, Norway.

Holliger, K. and Klemperer, S. L. (1989). A comparison of the Moho interpreted from gravity data and from deep seismic reflection data in the northern North Sea. *Geophys. J.*, **97**, 247–58.

Holliger, K. and Klemperer, S. L. (this volume). Gravity and deep seismic reflection profiles across the North Sea rifts.

Hospers, J. and Ediriweera, K. K. (1988). Mapping the top of the crystalline crust in the Viking Graben area, North Sea. *Norwegian Geol. Survey*, Special Publication, **3**, 21–8.

Hospers, J. and Ediriweera, K. K. (in press). Depth and configuration of the crystalline basement in the Viking Graben area, northern North Sea. *J. Geol. Soc.*

Hospers, J. and Finnstrøm, E. G. (1984). The gravity field of the Norwegian sector of the North Sea. *Norwegian Geol. Survey, Bull.*, **396**, 25–34.

Hospers, J., Finnstrøm, E. G., and Rathore, J. S. (1985). A regional gravity study of the northern North Sea (56–62°N). *Geophysical Prospecting*, **33**, 543–65.

Keen, C. E. (1987). Some important consequences of lithospheric extension. In *Continental extension tectonics* (ed. M. P. Coward, J. F. Dewey, and P. L. Hancock), *Geol. Soc. Special Publication*, **28**, 67–73. Oxford, Blackwell Scientific Publications.

Klemperer, S. L. (1988). Crustal thinning and the nature of extension in the northern North Sea from deep seismic reflection profiling. *Tectonics*,**7**, 803–21.

Latin, D. M., Dixon, J. E., and Fitton, J. G. (1989). Rift-related magmatism in the North Sea Basin. Abstract, *Terra*, **1**, 1, 34.

Le Pichon, X. and Sibuet, J. C. (1981). Passive margins: a model of formation. *J. Geophys. Research*, **86**, B5, 3708–20.

McKenzie, D. P. (1978). Some remarks on the development of sedimentary basins. *Earth and Planetary Science Letters*, **40**, 25–32.

McKenzie, D. P. and Bickle, M. J. (1988). The volume and composition of melt generated by extension of the lithosphere. *J. Petrol.*, **29**, 11–22.

Sellevoll, M. A. (1973). Mohorovičić discontinuity beneath Fennoscandia and adjacent parts of the Norwegian Sea and the North Sea. *Tectonophysics*, **20**, 359–66.

Solli, M. (1976). En seismisk skorpeundersøkelse Norge-Shetland. Unpublished thesis, University of Bergen, Norway.

Talwani, M. (1973). Computer usage in the computation of gravity anomalies. In *Methods in computational physics* (ed. B. A. Bolt), Vol. 13, New York, Academic Press.

Talwani, M. and Ewing, M. (1980). Rapid computation of gravitational attraction of three dimensional bodies of arbitrary shape. *Geophysics*, **XXV**, 1, 203–25.

Wernicke, B. (1985). Uniform-sense normal simple shear of the continental lithosphere. *Can. J. Earth Sciences*, **22**, 108–25.

Zervos, F. (1987). A compilation and regional interpretation of the northern North Sea gravity map. In *Continental extensional tectonics* (ed. M. P. Coward, J. F. Dewey and P. L. Hancock), *Geol. Soc. Special Publication*, **28**, 477–93. Oxford, Blackwell Scientific Publications.

Ziegler, P. A. (1982). *Geological atlas of western and central Europe*. Shell Internationale Petroleum Maatschappij, B.V.

5 Gravity and deep seismic reflection profiles across the North Sea rifts

K. Holliger and S. L. Klemperer

Abstract

The results of gravity modelling along 3000 km of deep seismic reflection profiles and along a refraction profile in the North Sea are discussed with study focusing on the Central Graben. In an attempt to reduce some of the inherent ambiguity of the gravimetric method, we used the crustal thickness as the only free parameter in our modelling. The geometry of the sedimentary pile was adopted from the depth-migrated seismic data and from published maps; its gravity effect was removed and the residual gravity anomaly was modelled in terms of variations in crustal thickness relative to a 32 km-thick reference crust. Our models show a minimum crustal thickness in the Central Graben of 23 km, corresponding to a β (stretching) factor of 2 for the crustal basement.

That our simplistic modelling approach leads to a good agreement between the reflection Moho and the gravity Moho, and to a good fit between the calculated and observed gravity anomalies, suggests that lateral variations in average basement velocity and density are, in general, relatively small (in spite of the local heterogeneity imaged by deep seismic reflection profiling), and thus that the regional isostatic compensation is of Airy type, not of Pratt type.

Gravity modelling of a refraction profile across the Central Graben of the North Sea shows that there is a density contrast of 400–500 kg/m^3 between the lower crust and the upper mantle of the North Sea. This rather high density contrast at the base of the crust favours an intermediate or anorthositic mean composition for the lower crust, and does not permit extensive basaltic underplating. This may not be true for the most severely extended parts of the Viking Graben and Central Graben, where our results favour limited addition of basaltic melt to the crust.

5.1. Introduction

Since 1984 the British Institutions' Reflection Profiling Syndicate (BIRPS) has participated in the collection of more than 3000 km of deep seismic reflection profiles across the Central Graben and Viking Graben (Fig. 5.1). These data have provided a wealth of structural information (Klemperer 1988; Klemperer and White 1989; Klemperer and Hurich, this volume) but conclusive information on seismic velocity within the basement cannot be extracted because of the short streamer length of 3 km. Consequently, lateral velocity and related density changes remain undetected (Nafe and Drake 1963; Anderson 1967; Ludwig *et al.* 1970; Woollard 1975). In the absence of coincident or at least nearby wide-angle seismic profiles, gravity modelling is the first choice to assist the interpretation of deep seismic reflection data (Holliger 1987; Holliger and Klemperer 1989).

In the deepest parts of the Central and Viking Grabens the low density Mesozoic sedimentary rocks reach thicknesses close to 10 km, which causes a negative gravity effect of at least 100 mGal (1 mGal $= 10^{-5}$m/s^2) with respect to the regions outside the grabens. However, the observed gravity anomalies in the North Sea have only low amplitudes (maximum peak-to-trough amplitudes are in the

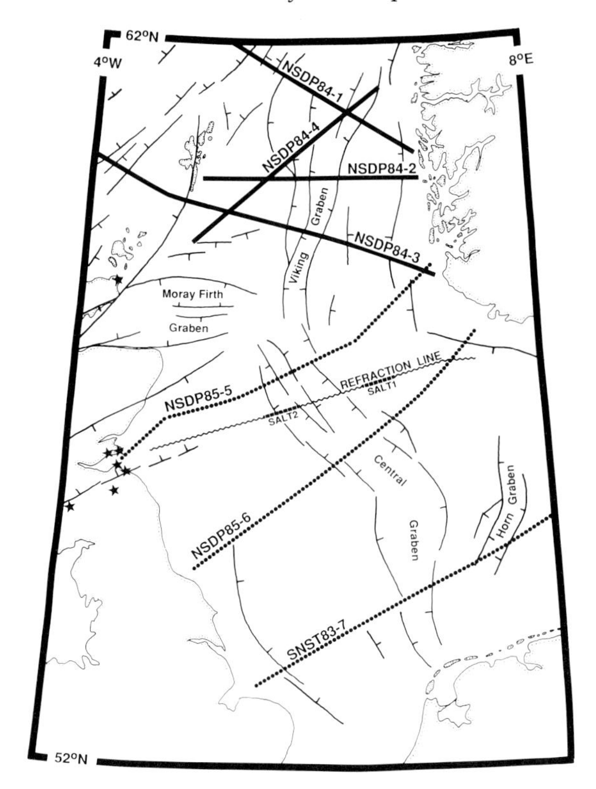

Fig. 5.1. Map of the North Sea area showing the location of the seismic profiles discussed in the text. Thin solid lines are principal fault structures. Stars indicate crustal-xenolith localities reported by Upton *et al*. (1983) and Hunter *et al*. (1984).

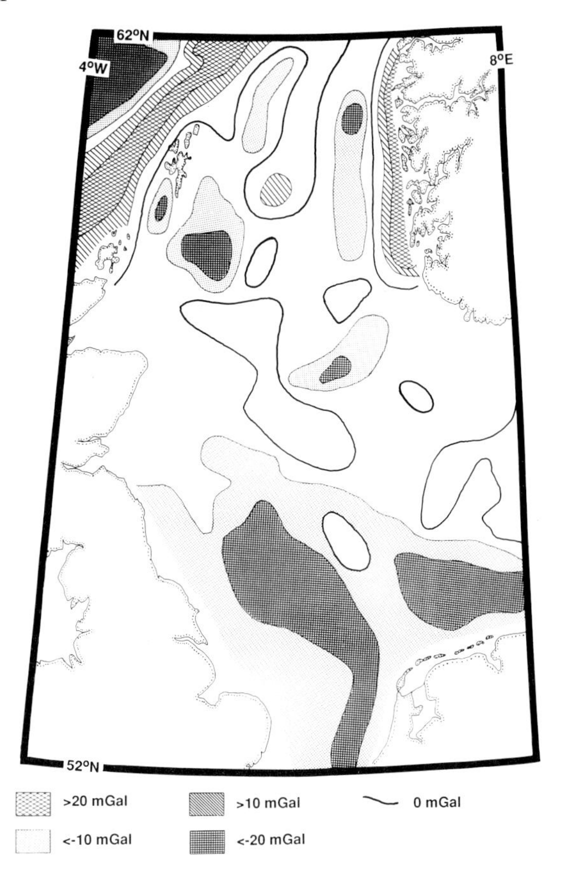

Fig. 5.2. Generalized free-air gravity map of the North Sea derived from SEASAT altimetry (Courtesy Cowan Geodata Services).

order of 50 mGal and are restricted to local features; average regional amplitudes are about 15 mGal); they show little or no correlation with the dominant tectonic trends (Fig. 5.2; Day *et al*. this volume). Obviously the mass deficiency caused by the low-density sediments must be compensated by excess mass at depth. Based on the assumption of isostatic compensation at a depth of approximately 30 km, Donato and Tully (1981) predicted a shallowing of the Moho to 20 km beneath the deepest parts of the Central and Viking Grabens. Their results were largely confirmed by a seismic wide-angle profile (Barton and Wood 1984) and by further gravity studies (Hospers *et al*. 1985; Zervos 1987).

The primary purpose of this paper is not to outline the Moho topography by gravity modelling but rather to compare the gravity Moho and the reflection Moho, and to discuss the resulting implications of such a comparison for crustal structure and the nature of the crust-mantle boundary. We made particular efforts to evaluate the average density contrast between the lower crust and the upper mantle, and also to test the hypothesis of a gravimetric Conrad discontinuity, i.e. a continuous boundary separating the lighter upper crust from the denser lower crust, such as has been invoked by other workers who have jointly modelled reflection and gravity data (Setto 1987; Setto and Meissner 1987; Dyment 1989; Sibuet *et al*. 1990).

Gravimetric interpretation of the NSDP84 profiles in the northern North Sea is given by Holliger and Klemperer (1989) and by Fichler and Hospers (this volume), and therefore in this paper emphasis is given to general results and to the profiles in the southern North Sea (NSDP85-5 and -6, and SNST83-7).

5.2. Deep seismic reflection data

In 1984 and 1985 a speculative deep seismic reflection survey, North Sea Deep Profiles (NSDP84 and NSDP85), was shot by the Geophysical Company of Norway (GECO) with the participation of the oil industry and BIRPS. Profile SNST83-7 in the southern North Sea is a 1983 speculative survey owned jointly by NOPEC(UK) and GECO. The acquisition and processing parameters of all these profiles are similar to those routinely used by BIRPS (Warner 1986; Klemperer 1989); details for the NSDP84 data are given by Holliger and Klemperer (1989). Post-stack reprocessing of all the deep profiles in order to enhance deep crustal reflectors was carried out using Merlin Profilers Seismic Kernel System. All these profiles were recorded to 15 s two-way time, sufficient to image the entire crust and uppermost mantle to a depth of about 50 km. Portions of the actual seismic data (NSDP84) are reproduced in papers by Klemperer (1988) and Holliger and Klemperer (1989), and (for NSDP85) by Klemperer and White (1989) and Klemperer and Hurich (this volume).

Line drawings were prepared from the seismic profiles at a scale of 1:200 000 and interpreted at this scale. Line drawings of seismic reflection data are intended as an objective reproduction of the principal reflections imaged, with noise, multiples and sideswipes excluded. The line drawings were digitized, and fed into a depth-migration routine based on inverse ray-tracing (Raynaud 1988). The sedimentary part of the seismic profiles is well constrained by extensive well-log information kindly provided by the oil industry. Based on the velocity logs from these wells, the following interval velocities were used for migration and depth conversion of the line drawings, as shown in this paper, of NSDP85 and SNST83-7 in the central and southern North Sea: water layer 1.5 km/s, Cainozoic 2.0 km/s, Cretaceous 3.4 km/s, Jurassic and Triassic 3.3 km/s, and Upper Palaeozoic 4.4 km/s. Appropriate velocities for the Viking Graben, northern North Sea, are: post-Palaeocene 2.1 km/s, Cretaceous 2.7 km/s, Jurassic and Triassic 3.2 km/s, and Upper Palaeozoic 4.5 km/s (data discussed by Klemperer 1988 and Holliger and Klemperer 1989).

Based on the results of seismic refraction experiments in the North Sea and adjacent areas (Smith and Bott 1975; Solli 1976; Bamford *et al.* 1978; Christie and Sclater 1980; Barton and Wood 1984) average velocities of 6.2 km/s and 8.0 km/s were assigned to the crustal basement and topmost mantle, respectively. The 'reflection Moho' (Klemperer *et al.* 1986), picked on the profiles as the base of the lower-crustal zone of high reflectivity (Barton *et al.* 1984) that is typical of the British shelf area (Matthews 1986), was taken as the base of the crustal layer of average velocity 6.2 km/s.

Probable errors in calculating the depth of the reflection Moho from the seismic reflection data are ± 1 km due to uncertainty in picking the travel-time to the reflection Moho, approximately ± 1 km due to uncertainty in the average basement velocity, and approximately ± 1 km due to uncertainties in the thickness of Upper Palaeozoic sedimentary rocks (Klemperer 1988). Thus, the total error in depth to the reflection Moho is unlikely to exceed ± 2 km in areas of good data quality. In areas of bad data quality (e.g. SNST83-7, 250 to 290 km) no meaningful estimate of depth to the reflection Moho can be made. Errors in estimates of the thickness of the pre-Mesozoic basement (the distance from the base of the Triassic to the reflection Moho) may have an additional ± 1 km error due to uncertainty in picking the base of the Mesozoic section.

In this study we have included Upper Palaeozoic sedimentary rocks in the gravity and reflection models, but we have regarded them as part of the crustal basement during the Mesozoic rifting of the North Sea. All our calculations of the degree of extension refer to the Mesozoic events and ignore any Palaeozoic extension.

5.3. Gravity modelling

5.3.1. *Gravity data*

At a regional scale, Bouguer gravity coverage is now complete for the entire North Sea area (Andersen 1978; Strang van Hees 1983; Hospers and Finnstrøm 1984; Tully and Donato 1985; British Geological Survey 1986; Goldschmidt 1986; Zervos 1987; Haagmans *et al.* 1988; Day *et al.*, this volume). For all data sets the Bouguer gravity anomalies (which in the case of marine gravity data simply consist of a replacement of the water layer) were calculated using a density of 2670 kg/m^3 for the Bouguer slab correction. The 'observed' Bouguer gravity profiles shown in Figs 5.3 and 5.5 were taken from the published maps at every contour (5 mGal in the Norwegian sector, 2 mGal elsewhere) and additionally by interpolation from the maps at regular intervals of 5 km. The water depth was taken every 5 km along the seismic profiles from the echo-sounder records and the density of sea water was taken to be 1030 kg/m^3.

5.3.2. *Densities*

Using available well information and following Donato and Tully (1981) the following densities were used: Cainozoic 2100 kg/m^3, Cretaceous 2350 kg/m^3 (northern North Sea) and 2450 kg/m^3 (southern North Sea), Jurassic and Triassic, and Permian in the southern North Sea, 2500 kg/m^3, Carboniferous and Devonian, and Permian in the northern North Sea, 2650 kg/m^3, crystalline basement 2750 kg/m^3, and upper mantle 3250 kg/m^3. Although reasonably well constrained, these density values should be looked at with caution. Common gravity modelling algorithms do not reproduce the earth's entire gravity field in a particular area but only calculate lateral deviations in gravity (i.e. gravity anomalies) relative to a given reference model. Therefore it is not absolute densities that are the key parameters in gravity modelling, but rather horizontal density contrasts with respect to a uniform reference density. In this study it was appropriate to relate the density contrasts of the sedimentary rocks and the upper mantle to the top of, and to the base of, the crystalline crust, respectively.

5.3.3. *Modelling philosophy*

The depth-migrated line drawings of the seismic profiles were simplified by subdividing them into a number of polygons representing the stratigraphic units listed above. In the southern and central North Sea the maps compiled at 1:2 500 000 by Day *et al.*(1981) served as the basis for the structural and stratigraphic interpretation of the seismic data down to the base of the Zechstein (Permian). Older sedimentary rocks were interpreted to be present in significant thickness beneath the western margin of the North Sea (100 km from the west end of NSDP85-5, Fig. 5.3a; 120 km from the west end of SNST83-7, Fig. 5.3c) based largely on the evidence of steep gravity gradients or short wavelength anomalies which require a shallow source. The polygons were digitized at approximately 10 km intervals and fed into a gravity modelling procedure based on the two-dimensional approach of Talwani *et al.* (1959).

Three-dimensional gravity modelling in the northern North Sea (Fichler and Hospers, this volume) confirms the results of previous two-dimensional models of the same area (Holliger and Klemperer 1989), and demonstrates that our two-dimensional approach is justified for profiles running more or less perpendicular to the dominant structural trends related to Mesozoic rifting. In the central and southern North Sea this assumption only breaks down at the east end of SNST83-7, where that profile is rotated more than 50° from a true dip-line to the southern Horn Graben (Fig. 5.1), so that the corresponding part of the 2-D gravity model (Figs 5.3c and 5.4, 450–500 km) should be treated with caution. For the purpose of gravity modelling the geological models were extended by 100 km on either side of the seismic profiles in order to avoid edge effects.

After removing the negative gravity effect of the sediments, the resulting residual gravity anomaly was modelled in terms of variations in Moho depth with respect to a 32 km thick reference crust (Sellevoll 1973; Solli 1976; Barton and Wood 1984). The basement density was held constant throughout the central and southern North Sea area discussed in this paper. In the northern North Sea the basement density was also held constant, except for the gravity highs associated with the Norwegian west coast, the Precambrian granulites of the Rona Ridge, and the mafic plutons of the Erland Terrace (Holliger and Klemperer 1989). For all these areas, previous workers had also interpreted higher-than-average basement velocities and densities (Sellevoll 1973; Smith and Bott 1975; Gatliff *et al.* 1984; Hospers and Finnstrøm 1984).

The main purpose of this study was to try to match the calculated with the observed gravity data, while keeping constant the sedimentary section as defined by the seismic reflection data, and using the Moho depth as the *only* free parameter. In general this approach allows us to reproduce the observed gravity anomalies to within their internal accuracy (4–5 mGal in the Norwegian sector, 2–3 mGal elsewhere). However, in a few places—especially where the observed gravity data contain short wavelengths (<20 km)—this approach has led to a 'poor' fit which could be improved by allowing more free parameters. The largest mismatch in the Viking Graben area is 8 mGal, on NSDP84-4; the largest mismatch in the central or southern North Sea is a 10 mGal misfit on NSDP85-6.

A number of other authors who have modelled gravity data along deep seismic reflection profiles have additionally assumed that reflectivity boundaries other than the reflection Moho are also density boundaries, and have produced good fits to the observed gravity with models based on this assumption (Setto 1987; Setto and Meissner 1987; Dyment 1989; Sibuet *et al.* 1990). However, there is no unique way to assign density to reflectivity patterns: for example, Setto and Meissner allow their unreflective upper crustal basement to vary laterally from 2600 to 2800 kg/m^3 along a single BIRPS profile. Additionally, intra-crustal reflectivity boundaries are typically less-well defined than the reflection Moho: for example, along a single BIRPS profile the boundary between the unreflective (less dense?) upper crust and the reflective (denser?) lower crust varies locally as much as 12 km vertically between the interpretation of Setto and Meissner, and the interpretations of Dyment, and Sibuet and co-workers. The assignment of differing densities to differing intra-crustal reflectivity patterns therefore introduces additional degrees of freedom into the gravity-modelling procedure; it is perhaps not surprising that such models closely fit the observed gravity. Because of these many degrees of freedom we do not believe that the existence of successful models of this type demonstrates that there is necessarily a relationship between crustal reflectivity and crustal density.

We prefer the approach used here, in which the interpretative aspects of gravity modelling are reduced to a minimum. Once the gravity effect of the sedimentary basin (calculated from independent data) has been stripped off, and a lower crust/upper mantle density contrast selected, our approach is essentially a one-parameter inversion (for the depth to the gravity Moho) rather than poorly constrained forward modelling.

However, because the wavelength of a gravity anomaly depends on its source depth, and gravity anomalies originating at the base of the crust can generally be expected to have a 5–10 times longer wavelength than those originating within the sedimentary section, our gravity models cannot fit the short-wavelength anomalies. For our purposes it was therefore appropriate to concentrate on a good fit for the intermediate to long wavelengths (>20 km) of the gravity field which contain most of the information on crustal structure. Further support for the validity of this approach comes from the good agreement of the 'SEASAT' Moho and the marine gravity and reflection Moho in the northern North Sea (Holliger and Klemperer 1989). (The gravity field resolved by SEASAT altimetry can be considered as a low-pass filtered version of the actual gravity field containing no anomalies with wavelength shorter than 30 km.)

5.3.4. *Accuracy*

We are aware of the limitations of the gravity-modelling philosophy outlined above, such as the omission from our models of lateral density changes observed in the sedimentary rocks, errors in interpretation and depth conversion of the seismic data for the sedimentary section, and possible unresolved Upper Palaeozoic sedimentary rocks. We expect these uncertainties lead to a probable error of ± 2 km in the depth to the gravity Moho with respect to our reference crust. This estimate was obtained both empirically (by changing the densities and geometries of the sedimentary pile within the likely ranges) and analytically (by applying the total differential to the gravity formula for a series of cylindrical slabs using average uncertainties of ± 50 kg/m^3 in slab density and ± 1 km in depth to the base of the Mesozoic).

5.4. Results

The output from our gravity modelling is the topography of an interface with a step-like density jump of $+500$ kg/m^3 relative to a reference depth of 32 km. This interface corresponds to Moho interpreted from gravity data and henceforward is referred to as 'gravity Moho' for simplicity.

Figure 5.3 shows the crustal structure inferred from gravity modelling along the three deep seismic reflection profiles in the southern North Sea (NSDP85 lines 5 and 6, SNST83-7). The gravity models of the NSDP84 profiles in the northern North Sea are discussed in full detail elsewhere (Holliger and Klemperer 1989); only a summary of the principle results of that modelling is given here.

5.4.1. *Moho topography and β-factors*

The amount of lithospheric extension is often quantified by the so-called β-factor, which corresponds to the ratio of lithospheric thickness before and after extension in uniform stretching models (McKenzie 1978). In this paper the β-factor is calculated as the ratio of the thickness of our reference crust (32 km) to the residual basement thickness as inferred from gravity modelling or from seismic-reflection data. These calculated β-factors represent the combined effects of all phases of post-Carboniferous stretching. As discussed earlier, our errors in basement thickness are unlikely to exceed ± 3 km, which leads to uncertainties of $+0.5$ and -0.3 for a typical β-factor of 2.0.

Along the axis of the Central Graben our minimum calculated depth to the gravity Moho decreases with increasing thickness of the sedimentary pile, from a minimum depth of 28 km beneath a maximum sedimentary thickness of 5.5 km in the south (Fig. 5.3c) to a minimum depth of 23 km beneath a maximum sedimentary thickness of 7.5 km in the north of the Central Graben (Fig. 5.3a). This corresponds to an increase in the maximum β-factors from 1.3 to 2.0, from south to north. The average β-factors inferred from gravity modelling along the three profiles NSDP85-5, NSDP85-6 and SNST83-7 are all close to 1.25.

The overall picture that emerges in the northern North Sea closely resembles that in the southern North Sea described above. However, maximum local extension and resulting crustal thinning were more severe, producing minimum gravity Moho depths of 20 km or even less, corresponding to maximum β-factors derived from gravity modelling of around 3 in the deepest parts of the Viking Graben (Holliger and Klemperer 1989) and average β-factors around 1.4 (Klemperer 1988).

The calculated β-factors for both the southern and the northern North Sea imply a fairly constant Mesozoic crustal extension of some 100 km because, in a general way, the maximum extension increases from south to north while the width of the basin (simplistically taken as the distance between opposing coastlines) decreases from south to north. The observed large-variations in local crustal attenuation might be due to differential exploitation during Mesozoic rifting of structures inherited from the early Palaeozoic Caledonian orogeny.

5.4.2. *Gravity Moho versus reflection Moho*

Figure 5.4 shows the depth-migrated line drawings of the deep seismic profiles in the southern North Sea, with the Moho topography derived from gravity modelling (Fig. 5.3) superimposed. The gravity Moho not only accurately reproduces the trend of the base of the reflective lower crust (the reflection Moho) after depth migration, but also generally agrees well within the given uncertainties (± 2 km for each method).

At several locations on the profiles shown in Fig. 5.4 there are reflectors several kilometres beneath the gravity Moho. Most of these occurrences of deep reflectors form dipping zones that have been mapped as being sub-parallel to the Viking Graben and Central Graben (west-dipping at 210–270 km on NSDP85-5, at 140–210 km on NSDP85-6, and at 10–60 km on SNST83-7; east-dipping at 500–540 km on NSDP85-5, and at 460–520 km on NSDP85-6). They have been interpreted as extensional shear zones within the upper mantle (Klemperer 1988; Klemperer and White 1989; Klemperer and Hurich, this volume).

The only area on the profiles in Fig. 5.4 where the diffuse pattern of crustal reflectivity apparently continues beneath the gravity Moho by more than the expected uncertainty is beneath the east margin of the Horn Graben (470 km to end of profile SNST83-7). Unfortunately, both the gravity modelling and the seismic data are suspect in this region, so that we are reluctant to draw any geological inference from the apparent discrepancy. The gravity Moho beneath the Horn Graben is suspect because the gravity model is a 2-D profile rotated 50° to the principal graben structures. The migration of the seismic data is suspect in this region because the profile is oblique to structure; in particular the brightest reflections apparently furthest beneath the gravity Moho (at 470 km on SNST83-7) are vertically beneath a 3 km-high salt wall (cf. sections in Best *et al.* 1983) that has probably seriously distorted the seismic image beneath the salt.

Away from the Horn Graben all three profiles show agreement within expected error between the gravity Moho and reflection Moho. There is an

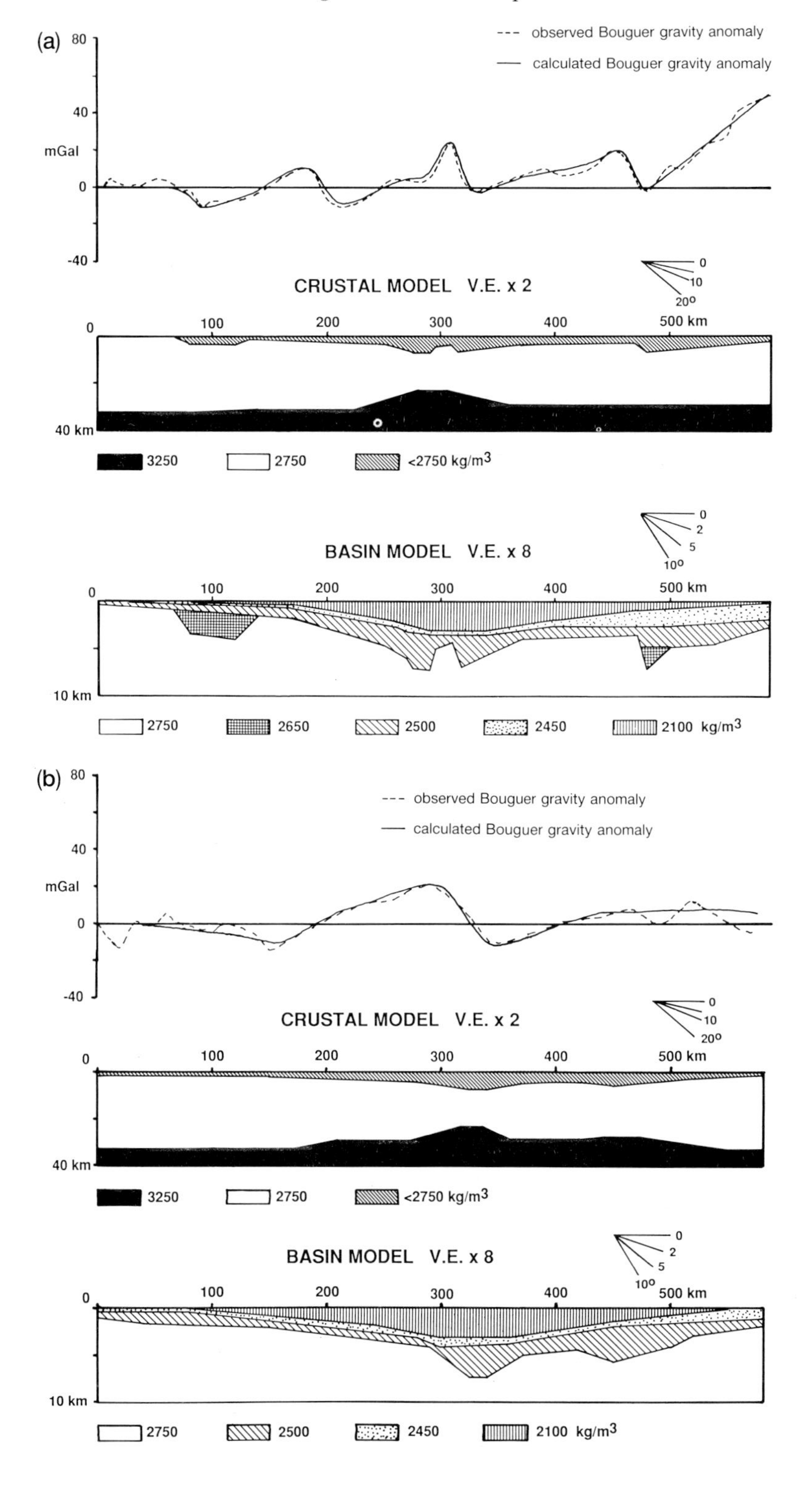
(a)
--- observed Bouguer gravity anomaly
— calculated Bouguer gravity anomaly
80
40
mGal
0
-40
CRUSTAL MODEL V.E. x 2
0
10
20°
0
100
200
300
400
500 km
40 km
3250
2750
<2750 kg/m3
BASIN MODEL V.E. x 8
0
2
5
10°
0
100
200
300
400
500 km
10 km
2750
2650
2500
2450
2100 kg/m3
(b)
--- observed Bouguer gravity anomaly
— calculated Bouguer gravity anomaly
80
40
mGal
0
-40
CRUSTAL MODEL V.E. x 2
0
10
20°
0
100
200
300
400
500 km
40 km
3250
2750
<2750 kg/m3
BASIN MODEL V.E. x 8
0
2
5
10°
0
100
200
300
400
500 km
10 km
2750
2500
2450
2100 kg/m3

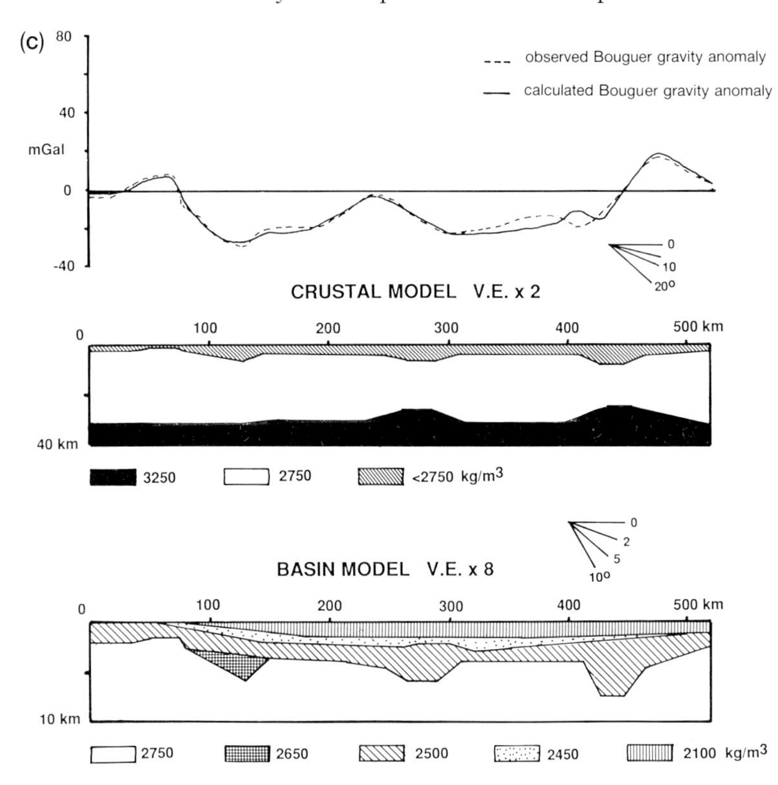

Fig. 5.3. Gravity models for **a.** NSDP85-5, **b.** NSDP85-6 and **c.** SNST83-7. For each profile is shown **top**, observed and calculated Bouguer gravity anomaly, **middle** the crustal structure (vertical exaggeration ×2) and **bottom** the details of the basin model (vertical exaggeration ×8) as adopted from the depth-migrated seismic reflection profiles and used in the gravity modelling.

apparent tendency for reflectivity within the Central Graben region on profiles NSDP85-5 and -6 to be slightly deeper than the gravity Moho, and though this discrepancy is beneath our resolution level for the reflection/gravity analysis, refraction data discussed below supports its existence.

The same agreement between gravity Moho and reflection Moho is present in the northern North Sea (Holliger and Klemperer 1989) with the single exception of the deepest part of the Viking Graben (profile NSDP84-2). Here the sedimentary pile reaches a thickness close to 10 km and the gravity modelling gives a Moho depth of 19–20 km, whereas the reflection Moho is interpreted to be at a depth of 25 to 26 km. Although in part of the area in question the reflection Moho is not unequivocally imaged, we consider the depth discrepancy to be outside the expected uncertainty and we tentatively suggest magmatic underplating in this most severely extended region of the North Sea rift system as a possible explanation (*see* Holliger and Klemperer 1989, for further discussion).

A new quantitative model relating lithospheric extension and basaltic melt production provides support for the idea of basaltic underplating and increased crustal density beneath the Viking Graben (McKenzie and Bickle 1988, Latin *et al.* this volume). McKenzie and Bickle (1988) predict that about 2 km of basaltic melt will be generated by stretching a 70 km-thick lithosphere (Stuart 1978) above an asthenosphere of normal potential temperature (1280°C) by a β-factor of 3.0. Basaltic intrusion could cause the increase in lower-crustal density and velocity required to remove the discrepancy between the reflection Moho and gravity Moho beneath the Viking Graben. Moreover, McKenzie and Bickle predict that little or no melt will be generated for β-factors smaller than 2.0 unless the lithosphere is substantially thinner than 70 km, or the potential temperature is greater than 1280°C.

This agrees well with our results in the rest of the North Sea, where β-factors interpreted from gravity modelling do not greatly exceed 2.0 and the reflection and gravity Moho coincide within the given uncertainties, and with the general paucity of rift-related volcanics in the North Sea basin (Latin *et al.*

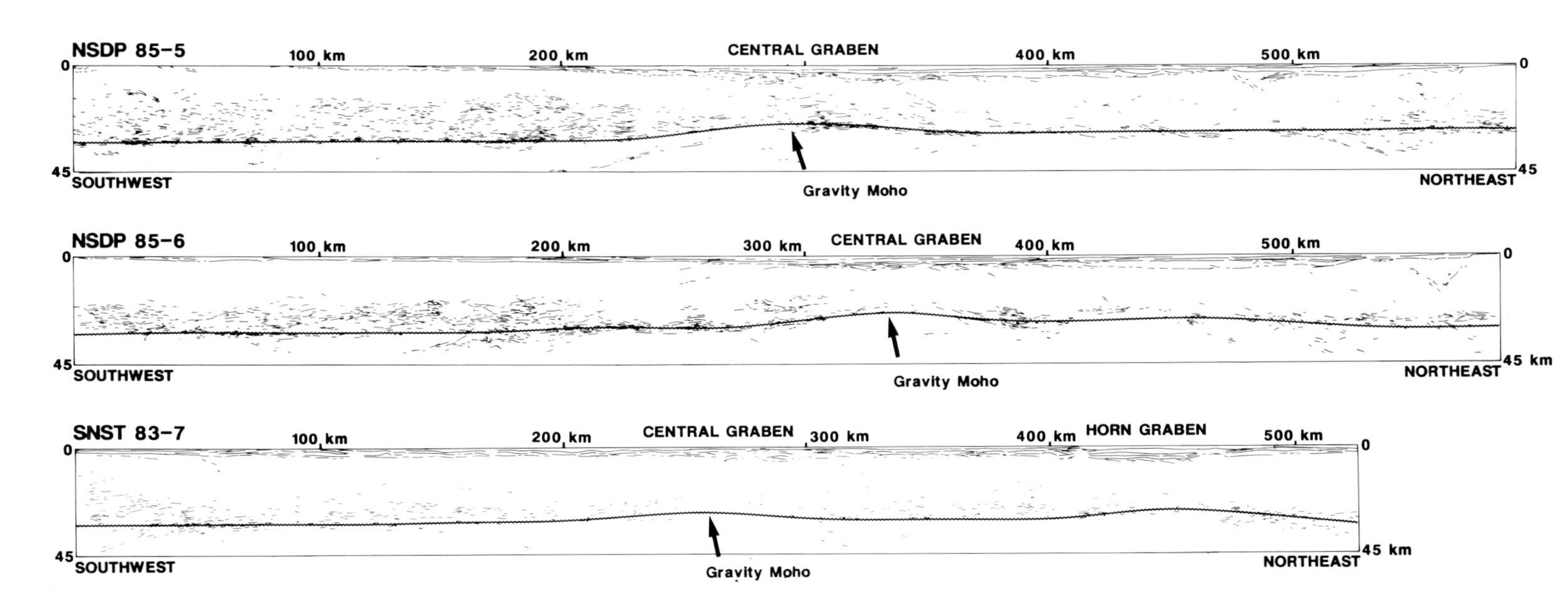

Fig. 5.4. Gravity Moho from gravity models in Fig. 5.3 (line) superimposed on depth-migration line drawings. Profiles are shown with no vertical exaggeration. **a.** NSDP85-5; **b.** NSDP85-6; **c.** SNST83-7.

this volume). Moreover, the good agreement between the reflection and the gravity Moho throughout the entire North Sea suggests that, averaged over its entire thickness, the Caledonian basement may be surprisingly homogeneous in terms of its average seismic velocity and density. In addition, there does not seem to be evidence for a correlation between the reflectivity and density of the lower crust, in contrast to interpretations of other regions of the British continental shelf (Setto and Meissner 1987; Dyment 1989; Sibuet *et al.* 1990).

5.5. Discussion

5.5.1. *Lateral density variations*

Because of the quasi-linear relationship between seismic velocity and density for crystalline rocks (the so-called 'Nafe-Drake' curve; Ludwig *et al.* 1970; Woollard 1975) one might think that the reflection Moho obtained by depth migration with a constant basement velocity is bound to be in good agreement with the corresponding gravity Moho based on a uniform basement density (Warner 1987). This is, however, only true as long as there are no (or only small) lateral density and velocity variations, because the magnitude of their effects on the depth of the gravity and reflection Moho differ considerably. Neither the interpreted reflection Moho nor the interpreted gravity Moho will be at the correct depth, and if the velocity–density change is sufficiently big there will be an observable discrepancy between the two interpreted Mohos.

How big do such velocity-density changes have to be in order to be detected? Local vertically-averaged seismic velocities for the basement beneath the North Sea can be expected to vary from 6.0–6.5 km/s (Sellevoll 1973; Solli 1976; Christie and Sclater 1980; Barton and Wood 1984). So if the basement velocity we used for depth migration (6.2 km/s) was too low by 0.3 km/s the interpreted reflection Moho will be too shallow by 1.5 km. The same velocity increase is expected to be associated with a basement density that is of the order of 75 kg/m^3 higher than average (Barton 1986*a*). Assuming isostasy and a lateral extent of approximately 20 km for this density anomaly would result in a calculated gravity Moho that is approximately 6 km too shallow. The resulting discrepancy of 4.5 km between the reflection and the gravity Mohos would exceed the sum of the expected uncertainties in the depth to gravity Moho and to the reflection Moho, so crustal heterogeneities of this magnitude should be detectable by a joint interpretation of reflection data and gravity data. The only discrepancy in the North Sea larger than 4 km is that in the deepest part of the Viking Graben (profile NSDP84-2) discussed above as possible evidence for localized mafic intrusion into the deep crust.

5.5.2. *The gravity Moho*

The Moho is one of the earth's few first-order seismic discontinuities for refracted and supercritically reflected waves. Steinhart (1967) defined it as the depth where the p-wave velocity discontinuously increases from values generally below 7.0 km/s to values of 7.6 to 8.6 km/s. Therefore the Moho is generally considered to correspond to the petrological crust-mantle boundary defined by the predominance, beneath the crust-mantle boundary, of high velocity/high density ultramafic rocks. Hence gravity modelling can be used to estimate changes in crustal thickness relative to a given reference crust, provided that the correct density contrast between the lower crust and upper mantle is chosen. In addition, a knowledge of the density contrast between the lower crust and upper mantle can provide constraints on the petrology of the lower crust and upper mantle.

The physical interpretation of the 'density contrast at the Moho' that we are able to estimate is that it represents the difference between the average density of the crust beneath a horizontal surface at the shallowest level of the gravity Moho, and the average density of the uppermost mantle above a horizontal surface at the deepest level of the gravity Moho (Fig. 5.6a). In this paper the depth range of the lower crust that we can study ranges from 20–32 km depth on profile NSDP84-2 to 26-32 km on profile SNST83-7. The density contrast actually present as a single step at the Moho could be less than the 'lower crust/upper mantle density contrast' we infer, if there are appropriate vertical and lateral density gradients within the lower crust or upper mantle over these depth ranges.

Uniform basement

In order to check the validity of our chosen density contrast at the gravity Moho, we applied the same gravity-modelling philosophy outlined above with a set of density contrasts ranging from 300–600 kg/m^3 to the only high-quality, seismic wide-angle profile in the North Sea (Fig. 5.1; Barton and Wood 1984). Along this profile it is thought that the 'true' Moho depth (i.e. the Moho *sensu stricto*, in this case a boundary between rocks with a seismic velocity ≤7.2 km/s and rocks with velocity ≥8.0 km/s) is known with an accuracy of ±1 km.

Figure 5.5 shows the gravity model of this refraction profile with the true Moho depth (as interpreted from the seismic wide-angle data) superimposed on the gravity Moho for a density contrast of 500 kg/m^3 (as used to model the reflection profiles in Fig. 5.4; the sedimentary part of this gravity model was taken from Day *et al.* 1981).

Outside the graben area the refraction Moho and gravity Moho agree most perfectly. In the axial parts of the graben the gravity Moho is too deep by 2 km. This discrepancy is, however, within the combined uncertainty of the two methods of ±3 km except on the eastern flank of the graben where the gravity Moho locally is too shallow by almost 4 km. In this region Barton and Wood (1984) and Barton (1986*a*) have interpreted unusually high velocities (6.9 km/s) and densities (3000 kg/m^3) near the base of the crust. Barton and her colleagues (1984) and Barton (1986*b*) found that the base of the lower crustal reflectivity corresponds to the refraction Moho along two short, seismic reflection profiles that are coincident with the wide-angle profile (Figs 5.1 and 5.5, SALT 1 and SALT 2). Along SALT 2, where the reflection Moho is reasonably well defined, the refraction Moho and the gravity Moho agree within 1.5 km.

This provides support for the hypothesis that, beneath the North Sea, the reflection Moho in general corresponds to the velocity and density jump

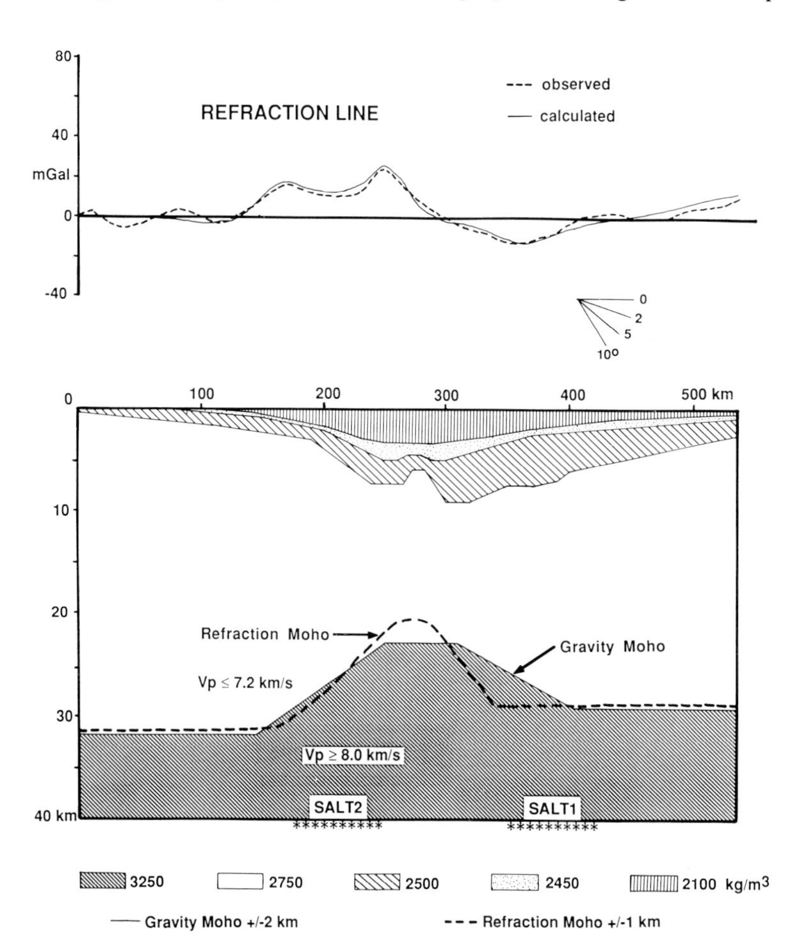

Fig. 5.5. Gravity model of the seismic wide-angle profile described by Barton and Wood (1984) (located in Fig. 1) with the 'true' Moho depth interpreted from the wide-angle seismic data superimposed on that derived from gravity modelling (vertical exaggeration ×8). Stars indicate the location of near-vertical seismic reflection profiles SALT-1 and SALT-2.

at the base of the crust, interpreted from seismic wide-angle data and gravity data. SALT 1 covers the region at the eastern margin of the Central Graben where the discrepancies between the refraction and the gravity Moho are most severe. However, along SALT 1 the imaging of the reflective lower crust is poor; so poor in fact that a definition of the reflection Moho with the desired accuracy is questionable (Barton 1986*b*).

However, 60 km along-strike from SALT 1, both to the north (at 310 km on NSDP85-5, Fig. 5.4a) and to the south (at 390 km on NSDP85-6, Fig. 5.4b), there is a suggestion that the reflection Moho beneath the eastern margin of the Central Graben is 0–3 km deeper than the gravity Moho. This is a smaller discrepancy than our estimated combined uncertainties. However, taken in conjunction with the evidence from the wide-angle seismic profile that the refraction Moho (interpreted by Barton and Wood 1984, is 0–4 km deeper than our gravity Moho (at 350 km on the refraction line, Fig. 5.5), we tentatively suggest that dense, mafic material has been added to the crust in this region.

Barton's (1986*a*) gravity model, along her wide-angle seismic profile, allowed basement density to vary laterally and vertically in order to fit the observed gravity, but was constrained to have a density jump at the refraction Moho, included a 0–7 km thick pillow of crust of density of 2950–3100 kg/m^3 beneath the eastern margin of the Central Graben. Just as our proposed location of minor magmatic underplating of the northern North Sea is beneath the deepest part of the Viking Graben, so is the region between profiles NSDP85-5 and -6 the deepest part of the Central Graben (base Zechstein at >7 km depth), and the location of maximum thermal subsidence (⩾3 km Cainozoic strata; Day *et al.* 1981); it is therefore the most plausible region for intrusion of mantle melts.

To summarize the results of Fig. 5.5, for a density contrast of 500 kg/m^3 at the base of the crust on the wide-angle profile, the discrepancies between refraction Moho and the gravity Moho are very small (⩽1 km) over 65 per cent of the length of the profile, acceptable (between 1 and 3 km) over 30 per cent, and unacceptable (greater than our estimated uncertainty of 3 km) over 5 per cent of the length.

The physical significnce of the 'density contrast' as illustrated in Fig. 5.6a is the change from the average density ρ_2 of the lower crust, between the shallowest Moho level z_2 and the Moho, to the average density ρ_3 of the upper mantle between the Moho and the (deepest) regional Moho level z_3. Figure 5.6b shows a comparison of the refraction Moho and the gravity Mohos for density contrasts across the gravity Moho (i.e. ρ_3–ρ_2 in Fig. 5.6a) of 300, 400, 500 and 600 kg/m^3 (the modelling procedure that was used is the same as for Figs 5.3 and 5.5). Although perfect agreement cannot be achieved over the entire length of the profile, the gravity Moho for a density contrast of 500 kg/m^3 reproduces the refraction Moho better than the gravity Mohos for density contrasts of 300, 400, and 600 kg/m^3. Density contrasts of 400 and 600 kg/m^3 do produce an acceptable fit (less than our expected uncertainy of 3 km) over most (>90 per cent) of the profile's length, whereas the fit becomes largely unacceptable (>3 km) for a density contrast of 300 kg/m^3 (acceptable fit over <50 per cent of the profile).

The models in Figs 5.5 and 5.6b used a laterally-constant density contrast at the Moho. However, the Barton and Wood (1984) refraction interpretation shows average velocities in the lower crust (between 20 km depth and the Moho) of around 6.5 km/s in the western North Sea but up to 6.9 km/s in the eastern North Sea. Given this evidence for lateral heterogeneity in the lower crust, it is beyond the resolution of the analysis presented in Fig. 5.6b to

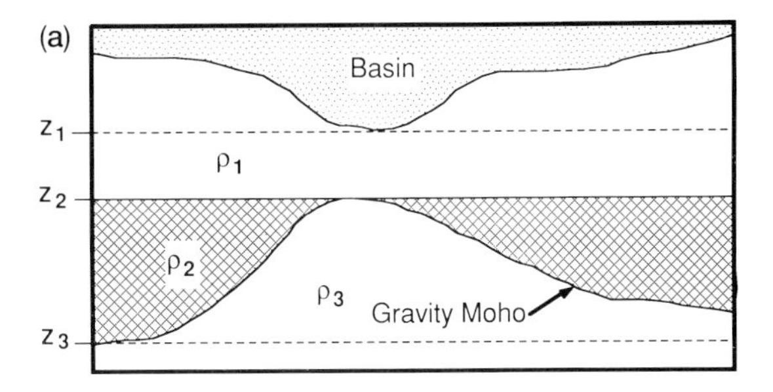

Fig. 5.6.a. Cartoon to illustrate the physical significance of the lower crust/upper mantle density contrast ($\rho_3 - \rho_2$). The depths between which these average densities pertain are z_2, the shallowest Moho depth along each profile; and z_3, the greatest Moho depth on each profile (the regional Moho depth). Note that ρ_1, the average crustal density between the greatest basin depth z_1 and the shallowest Moho depth z_2, is not determined in this study, and could for example be equal to ρ_2. For the profiles illustrated in this paper and in Holliger and Klemperer (1989) z_1 ranges from 5–10 km, z_2 ranges from 19–28 km, and z_3 is 32 km.

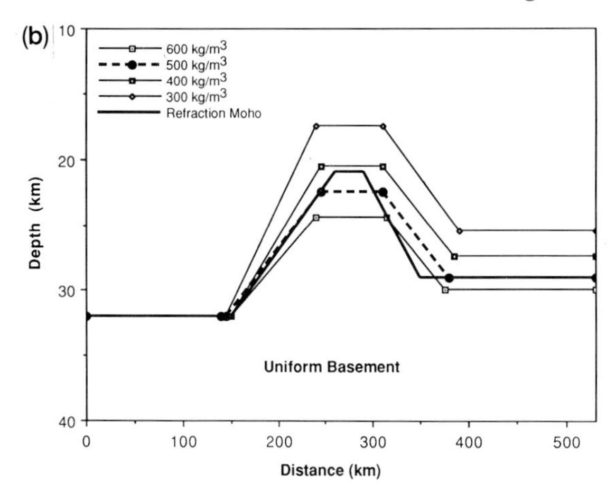

Fig. 5.6.b. Comparison of the depth to Moho inferred from modelling of wide-angle seismic data with the depths to the gravity Moho for a uniform basement density and density contrasts ($\rho_3 - \rho_2$ in Fig. 5.6a) of 300, 400, 500 and 600 kg/m^3 at the gravity Moho.

distinguish between mean density contrasts at the Moho of 400 kg/m^3 and 500 kg/m^3. In fact, the preferred gravity model of Barton and Wood shows a lateral variation in the contrast between the mean lower crustal density and upper mantle density from about 500 kg/m^3 in the western North Sea to about 400 kg/m^3 in the eastern North Sea.

Conrad discontinuity

As will be discussed below, a density contrast of 500 kg/m^3 between the lower crust and upper mantle is higher than inferred by previous workers; it also imposes severe constraints on the petrological interpretation of the lower crust. We therefore next test whether it is possible to lower the required density contrast between the lower crust and the upper mantle by introducing a Conrad-type boundary separating a less-dense upper crust from a denser lower crust, and still reproduce the refraction Moho as accurately as with a uniform-basement model. Setto (1987), Setto and Meissner (1987), Dyment (1989) and Sibuet and his colleagues (1990) have used this approach to model the gravity field along the BIRPS SWAT profiles, taking the top of the lower crustal zone of high reflectivity as the Conrad-type (density) discontinuity, even though Wever (1989) suggests that the Conrad (velocity) discontinuity is normally more than 2 km beneath the top of the reflective lower crust.

Assigning a higher density to highly reflective basement areas is arbitrary, and implicitly favours the presence of dense mafic material as the source of the reflectivity. Although basaltic sills certainly can increase the reflectivity of the basement there are other sources of reflectivity (shear zones, zones of fluid-filled porosity: Warner, 1990) that need not materially affect the density. Moreover, except in the areas of strongest reflectivity, the top of the reflective lower crust is difficult to pick out precisely. Finally, constraints for the density contrast between the transparent upper and the reflective lower crust are virtually non-existent: Setto and Meissner (1987) used a laterally-varying density contrast of 0–300 kg/m^3 (most frequently 200 kg/m^3) whereas, for the same data set, Sibuet and co-workers (1990) used a density contrast of 100 kg/m^3. There is certainly no widespread agreement on what the Conrad discontinuity represents, or even whether it has any geological significance (Litak and Brown 1989).

Although the Barton and Wood (1984) wide-angle interpretation did not include any first-order seismic interfaces (e.g. Conrad) between the sedimentary basin and the Moho, we fixed the top of the dense lower crustal layer for our reference model at a depth of 16 km (half the reference thickness); this was based on observations of the Conrad discontinuity at 16 to 18 km depth in the western North Sea, 100 km further south, on the Caledonian Suture Seismic Profile (Bott *et al*. 1985).

Since in our case geometrical constraints for the hypothetical Conrad discontinuity are lacking along most of the profile's length, three different modelling schemes were used:

(1) the Conrad is always parallel to, and 16 km shallower than, the Moho;
(2) the Moho topography was adopted from the interpretation of the wide-angle seismic data, and the depth to the Conrad was used as the only free parameter;
(3) the Conrad is always in the middle of the crystalline basement, as would result from applying uniform stretching (McKenzie 1978) to an initially laterally homogeneous two-layer crust.

All three models were calculated with density contrasts of both 100 and 200 kg/m^3 between the upper and lower crust, and corresponding density contrasts of 400 and 300 kg/m^3 at the Moho.

Approaches (1) and (2) required more than 7 km

vertical relief on the Conrad in order to fit the observed gravity profile, and therefore resulted in the Conrad penetrating the sedimentary section. This we considered unrealistic, so we discarded these models. Approach (3) only leads to minor deviations (≤2 km) of the Conrad from its reference level. Consequently the Conrad does not contribute significantly to the calculated gravity field (Fig. 5.6c) and the resulting Moho depths closely resemble those of the uniform basement models with density contrasts of 300 and 400 kg/m^3 (Fig. 5.6b). A comparison of the seismic Moho, the gravity Moho for a uniform basement with a density contrast of 500 kg/m^3, and the gravity Moho for crust in which the 500 kg/m^3 contrast is partitioned between the Conrad and the Moho (Fig. 5.6c) shows that the best fit is obtained with the uniform basement model and the large 500 kg/m^3) density contrast at the Moho. This conclusion is only altered if a density contrast larger than 200 kg/m^3 is allowed at the mid-crustal boundary, or if the mid-crustal boundary is allowed a more arbitrary depth variation.

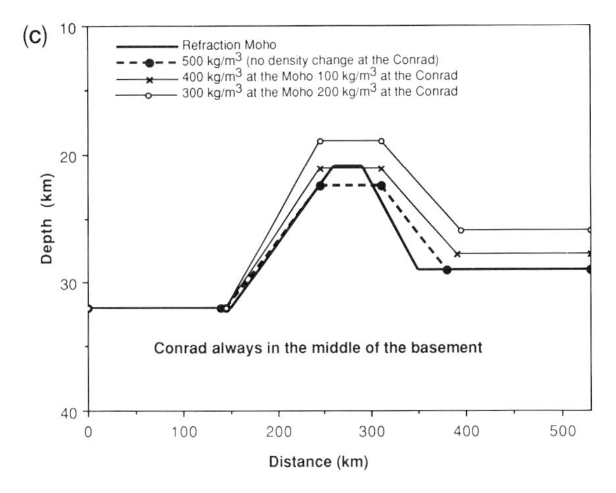

Fig. 5.6.c. Comparison of the depth to Moho inferred from modelling of wide-angle seismic data with the depth to the gravity Moho for a uniform basement density and a density contrast ($\rho_3 - \rho_2$) of 500 kg/m^3 at the gravity Moho, and with the depths to the gravity Mohos for a Conrad-type crust with density contrasts ($\rho_3 - \rho_2$) of 300 and 400 kg/m^3 at the gravity Moho and of 200 and 100 kg/m^3 between the upper and lower crust (not indicated in the cartoon Fig. 5.6a). The Conrad discontinuity was placed in the middle of the basement along the entire profile (see text).

5.5.3. *Lower crust upper/mantle density contrast and the bulk composition of the lower crust*

Because of the limited compositional range of potential upper-mantle rocks the density contrast at the base of the crust drastically restricts the average compositional range of the lower crust. The most probable upper-mantle composition is peridotitic (density about 3320 kg/m^3), but some regions of the continental upper mantle may be eclogitic (density about 3500 kg/m^3) (Ringwood 1975). However, of several thousand deep-provenance xenoliths collected in the British Isles, none are mantle eclogites, while peridotites are common (Upton *et al.* 1983). Assuming a normal peridotitic upper mantle beneath the North Sea rift, the density contrast at the base of the continental crust is likely to be in the following range: for a granitic or anorthositic lower crust, about 500 kg/m^3; for diorite, about 400 kg/m^3; and for gabbro, about 250 kg/m^3 (Herzberg *et al.* 1983; Meissner 1986).

The density contrast of 500 kg/m^3 used in this study is compatible with the average velocity jump at the Moho interpreted by Barton and Wood (1984). However, Barton (1986*a*) showed that the direct conversion of seismic velocities interpreted from seismic wide-angle data into densities is far from unique. Such a density contrast between the lower crust and the upper mantle is high; Woollard (1966) estimated an average of 390 kg/m^3 by correlating surface topography and crustal thickness in regions with no significant isostatic gravity anomalies. This favours a granodioritic to quartzdioritic, or gabbro-anorthositic, bulk composition of the lower crust beneath the North Sea (Herzberg *et al.* 1983; Meissner 1986). Although minor additions of basaltic melt beneath the level of resolution of the gravity method may have occurred, extensive basaltic underplating is unlikely, because igneous rock formed by extension and decompression melting of the mantle has a density of at least 3000 kg/m^3 (White and McKenzie 1989).

An additional constraint on lower-crustal composition, that of deep-provenance xenoliths, is available, and is applicable particularly to the western central North Sea. Of the many lower-crustal xenoliths found throughout Scotland and on the western shore of the North Sea, plagioclase-rich metagabbroic granulites, including some monomineralic anorthosites, form the major component;

intermediate-composition, high-grade xenoliths are scarce (Upton *et al.* 1983; Hunter *et al.* 1984). The plagioclase-rich mafic lower crust implied by these xenoliths would have a density compatible with our estimates. In the compilation of Christensen (1982), ten out of twelve anorthosites and gabbroic anorthosites have densities less than 2920 kg/m^3 (implying a lower crust/upper mantle density contrast at the Moho with 3320 kg/m^3-peridotite of greater than 400 kg/m^3) and six out of twelve samples have densities less than 2820 kg/m^3 (implied density contrast greater than 500 kg/m^3).

However, the low mean seismic velocities found by Barton and Wood (1984) for the lower crust of the western North Sea, about 6.5 km/s, are not consistent with these mafic lower-crustal xenoliths being an unbiased sample of a dry lower crust: in the same compilation by Christensen (1982) the average 6 kb velocity of the twelve anorthosites and gabbroic anorthosites is 6.9 km/s, and none of these rocks had 6 kb velocities less than 6.7 km/s. Hyndman and Klemperer (1989) suggest that a fluid-filled porosity of 1–2 per cent within a mafic lower crust could account for the low observed velocity. A crustal porosity of 2 per cent would reduce the density contrast to be fit with the measured, zero-porosity, densities by 2 per cent at the most, i.e. from 500 to 490 kg/m^3, which is compatible with an anorthositic (but not with a gabbroic) bulk composition of the lower crust.

Thus our inferred lower-crustal densities, taken in conjunction with previously reported seismic velocities, are consistent with the reported xenolith compositions if the lower crust is wet. Alternatively, it is possible that the average seismic velocity of the lower crust is reduced by the presence of a substantial component of intermediate to silicic composition. If this is so, the reported xenolith suites must be biased towards the mafic component, and must undersample the more felsic lower crustal rocks.

5.5.4. *Isostasy and elastic thickness*

The gravity models presented in this paper imply the operation of Airy isostasy at a broad wavelength—that is, the North Sea basin is largely compensated by the Moho anti-root vertically beneath the basin, as first deduced by Collette *et al.* (1965) from the absence of large negative gravity anomalies over the North Sea graben system. However, our models are not in strict *local* isostatic equilibrium. Vertical columns from the surface to the regional Moho depth (32 km) exert loads at this depth that increase from the basin margins to the basin centre by up to 350 MPa (3.5 kb). Though local isostatic equilibrium is often made a requirement of acceptable gravity models (Dyment 1989; Sibuet *et al.* 1990) this is to assume that the crust has zero lateral strength and zero elastic thickness. For the North Sea this is not correct. Barton and Wood (1984) found that a lithospheric elastic thickness of about 5 km best fitted the observed gravity field of the North Sea.

This result is borne out by the apparent lack of compensation of short-wavelength loads in the North Sea. As an example, the paired high-low gravity anomalies at the western ends of the NSDP85-5 and SNST83-7 lines (Figs 5.3a, and c) can be interpreted either as granitic intrusions (Tully and Donato 1985) or, based on indications on the seismic data, as Upper Palaeozoic sedimentary basins (our interpretation). The short wavelengths and associated steep gradients of these anomalies require a source depth no greater than 10 km. The caluclated anomalies for these basin-footwall pairs can fully account for the observed residual anomalies after stripping off the gravity effect of the Mesozoic sediments. This implies that these Palaeozoic basins are entirely supported by the strength of the lithosphere. The same is true for the horst structures in the Central Graben (NSDP85-5, the Forties-Montrose High at 300 km on Fig. 5.3a) and Viking Graben (Holliger and Klemperer 1989), and for the Permo-Carboniferous Egersund Basin at the eastern end of NSDP85-5 (at 480 km on Fig. 5.3a).

Using standard elastic theory (Turcotte and Schubert 1982) a minimum elastic thickness of the lithosphere of 2–3 km is required to support the vertical stress exerted by the above features (about 10 MPa, 0.1 kb), in agreement with the Barton and Wood (1984) result. All the Palaeozoic loads (whether granites or basins) modelled in this study or by Holliger and Klemperer (1989) are uncompensated, in contrast to the large degree of compensation found for most of the Mesozoic structures. Donato and Tully (1981), from their more regional study of the North Sea gravity data, also concluded that 'the compensation mechanism is associated with the younger Mesozoic development of the North Sea basins and does not affect the Palaeozoic basins'.

The existence of flexural strength and non-zero

elastic thickness permits departures from strict local isostasy. Figure 5.7 demonstrates that the observed gravity field is not satisfied by models in which complete Airy compensation occurs at the Moho to support the sedimentary-basin load. For the profile shown in Fig. 5.5 we show in Fig. 5.7 the 'isostatic gravity Moho', calculated to give the same total load everywhere along the profile at 32 km depth, using our preferred lower crust/upper mantle density contrast of 500 kg/m^3. Though this isostatic Moho is close to the gravity Moho and the refraction Moho beneath the basin margin, in the graben area it is too deep by about 5 km. The gravity anomaly resulting from this model is about 40 mGal lower than the observed gravity field over the basin axis. This is the result of the $1/r^2$ dependence of the gravitational field from the negative load of the basin near the surface and the positive load at depth, due to the shallow mantle beneath the graben (the 'proximity effect' of Meissner, 1986). It is clear from Fig. 5.7 that models with local Airy isostasy cannot satisfy

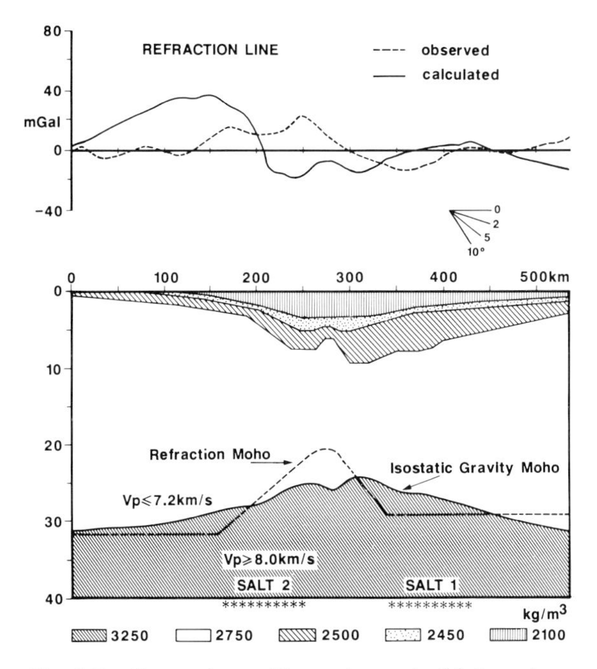

Fig. 5.7. Comparison of isostatic gravity Moho calculated to give perfect isostatic compensation at the Moho, for a lower crust/upper mantle density contrast of 500 kg/m^3, with the true Moho depth interpreted from the wide-angle seismic data superimposed. Note the poor fit of the observed and calculated gravity fields. Compare with Fig. 5.5.

the observed gravity field, even though regional Airy isostasy is responsible for compensating the basin; it is therefore reasonable to present models (Figs 5.3, 5.5 and 5.6) which are not in strict local isostatic equilibrium.

5.6. Conclusions

Coupled changes of 0.3 km/s and 75 kg/m^3 or more in average basement velocity and density are likely to be detected by a joint interpretation of deep seismic reflection and gravity data. The good agreement between the gravity Moho and reflection and refraction Mohos implies that, averaged over its thickness, the crystalline crust in the basin area of the North Sea is remarkably homogeneous, and that the reflection Moho corresponds to the density-velocity jump at the base of the crust imaged by the gravimetric and the seismic wide-angle method. That the density, velocity, and reflectivity of the lithosphere all change at the Moho seems inconsistent with the recent iconoclastic suggestion that the continental Moho is not the crust-mantle boundary (Griffin and O'Reilly 1987). The gravity data do not require a Conrad-type, mid-crustal, density contrast, but cannot rule out its existence.

A comparison of the gravity Moho and refraction Moho shows that an average density contrast of 400–500 kg/m^3 between the lower crust and upper mantle is a realistic average value, which favours an anorthositic or intermediate bulk composition for the lower crust. Unless the pre-Mesozoic composition of the lower crust beneath the North Sea was granitic, the high inferred density contrast does not permit substantial additions of basaltic melt to the crust during rifting, with the possible exception of the most strongly extended parts of the graben system.

An average intermediate composition for the lower crust is also consistent with the low seismic velocities, but not with the dominantly-mafic Scottish xenolithic population. An anorthositic lower crust is consistent with the xenoliths but not the seismic velocities for the western North Sea, unless the lower crust is wet.

The success of gravity modelling in outlining the topography of the Moho is due to the most variations in average basement density. Regional compensation is of Airy-type, though strict local isostasy is not achieved, with Palaeozoic basins and intra-graben horsts being supported by the flexural strength of the crust.

Acknowledgements

Special thanks to Duncan and Sheila Cowan (Cowan Geodata Services) for kindly providing the SEASAT data. This project was assisted by other members of the BIRPS group, which is funded by the National Environment Research Council (NERC), and by Penny Barton, Dave Snyder, and Edi Kissling. The comments of two anonymous reviewers helped to improve the quality of this paper. The NSDP programme was shot and processed by GECO as a speculative survey with participation by BIRPS. Profiles NSDP84-1, -2, 3, and -4, and NSDP 85-5 and -6 are available from Merlin Geophysical Ltd, Merlin House, Boundary Road, Woking, Surrey GU21 5BX, UK. NOPEC very kindly allowed access to their proprietary deep reflection seismic profile SNST83-7. BIRPS profiles SALT-1 and SALT-2 are available from Marine Geophysics Programme Manager, British Geological Survey, West Mains Road, Edinburgh EH9 3LA, UK. Well-log information was generously provided by Amoco Norway, Amoco UK Exploration, Conoco Norway, Murphy Petroleum, Norske Shell, Occidental Petroleum (Caledonia), Phillips Petroleum (Norway), Phillips Petroleum (UK), Ranger Oil (UK), Shell UK with Esso Exploration and Production UK, Texaco, and Unocal Norge. Without the generous assistance of all the above companies, this study would not have been possible. K.H. was supported by a NERC/BIRPS visiting fellowship. S.L.K. is supported by a University Research Fellowship of the Royal Society.

Contribution No. 608, ETH-Geophysics, Zurich, Switzerland; Cambridge Earth Sciences Contribution No. 1451.

References

Andersen, O. B. (1978). Bouguer anomaly map of Denmark and surrounding waters. *Geodaetisk Instituts Skrifter*, **3**, 44.

Anderson, D. L. (1967). A seismic equation of state. *Geophys. J. Roy. Astron. Soc.*, **13**, 9–30.

Bamford, D., Nunn, K., Prodehl, C., and Jacob, B. (1978). LISPB-IV. Crustal structure of northern Britain. *Geophys. J. Roy. Astron. Soc.*, **54**, 43–60.

Barton, P. J. (1986*a*). The relationship between seismic velocity and density in the continental crust—a useful constraint? *Geophys. J. Roy. Astron. Soc.*, **87**, 195–208.

Barton, P. J. (1986*b*). Comparison of deep reflection and refraction structures in the North Sea. In *Reflection seismology: A global perspective* (ed. M. Barazangi and L. Brown). *Am. Geophys. Union*, Washington DC, Geodynamics Series, **13**, 297–300.

Barton, P. and Wood, R. (1984). Tectonic evolution of the North Sea basin: stretching and subsidence. *Geophys. J. Roy. Astron. Soc.*, **79**, 987–1022.

Barton, P., Matthews, D. H., Hall, J., and Warner, M. (1984). Moho beneath the North Sea compared on normal incidence and wide-angle seismic records. *Nature*, **308**, 55–6.

Best, G., Kockel, F., and Schöneich, H. (1983). Geological history of the southern Horn Graben. *Geol. en Mijnbouw*, **62**, 25–33.

Bott, M. H. P., Long, R. E., Green, A. S. P., Lewis, A. H. J., Sinha, M. C. and Stevenson, D. L. (1985). Crustal structure south of the Iapetus suture beneath northern England. *Nature*, **314**, 724–7.

British Geological Survey (1986). *Bouguer gravity anomaly map*, 1:1 000 000 Series, Southern North Sea sheet.

Christensen, N. I. (1982). Seismic velocities. In *Handbook of physical properties of rocks* (ed. R. S. Carmichael), vol. 2, 1–228. Boca Raton, Florida, USA, C.R.C. Press.

Christie, P. A. F. and Sclater, J. G. (1980). An extensional origin for the Buchan and Witchground Graben in the North Sea. *Nature*, **283**, 729–32.

Collette, B. J., Lagaay, R. A., and Ritsema, A. R. (1965). Depth of the Mohorovičić discontinuity under the North Sea. *Nature*, **203**, 688–9.

Day, G. A., Cooper, B. A., Andersen, C., Burgers, W. F. J., Rønnevik, H. C., and Schöneich, H. (1981) Regional seismic structure maps of the North Sea. In *Petroleum geology of the continental shelf of north west Europe*, (ed. L. W. Illing and G. D. Hobson), 76–84. London, Institute of Petroleum, Heyden and Son.

Day, G. A., Anderson, O. B., Ensager, K., Finnstrøm, E. G., Hospers, J., Liebe, T., Makris, J., Plaumann, S., Strang van Hees, G., and Walter, S. A. (this volume). North Sea gravity map.

Donato, J. A. and Tully, M. C. (1981). A regional interpretation of North Sea gravity data. In *Petroleum geology of the continental shelf of north west Europe*, (ed. L. W. Illing and G. D. Hobson), 65–75. London, Institute of Petroleum, Heyden and Son.

Dyment, J. (1989). SWAT et les bassins Celtiques: relations avec la croûte hercynienne, néoformation du Moho. *Bull. de la Société Géologique de France*, **(8)**, **V**, **3**, 477–87.

Fichler, C. and Hospers, J. (this volume). Gravity modelling in the Viking Graben area, North Sea.

Gatliff, R. W., Hitchen, K., Ritchie, J. D., and Smythe, D. K. (1984). Internal structure of the Erland Tertiary volcanic complex, north of Shetland revealed by seismic reflection. *J. Geol. Soc. London*, **141**, 555–62.

Goldschmidt, A. (1986). Das Schwerefeld der Nordsee und eine erste Interpretation. Unpublished thesis, University of Hamburg.

Griffin, W. L. and O'Reilly, S. Y. (1987). Is the continental Moho the crust-mantle boundary? *Geology*, **15**, 241–4.

Haagmans, R. H. N., Husti, G. J., Plugers, P., Smit, J. H. M., and Strang van Hees, G. L. (1988). *Netherlands Geodetic Commission, Publications on Geodesy*, New Series, **32**.

Herzberg, C. T., Fyfe, W. S., and Carr, M. J. (1983). Density constraints on the formation of the continental Moho and crust. *Contributions to Mineralogy and Petrology*, **84**, 1–5.

Holliger, K. (1987). Crustal structure of the North Sea based on the analysis of SEASAT altimetry, deep seismic reflection data and shipborne Bouguer gravity anomalies. Unpublished thesis, ETH Zürich.

Holliger, K. and Klemperer, S. L. (1989). A comparison of the Moho interpreted from gravity data and from deep seismic reflection data in the northern North Sea. *Geophys. J.*, **97**, 247–58.

Hospers, J. and Finnstrøm, E. G. (1984). The gravity field of the Norwegian sector of the North Sea. *Norges Geologiske Undersøkelse Bulletin*, **396**, 25–34.

Hospers, J., Finnstrøm, E. G., and Rathore, J. S. (1985). A regional gravity study of the northern North Sea (56–60°N). *Geophysical Prospecting*, **33**, 543–66.

Hunter, R. H., Upton, B. G. J., and Aspen, P. (1984). Meta-igneous granulite and ultramafic xenoliths from basalts of the Midland Valley of Scotland: petrology and mineralogy of the lower crust and upper mantle. *Trans. Roy. Soc. of Edinburgh: Earth Sciences*, **74**, 75–84.

Hyndman, R. D. and Klemperer, S. L. (1989). Lower-crustal porosity from electrical measurements and inferences about composition from seismic velocities. *Geophys. Research Letters*, **16**, 255–8.

Klemperer, S. L. (1988). Crustal thinning and nature of extension in the northern North Sea from deep seismic reflection profiling. *Tectonics*, **7**, 803–21.

Klemperer, S. L. (1989). Processing of BIRPS deep seismic reflection data: a tutorial review. In *Digital seismology and fine modeling of the lithosphere* (ed. R. Cassinis, G. Nolet, and G. F. Panza), vol. 42, 229–57, Ettore Majorana International Science Series (Physical Sciences). New York, Plenum Press.

Klemperer, S. L. and Hurich, C. A. (this volume). Lithospheric structure of the North Sea from deep seismic reflection profiling.

Klemperer, S. L. and White, N. J. (1989). Coaxial stretching or lithosphere simple shear in the North Sea? Evidence from deep seismic profiling and subsidence. In *Extensional tectonics and stratigraphy of the North Atlantic margin* (ed. A. J. Tankard and H. R. Balkwill). *Am. Assoc. Petrol. Geol. Memoir*, **46**, 511–22.

Klemperer, S. L., Hauge, T. A., Hauser, E. C., Oliver, J. E., and Potter, C. J. (1986). The Moho in the northern Basin and Range Province, Nevada, along the COCORP 40°N seismic reflection transect. *Geol. Soc. of America Bulletin*, **97**, 603–18.

Latin, D. M., Dixon, J. E., and Fitton, J. G. (this volume). Rift-related magmatism in the North Sea Basin.

Litak, R. K. and Brown, L. D. (1989). A modern perspective on the Conrad discontinuity. EOS, *Transactions American Geophysical Union*, **70**, 713, 722–5.

Ludwig, J. W., Nafe, J. E., and Drake, C. L. (1970). Seismic refraction. In *The Sea* (ed. A. E. Maxwell), vol. 4, 53–84. New York, Wiley.

McKenzie, D. (1978). Some remarks on the development of sedimentary basins. *Earth and Planetary Sciences Letters*, **40**, 25–32.

McKenzie, D. and Bickle, M. J. (1988). The volume and composition of melt generated by extension of the lithosphere. *J. Petrol.*, **29**, 625–79.

Matthews, D. H. (1986). Seismic reflections from the lower crust around Britain. In *The nature of the lower continental crust* (ed. J. B. Dawson, D. A. Carswell, J. Hall and K. H. Wedepohl). *Geol. Soc. London Special Publication*, **24**, 11–22.

Meissner, R. (1986). *The continental crust—a geophysical approach*. London, Academic Press, International Geophysics Series, vol. 34.

Nafe, J. E. and Drake, C. L. (1963). Physical properties of marine sediments. In *The Sea* (ed. M. N. Hill), vol. 3, 793–815. New York and London, Interscience Publishers.

Raynaud, B. (1988). A 2-D, ray-based, depth migration method for deep seismic reflections. *Geophys. J.*, **93**, 163–171.

Ringwood, A. E. (1975). *Composition and petrology of the earth's mantle*. New York, McGraw-Hill.

Sellevoll, M. A. (1973). Moho beneath Fennoscandia and adjacent parts of the Norwegian Sea and the North Sea. *Tectonophysics*, **20**, 359–66.

Setto, I. A. E. R. S. A. (1987). The use of seismic refraction and reflection data for gravity modelling of crustal structures. Unpublished PhD thesis, University of Kiel.

Setto, I. and Meissner, R. (1987). Support from gravity

modelling for seismic interpretation. *Annales Geophysicae*, **5B**, 395–402.

Sibuet, J.-C., Dyment, J., Bois, C., Pinet, B., and Ondreas, H. (in press). The structure of the Celtic Sea and Western Approaches from gravity data and deep seismic profiles: constraints on the formation of continental basins. *J. Geophys. Research*.

Smith, P. J. and Bott, M. H. P. (1975). Structure of the crust beneath the Caledonian foreland and the Caledonian belt of the North Scotland shelf region. *Geophys. J. Roy. Astron. Soc.*, **40**, 187–205.

Solli, M. (1976). En seismik skorpeundersokelse Norge-Shetland, Unpublished M.Sc. thesis, University of Bergen.

Steinhart, J. S. (1967). Mohorovičić discontinuity. In *International dictionary of geophysics* (ed. K. Runcorn), vol. 2, 991–4. Oxford, Pergamon Press.

Strang van Hees, G. L. (1983). Gravity survey of the North Sea. *Marine Geodesy*, **6**, 167–81.

Stuart, G. W. (1978). The upper mantle structure of the North Sea from Rayleigh wave dispersion. *Geophys. J. Roy. Astron. Soc.*, **52**, 367–82.

Talwani, M., Worzel, J. L., and Landisman, M. (1959). Rapid gravity computations for two-dimensional bodies with application to the Mendocino submarine fracture zone. *J. Geophys. Research*, **64**, 49–59.

Tully, M. C. and Donato, J. A. (1985). 1:1,000,000 northern North Sea Bouguer anomaly gravity map. *Report of the British Geological Survey*, **16**, (6).

Turcotte, D. L. and Schubert, G. (1982). *Geodynamics: application of continuum physics to geological problems*. New York, Wiley.

Upton, B. J. G., Aspen, P., and Chapman, N. A. (1983). The upper mantle and deep crust beneath the British Isles: evidence from inclusions in volcanic rocks. *J. Geol. Soc.*, **140**, 105–21.

Warner, M. R. (1986). Deep seismic reflection profiling the continental crust at sea. In *Reflection seismology: A global perspective* (ed. M. Barazangi and L. Brown). Washington DC, *Am. Geophys. Union*; Geodynamics Series, 13, 281–6.

Warner, M. R. (1987). Seismic reflections from the Moho—the effect of isostasy. *Geophys. J. Roy. Astron. Soc.*, **88**, 425–35.

Warner, M. R. (1990). Basalts, water or shear zones in the lower continental crust? *Tectonophysics*, **173**, 163–74.

Wever, T. (1989). The Conrad discontinuity and the top of the reflective lower crust—do they coincide? *Tectonophysics*, **157**, 39–58.

White, R. and McKenzie, D. (1989). Magmatism at rift zones: the generation of volcanic continental margins and flood basalts. *J. Geophys. Research*, **94**, 7685–729.

Woollard, G. P. (1966). Regional isostatic relations in the United States. In *The Earth beneath the continents* (ed. J. S. Steinhart and G. P. Smith). Washington, DC, *Am. Geophys. Union*: Geophysical Monograph No. 10, 557–94.

Woollard, G. P. (1975). Regional changes in gravity and their relation to crustal parameters. *Bureau Gravimétrique International, Bulletin d'Information*, **36**, 106–10.

Zervos, F. (1987). A compilation and regional interpretation of the northern North Sea gravity map. In *Continental extensional tectonics* (ed. M. P. Coward, J. F. Dewey, and P. L. Hancock). *Geol. Soc. Special Publication*, **28**, 477–93.

6 Rift-related magmatism in the North Sea basin

D. M. Latin, J. E. Dixon, and J. G. Fitton

Abstract

Mesozoic rifting and associated decompression melting in the North Sea basin gave rise to igneous rocks which vary from mildly to highly undersaturated in character. Recent theoretical developments based on a parameterization of melting experiments on dry peridotite (McKenzie and Bickle 1988) suggest that there should be quantitative relationships between the compositions and volumes of magmas generated in rift environments, degrees of partial melting of the asthenospheric mantle, mantle temperature, and amounts of extension. The North Sea is a good place to assess theoretical models because a large number of the required parameter values have already been constrained (e.g. the degree of extension, pre-rift lithosphere thickness, rifting duration). In order to test the quantitative model of McKenzie and Bickle, observed and expected major element compositions are compared, and trace element models are used to make estimates for degrees of partial melting represented by the North Sea rocks. These models suggest that if the source was dry peridotite able to produce MORB (mid-ocean ridge basalt) on melting by 15–20 per cent then less than 0.5 per cent melting in a single-stage melting event is required to explain the least undersaturated Mesozoic rocks (Forties alkali basalts). However, many of the North Sea rocks have been LILE (large ion lithophile element) enriched by some mechanism and can *only* be produced by melting enriched mantle either alone or by mixing with melts from a depleted MORB-source asthenosphere. Indeed even the alkali basalts of Forties appear to require some enrichment, in terms of K_2O, TiO_2, and H_2O, compared to the predicted smallest melts from dry peridotite. The relatively minor volatile content of the alkali basalts, the thin mechanical boundary layer at the time of rifting, and prior extensional events, make the lithosphere an unattractive source for the enrichment; a streaky asthenosphere source which could produce both MORB and OIB (ocean island basalt) at different degrees of melting is considered most likely. The small melt fraction values (<0.5 per cent) obtained for a depleted MORB source can be used, in conjunction with other observational data from the North Sea, as input parameters for the McKenzie and Bickle model, to place a lower limit on the required amounts of upwelling at different potential temperatures, given dry melting conditions. The model predicts no melting in the North Sea for normal (1280 °C) potential temperatures. The observations may be reconciled with volatile-induced melting at temperatures about 100 °C lower than the dry solidus, or larger amounts of extension, at normal potential temperatures. Alternatively, but less probably, potential temperatures of 1380 °C will produce the alkali basalts from dry peridotite at the observed extension factors. This analysis suggests that such models have at present extremely limited application when melt fractions are small, except to provide upper limits on combinations of asthenosphere temperature and extension. Some quantification becomes possible when magmas capable of being derived from the dry peridotites used in the experiments are erupted, as happened in the most stretched part of the North Sea; models may well be useful in other more extended regions. Where enriched, volatile-bearing lithosphere dominates magma generation no reliable framework yet exists to relate compositions to stretching.

6.1. Introduction

This paper concentrates on igneous rocks of Jurassic and Cretaceous age, relating their occurrence, composition, and volume to the degree of lithospheric attenuation resulting from Mesozoic extension. Igneous rocks of Lower Permian age are not discussed in detail, but are introduced for comparison. The work described here builds on an earlier study by Dixon and colleagues (1981) which established the undersaturated character of North Sea rift magmatism, and drew attention to its compositional similarity to other rift provinces. Since 1981 a number of new wells have sampled igneous rocks. These have now been studied, and the original samples have been re-analysed with samples from the Forties volanic province (Howitt *et al.* 1975; Woodhall and Knox 1979; Fall *et al.* 1982) and are interpreted in the light of new theory (McKenzie and Bickle 1988). In recent years the theoretical framework governing the generation of magmas in rift zones has become quantitative. The main aim of this study is to test quantitative models for magma generation using the North Sea rift system to provide the observational data. Discussion of the theory and its relationship to the observations forms the main body of the paper.

Areas of lithospheric extension are often sites of volcanism; it has long been recognized that magma generation may be linked to partial melting of upwelling asthenosphere which undergoes adiabatic decompression during the stretching process. However, although Dixon and co-workers (1981) discussed a qualitative relationship between stretching and magma composition, until recently little attempt has been made to make quantitative estimates of the volumes and compositions of magmas produced in this way (Foucher *et al.* 1982; McKenzie and Bickle 1988).

Once this theoretical problem had been overcome, it became apparent that the North Sea should be an excellent area for looking at the relationship between extension and magmatism; the amount of extension is well constrained and estimates made by different methods (subsidence modelling, fault block restoration, seismic reflection, and gravity models) converge on similar values (Sclater and Christie 1980; Barton and Wood 1984; N. J. White 1988; Holliger and Klemperer 1989; Klemperer and White 1989). Most important is the fact that the North Sea is a rift which is old enough to show the thermal-subsidence (post-rift) stage of rift evolution and where there is enough data to make extension estimates from thicknesses of thermal-subsidence stage sediments (Sclater and Christie 1980; Barton and Wood 1984). For instantaneous stretching the amount of thermal subsidence (caused by conductive cooling) is directly related to the size of the rift-induced thermal anomaly, and hence to the amount of upwelling of the asthenosphere.

In the first part of this paper we will consider the theory which relates the composition and volumes of magmas produced during asthenospheric upwelling to amounts of instantaneous extension (McKenzie and Bickle 1988), ideas which we set out to test in the North Sea. We will then review what is known about the spatial, temporal, petrographic, chemical, and volumetric aspects of Mesozoic magmatism in the North Sea. We hope to show that, although severe alteration makes interpretation of chemical data difficult, estimates can still be made of the degree of partial melting (of the asthenosphere) represented by the North Sea rocks. These estimates turn out to be very small for all reasonable asthenosphere source compositions but, crucially, some North Sea rocks have element ratios which cannot be modelled by any single stage partial melting process, and element concentrations which require some 'enriched' region to be sampled. We then consider whether the theoretical framework provided by McKenzie and Bickle, which works well for larger degrees of melting characteristic of MORB (mid-ocean ridge basalt), can account in a quantitative way for any of the Mesozoic igneous rocks alone, or by way of providing a component of the final melt. We find that this is only possible for a limited set of rocks, and only then if the asthenosphere temperature is set at a value which seems unacceptably high on independent grounds. Even if this is accepted many of the North Sea magmas still cannot be modelled.

This is clearly not satisfactory, and in the concluding sections we attempt to show how the theoretical framework will need to be refined if it is to provide quantitative information in situations for which it was not designed. Our analysis does, nevertheless, support a model of normal-temperature passive propagating-rift extension for the North Sea rift system. It also shows that the theory is potentially able even now to constrain extension estimates in other basins, given appropriate magma compositions and independent control on asthenosphere temperature from uplift patterns.

6.2. Theoretical aspects

Within-plate extension can be viewed as part of a process of lithospheric stretching which may ultimately lead to the formation of an ocean basin and spreading centre. The production of magma beneath ocean ridges and its subsequent extrusion as MORB (mid-ocean ridge basalt) can be seen as an extreme example of the within-plate process. The key difference is that the mid-ocean ridge is a steady-state upwelling system under zero-thickness lithosphere and is essentially time-invariant, whereas the behaviour of stretching lithosphere and underlying asthenosphere varies according to how far and how fast the stretching has proceeded. The steady-state process is nevertheless the key to the quantitative model.

In order to discuss magma genesis it is first necessary to consider whether or not melting occurs everywhere under old oceanic or continental lithosphere under 'normal' conditions. Consider the geothermal gradient through old oceanic lithosphere which has achieved steady-state heat flow (Fig. 6.1). The upper part of the lithosphere (the 'plate' itself, or mechanical boundary layer, MBL) has a more or less linear temperature gradient governed by conductive heat-loss. The lithosphere must overlie asthenosphere in which heat transport is advective and which therefore has an adiabatic (and thus near vertical) geotherm. Between these two regions lies a thermal boundary layer (TBL) in which heat is transported both by conduction and advection, and which is inferred to be periodically incorporated into the asthenosphere circulation. The conductive and advective geotherms will intersect at some point within the TBL, and this can be used to define the base of the lithosphere (Parsons and McKenzie 1978; McKenzie and Bickle 1988; R. S. White 1988a).

In order for melting to occur the geotherm must

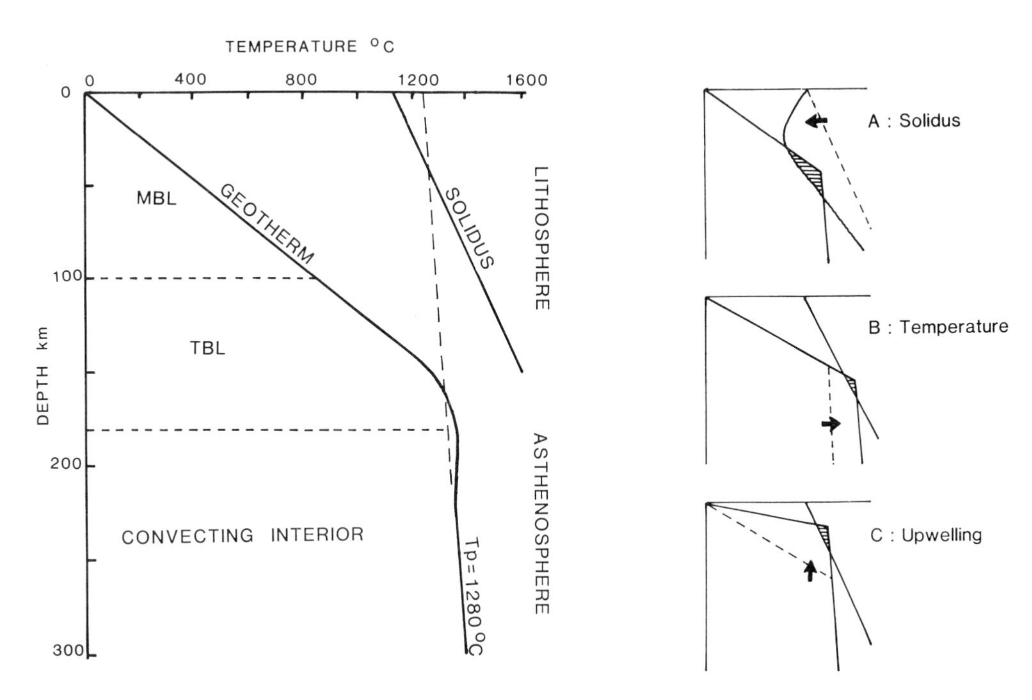

Fig. 6.1. The horizontally-averaged thermal structure of the lithosphere for a potential temperature of 1280 °C, a mechanical boundary layer thickness of 100 km, and an interior viscosity of 2×10^{17} m^2 s^{-1} following McKenzie and Bickle (1988). The corresponding adiabatic upwelling curve (assuming no melting) is shown dashed. The solidus position for dry peridotite is also shown. MBL = mechanical boundary layer; TBL = thermal boundary layer (see discussion in text). The highly schematic cartoons to the right of the main diagram show three end-member melting mechanisms: A. melting due to a change in solidus position, B. melting by raising the potential temperature, and C. melting by adiabatic upwelling.

intersect the mantle melting curve (solidus). Under 'normal' conditions the geotherm is a long way below the solidus for dry peridotite; thus if this solidus controls melting, it will not occur. If however, the solidus drawn in Fig. 6.1 is not the true initiating of melting line but represents only the line of significant (or experimentally-visible) melt generation (5–10 per cent) then small melt fractions could occur widely in the mantle at significantly lower temperatures (for a given pressure). This concept has been given new impetus by recent theoretical work on the physical properties of small melt fractions and will be returned to in a later section. It is none the less clear that for more than 5–10 per cent melting to occur the system must be perturbed in one of three ways:

(1) by changing the position of the solidus, for example by the addition of water, as in the case of melting below island arcs;
(2) by increasing the temperature of the asthenosphere and so moving the geotherm to the right where it may intersect the solidus; this appears to be the case for ocean island volcanoes (e.g. Hawaii) located over 'hot spots' (plumes), and may also be important below some segments of ocean ridges (Iceland) and continental rifts (East Africa);
(3) by raising the base of the lithosphere closer to the surface, thus causing adiabatic upwelling of the asthenosphere below; this is the process operating at normal ocean ridges and is implied by the passive stretching model (McKenzie 1978) for rift systems like the North Sea (Dixon *et al.* 1981).

In the following sections we review current ideas on the melt generation process, models for lithospheric attenuation, estimates of the amount of Mesozoic extension in the North Sea, and the compositions of magmas produced in rift-related environments.

6.2.1. *Melting by upwelling*

It has long been believed that large quantities of melt are generated at ocean ridges as a response to adiabatic upwelling of the asthenosphere (Verhoogen 1954; Green and Ringwood 1967). Several recent studies have calculated the extent of melting of the mantle during decompression (Ahern and Turcotte 1979; McKenzie 1984). These calculations have then been used in conjunction with variants of the passive stretching model (McKenzie 1978) in order to estimate the volumes (Foucher *et al.* 1982; McKenzie and Bickle 1988) and compositions (McKenzie and Bickle 1988) of melts generated during extension.

McKenzie and Bickle pointed out that the generally uniform thickness of oceanic crust (6–7 km), regardless of spreading rate, must mean that the asthenosphere temperature must be uniform everywhere at a given depth. The spreading apart of plates at a ridge must be exactly balanced by the influx (and upwelling) of asthenosphere. As McKenzie and Bickle showed, the further the geotherm rises above the solidus the greater is the degree of melting, and the greater is the total amount of melt produced in a progressively rising column (Fig. 6.2). Because the geometry of the process is constant, the amount of upwelling is fixed for a given amount of spreading, independent of rate; thus to produce constant melt (i.e. crust) thickness, the solidus must be crossed by the same amount everywhere. This can only be achieved if the asthenosphere temperature is constant for a given depth.

Anomalously thick crust in regions such as Iceland (27 km), astride spreading ridges, is thus most readily explained by higher than normal asthenosphere temperature below the ridge, causing the solidus to be encountered at much greater depth in the upwelling path (Fig. 6.2). Asthenosphere temperature is best expressed by the 'potential temperature' (Tp), the temperature on the adiabatic gradient projected to surface pressure (1 bar). Under normal lithosphere the Tp is inferred to be 1280 °C; in plume situations, such as under Iceland, it is some 200 °C hotter (R. S. White *et al.* 1987; R. S. White 1988*b*; R. S. White and McKenzie 1989; McKenzie and Bickle 1988). Many readers may be familiar with the figure of 1333 °C for the temperature of the mantle (Parsons and Sclater 1977; McKenzie 1978). This value is approximately equal to the temperature on the 1280 °C adiabat at the base of the oceanic lithosphere (about 125 km).

McKenzie and Bickle also showed that the extent of melting is the dominant control on the composition of melts that reach the surface, even after the effects of low-pressure fractional crystallization and contamination are allowed for. The constancy of MORB composition reflects its generation by between 15 per cent and 20 per cent partial melting of homogeneous asthenosphere.

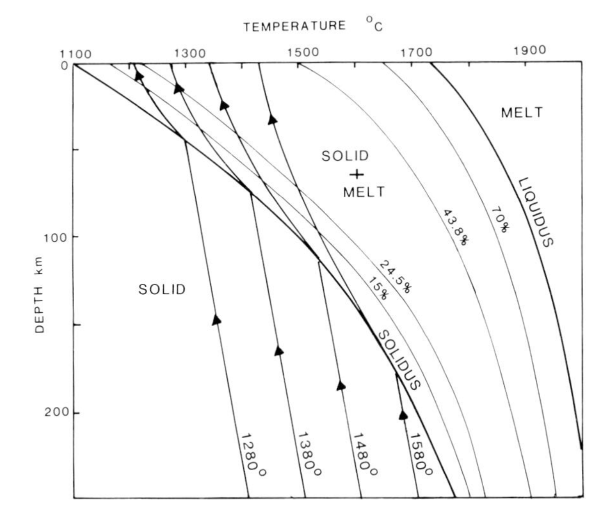

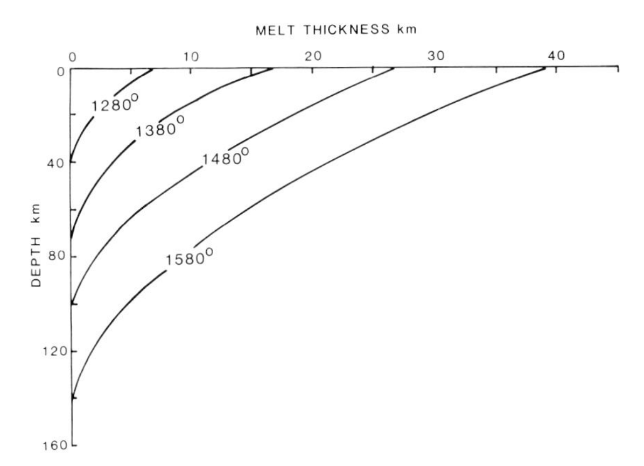

Fig. 6.2. The upper diagram shows adiabatic decompression paths for different potential temperatures. The curves between the solidus and liquidus are labelled with per cent melting of dry peridotite. The lower diagram shows the total thickness of melt generated below a given depth plotted as a function of depth and potential temperature. Both diagrams are taken from McKenzie and Bickle (1988).

In both the 'normal' ridge and the 'plume + ridge' cases mentioned above, the amount of melt produced and thus its composition, is governed purely by the potential temperature, because upwelling continues all the way to the surface as the plates separate. In the case of a continental rift, such as the North Sea, the amount of extension is finite and so controls the degree of upwelling of mantle for any potential temperature. However, the initial thickness of the lithosphere (prior to stretching) is also important. Clearly for thinner initial thicknesses the convecting asthenosphere is already closer to the solidus and so for a given amount of extension will not need to rise as far before it starts to melt. Slow rates of extension will reduce the amount of melting if thermal conduction can keep pace with upwelling, causing asthenosphere to become lithosphere and follow a conductive cooling path, thus adding to lithospheric thickness.

Less clear is the effect of the composition of the mantle undergoing melting. A first order observation is that mantle portions of different composition may produce melts of different composition. The position of the solidus defined by McKenzie and Bickle is based on a parameterization of published melting experiments carried out on garnet peridotite under 'dry' conditions. This solidus should not be regarded as unique, and any change in its position will affect the quantity and composition of melts produced. The presence of carbon dioxide and water (even in very small amounts) has long been known to have a dramatic effect on the position of the solidus (Wyllie 1987; Eggler 1976, 1978; Edgar 1987; Wallace and Green 1988; Green and Wallace 1988; Hunter and McKenzie 1989). It seems likely that there will be a number of solidi below the McKenzie and Bickle solidus in pressure–temperature space which may be important when considering very small amounts of partial melting (less than 5 per cent), and hence small degrees of upwelling. These are relatively unimportant in the mid-ocean ridge situation as small degree melts are overtaken and amalgamated with large degree melts in the progressive upwelling.

Finally the melt generated in any given situation has to be extracted from the residue. The nature and efficiency of the extraction process has implications for melt composition and quantity (McKenzie 1984). For melts to reach the surface following stretching of the continental lithosphere they must pass through the continental lithospheric mantle and crust, which both act as sinks for heat and potential sources of contamination.

To summarize, the amounts and composition of partial melts produced by upwelling are controlled by seven fundamental parameters:

(1) the potential temperature of the convecting mantle;
(2) the amount of extension;

(3) the initial thickness of the lithosphere;
(4) the composition of asthenospheric mantle;
(5) the rates of extension;
(6) the nature of the extraction processes, and
(7) the extent to which magmas interact with the lithosphere.

6.2.2. *Models of lithospheric attenuation*

It is not our intention here to review all models for the formation of sedimentary basins, but rather to concentrate on the fundamental principles of a number of models, in particular the passive stretching model (McKenzie 1978) and its application to the North Sea.

By analogy with ocean ridges, Sleep (1971) considered that injection of hot low-density asthenospheric melts into the continental lithosphere would cause uplift, with subsequent cooling and contraction giving rise to subsidence. This model requires considerable amounts of erosion to occur during the uplift stage if the area is to subside below its original level to form a basin.

McKenzie (1978) argued that there is no convincing evidence for such large-scale erosion (in the order of 2 km or more) in the North Sea (in direct conflict with Ziegler 1975, 1981, 1982*a*, *b*; and Ziegler and Van Hoorn 1989) and so proposed a simple one-dimensional model in which instantaneous stretching of the lithosphere causes thinning, bunching of the isotherms, and a rise in the geothermal gradient due to the resulting passive upwelling of the asthenosphere from below. The lithosphere is compensated isostatically before and after extension, and so this stage of basin formation is associated with an initial fault-controlled subsidence in the crust, with plastic flow lower down. With time, the lithosphere thickens by conduction of heat to the surface, and the geotherm (elevated during extension) relaxes, causing a further thermal subsidence which decays exponentially with time. Thermal subsidence and surface heat flow depend only on the amount and rate of stretching. The total subsidence is the sum of the initial (isostatic) and the thermal subsidence.

The McKenzie model successfully accounts for the subsidence history of the North Sea (Sclater and Christie 1980; Barton and Wood 1984) and the Bay of Biscay passive margin (Le Pichon and Sibuet 1981). However, studies of the eastern Canadian margin (Royden and Keen 1980) and the intra-Carpathian basins (Sclater *et al.* 1980) have shown differences between the subsidence histories predicted (using McKenzie's model) and those observed. In order to explain the lower initial subsidence and more rapid thermal subsidence, Hellinger and Sclater (1983) proposed a two-layer heterogeneous stretching model in which the crust and sub-crustal lithosphere could be stretched by different amounts. The main problem with two-layer models is that of mass conservation (maintaining compatibility in the section). N. J. White and McKenzie (1988) show how the 'steer's-head' geometry of many sedimentary basins (Dewey 1982) can be produced by constraining crustal deformation to a narrow rift zone and allowing the sub-crustal lithosphere to extend over a wider area, whilst keeping the overall amount of extension in both parts the same (Dixon *et al.* 1981).

The simple model of stretching (McKenzie 1978) is based upon instantaneous extension. Jarvis and McKenzie (1980) noted that, if the stretching period was protracted, then there would be vertical diffusion of heat away from the upwelling asthenosphere during extension. This would act to superimpose part of the thermal subsidence on the initial subsidence stage, and would lessen the amount during the later (strictly thermal) stage. However, they conclude that heat loss in the initial stage is not a real problem in the North Sea, provided that the duration of stretching is less than ~20 Ma. Models for the North Sea now commonly incorporate finite duration stretching (usually 60 Ma) in the calculations (Barton and Wood 1984; Wood 1982; N. J. White 1988).

Until recently one of the main arguments against the simple model was that degrees of extension calculated from the restoration of fault blocks were significantly less than would be predicted by the model and the thermal subsidence (Ziegler 1983; Ziegler and Van Hoorn 1989). By careful examination of fault block geometries, and using a kinematic model for fault block restoration, N. J. White (1988, 1989) has shown that extension calculated from the faults in part of the Viking Graben is very similar to that predicted from the subsidence, so eliminating the extension discrepancy.

Ziegler (1982*b*) appears to favour a mechanism similar to that of Sleep (1971), for the North Sea, citing as evidence large scale (of the order of 2 km) regional uplift and associated magmatic activity during the Mid-Jurassic, and a discrepancy between fault-controlled and thermal subsidence which is not to be expected from the McKenzie (1978) model. However, although sedimentary facies patterns may also support the emergence of land masses in the central North Sea during that period (Eynon 1981), they do not necessarily require uplift of more than 100 m or so according to Leeder (1983).

The main problem with Ziegler's approach is the mechanism. If the lithosphere is to be injected with melts then it must be as a result of one of the mechanisms discussed earlier. R. S. White and McKenzie (1989) show that production of melt during passive stretching over asthenosphere with a range of potential temperature is not enough in itself to generate positive relief; the reason is that the elevation due to presence of melt is always less than the isostatic subsidence due to the stretching which is creating it! The obvious way to produce uplift of the order of several kilometres is to invoke a mantle plume which will create a large thermal anomaly and lower the density of asthenosphere and lithosphere. It will exert dynamic forces on the base of the lithosphere, and will generate low density melts from the asthenosphere. The mantle-plume model for the North Sea is not one which can be dismissed lightly. The presence of magmatism, a trilete rift system, and domal uplift are taken to be the classic characteristics of plume-generated triple junctions (Burke and Dewey 1973) and have led to strong support for the plume model in the literature (Kent 1975; Whiteman *et al*. 1975; Ziegler 1982*b*; Ziegler and Van Hoorn 1989). We discuss this further in a later section.

Another model in common use for North Sea extension is the lithospheric simple shear or 'Wernicke' model (Wernicke 1985; Gibbs 1987*a*, *b*, 1989*a*, *b*). There appear to be three major drawbacks to this model; it requires a detachment plane to pass obliquely through the entire lithosphere, which seems unlikely on rheological grounds and is not observed on deep seismic profiles (Klemperer 1988; Klemperer and N. J. White 1989); thermal subsidence should be significantly different in extent and location from that predicted by the McKenzie model and, more importantly, the ratio of initial to thermal subsidence will be much greater. The observed subsidence fits the McKenzie model (N. J. White 1988; Klemperer and N. J. White 1989; N. J. White 1989). In offsetting the asthenospheric upwelling and crustal subsidence it predicts that magmatic activity would not be centred on the regions of greatest subsidence (Bosworth 1987) whereas, as we show later, the two coincide closely. Because the upwelling is distributed over a wide area in the lithospheric simple shear model, it is difficult to produce melt even at large amounts of crustal extension or elevated temperatures. Latin and N. J. White (1990) show that the lithospheric simple shear model fails to account for the existence or the location of magmatism in the North Sea, with or without a mantle plume.

Whether or not a mantle plume was present at the time of rifting (Mid-Jurassic) in the North Sea will have important implications for the volumes and compositions of magmas generated. Study of rift magmatism should therefore point us in the direction of the more likely model. The magmatic evidence is discussed in later sections.

6.2.3. *Estimates of extension in the Mesozoic North Sea*

Although a number of different techniques have been used to estimate the amount of extension in different parts of the North Sea (subsidence modelling, fault restoration, gravity models, and seismic refraction and reflection) we are primarily interested in those estimates obtained from subsidence records as these are most likely to reflect the amount of subcrustal lithospheric attenuation and hence the amount of asthenospheric upwelling. The amount of extension will be described by use of the parameter β equal to the ratio of pre-stretched to stretched thickness (McKenzie 1978).

The simple stretching model (McKenzie 1978) was first tested by Christie and Sclater (1980) using a seismic refraction line and subsidence data calculated from well logs across the Witch Ground Graben in the Moray Firth. Wood (1982) and Barton and Wood (1984) carried out a similar but much more detailed study of the 'triple junction' and Central Graben. All of these studies concluded that the amount of stretching (during the Mid-Jurassic to Cretaceous period) calculated by backstripping the subsidence data, agrees well with that measured by crustal thinning. All obtained similar estimates for the maximum amounts of extension; $\beta = 1.8$ in the area of Forties (Christie and Sclater 1980) and $\beta = 1.6$ in the Central Graben (corresponding to about 70 km of extension in the North Sea), and $\beta = 1.4$–1.6 in the Forties area (Wood 1982; Barton and Wood 1984). The values given by Wood and

Barton and Wood are obtained from the subsidence data, using best-fit theoretical curves with finite (60 Ma) rates of extension and taking account of pre-Middle Jurassic extension of approximately $\beta = 1.25$, and hence reduced initial crustal thicknesses. In a later section we will compare β values with melt compositions in different locations.

The values we have chosen to use are those of Barton and Wood: Forties/Central Graben, β = up to 1.6; Central Graben flanks, $\beta = 1.35$; Egersund Basin, $\beta = 1.2$. It should be noted that extension during the Triassic was probably less than 1.25, and that the thermal anomaly should have largely decayed away by Middle Jurassic times. However, any remanent thermal anomaly is unlikely to increase effective Jurassic β values to greater than 1.8 in the most stretched areas, and a maximum upper limit of $\beta = 2.0$ is used later in the paper.

Ziegler (1983) and Ziegler and Van Hoorn (1989) have pointed out that in the Central Graben the amount of extension measured across faults is considerably less (by about a factor of two) than that calculated from the subsidence. However, according to N. J. White (1988) the faulting is complex and difficult to interpret in the Central Graben; this discrepancy in the extension is not observed in the rather simpler Shetland Terrace, where faults and subsidence calculations agree well giving $\beta = 1.5$ (N. J. White 1988, this volume) for the Mid-Jurassic to Cretaceous event.

6.2.4. *Melt compositions in rift systems*

The alkaline nature of magmatic provinces associated with rifting is well known. Much early work concentrated on the East African rift system, and it remains the classic model. Several authors (Baker 1987) have commented on two distinct associations found there, a strongly alkaline basanite/nephelinite and phonolite/carbonatite association, and a transitional or mildly alkaline olivine basalt, trachyte, rhyolite association. This bimodality was identified by Mohr (1982) as a common feature of several other continental rift provinces.

Baker (1987) noted that in the Kenya Rift, magmas became increasingly undersaturated at any time with distance away from the rift axis, and silica undersaturation decreased with time. In the Ethiopian section of the East African system, transitional to mildly alkaline rocks dominate, with more strongly alkaline silica-undersaturated rocks occurring off-axis on the plateau (Barberi *et al.* 1982). Similar trends have been observed in a number of other areas. Dixon and co-workers (1981) comment on the similarity (in terms of composition and location) between the Mesozoic volcanism in the North Sea and that seen in the Kenya Rift.

Dixon and colleagues also related the magmatic activity observed in the North Sea rift system to variants of McKenzie's (1978) simple stretching model. They suggested that the shape of the stretching envelope (that is the way in which strain is partitioned within the lithosphere) exerted an important control on the depth of melting, and thus on the magma compositions. Using the work of Yoder (1976) they suggested for simplicity that the greater the extension, the shallower the depth of melting, and that these melts would be more silica-saturated than melts from greater depths where the stretching was less. Thus one might intuitively expect to find the most silica-saturated melts along the rift axis, with more highly alkaline and undersaturated melts occurring on the flanks.

The early ideas lacked the additional insight given by more recent work on the problems of melt extraction. Physical modelling of melt generation and its segregation in the mantle suggests that an interconnecting porosity is established at very small degrees of melting. This melt is extracted from the matrix by buoyancy and surface energy effects, and the rates of extraction are geologically fast (Ahern and Turcotte 1979; Stolper *et al.* 1981; McKenzie 1984; Scott and Stevenson 1986; Ribe 1985; Ribe 1987; Spiegelman and McKenzie 1987). The ability of melt to separate from the residue is also dependent on the viscosity of the melt. Melts of low viscosity, such as carbonatites, can separate very rapidly at very small melt fractions (Hunter and McKenzie 1989). The first formed melt fractions in the mantle can theoretically be extracted when amounts are as small as 10^{-2} per cent (McKenzie 1989). Extraction may therefore no longer be a limiting factor, but passage through the lithosphere to the surface may still be governed by total melt volume and the prevailing structural style.

We can now consider on/off axis compositional relationships of rift valley magmas in a slightly different way from that of Dixon and his colleagues. Rather than perceiving depth of melting as being the fundamental control on alkalinity and undersaturation, it is now thought that this control is the extent of melting (which is itself very closely related to depth).

Small degree (highly undersaturated) melts are unlikely to be generated at shallow depths (<40 km), because by the time mantle of normal potential temperature has risen to such levels it will have already undergone a considerable amount of partial melting. The converse is also true. Large degree melts (less silica undersaturated) can only be produced if the mantle rises far enough over its solidus; for a given temperature, they require greater upwelling, and hence are produced at shallower levels.

The general conclusion is that the compositions and volumes of magmas generated in rift environments, such as the North Sea, should be closely related to the degree of partial melting of the mantle, and hence to the amounts of stretching and upwelling. However, in the application of this principle to the North Sea we make two fundamental assumptions. First, that any melt observed at a particular locality is the product of lithospheric attenuation directly below, and has migrated upwards vertically. Secondly, that the melts produced reflect something close to the total amount of sub-crustal extension at any locality during the extensional phase, i.e. that melts do not permanently lag behind extension.

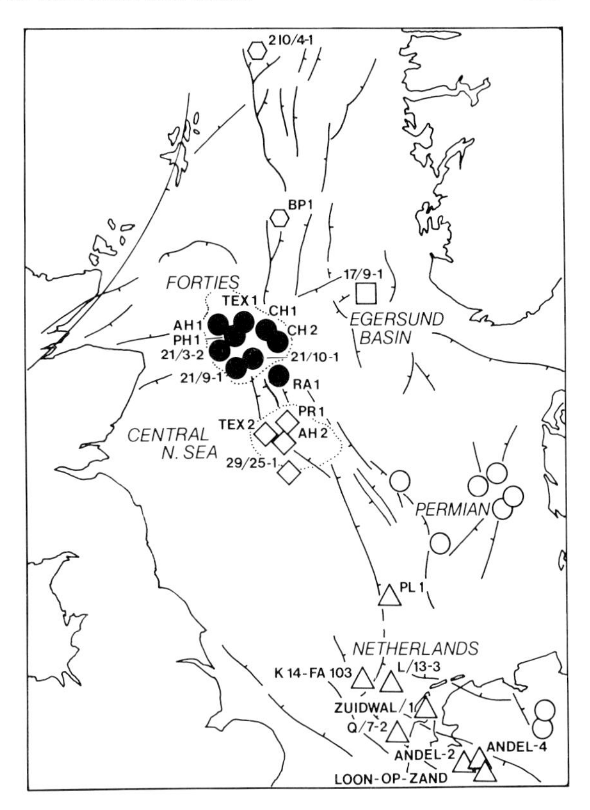

Fig. 6.3. Location of Mesozoic and Lower Permian igneous rocks in the North Sea rift system. Labels correspond to well locations, but some wells are unreleased and so are given code names (refer to text). ● Forties Province rocks; ◇ Central North Sea Mesozoic rocks; ⬡ Viking Graben occurrences; △ Netherlands on-shore and off-shore Mesozoic rocks; ○ Permian occurrences.

6.3. North Sea magmatism

In this section we concentrate on the nature and generation of the magmas and hence primarily on their chemistry. The petrography of the rocks reflects their chemistry before and after alteration, and so is only of secondary interest here; it is nevertheless a crucial component in the study. It establishes the scale and character of alteration effects; it effectively establishes the petrological and major-element character of the rocks by analogy with fresh analogues elsewhere in the world, despite their alteration; it can provide important evidence that selected enrichments are of primary magmatic origin and are not due to alteration, for example, from the presence of primary potassic micas and amphiboles in some K_2O rich rocks; and finally, it can occasionally hint at mantle source mineralogy, as in the resorbed xenocrysts of phlogopite and amphibole in samples from 29/25-1 and PL1 respectively.

Here we review the North Sea occurrences, area by area (Fig. 6.3). Unreleased wells are given code names in Fig. 6.3 and these are used throughout the paper. Radiometric dates are reviewed in Table 6.1.

For petrographic details the reader is referred to the literature on the Forties volcanic province (Howitt *et al*. 1975; Gibb and Kanaris-Sotiriou 1976; Woodhall and Knox 1979; Fall 1980; Fall *et al*. 1982), and to Dixon *et al*. (1981) for the remaining occurrences known at that time. We will not discuss the petrography of Lower Permian occurrences here as there is nothing new to add to the work of Dixon and his colleagues. The petrology of the rocks from newly released wells will be presented elsewhere, as will the results of our re-investigation of the material from the Egersund Basin and Zuidwal.

Table 6.1. Age data for Mesozoic igneous rocks

Well	Date	Technique	Source
210/4-1	152 ± 3	K-Ar	Dixon *et al.* 1981
21/3-1a	169 ± 4[I]	K-Ar	Howitt *et al.* 1975
21/10-1	147 ± 5[I]	K-Ar	Howitt *et al.* 1975
	164 ± 12[I]	K-Ar	Howitt *et al.* 1975
21/9-1	109 ± 2	K-Ar	Howitt *et al.* 1975
21/3b-3	158 ± 8	K-Ar	Ritchie *et al.* 1988
	148 ± 2	$^{40}Ar/^{39}Ar$	Ritchie *et al.* 1988
	154 ± 2[II]	$^{40}Ar/^{39}Ar$	Ritchie *et al.* 1988
15/21-1	155 ± 6[I]	K-Ar	Howitt *et al.* 1975
RA1	188 ± 10	K-Ar	unpublished (Ranger)
17/9-1	177–180	K-Ar	Furnes *et al.* 1982
29/25-1	138 ± 4	$^{40}Ar/^{39}Ar$	Dixon *et al.* 1981
PL1	99 ± 5	K-Ar	unpublished (NAM)
Zuidwal-1	145	K-Ar	Harrison *et al.* 1979
	144 ± 1	$^{40}Ar/^{39}Ar$	Dixon *et al.* 1981
	152 ± 3	$^{40}Ar/^{36}Ar–^{40}K/^{36}Ar$	Perrot and Van der Poel 1987
L/13-3	101 ± 1	$^{40}Ar/^{39}Ar$	Dixon *et al.* 1981
Q/7-2	95 ± 2 to 106 ± 2	$^{40}Ar/^{39}Ar$	Dixon *et al.* 1981
E/6-1	161 ± 4	K-Ar	unpublished (NAM)
Loon-op-Zand-1	132 ± 3	$^{40}Ar/^{39}Ar$	Dixon *et al.* 1981
Andel-4	133 ± 2	$^{40}Ar/^{39}Ar$	Dixon *et al.* 1981

[I] Revised using the decay constants of Steiger and Jäger (1977).
[II] Does not meet the % Ar-release criteria of Fleck *et al.* (1977).

6.3.1. *Whole-rock chemistry and alteration*

All major- and trace-element analyses presented in this paper were determined using X-ray fluorescence techniques in Edinburgh (as in Fitton and Dunlop 1985, except that a Rh-anode X-ray tube was used). In view of the often highly altered nature of rocks in the North Sea, all powders used for major element determinations were dried at 110°C and then ignited at 1100°C for one hour to remove volatile material. Percentage weight loss on ignition (LOI) was calculated, and the ignited material used for the preparation of fused glass discs. Major-element concentrations are therefore presented on an anhydrous basis. Trace element concentrations were determined using pressed-powder pellets, and are also expressed on an anhydrous basis.

Because of the marked effects of alteration on the more mobile elements (e.g. K, Na, Ca, Si, Mg, Sr, Rb, Ba, and Pb), conventional chemically-based classification can only be used on rare fresh rocks. The classification of the rocks in each area is therefore based on a combination of chemistry, petrography, and relict mineral compositions.

The effects of alteration on major element concentrations in a 15 m ankaramitic flow from well AH1 in the Forties province are shown in Fig. 6.4. The flow appears to be homogeneous petrographically, but shows large variations in chemistry towards the top where alteration is most marked. The degree of alteration at any point is measured by the wt% LOI (the volatile components lost on ignition of the sample at 1100°C), which shows a marked increase near the top of the flow where it reaches 20 per cent. As LOI increases, Na_2O, K_2O, as well as all the elements traditionally regarded as immobile, increase, while SiO_2, MgO, and CaO decrease in abundance. The increase in the alkalis might be explained by their introduction, whereas the increases observed in the immobile elements are

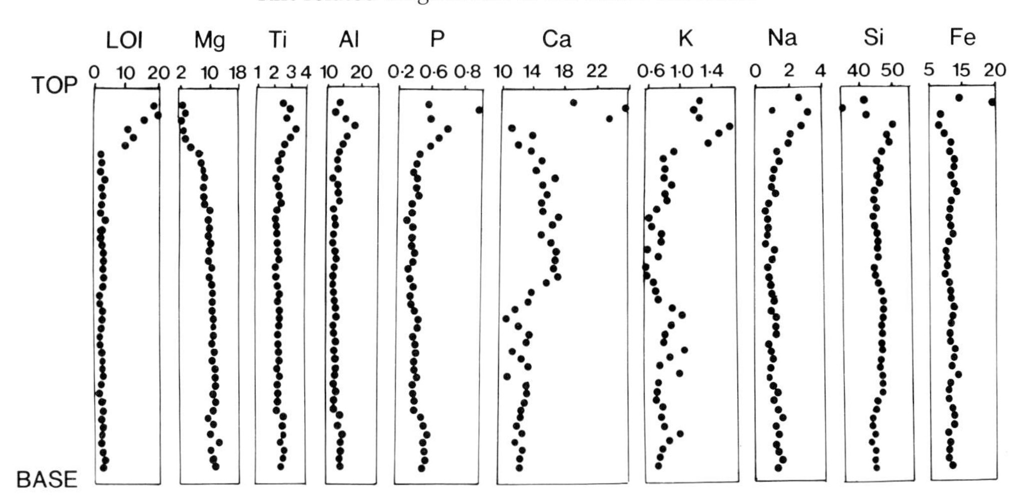

Fig. 6.4. Major element oxides (wt%) plotted as a function of depth through an ankaramitic alkali basalt flow (core from well AH1) with a severely altered top. Total thickness of the flow is approximately 15 metres. LOI = loss on ignition (*see* text for details).

thought to reflect their concentration in the residue—a 'lateritization' effect.

Table 6.2 summarizes the observed effects of alteration on two ankaramitic flows (from wells AH1 and 21/3-1A) and those noted by Fall (1980) for the Forties province as a whole, and Hart (1973) for low temperature alteration by sea-water.

Table 6.2. Alteration effects on major element oxides in alkali basalts from Forties

OXIDE	A	B	C	D
MgO	–	–	+	–
SiO_2	+	–	–	–
Al_2O_3	*	*	*	*
Fe_2O_3	–	*	*	*
CaO	–	–	–	–
Na_2O	+	+	+	+
K_2O	+	+	+	+
TiO_2	*	*	*	*
P_2O_5	*	*	*	*

\+ elements added to the rock during alteration.
– elements leached from the rock during alteration.
* immobile elements during the alteration process.

A: Samples from well AH1.
B: Samples from well 21/3-2 (Total).
C: Effects noted by Fall (1980) for Forties basalts.
D: Low temperature alteration by sea water (Hart 1973).

In a chemical study of altered rocks the immobile elements are clearly the only reliable sources of information. Although immobile elements may increase in concentration during alteration, ratios between such elements are unaffected. Figure 6.5 shows (as an example) that Zr/Nb and Ce/Y (in the core from AH1) are the same for samples with 1 per cent LOI as they are for samples with 20 per cent LOI. Differences in such ratios and ratios between ratios are, therefore, independent of the extent of alteration in this core; this result is taken to apply to the low-temperature alteration in the whole suite.

We show in section 6.4 (partial-melt modelling) that Zr/Nb and Ce/Y are ratios which also vary systematically with the extent of melting, and are insensitive to the extent of subsequent fractional crystallization. The following discussion of the North Sea rocks often relates differences to the values of such ratios.

6.3.2. *Mid-Jurassic to Mid-Cretaceous occurrences*

The Forties Province

Location and age The preponderance of volcanic activity in the area of the Moray Firth—Central Graben—Viking Graben triple junction is the most striking feature of the magmatism during this period. The volcanic province was first discovered in 1970 when exploratory wells were drilled in what is now

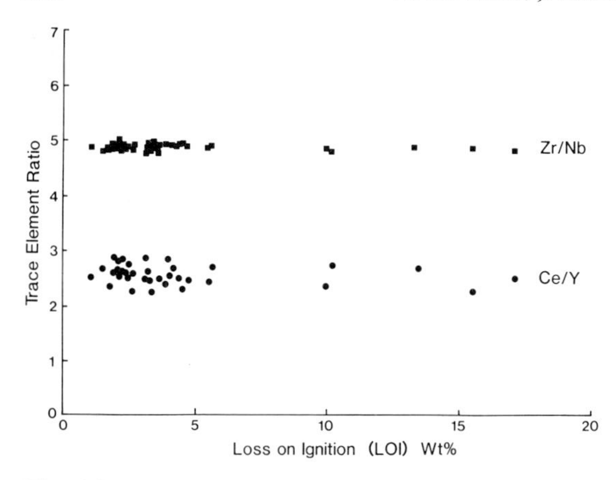

Fig. 6.5. Ce/Y and Zr/Nb plotted as a function of LOI (wt% loss on ignition, *see* text) for samples of ankaramitic alkali basalt from well AH1.

the Forties Field (Howitt *et al.* 1975). We now know of at least ten wells in which igneous rocks have been cored, and many more from which cuttings have been recovered. In terms of lithostratigraphy the province is termed the Rattray Volcanic Formation, whose lateral equivalent is the Pentland Formation (mainly sands). Both formations comprise the Fladen Group.

The igneous rocks are overlain by 2–3 km of younger Mesozoic and Cainozoic sediments, and overlie barren red beds of inferred Triassic age. In most of the wells we have studied the sediments directly above the igneous rocks are Oxfordian or younger, thus providing a minimum age constraint. A Middle Jurassic (Bathonian/Bajocian) age has been determined from microfossils (foraminifera, ostracodes, spores, microspores, and dinoflagellates) found in interbedded sediments. However, questions have been raised as to the validity of the microfossil evidence (Ritchie *et al.* 1988), as palynomorphs used for dating may have been reworked.

Severe alteration makes radiometric dates unreliable. The radiometric dates are reviewed in Table 6.1. A Middle Jurassic minimum age was given by Howitt and co-workers on the basis of conventional K/Ar dating on whole-rock samples of volcanic rocks. These dates, revised by Ritchie and co-workers using more recent IUGS constants (Steiger and Jäger 1977), are given as 169 ± 4 (Total, 21/3-1a), 155 ± 6 (Monsanto, 15/21-1), 147 ± 5 and 164 ± 12 (BP, 21/10-1). The oldest of these dates is Bajocian on most time scales, but the overall spread is from Bajocian to Oxfordian (on the scales of Hallam *et al.* 1985, and Van Hinte 1976) or Tithonian (on the scales of Harland *et al.* 1982, and in press). Ritchie and colleagues have challenged the minimum-age hypothesis of Howitt *et al.* (1975) on the grounds that the K/Ar age may be reduced *or increased*, depending upon the relative proportion of radiogenic Ar lost to the proportion of K lost or gained during the various alteration processes.

Ritchie and colleagues (1988) have produced what are probably the most reliable dates for the province. They carried out a detailed $^{40}Ar/^{39}Ar$ stepwise degassing study of rocks from well 21/3b-3 (Occidental). This is one of the few ways of producing reliable radiometric dates from altered rocks (another being U/Pb studies on zircons). The study produced two $^{40}Ar/^{39}Ar$ dates of 148 ± 2 and 154 ± 2. Of these dates, only the former meets the somewhat arbitrary 50 per cent Ar-loss constraint of Fleck *et al.* (1977), but both are probably reliable estimates. The Ritchie *et al.* dates suggest that the province could be as young as Tithonian, or as old as Callovian (depending on the time scale used). Because the province is overlain by sediments of Oxfordian age, we feel that a Callovian or older (Bathonian) age is justified. This may change if further $^{40}Ar/^{39}Ar$ studies are carried out.

Volumes The total amount of igneous material in the province is still uncertain. In the northern part of the province, wells have been drilled through the entire igneous sequence, where it is rather thin (often less than 200 m thick). Woodhall and Knox (1979) report a total of 580 m of volcanics penetrated in a well 'close to latitude 58°N', and an adjacent well penetrated 1100 m without reaching the base. According to Ritchie *et al.*, the thickest sequence encountered to date is 1492 m recorded from 21/3b-3 (Occidental). Howitt *et al.* (1975) suggested that total thicknesses might reach 1.5 km. Woodhall and Knox (1979) integrated the evidence from well stratigraphy, seismic, aeromagnetic, and gravity surveys, to produce an isopach map for the province, which indicates a maximum thickness of 3–4 km in some places. Recent geophysical modelling (J. L. Swallow in prep.) suggests that the thickness rarely exceeds 2.5 km.

Estimates of volume from the isopach map (Woodhall and Knox 1979) lead us to suggest that there is at most 9000 km^3 of igneous material in the province. When this is averaged over the area in which igneous rocks are found (in the order of 12 000 km^2 or more) it gives a maximum melt thickness (using the terminology of McKenzie and Bickle, 1988) of approximately 0.5 km. These estimates may, however, only give a minimum value because they cannot account for any underplated material. We are confident therefore, that at least 0.5 km of melt was produced in Forties.

Petrography and mineralogy The dominant rock type in the Forties province is an extrusive, mildly undersaturated, porphyritic alkali basalt. Typically these basalts are strikingly rich in large euhedral olivine and titanaugite phenocrysts and may, therefore be termed ankaramite. The olivines are almost invariably completely transformed to chlorites, clays, carbonate, or zeolite, but Fall *et al.* (1982) found one core (from 15/22-1) with fresh olivine, and we can add one more (from well AH1, close to the Halibut Horst; see Plate 6.1a). Pyroxenes in contrast are commonly well preserved and show concentric, oscillatory and hour-glass zoning, all features typical of ankaramitic basalts from ocean islands and rift provinces. Calcic plagioclase, clinopyroxene, olivine (pseudomorphed), opaque spinel, and apatite make up the groundmass. Brown amphibole (kaersutite) and, more rarely, biotite are locally present as groundmass phases.

More evolved less mafic rocks with fewer ferromagnesian phenocrysts and a more sodic plagioclase also occur. These are hawaiites and mugearites. As feldspar increases in abundance the rock can acquire a marked trachytic or flow-aligned texture of plagioclase laths, as in samples from wells CH1 and CH2 (Plate 6.1b).

Ocelli, sub-spherical patches up to 10 mm in diameter with a modal mineralogy and texture sharply distinct from the enclosing host rock, are locally common (e.g. the ankaramites from wells 21/10-1, 21/3-2 and 21/3-1a; Plate 6.1c). They are richer in feldspar, kaersutite, magnetite, and apatite than the host groundmass and may be vesicular. They are usually interpreted as evidence of the separation of an immiscible volatile-rich silicate liquid (Fall 1980) and are quite common in alkali basalts. In this instance they attest to some enrichment in both H_2O and K_2O in the parental Forties magma as their host rocks have not evidently undergone much low-pressure crystal fractionation. This supports our questioning, later in this paper, of the validity of the dry peridotite solidus as a control on Forties basalt generation.

Textures: evidence of eruptive mode The majority of Forties volcanic rocks appear to be extrusive. A few minor intrusives were recognized by Fall *et al.* (1982) in 21/3-1a. Individual flows recognized by Fall (1980) in 21/9-1 had reddened (oxidized) tops, probably indicative of sub-aerial eruption. Flow-front breccias are locally abundant and contain glassy, variolitic or other textural evidence of rapid cooling. Volcaniclastics have often been extensively reworked and are incorporated into sediments of deltaic facies. There is however no evidence of eruption into large bodies of water or into wet sediment. If water did enhance solidification it may have been as ephemeral pools, rivers, or shallow lakes in a generally subaerial setting.

Chemistry We have analysed 135 samples from eight wells (B.P. 21/9-1, B.P. 21/10-1, Total 21/3-2; AH1, PH1, TEX1, CH1, and CH2). Our conclusions are in accord with those of Fall *et al.* (1982) and with the petrographic results discussed above. Representative analyses are shown in Table 6.3. All the rocks are members of the alkaline series. The CIPW Norm is unreliable for all except the rare fresh rocks such as the mid-flow samples of AH1, which have moderate *ne* and abundant *ol*.

On the TiO_2 *vs* P_2O_5 diagram (Fig. 6.6), rocks

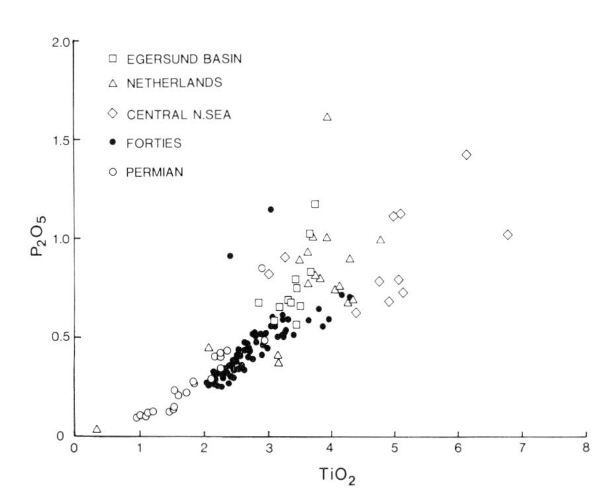

Fig. 6.6. TiO_2 *vs* P_2O_5 (wt%) in Mesozoic and Lower Permian basic igneous rocks (>4wt% MgO) from the North Sea Basin.

Table 6.3. Representative analyses of mafic igneous rocks from the Mesozoic North Sea

	A	B	C	D	E	F	G
SiO_2	47.81	50.72	45.00	42.68	43.62	40.13	43.44
Al_2O_3	11.90	17.63	15.07	12.44	117.14	14.45	18.68
Fe_2O_3T*	12.51	7.97	11.97	14.13	112.51	14.64	14.06
MgO	9.63	2.57	8.99	10.07	7.49	10.74	7.74
CaO	12.43	11.30	11.10	12.14	5.36	8.78	9.27
Na_2O	2.02	2.94	2.19	2.49	1.81	0.78	1.83
K_2O	0.90	1.65	1.03	0.50	5.24	4.18	0.81
TiO_2	2.19	3.32	3.26	3.66	4.90	4.30	3.91
MnO	0.11	0.11	0.40	0.21	N.D	0.24	0.12
P_2O_5	0.32	0.50	0.52	0.83	0.68	0.69	1.00
TOTAL	99.81	98.70	99.55	99.16	98.75	98.93	100.88
LOI	2.60	10.20	2.90	6.30	9.4	3.20	9.80
Ni	236	189	65	63	98	46	151
Cr	815	336	132	245	69	32	312
V	396	571	360	451	523	501	439
Sc	50	57	32	42	49	36	38
Cu	57	56	72	110	32	69	58
Zn	107	36	108	96	166	94	79
Sr	481	705	905	519	1043	883	823
Rb	32	64	15	30	97	63	24
Zr	166	255	273	354	360	318	311
Nb	37	61	78	154	106	94	109
Ba	502	1022	593	293	1040	760	905
Pb	N.D	N.D	1	5	14	2	4
Th	1	5	2	10	8	6	8
La	29	41	56	95	84	67	92
Ce	64	81	130	208	171	160	157
Nd	27	34	56	70	64	67	57
Y	21	22	30	23	22	27	24

* Total iron calculated as Fe_2O_3.
Note: Major element oxides are given as wt%; trace elements are in ppm.
A: Ankaramitic alkali basalt from well AH1, Forties Province.
B: Altered ankaramitic alkali basalt from well AH1, Forties Province.
C: Alkali Basalt from well 21/9-1 (BP), Forties Province.
D: Nephelinite from well 17/9-1 (Esso Norway), Egersund Basin.
E: Monchiquite from well 29/25-1 (Shell/Esso), Central Graben.
F: Lamprophyric basanite from well PL1, Netherlands.
G: Olivine nephelinite/basanite from well Andel-2 (NAM), Netherlands.

with >4wt% MgO plot as a distinct group, intermediate in concentration between the other Mesozoic samples and the Lower Permian basalts from the Mid North Sea High and the Netherlands (see Fig. 6.3). The spread is probably a combination of variable melting at source, low-pressure fractionation (and crystal accumulation) of olivine and clinopyroxene, and alteration effects.

On the incompatible element ratio plot (Fig. 6.7) with the low-pressure fractionation and alteration effects removed, the Forties basalts are more tightly grouped. The implications of their position on this diagram are discussed later in the paper, but the general conclusion from petrography and chemistry is that the Forties province has a very restricted range of parental alkali basalts representing, presumably, a correspondingly restricted range of parameters controlling magma generation.

Central Graben occurrences

Location, age and volumes The rocks described in this group are restricted to quadrants 29 and 30 in

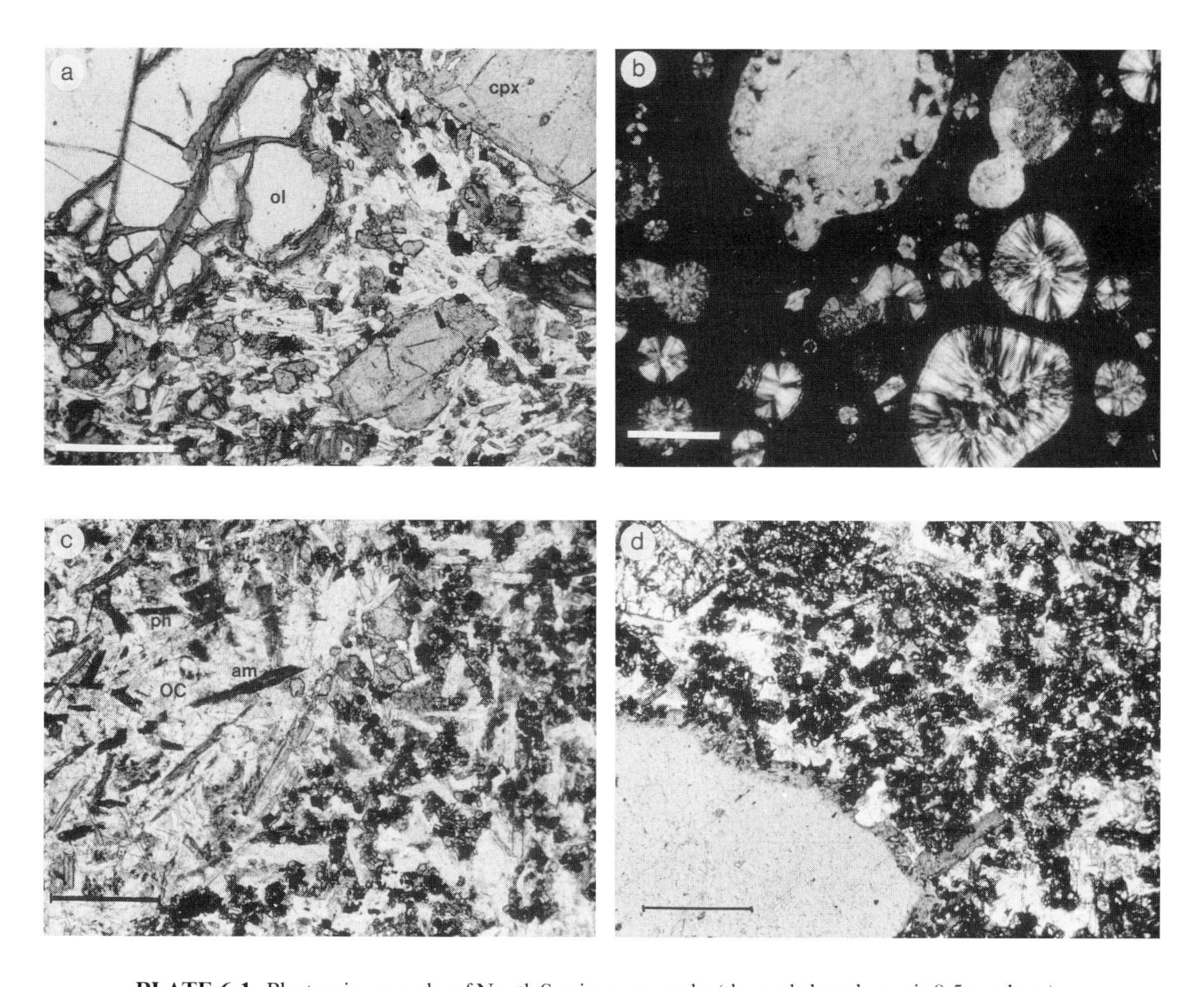

PLATE 6.1 Photomicrographs of North Sea igneous rocks (the scale bar shown is 0.5 mm long).

a Ankaramitic basalt. AH1. Forties province (ppl). Phenocrysts of partially altered *olivine* (ol), *titanaugite* (cpx) in a groundmass of *clinopyroxene*, *plagioclase*, and opaque *spinel*, with minor *olivine* microphenocrysts.

b Amygdaloidal, glassy hawaiite. CH1. Forties province (xpl). Amygdales are filled with radiating *zeolite* prisms or vuggy *carbonate* (top). The groundmass is turbid brown *glass* with *feldspar* microlites and altered *clinopyroxene* microphenocrysts.

c Ocellus in ankaramite. 21/9-1 (BP). Forties province (ppl). Part of an ellipsoidal 2 cm long *ocellus* is shown in the left half of the plate (oc). It is composed of blades of *kaersutitic amphibole* (am) and *phlogopite* plates (ph) set in *sodic plagioclase* with abundant *apatite* needles. The adjacent groundmass to the right has *titanaugite* microphenocrysts, and groundmass *clinopyroxene* and brown *hornblende*, *plagioclase*, and opaque *spinel*.

d Monchiquite. 29/25-1 (Shell). Central North Sea (ppl.). A large, partially resorbed *phlogopite* ?xenocryst with fringing overgrowths of darker brown *biotite* set in a fine-grained groundmass of *carbonate*+*ore* pseudomorphs after *titanaugite*, minor brown *hornblende*, and *alkali feldspar*.

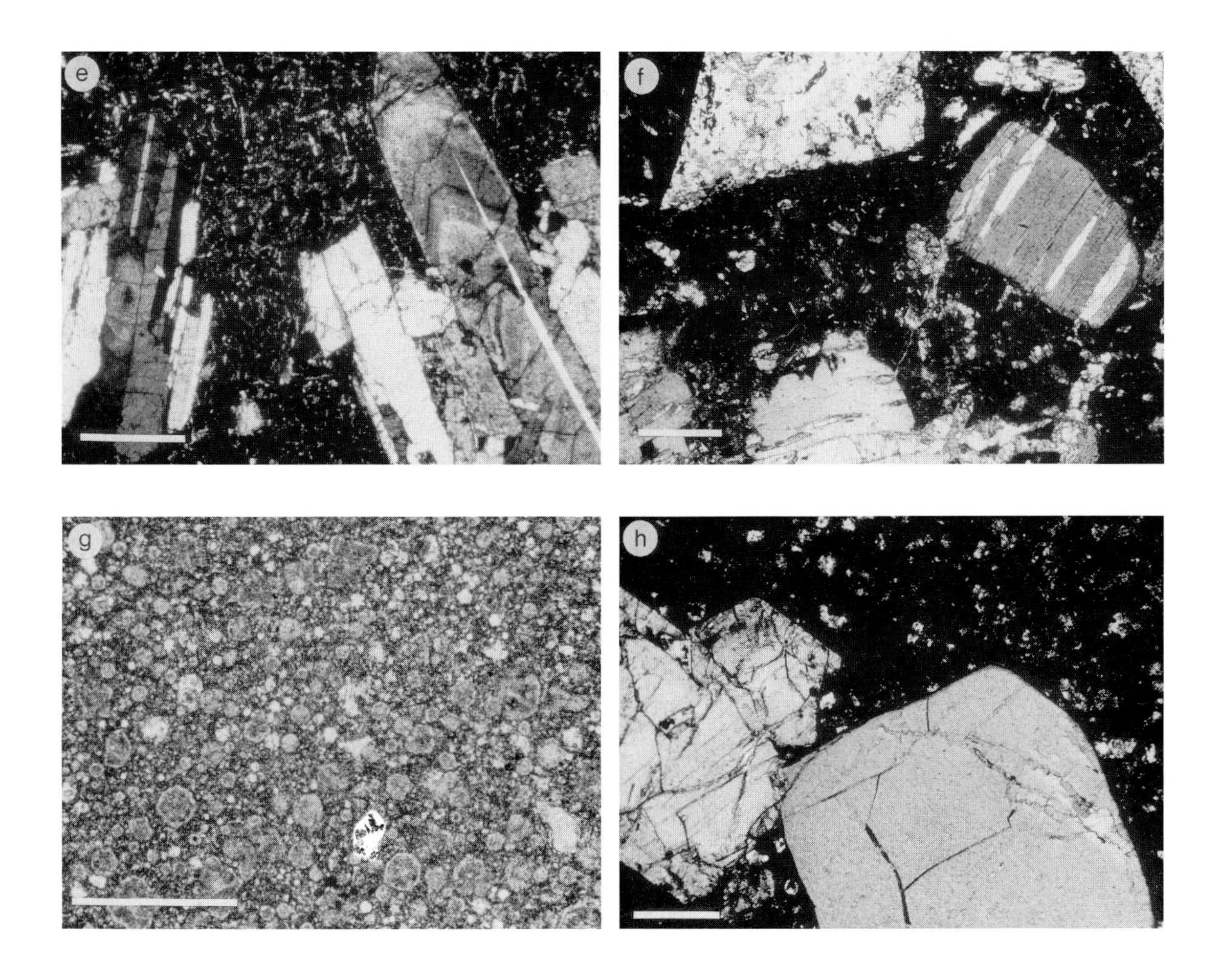

e Nephelinite. 17/9-1 (Esso Norway). Egersund Basin (xpl). Prominent twinned and zoned *titaniferous clinopyroxene* phenocrysts are set in a near isotropic groundmass of *feldspathoid*, *zeolite* after *glass*, opaque *spinel* and microlites of birefringent *clinopyroxene*.

f ?Alnöite. 17/9-1 (Esso Norway). Egersund Basin (ppl). Rounded phenocrysts/xenocrysts of *phlogopite*, partially carbonated along cleavage planes, with large, euhedral *carbonate* pseudomorphs after *clinopyroxene*, and pseudomorphed *olivine* microphenocrysts. The groundmass is turbid carbonate, probably after feldspathoid.

g Leucitite. Zuidwal-1 (Petroland). North Netherlands (ppl). The rock is entirely composed of euhedral *carbonate* pseudomorphs of *leucite* icositetrahedra interspersed with small hexagonal crystals of *hydronepheline* after *nepheline*.

h Lamprophyric basanite. PL1. Netherlands Sector, southern North Sea (ppl). Euhedral phenocrysts of *titaniferous clinopyroxene* (upper left) adjacent to a rounded (partially resorbed) phenocryst/xenocryst of *kaersutitic hornblende*. The groundmass is dominantly *clinopyroxene*, brown *hornblende* and opaque *spinel* with very minor laths of *plagioclase* and some interstitial *alkali feldspar*.

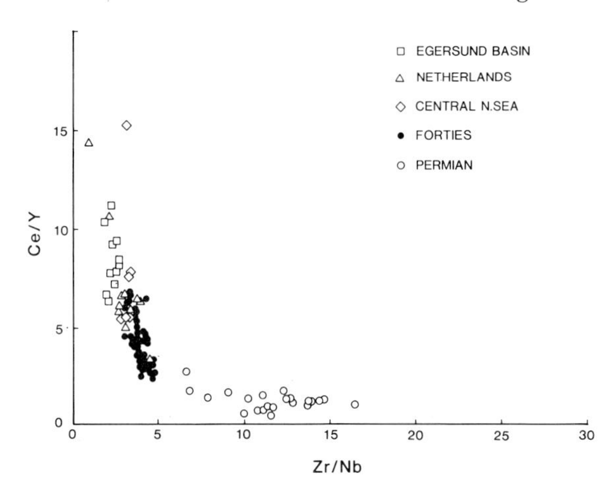

Fig. 6.7. Ce/Y *vs* Zr/Nb in Mesozoic and Lower Permian basic igneous rocks (>4wt% MgO) from the North Sea Basin.

the Central Graben area. This part of the North Sea has yet to be fully investigated for igneous rocks, but appears to be quite a major area of magmatic activity, perhaps second in scale to the Forties province.

Dixon *et al*. (1981) described two sequences of intrusive rocks from the Zechstein sequence in well 29/25-1 (Shell) on the edge of the Mid North Sea High (Fig. 6.3), flanking the Central Graben to the west, one in the Upper Permian, the other a few hundred metres lower in the Lower Permian, both very similar texturally. $^{40}Ar/^{39}Ar$ age spectra showed a plateau oscillating about 135 Ma, with a best estimate age of 138 ± 4 Ma and outside limits of 128 and 157 Ma (the high age probably due to a gain in ^{40}Ar from the surrounding salts). Both sequences are interpreted as parts of a single intrusion or as two coeval sills or dykes of latest Jurassic (Tithonian) or earliest Cretaceous (Berriasian/Valanginian boundary) age depending on the time scale used. A similar intrusion (not cored) is seen in the Zechstein in well 29/19-2 (Shell).

We now know of a number of wells in the area which have penetrated igneous rocks reported to be within the Mid-Jurassic part of the sequence, where these have been attributed to the Rattray Volcanics Formation. These wells lie on the edge of or within the graben itself, and the rocks may be somewhat older than those in 29/25-1. We have studied cores (wells AH2, PR1, TEX2) of volcaniclastic and epiclastic material, in each case overlain by Upper Jurassic sediments. A lava flow was cored in a production well in the Auk field (Shell, 30/16-A11; not on the map in Fig. 6.3) where the extrusive rocks are in the Upper Jurassic/Early Cretaceous part of the sequence. Upper Jurassic/Early Cretaceous ages are supported by K/Ar dates of 127.4 ± 2.6 Ma on these rocks (A. Heward, pers. comm. 1989). The area of igneous rocks outlined on Fig. 6.3 was compiled from unreleased wells. Volume estimates cannot yet be made.

Well AH2 penetrates volcanic and tuffaceous rocks in the Jurassic part of the sequence (none of which is cored). Core 11 from this well contains a conglomerate of well-rounded cobbles of lava (up to 5 cm in diameter) of moderate sphericity, supported by a matrix of bioclastic sandstone. The unit is interpreted as representing shelf edge conditions, and probable debris-flow deposition.

Petrography and mineralogy Both cored sequences encountered 'basaltic' rocks which are more undersaturated and enriched in potassium and incompatible trace-elements than the Forties alkali basalts. They have primary biotite and amphibole, as well as large rounded phlogopite xenocrysts (Plate 6.1d). In 29/25-1 (described in detail by Dixon *et al*. 1981), pseudomorphs after a feldspathoid phase were tentatively identified, which would in turn identify the rock as a tephrite or mafic phonolite. Intrusive rocks with the same mineralogy (olivine, clinopyroxene, alkali feldspar, plagioclase, leucite, or sodalite, biotite ± amphibole) are termed monchiquites. Well AH2 penetrated a volcanic debris-flow or conglomerate containing large rounded clasts of similar highly altered undersaturated potassic rocks to those in 29/25-1, implying prior eruption to the surface in this case.

Chemistry Alteration masks primary features but clearly, from the mineralogy, the high K_2O content is original. The 7 per cent *ne* and 26 per cent *lc* in the norm of rocks from 29/25-1 are probably close to real values, and support the petrographic identification. Incompatible elements are enriched relative to Forties in absolute terms (Fig. 6.6) and match those in the fresh nephelinites from the Egersund Basin (well 17/9-1) described below, with the exception of K_2O which is nearly ten times higher (Table 6.3). The values of Ce/Y and Zr/Nb (Fig. 6.7) are generally intermediate between those of the nephelinites and the Forties basalts.

The Egersund Basin

Location, age, and volumes Two sequences of igneous rocks were cored in well 17/9-1 (Esso) in the Egersund Basin in the Norwegian sector of the North Sea. To our knowledge no other igneous rocks have been cored in this area. The upper sequence in well 17/9-1 rocks comprises approximately 400 m of volcanic rocks and interbedded sediments which are overlain by claystones of Oxfordian age, and which overlie Lower Jurassic shales. These rocks have been dated (Table 6.1) using conventional K/Ar techniques (Furnes *et al.* 1982) and give Bajocian ages (179, 180, 177 Ma). The lower sequence is intrusive into Lower Jurassic shales and ?Triassic sands. The intrusive rocks also give Bajocian (177, 178 Ma) K/Ar dates (Furnes *et al.* 1982). Both igneous sequences are pervasively altered, and this makes us question the reliability of dates determined by standard K/Ar techniques. However, the sequences are well constrained stratigraphically, interbedded sediments are described as Dogger (Bathonian-Bajocian) in age (Dixon *et al.* 1981). Mid-Jurassic dates are probably reasonably close to the true ages. We have no information from seismic reflection data or other wells to constrain volumes.

Petrography, mineralogy, and eruptive character The lava samples studied are relatively fresh porphyritic nephelinites (Plate 6.1e) with either a fine grained holocrystalline, or a turbid brown glassy sometimes brecciated, groundmass. The samples are highly vesicular and many contain chlorite-filled amygdales. At rapidly chilled flow tops, streamers of vesicles are observed. The textures suggest rapid cooling of flows some of which may have been erupted into water. All contain prominent rectangular phenocrysts of a brown clinopyroxene up to 6 mm in length, which shows oscillatory zoning, multiple twinning, and anomalous birefringence (Plate 6.1e). Microprobe analysis (Dixon *et al.* 1981) shows them to be titaniferous salites typical of nephelinites. Rare larger euhedral serpentine pseudomorphs after olivine occur in a number of sections. Altered nepheline, clinopyroxene, and ore make up the groundmass. A few samples contain 6- to 8-sided phenocrysts and groundmass crystals after another feldspathoid phase, now analcime, but probably a sodalite group mineral originally. Contrary to Furnes and colleagues (1982) we have found no samples containing any feldspar in 20 thin sections, and so are confident that these rocks are extrusive nephelinites.

Samples from the intrusive sequence show abundant large pseudomorphed euhedral phenocrysts of clinopyroxene and olivine, both now replaced by similar aggregates of carbonate and green chlorite. Large, slightly rounded phlogopite phenocrysts are prominent (Plate 6.1f) and are only partially altered along cleavage traces. A single distinctive chlorite pseudomorph rimmed by phlogopite and magnetite is thought to be a partially resorbed orthopyroxene xenocryst subsequently altered hydrothermally. The rock's texture is striking, arising from abundant euhedral microphenocrysts of phlogopite and altered clinopyroxene and highly irregular fractured larger grains, suggesting transport in a rapid moving volatile-rich matrix which was then partly extracted, concentrating the phenocrysts. Such textures are common in kimberlites and related mafic alkaline dyke rocks, the alnöites.

Chemistry The few fresh samples are strongly *ne*-normative. Absolute levels of incompatible trace-elements (Fig. 6.6 and Table 6.3), Ce/Y and Zr/Nb ratios (Fig. 6.7), are all comparable to those of nephelinites from the literature. Ce/Y ratios are twice those of the Forties alkali basalts, and it will be argued later that these rocks cannot be derived directly by small amounts of melting of a homogeneous asthenosphere source (i.e. one that would yield MORB if melted to 20 per cent). In many ways they encapsulate the problem of rift-magmatism. They require an enriched source or sources, they are identical in chemistry to oceanic nephelinites, and here they occur in a region which has undergone only slight extension.

Netherlands, on-shore and off-shore: Zuidwal

Location and age Zuidwal-1 is a well located close to the island of Texel in the Waddensee (Fig. 6.3). As Dixon and co-workers (1981) noted, there is excellent agreement between the $^{40}Ar/^{39}Ar$ age of 144 ± 1, the K-Ar age quoted in Harrison *et al.* 1979 (145 Ma), and the stratigraphic evidence of overlying Valanginian sediments. More recently Perrot and Van der Poel (1987) quote a slightly older age of 152 ± 3 from $^{40}Ar/^{36}Ar—^{40}K/^{36}Ar$ dating (Table 6.1).

Volumes The Zuidwal volcano was evidently a major edifice (Cottençon *et al.* 1975; Perrot and Van der Poel 1987) perhaps comparable to the off-axis volcanoes of East Africa such as Mount Elgon. However, no reliable estimate of the volume of basaltic parental magma can be made from the available data. All the volcanic material occurs as clasts in agglomerate as Zuidwal-1 evidently scored a direct hit on the volcano's central conduit.

Petrography, mineralogy, and chemistry The Zuidwal core encountered a great range of different rock types, from feldspathoidal basalts through to evolved types rich in feldspathoid alone (e.g. leucitite; Plate 6.1g), feldspathoid and feldspar together (phonolite) or feldspar alone (trachyte), as reported by Jeans *et al.* (1977). Some of the basic rocks have abundant modal nepheline (now pseudomorphed), others have pseudomorphed leucite microphenocrysts. The petrographic nomenclature gets rather out of hand in such rocks, but the terms basanite, tephrite, and leucite-basanite or leucite-tephrite would commonly be used. The key feature of Zuidwal is the implied, but not necessarily large, variability in the ratios of Na_2O to K_2O and of total alkalis to alumina and silica in the parental basaltic undersaturated magmas. Well documented low-pressure fractionation operating on these parents will then exploit what may be small bulk-chemical differences, to generate a wide range of evolved products. Such implied variation in magma character from source is difficult to reconcile with melting of a uniform, well-mixed asthenosphere.

Dixon and colleagues (1981) showed that the relict pyroxenes in a Zuidwal mafic phonolite were typical of highly undersaturated feldspathoidal magmas. In the absence of any fresh rocks, major element chemistry is useless. The Zuidwal samples on the immobile element ratio plot (Fig. 6.5) have been filtered to remove evolved samples with less than 4wt% MgO. Like the Netherlands suite in general, Zuidwal occupies a position comparable to the central North Sea rocks, intermediate in implied *ne*-normative character between the Forties alkali basalts which have no modal feldspathoids, and the Egersund Basin rocks with no modal feldspar.

In addition to the volcanics, a recognizable but highly altered peridotite nodule has also been sampled by the Zuidwal-1 core. This contains relics of clinopyroxene but has not yet been analysed.

Netherlands, on-shore and off-shore: other occurrences

Location, age, and volumes In addition to Zuidwal, igneous rocks have been found in three on-shore wells (Andel-2 and -4; Loon-op-Zand 1) and four off-shore sites (PL1, K/14-FA103, L/13-3, and Q/7-2; see Fig. 6.3). These rocks are all rather younger than the other Mesozoic igneous occurrences (Table 6.2 and Fig. 6.15). Dixon *et al.* (1981) give $^{40}Ar/^{39}Ar$ dates of 133 ± 2 and 132 ± 3 for the Andel-4 and Loon-op-Zand 1 respectively. The off-shore rocks all give ages close to 100 Ma by whatever method (Table 6.1). All of the rocks appear to be intrusive in character, and none intrudes rocks younger than Middle Jurassic age. As yet we have no information relating to volumes.

Petrography, mineralogy, and chemistry All of the rocks are strongly undersaturated, but mafic in character. They are more undersaturated than the Forties alkali basalts, and show similarities with both the Central Graben (PL1, K/14-FA103) occurrences and the Egersund Basin nephelinites (Loon-op-Zand 1, Andel-2 and -4, L/13-3, Q/7-2).

Primary magmatic potassic phases are common in rocks from PL1 and K/14-FA103. Lamprophyres (more specifically, lamprophyric basanites) from PL1 show abundant amphibole and biotite (both as phenocrysts and in the groundmass), and this is reflected by high K_2O contents (Table 6.3, Plate 6.1h). Enrichment in TiO_2 appears very characteristic of both PL1 and 29/25-1 (Central Graben) rocks (Fig. 6.6) and serves to emphasize their close similarity. K/14-FA103 contains moderately evolved rocks belonging to the basanite to trachyte/phonolite suite, which contain ocelli rich in acicular amphibole pseudomorphs. The petrography of both PL1 and K/14-FA103 rocks attests to a rather volatile and potassium-rich primary magma. The rocks from PL1 also contain resorbed xenocrysts of amphibole which provide further evidence for a hydrous primary magma.

The other rocks from this Netherlands group are closer in character, though less markedly undersaturated, to the nephelinites from 17/9-1. They contain pseudomorphed olivine and clinopyroxene phenocrysts set in a glassy base rich in the same phases, but also containing kaersutitic hornblende, biotite, and apatite. Such strongly undersaturated, feldspar-free, glassy basaltic rocks are termed limburgites. Analysis of rare unaltered pyroxene cores from Loon-op-Zand samples showed them to be chemically indistinguishable from the titaniferous salites seen in the Egersund Basin rocks (Dixon *et al.* 1981). Some Andel-2 samples contain possible pseudomorphs after nepheline, and minor amounts of feldspar are present in rocks from L/13-3.

On both the TiO_2 *vs* P_2O_5 and Ce/Y *vs* Zr/Nb diagrams (Figs 6.6 and 6.7) the Netherlands field overlaps that of the Central Graben rocks, with which they have been grouped in later discussions (Fig. 6.11).

Other occurrences

In the Viking Graben there are two locations where igneous rocks have been reported. In quadrant 9, close to the Beryl field, well BP1 shows some small amounts of volcanic debris in sediments from the Middle Jurassic part of the sequence. Other wells in this area may also contain volcanic material, but as far as we know no volcanic flows, intrusive rocks, or tuffs have been cored in this area. Dixon and colleagues (1981) reported igneous rocks intrusive in Permian sediments (probably hawaiites or mugearites) from 210/4-1 in the northern part of the Viking Graben. The rocks from this well were dated as Kimmeridgian (152 ± 3 Ma) by conventional K/Ar methods. No material suitable for analysis has yet been obtained from these areas.

Dixon and colleagues also drew attention to a number of other locations, not in the North Sea but in close proximity, where igneous activity occurred during the Mesozoic. The Sunnhørdland area of West Norway has a suite of northerly-trending alkali basaltic dykes intruded over a long period from Permian to Middle Jurassic (Faerseth *et al.* 1976); these have chemical characteristics similar to the Forties basalts. In the Skåne district of southern Sweden occur alkali olivine basalt flows and plugs which range in age from Middle Jurassic to Early Cretaceous; they may be similar to the Forties rocks. Wolf Rock (south west of Cornwall) is a nosean phonolite of similar age and character to rocks from the Netherlands group in the North Sea. The Rhine Graben shows olivine nephelinite magmatism which dates back to the Early Cretaceous, and might have some connection with the magmatism in the North Sea.

6.4. Partial-melt modelling

In the first section of this paper we concluded that the compositions and volumes of basic magmas generated in rift environments, such as the Mesozoic North Sea, should be closely related to the degree of partial melting of the mantle, and hence to the amounts of stretching and upwelling. In this section we look at ways in which the chemistry of the igneous rocks in the North Sea can be used to estimate the degree of partial melting that gave rise to them.

In order to estimate degrees of partial melting it is first necessary to constrain as many of the variables which influence magma composition as possible: the nature of the source (i.e. mantle composition), and the history of fractional crystallization and crustal contamination. We have shown in the previous section that the effects of alteration can be eliminated by the use of ratios such as Zr/Nb and Ce/Y.

6.4.1. *Fractional crystallization*

Fractional crystallization is usually the most important process which operates to change the composition of magma, once it has segregated from its mantle source. Its effects can be described with reference to TiO_2 and P_2O_5, which are immobile, but which will be subject to enrichment in highly altered rocks.

Small degrees of melting give rise to primary magmas with higher TiO_2 and P_2O_5 concentrations than do large degrees of melting, since both Ti and P are incompatible in mantle phases (they prefer to reside in any melt formed; see the next section on melting). The dispersion of the different North Sea groups in terms of TiO_2 and P_2O_5 (Fig. 6.6) might, at first sight, be explained in terms of differing degrees of partial melting. However both elements are involved in the subsequent crystallization of these melts. Figure 6.8 shows the extent of Ti and P variation attributable firstly to varying degrees of melting of a simple MORB-type source and, in addition, to the effects of progressive crystallization and extraction of the expected mineral phases from the basaltic parent. Varying the source composition would introduce still more dispersion. Removal of olivine, with no Ti or P, concentrates both elements in the residual liquid in much the same way that alteration concentrates immobile elements. Clinopyroxene contains significant Ti but no P. Magnetite and apatite contain abundant Ti and P respectively. The net effect is to generate a wide range of TiO_2,

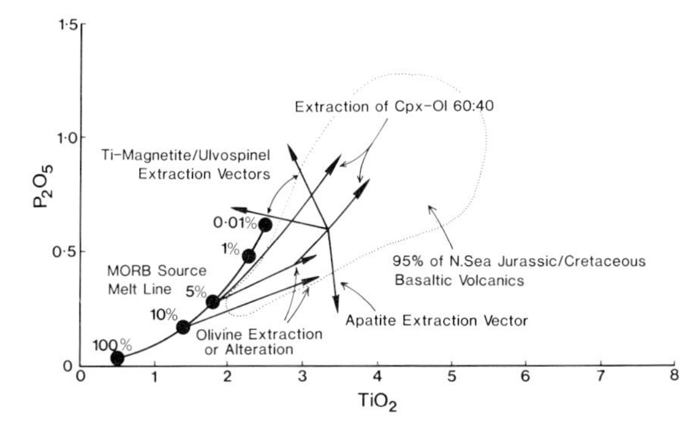

Fig. 6.8. The dotted line shows the spread of Mesozoic basic igneous rocks on the TiO_2 *vs* P_2O_5 diagram of Fig. 6.6. A simple equilibrium partial melting path (from a MORB source) is shown by the curved line, and the per cent melting at different positions on the line are given. Vectors for the effects of alteration and fractional crystallization of various phases on TiO_2 and P_2O_5 concentrations are also drawn.

P_2O_5, and TiO_2/P_2O_5 values in rocks still broadly basaltic (>4wt% MgO), thus completely obscuring the values in the parent magmas generated on partial melting of the source.

What is needed are elements which will not enter the crystallizing phases (strictly that they are equally incompatible in these phases), and that are immobile during alteration. The ratios between such elements will not change and will therefore reflect the mantle source composition and may be related to the degree of partial melting. We will show below that the ratios of Zr/Nb and Ce/Y meet these criteria. They are unaffected by either alteration or fractional crystallization for rocks with MgO >4 per cent. We now need to consider how the compositions of primary magmas reflect both their source composition and the extent of melting of that source.

6.4.2. *Partial melting*

Melting theory

The degree and depth of partial melting of the mantle source and its composition are the main controls on the chemical composition of the primary melt formed. Different elements partition differently between the melt phase and the solid phase. Incompatible elements (many trace elements, and some minor elements such as Ti, P, K) prefer to reside in the melt phase and thus, when there is only a small amount of melt present (after a small degree of melting), the melt will be strongly enriched in these elements relative to their concentration in the mantle source. The concentration of any element in the melt compared to its concentration in the solid residue depends upon a coefficient (D, the bulk distribution coefficient) which describes partitioning of the element between the solid and the liquid. D = concentration of the element in the solid/concentration of the element in the liquid. For incompatible elements $D < 1$, whereas for compatible elements (most major elements and trace elements, such as Ni and Cr) $D > 1$. The compatibility of any element will depend on the proportions of individual minerals comprising the solid phase. Thus Ni and Cr are compatible in olivine and clinopyroxene respectively, but not in plagioclase for which they may be considered incompatible. In the cases investigated in this paper, the solid is a multi-phase assemblage of primary olivine, orthopyroxene, clinopyroxene, and an aluminous phase (garnet or spinel). When the term incompatible is used here, therefore, it is being applied only to these phases.

The value of D thus controls the distribution of the elements between the solid and coexisting melt. If two elements such as Ce and Y have different values of D, then the ratio between the two in the solid before melting will be different from the Ce/Y value of any small-degree melt formed from that solid. D is smaller for Ce than it is for Y in mantle phases, and therefore Ce/Y will be higher in the melt than in the starting solid because Ce enters the melt more readily than does Y. The difference between this ratio and that of the solid will be greatest when only a very small amount of melting has occurred. As melting progresses, the ratio in the melt will approach that in the original solid, providing equilibrium is maintained.

It follows that, if we know the D values for various elements in the mantle, the initial concentrations of these elements in the mantle, and how melting takes place (fractional or equilibrium), then we might use the ratios of such elements in a primary magma to predict the degree of melting that it represents. To reiterate the conclusions of the preceding sections, if ratios of these elements are also unaffected by alteration and fractional crystallization, we may also be able to do this for variably evolved and altered rocks.

We have already shown (Fig. 6.5) that ratios of immobile trace elements such as Zr/Nb and Ce/Y are unaffected by low temperature alteration of the sort observed in the basalts from well AH1. Figure 6.9 depicts Zr/Nb and Ce/Y plotted against wt% MgO for 116 analyses (J. G. Fitton, unpublished data) of unaltered, variably evolved alkaline rocks (similar in character to those in the North Sea) from the ocean island of São Tomé, part of the Cameroon line of West Africa (Fitton 1987). Note that for rocks which have suffered little or only moderate degrees of fractional crystallization (MgO > 4 per cent) the ratios are remarkably constant, a sign that the elements concerned are equally incompatible in the fractionating phases. In the more evolved rocks (MgO < 4 per cent) the ratios start to increase and become highly variable. Clearly the elements concerned are not equally incompatible during this late stage in the crystallization history.

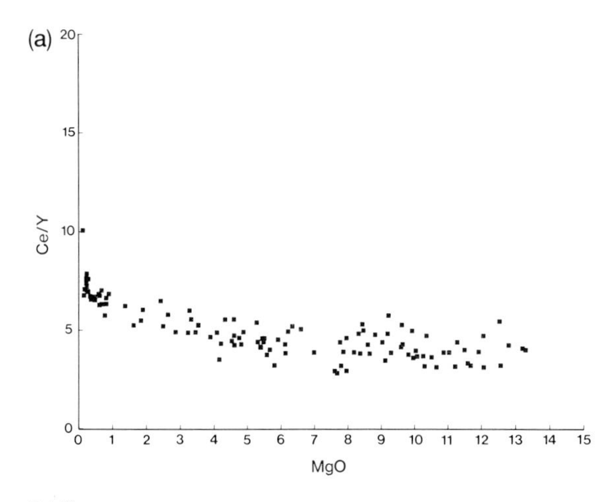

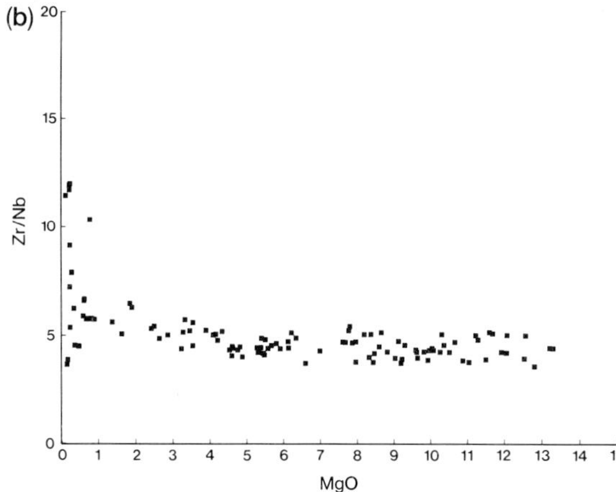

Fig. 6.9. Variations in trace element ratios in a suite of alkaline lavas from the ocean island of São Tomé (Cameroon line, West Africa) produced as a result of fractional crystallization. **a.** Variation of Ce/Y with wt% MgO; **b.** Zr/Nb against wt% MgO. Analyses were produced by XRF analysis in Edinburgh, and are from the unpublished data set of J. G. Fitton.

From this study of fresh material we conclude that the ratios Zr/Nb and Ce/Y can be used to 'see through' both the alteration and fractional crystallization of igneous rocks, provided that only data from moderately or unevolved (MgO > 4 per cent) samples is used. Values of Ce/Y and Zr/Nb from such samples should thus only depend on the composition of the mantle source and the degree of partial melting which produced the primary magma. The concentrations of the individual elements will be clearly subject to strong enrichment by fractional crystallization and by alteration, or both.

Melt compositions

Smaller-degree melts are likely to be more undersaturated and form at greater depths than larger degree melts. From the discussion above it would therefore seem likely that differences in the ratios of incompatible trace elements should correlate directly with differences in the degree of undersaturation. Higher values of Ce/Y ($D_{Ce} < D_Y$) and lower values of Zr/Nb ($D_{Zr} > D_{Nb}$) would be expected in smaller degree melts, which should also be more undersaturated. We might also expect Ce/Y to be especially high when garnet is a residual phase (at depths >60 km or so; pressures >2.0 GPa), because Y tends to reside in the garnet.

In order to test these ideas we devised an index of silica saturation (S.I. $= 2Si - (11(Na + K) + 3Ca + Al + Fe^{II} + Mg)$ against which we have plotted the Ce/Y and Zr/Nb values of over 800 analyses of fresh, mafic (> 4wt% MgO) ocean island basalts (J. G. Fitton and D. James, unpublished data). Increasingly negative values of S.I. correspond to increasing degrees of undersaturation away from the critical plane of silica-undersaturation in the basalt tetrahedron, where S.I. = 0; in contrast, positive values reflect an excess of silica, and increase as the rock becomes progressively more silica-saturated. The results of this analysis are encouraging. The ocean island basalt data set shows a marked negative correlation of Ce/Y with S.I. and a positive correlation of Zr/Nb with S.I. (Fig. 6.10). The spread of the data is most easily explained in terms of variable amounts of crystal fractionation, largely of olivine, affecting the value of S.I.

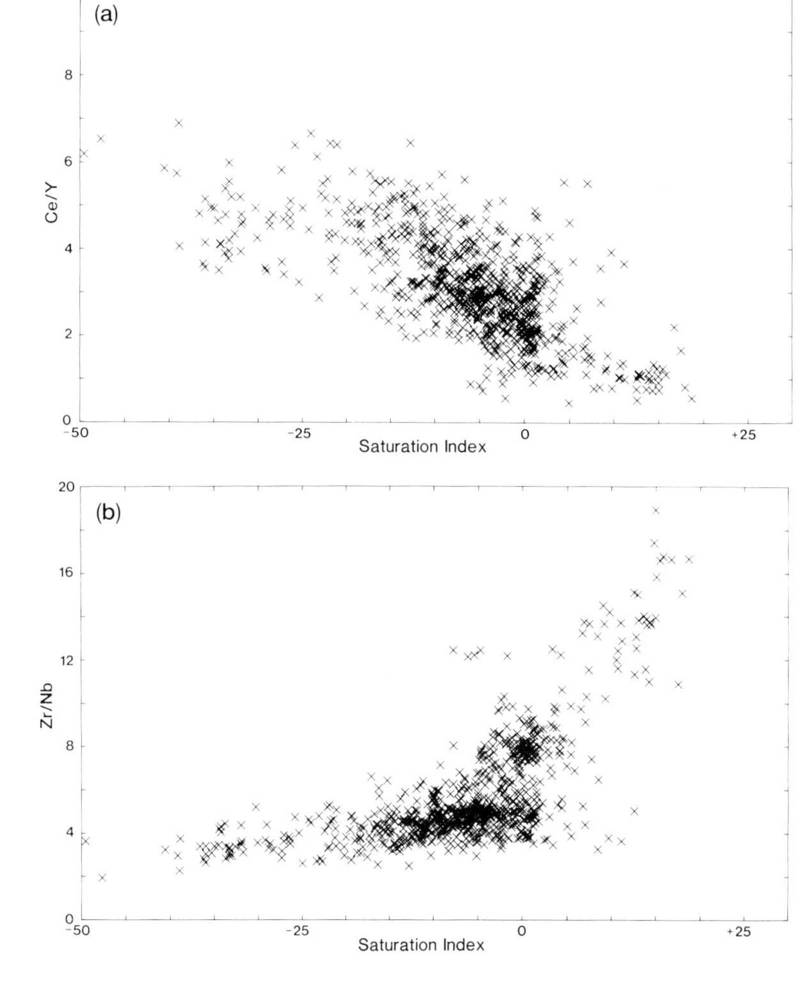

Fig. 6.10. Variation in trace element ratios of OIBs with silica saturation as defined by the Saturation Index (S.I. = 2Si − (11(Na + K) + 3Ca + Al + Fe^{II} + Mg)). ***a.*** Ce/Y *vs* Saturation Index, and **b** Zr/Nb *vs* Saturation Index. Negative values of the Saturation Index indicate undersaturated rocks. The OIBs plotted are all basic volcanic rocks (>4wt% MgO) from the unpublished data set of J. G. Fitton and D. James.

Figure 6.10 implies that in altered basalts, where the saturation index is useless, ratios of such immobile trace elements may be used to give an indication of the degree of saturation of the primary melt. Our diagram of Ce/Y *vs* Zr/Nb (Fig. 6.7) can therefore be used to study the relative degrees of undersaturation of altered mafic rocks. The North Sea rocks with highest Ce/Y and lowest Zr/Nb (the Egersund Basin nephelinites) could therefore be considered the most undersaturated rocks in the region, and probably reflect the smallest degrees of partial melting. Following the same logic, the Forties Province alkali basalts would be regarded as the least-undersaturated, largest-degree melts of all of the Mesozoic rocks, and the Lower Permian basalts the most silica-saturated, largest-degree melts studied so far. This treatment of undersaturation confirms and extends the conclusions of Dixon *et al.* (1981) which were based on the major element chemistry of rare fresh rocks in each region and on compositions of relict pyroxenes, which are well known to reflect the degree of saturation of the host magma.

Quantitative modelling

The McKenzie and Bickle model is the only quantitative framework available for relating solidus overstep to composition to melt fraction. The model is founded on a parameterization of the results of melting experiments on dry peridotite. While these experiments yielded good agreement on solidus position at a given temperature and pressure, the smallest observed melt fraction (X) for which compositional data were incorporated was 0.14 (14 per cent). Despite the lack of compositional data for extents of melting <0.14, the parameterization generated a relationship between melt fraction and T′ (a dimensionless temperature above the solidus, expressed as a function of the solidus to liquidus temperature interval, i.e. the effective solidus overstep) in the interval from the solidus, where $T' = -0.5$ and $X = 0$, to $T' = -0.4$ where $X = 0.15$ (see McKenzie and Bickle 1988, p. 641). The T′ interval where X rises rapidly from zero, at the solidus, to 0.15 (15 per cent melting) is, therefore, not well constrained. Na_2O falls rapidly away from the solidus, but little reliance can be placed on values for either T′ or X for a given composition generated in the melting experiments when less than about 10 per cent melt was present.

Nevertheless we shall argue in this section that if any of the North Sea basalts match compositions generated in these experiments, then important constraints are placed on the pressure-temperature point reached by the upwelling asthenosphere, *even with the low melt fraction uncertainties* and *even if the asthenosphere has a different, more enriched composition* than that of the peridotites used in the experiments. To establish the match we consider the most reliable major element analyses from the region of greatest melting in the North Sea (the Forties province) and then, as an independent check, use the immobile trace-element ratios to assess whether a source capable of producing MORB at 15–20 per cent melting will yield a magma with the ratios in the proposed match at a melt fraction within the band of uncertainty, i.e. between 0 and 10 per cent melting. We would expect to find that the trace elements constrain the degree of melting more tightly than the major elements (which are unable to distinguish 1 per cent from 10 per cent using the parameterization), even with the uncertainties in distribution coefficients.

The obvious next question is, can all of the North Sea rocks be accounted for by variable degrees of melting of the same dry MORB-type source? If, as the mineralogy of some suggests, they cannot, because of the extreme and variable enrichment in incompatible elements and volatiles, how can they be explained? Can a model for their generation be found which is still consistent with the constraints imposed by the fit of a subset of these rocks to the dry MORB source model? We conclude that it can, but emphasize that the characterization of magma sources in continental magmatic provinces requires a multi-system isotopic approach, combined with geochemistry. Rb-Sr, Sm-Nd, and U-Pb isotopic work is in progress. Our aim in the interim is to clarify the constraints on the upwelling paths that the geochemical data alone impose.

Major elements With the exception of TiO_2 and K_2O (which are both too high) the major element compositions of many of the freshest Forties alkali basalts (e.g. A and C in Table 6.3) can be produced by the smallest extents of melting in the parameterization (McKenzie and Bickle 1988, table A1). The compositions, including TiO_2 and K_2O, are also in good agreement with the melts produced close to the solidus ($X = 0.10$ or less) in the melting experiments of Jaques and Green (1980) and of Takahashi and Kushiro (1983) which used Hawaiian basalts (relatively high in TiO_2 and K_2O), probably containing a component from a source more enriched in incompatible elements than the dry MORB source in the starting materials. The parameterization fails to predict the correct TiO_2 and K_2O contents for the Forties basalts because it was produced from a wide experimental data base, including many experiments that used starting materials which were relatively low in TiO_2 and K_2O. The original relatively high levels of TiO_2 and K_2O in the Forties basalts are supported by the presence of biotite and hornblende in the groundmass which, with the ocelli, also suggests the presence of minor amounts of volatiles in the primary magma. The presence of K-bearing hydrous phases is evidence for input from a source more enriched in incompatible elements than the dry MORB source, an observation with rather profound implications, which are discussed further in a later section.

The major-element compositions of the other Mesozoic rocks are not observed in the melting experiments on dry peridotite. For the nephelinites and alnöites from the Egersund Basin (Table 6.3, example D) it would be generally agreed that melting of garnet peridotite at depth in the presence of volatiles (notably CO_2) would be required (Eggler 1976; Edgar 1987; Wyllie 1987). The highly potassic, volatile-enriched nature of the Central Graben rocks (Table 6.3, example E) and the rocks from PL1 (example F) as well as others from the

Netherlands, requires melting of a volatile-saturated peridotite enriched in both TiO_2 and K_2O which, as we discuss later, is only likely to reside in the lithosphere.

We conclude that the major element compositions of the Forties basalts can be explained by melting, in the less than 10 per cent region, of dry peridotite of that type that would also produce MORB at 15–20 per cent melting, but that some input from a wetter more enriched source is also required. The other North Sea rocks are all more extreme both in terms of their volatile contents and of incompatible major elements (notably Ti and K), and they could not be produced by dry melting of MORB-source peridotite.

Trace elements For our trace element model we use the bulk mantle D values presented by Fitton and Dunlop (1985). These values ($D_{Ce} = 0.012$, $D_Y = 0.183$, $D_{Zr} = 0.049$, $D_{Nb} = 0.0057$) were produced by using a low-degree partial-melting model to generate basalts from a mantle source composition which would produce the range of MORB basalts at higher degrees of melting (15–20 per cent). Because of the critical dependence of melt fraction on D at very small degrees of melting, small changes in the value of D lead to large differences in melt fractions calculated during the modelling. We use the values of Fitton and Dunlop because they produce reasonable melt fraction estimates in the calculations for alkali basalts (<5 per cent melting; Gast 1968) and MORB (15–20 per cent), and because they are consistent with experimentally-determined values for mantle phases. The D values are appropriate for a garnet lherzolite source containing a small amount of a K-rich phase such as the amphibole, richterite. It must be stressed that these D values are model-dependent, and may not be accurate absolute values. They will, however, be internally consistent and may therefore be used to calculate relative degrees of melting.

We use the MORB source of Fitton and Dunlop which is depleted in highly incompatible, large ion lithophile elements (LILEs, e.g. Rb, Ba, K, La, Nb) due to repeated melt extraction through time. This depleted MORB source has Ce/Y equal to 0.27, and Zr/Nb equal to 28.1, and produces the ratios commonly found in MORB (0.3–0.4 and 25–26 respectively) as a result of 15–20 per cent partial melting. Because this is essentially the same as the source used by McKenzie and Bickle (1988) we can compare partial melting estimates generated with it directly with those predicted by their model (i.e. <10 per cent melting).

As an example of a relatively more enriched asthenosphere, we also use a Bulk Silicate Earth (BSE) source (James 1987; Kostopoulos and James, in preparation) which is considerably less depleted in LILEs than is the MORB-source and therefore would produce MORB-like melts at much larger degrees of melting (30 per cent). This source has Ce/Y = 0.4 and Zr/Nb = 19.2, and is used to assess the effects of bulk-mantle enrichment with respect to the depleted MORB source. It should be noted that the far greater degrees of melting required to produce anything like MORB from the BSE source are perhaps unreasonably large when the constraint of oceanic crustal thickness is concerned.

All melting models have the property that, as the degree of melting (F) tends to zero, then C_l/C_o (concentration in liquid/concentration in the source) tends to $1/D$, and so will give similar values for the liquid composition. All models are also constrained such that when $F = 1$ (100 per cent melting), then $C_l = C_o$. Because of these constraints the computed concentration of a given element in the liquid for high and low degrees of melting of the mantle will vary little between the different models. We have used a simple form of the equilibrium melting equation (Shaw 1970) in our calculations:

$$C_l = C_o/(F + D - (FD))$$

This model assumes that melt, once formed, remains in contact with, and therefore in equilibrium with, the solid residue, until it is extracted in one step—i.e. it is a batch melt. We feel justified in the use of this model for two reasons. First, there is little or no difference between results generated with this model and those produced by other more complex models. Secondly, the exact nature of the melting process in the mantle is not yet fully understood. Simple equilibrium melting cannot strictly be the true process (which is more likely to be somewhere between equilibrium and fractional melting) but it provides a sufficiently accurate description for our present purposes.

The equilibrium melting model was used to calculate expected melt compositions during progressive partial melting of the two mantle sources described above. These compositions were used to calculate Ce/Y and Zr/Nb values at different degrees of melting. The same technique was used for

TiO_2 and P_2O_5 in Fig. 6.8 ($D_{Ti} = 0.11$, $D_P = 0.053$; Fitton and James 1986) but only with the depleted MORB source (Fitton and Dunlop 1985).

Model results and interpretation

The results of the equilibrium partial melting model are displayed in Fig. 6.11. On this diagram there are two non-linear trends of decreasing Ce/Y and increasing Zr/Nb with increasing degrees of partial melting, one trend for each of our two sources. The trends are labelled for degrees of melting along their length. Notice that, at large degrees of melting (20 per cent), there is a significant positional discrepancy between the two paths, the more enriched BSE-source producing much lower Zr/Nb values (and slightly higher Ce/Y) than the melts from the depleted MORB-source. However, at small degrees of melting (<2 per cent) the discrepancy is not so large and tends to be dominated by Ce/Y rather than Zr/Nb. Figure 6.11 also shows fields for the different groups of Mesozoic North Sea rocks, the Lower Permian basalts, and ocean island basalts (OIB; J. G. Fitton and D. James, unpublished data). The majority of the North Sea rocks coincide with the OIB field. The BSE source will not produce the ratios required for MORB (Ce/Y = 0.3–0.4, Zr/Nb = 25–26) at any degree of melting, nor will it produce many OIBs.

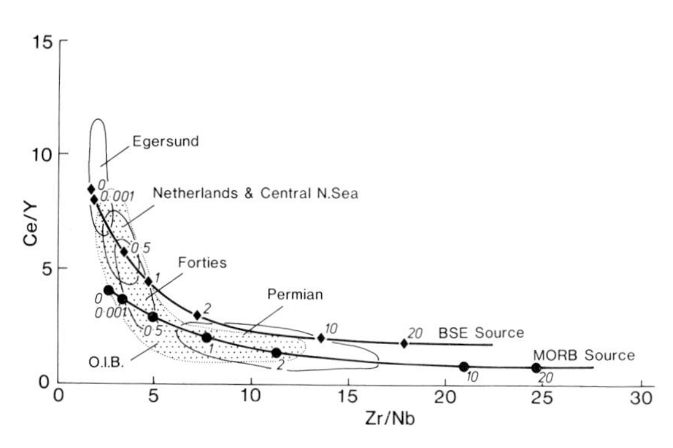

Fig. 6.11. Equilibrium partial melting lines calculated for Ce/Y *vs* Zr/Nb from a MORB source (Fitton and Dunlop 1985) and a BSE source (Kostopoulos and James 1989). The melting lines are marked for per cent melting and were calculated assuming simple single-stage equilibrium partial melting using the *D* values calculated by Fitton and Dunlop (1985). Fields for basic igneous rocks (>4wt% MgO) from the North Sea basin are outlined and the OIB field is dotted (unpublished data of J G. Fitton and D. James).

The Forties Province alkali basalts, as was suspected, give the largest melting estimates of all the Mesozoic groups with 1 to 0.5 per cent for the BSE source and 0.5 to 0 per cent for the depleted MORB source, for those with the lowest Ce/Y (around 3.5–4). All other Mesozoic groups represent less than 1 per cent melting of the BSE source and less than 0 per cent melting of the depleted MORB source! The observation that it is possible to generate at least some of the Forties alkali basalts from a depleted MORB source agrees with the major element evidence, and the figure for the extent of melting (<0.5 per cent) is more precise than the 0 to 10 per cent range given by the parameterization of experimental data. That the other Mesozoic rocks can only be derived from a more enriched source is no surprise, as it was already suspected from the petrography and major element chemistry. The mildly alkaline basalts of the Lower Permian have Ce/Y and Zr/Nb values suggestive of much larger degrees of melting; 5 to 0.5 per cent from the depleted MORB source, and 15 to 2 per cent from the BSE source.

The key conclusion from the analysis above is that at least some of the Forties basalts can be produced by small amounts of melting of dry-peridotite (depleted-MORB source) of the sort used in the experiments that were parameterized by McKenzie and Bickle (1988), peridotites which liberate MORB at 15–20 per cent melting but which will not generate the full range of OIB or the North Sea rocks at smaller amounts of melting. Where the major elements only limit the extent of melting to <10 per cent from such a source, the trace element ratios tell us that it was likely to have been less than 0.5 per cent.

6.4.3. *Alternative sources: asthenosphere and lithosphere*

In an earlier section we presented a definition of the lithosphere, which included the MBL (mechanical boundary layer) and the upper part of the TBL (thermal boundary layer). Here we note that because the MBL is the part of the lithosphere that is separated from mantle convection, then for geochemical and isotopic studies the two terms (lithosphere and MBL) are often understood to mean the same thing. In the rest of the paper, when we consider permanent enrichment of the lithosphere, the terms lithosphere and MBL are taken to be synonymous unless it is stated otherwise.

The boundary layer theory used to compute the thermal structure of the Earth from the surface downwards (McKenzie and Bickle, 1988, Appendix B) yields a largely conductive thermal gradient through the TBL (Fig. 6.1). The base of the MBL will always be at a significantly lower temperature than the 'knee' in the geotherm at the top of the convecting asthenosphere. This geometry means that any solidus curve crossed by the base of the MBL during a thinning event will already have been crossed much further by the TBL and the convecting part of the asthenosphere. In the simplest case, if no compositional difference exists across the MBL/TBL boundary it will be impossible to melt the MBL without melting the asthenosphere/TBL more extensively in the same event. Likewise it would only be possible to derive melts from the MBL alone in a decompression event if separate and sufficiently well spaced solidi applied to the MBL and the TBL/asthenosphere rocks, because of compositional and volatile-content differences between the two.

Our concern in this paper is to establish whether the quantitative framework provided by McKenzie and Bickle is relevant to North Sea magmatism. We have seen how a depleted source capable of producing MORB at 15–20 per cent melting, when melted to <0.5 per cent, could yield the Forties alkali basalts. The modelling has also shown that this same source cannot produce all North Sea rocks. Some are too enriched in incompatible elements to be produced by even infinitely small degrees of melting of such a depleted source. In this respect they overlap in their geochemistry with OIB, as do alkaline rocks from other continental rifts. This general identity is an important observation, because it strongly suggests that a major component of OIB/rift magmas must come from the same widespread shallow-level asthenosphere source from which MORB is derived by more extensive melting, and that this is not the depleted MORB source discussed above. To attribute the enriched component in both to a deep mantle, plume source (e.g. Hofmann and White 1982; Allègre 1982) would imply that such sources are always present everywhere beneath magmatic rifts, which seems unlikely, even if plumes do occur beneath segments of some rifts.

Conversely, to appeal to a lithosphere source enriched in some way, to account for all alkaline rocks more extreme in composition than alkali basalt runs into a potentially serious but admittedly unquantified problem. Lithosphere needs time to acquire enrichment. Therefore young oceanic lithosphere or recently mobilized continental lithosphere is not a likely source for enriched magmas, whereas old cratonic lithosphere might well be. It should be noted that a source in the lithosphere for the volatile-rich highly-potassic rocks, found in the Central Graben and parts of the Netherlands and common to many continental rifts, is probably inescapable because they have no oceanic counterparts.

A third possibility for generating enriched magmas in oceanic islands and in rifts is the 'streaky mantle' or 'marble-cake' model (Sleep 1984; Fitton and Dunlop 1985; Allègre 1987) in which fusible streaks, rich in garnet and clinopyroxene, provide enriched alkaline rocks (such as nephelinites) at low degrees of melting. The streaks reside in a host which is more depleted than the depleted-MORB source. When this streaky MORB source is melted 15–20 per cent it yields MORB because, at such high degrees of melting, the effect of the streaks is swamped by melting of the depleted host. The universality of such a streaky MORB-source model is obviously appealing, and it gets round the awkward dependence on either plumes or enriched lithosphere in situations such as the Cameroon line (West Africa) where coeval compositionally identical volcanoes occur in a line straddling a continent-ocean boundary (Fitton and Dunlop 1985).

Our aim in this section is to consider the implications of the two end-member non-plume models for generating enriched melts, streaky MORB source mantle and lithosphere enrichment, in the light of our previous result, that dry melting of a depleted-MORB source asthenosphere could produce Forties alkali basalt. Are these two models capable of generating the appropriate trace element signature? Is the enriched-lithosphere model capable of generating the alkali basalts unaided? Under what circumstances can we still use the dry solidus result to constrain upwelling? A full discussion of possible sources for North Sea magmas must await the results of Rb-Sr, Sm-Nd, and U-Pb isotopic work currently in progress.

We will take the process of lithosphere enrichment by volatile- and incompatible-element enriched small melt fractions as a fact, because it is now accepted as the only mechanism to account for ultra-potassic and kimberlitic magmatism within continental plates and is also very strongly supported by the character of metasomatized peridotite xenoliths (Brooks *et al.* 1976; Harte 1983; Cox 1983; Hawkesworth *et al.* 1987; McKenzie 1989). We will not consider for the moment the actual location of appropriate solidi in pressure-temperature space, but we mention again the basic principle that the base of the MBL is much cooler than the 'knee' in the geotherm, and is therefore inherently further from any given solidus than the uppermost part of the asthenosphere (Fig. 6.1).

Because the lithosphere can be enriched by a volatile-rich phase, considered by most authors to be a melt, it is reasonable to expect melting of enriched or metasomatized lithosphere also to yield a volatile-rich melt. Its behaviour will not, therefore, be governed by dry melting solidi. Whether the liberated melt is as volatile-rich as the enriching melt will depend on the behaviour of the enriching melt as it cools and solidifies within the MBL. It is appropriate to point out that such lithosphere-enrichment models assume that a volatile-bearing small melt fraction is liberated from the asthenosphere either continuously or as a result of discrete events: upwelling or plume-passage. The experimental dry peridotite solidus for the asthenosphere upon which the argument rests at present is not strictly compatible with this principle. However, if the steady-state volatile content of the asthenosphere is small and these volatiles are largely incorporated in the earliest small melt fraction, whether continuously or not, the asthenosphere may be considered as effectively dry, certainly at moderate amounts of melting. Such adherence to the dry solidus is unlikely to be true of the enriched lithosphere when remelted, and its volatile content must therefore be taken in to consideration.

The potential sources of the North Sea magmas are thus:

(1) the MBL (permanent lithosphere), which would have to be enriched and volatile-bearing for it to play any role in melt generation at low degrees of extension, because of its low temperature; or
(2) the asthenosphere and TBL (both convective and therefore non-permanent), which may be depleted but possess a very minor volatile content (as solid phases or in-transit melt) or which may be 'streaky', and so capable of yielding enriched magmas on its own.

We can now define conditions for our <0.5 per cent melting result for the Forties basalt to be still useful, if we consider what might be generated in either of these enriched sources at the same degree of perturbation. If neither source can have actually produced alkali basalt by this point, but only more enriched magmas, then more melting has to take place to reach alkali basalt compositions. In fact a greater degree of melting, and thus a larger temperature overstep (above any solidus), must occur if any more enriched magmas are produced for incorporation in the final melt, in order to lower the levels of its incompatible elements to those typical of alkali basalt. The 0.5 per cent result will only be invalidated if volatiles play a major role in the melting process, either by allowing enriched lithosphere to generate alkali basalt before asthenosphere melting will do so, or by significantly changing the behaviour of melting in the asthenosphere at small extents of melting.

The lithosphere (MBL): an exclusive source for alkali basalt?

The circumstances which might permit alkali basalt generation entirely within the lithosphere are going to be those in which the metasomatized base of the MBL is on or very close to its volatile-dependent solidus, so that any perturbation can mobilize a melt. If 'events' led to the enrichment of the lithosphere, thermal relaxation will have stranded any solidified metasomatizing/enriching melts higher up in the cooler parts of the MBL, rendering them more inaccessible and decreasing the available temperature interval before the asthenosphere is also involved in melting. The question to be asked is whether it is possible to generate a volatile-bearing batch melt of alkali basalt composition from enriched lithosphere, with a liquidus temperature (the temperature overstep of the appropriate solidus that is required to produce the melt) which is low enough to be intersected by the MBL before the asthenosphere starts to melt to yield a similar magma.

Model calculations (Fig. 6.12) show that the chosen immobile trace-element ratios of the mafic North Sea rocks can be generated by various routes, which include the remelting (by 10 per cent) of a depleted MORB-type lithosphere source previously enriched by small melt fractions (0.1 per cent melts of depleted MORB-type asthenosphere) in the form of veins to a density of 5 per cent. However, the model must pass the same test for all incompatible elements and for volatiles. Sufficient amounts of

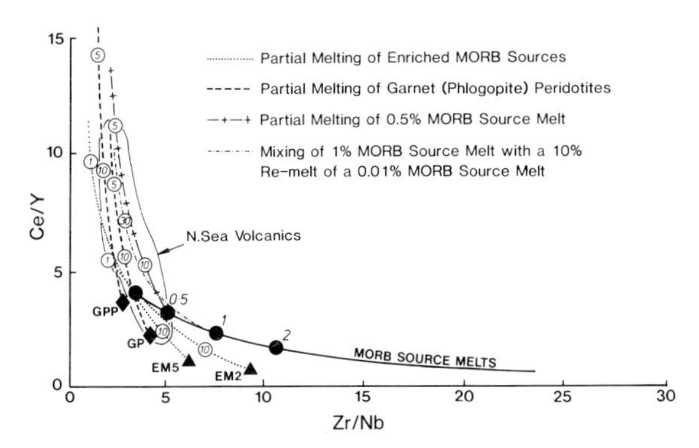

Fig. 6.12. Ce/Y *vs* Zr/Nb relationships for single and multi-stage melting models. The bold curved line shows Ce/Y *vs* Zr/Nb values for different degrees of partial melting of a MORB source (Fig. 6.11, and text). The other curved lines refer to different hypothetical melting models which could lead to the enriched values shown by the Mesozoic rocks (Figs 6.5 and 6.11) of the North Sea. Each of the lines relates to a different source composition, and each is marked with values relating to the per cent of partial melting of that particular source. It is important to note that these values should be treated with caution because the *D* values used in the calculations were intended for use with a MORB source and may not be strictly applicable for these sources.

The purpose of the diagram is simply to show that it is possible to produce the enriched values shown by the North Sea data (outlined by the fine continuous-line) through melting of sources other than the MORB source. EM2 and EM5 are enriched MORB sources produced by mixing 98 per cent and 95 per cent MORB source respectively with 2% and 5 per cent of a 0.1 per cent melt of the MORB source. GP is a garnet lherzolite xenolith composition from Kimberley, South Africa (Erlank *et al.* 1987). GPP is an average garnet phlogopite lherzolite composition from the data of Erlank *et al.* for GPP xenoliths from Kimberley, South Africa.

lithospheric olivine and pyroxene have to dissolve into the remobilized small melt fraction to dilute the incompatible elements down to the values of alkali basalt, but at the same time leave the volatile content still high enough to depress the liquidus temperature. An asthenosphere-derived volatile-bearing small melt fraction may be added in at the time of the perturbation, but it will bring its own quota of incompatible elements and so will not help. If all of the volatiles in the enriched lithosphere are partitioned into the final melt, this will have a volatile content equal to that of the original small melt fraction divided by the subsequent dilution factor. For example, a 10 per cent melt of dry lithospheric mantle which has been enriched to 5 per cent (by volume) with veins of melt containing 5wt% of volatiles would generate an alkali basalt with 2.5wt% of volatiles. A volatile content of 2.5wt% might lower the alkali basalt liquidus temperature by some 100–200 °C, below that of the liquidus when melting dry peridotite, at any pressure (Edgar 1987).

Our interim conclusion is that melts from metasomatically enriched lithosphere could have the required major- and trace-element contents for alkali basalt and could also have high volatile contents if volatiles are effectively handed on in the enrichment/mobilization process. The MBL enrichment model itself is based on the premise that the enriching melts have their liquidus exceeded by the steady state geotherm, and so migrate into the MBL until reaching their solidus (see Fig. 6.14). If perturbing the geotherm can remelt them and incorporate sufficient of the host olivine and pyroxene to move the composition to alkali basalt, before the asthenosphere melts, then all the quantitative arguments based on the dry alkali basalt liquidus break down, except as an upper limit on mantle temperature and degree of extension of the lithosphere.

At present we feel that such a model cannot be totally excluded, but we find three arguments which make it unattractive. The volatile content of the Forties alkali basalts was evidently not generally high, probably less than 1wt%. They are rarely vesicular or fragmental, and they only contain minor amounts of volatile phases, unlike some of the other North Sea occurrences. The other two objections are discussed further below in our summary of probable behaviour. In essence, for a relatively thin MBL (~100 km) prior to Jurassic extension, such as that inferred for the North Sea area, the base of the MBL is too cool to intersect even a volatile-saturated solidus, and is at such a shallow depth that any solidus curves that might be accessible are steepening and curving away towards higher temperatures at shallower depths. Perturbations due to extension simply do not stand much chance of causing a shallow MBL to overstep the volatile-undersaturated liquidus curve, for alkali basalt, down pressure. Another important consideration concerns extensional events during the Permian and Triassic in the North Sea, which would have caused upward migration of enrichment to levels where the

MBL is too cold ever to re-intersect their solidus curves, thus leaving them stranded and inaccessible to even the volatile saturated solidus during the Jurassic event.

A further complication, rather than an objection, posed by the North Sea magmas (as well as those of other rifts) is that no single small melt fraction can account for lithosphere enrichment that will generate the low K_2O Egersund nephelinites, the high K_2O Central Graben monchiquites, and the Forties alkali basalts by variable degrees of remelting. At least two end members are needed, one high K (kimberlitic) and one low K (nephelinitic?) to account for the extreme rocks alone.

Finally, if these objections can be overcome, the lithosphere would still be required to be enriched so that perhaps half of the Forties magma were derived from the introduced metasomatic enrichment. If 0.5 km of magma (and there is at least 0.5 km in Forties) represents a 10 per cent melt of a layer of lithosphere enriched by 5 per cent with 0.1 per cent small melt fractions from a depleted MORB asthenosphere, then 0.25 km represents this original 0.1 per cent melt. The enriching melt must therefore have come from a column of asthenosphere 250 km high. Net upwelling (due to mantle-wide convection) at a rate of 1 mm/year over 250 Ma would be sufficient to create such a layer (enriched to 5 per cent) 5 km thick in the overlying lithosphere—about the time available since the Caledonian tectonism. Such rates of turnover may not be unreasonable. However, if the enriched layer had been displaced upwards, if not erupted, during the Early Permian and Triassic events, then the prospect of replacing it before the Middle Jurassic would seem slight.

The 'streaky' asthenosphere as a source

We turn now to the other end-member model for deriving enriched magmas—the streaky mantle or marble cake model (Sleep 1984; Fitton and Dunlop 1985; Allègre 1987). This model envisages a heterogeneous asthenosphere with a depleted host containing incompatible element-enriched, fusible streaks rich in garnet and clinopyroxene. Unpublished calculations by J. G. Fitton suggest that such a model can generate the range of OIB trace element ratios at low degrees of melting and still yield MORB at around 15–20 per cent melting.

If the Forties alkali basalts could have been produced from a depleted MORB-source, then, if the source was less depleted by virtue of its streaks, more of the depleted host must have melted to compensate for the enriched melt from the streaks, in order to yield an alkali basalt. In other words the alkali basalt liquidus for the depleted MORB-source must be further overstepped by the perturbed geotherm if the source is enriched. The alkali basalts may well represent something closer to a 2 per cent melt of an enriched source. The solidus for the streak mantle may well be at a lower temperature than that for the depleted MORB-source and the relations between T′ (degree of solidus overstep), X (melt fraction), and composition may be different from those in the experiments in the low melt fraction range. Such differences are not going to be apparent at the large extents of melting characteristic of MORB production. The same argument concerning the liquidus overstep also applies to mixing asthenosphere melts with enriched lithosphere melts. There may however, be another effect on the solidus temperature due to volatiles (as was discussed earlier) whereby the streaks yield a volatile-bearing magma on a lower solidus, which produces a slight lowering of liquidus temperatures in the alkali basalt range relative to the dry experiments. The size of liquidus drop is not likely to be as pronounced as in the MBL because the convecting asthenosphere (and TBL) is probably losing its lowest melt fraction continually, and because there can be no long-term accumulation of volatiles. Nevertheless we feel that some lowering of liquidus temperatures is likely at low melt fractions relative to the dry experiments, and for this reason we are not willing to be categoric, in section 6.5., about the operative asthenosphere potential temperature in the North Sea at the time of rifting.

Conclusions concerning source

Our conclusion is that the most plausible source for the highest degree (Forties) melts is the streaky asthenosphere. If this is so, then the dry solidus position must have been crossed because the alkali basalt liquidus in a dry, depleted MORB source is more easily attained than is the same liquidus in a dry, enriched, streaky MORB source. This conclusion is made with the provisos concerning volatiles noted in the previous paragraph which might lower the solidus position by as much as 100–200°C. A strongly volatile-influenced lithosphere source for the alkali basalts, and therefore by implication for all of the North Sea rocks, is nevertheless still a real alternative. We are reluctant to endorse it strongly because it carries the profound implication that either all oceanic alkaline rocks are produced from deep enriched mantle sources (plumes), or that lithosphere enrichment can occur very effectively in tens of millions of years.

6.4.4. *Crustal effects*

The highly LILE-enriched nature of many of the North Sea rocks, as for example in K_2O which can reach over 5wt% in rocks which still have over 6wt% MgO (see Table 6.3), implies an enriched, perhaps 'streaky', mantle source or an enriched layer in the lithosphere. However it might also be a crustal contamination effect. If the contaminant were crustal, then in order to generate a magma with 4wt% K_2O from a depleted MORB source melt containing only 0.25wt% we would need to add 60–70 per cent by mass of granite (with 5wt% K_2O). Such large amounts of assimilation would drive the basalt composition towards that of a granodiorite (60 wt% SiO_2), and are also highly unlikely on thermal grounds. Crustal contamination is unlikely to be an important process.

6.4.5. *Summary*

In this section we have shown that, even in highly altered basaltic rocks, it is possible to make semi-quantitative estimates for degrees of partial melting of the asthenosphere (and the TBL). Alteration effects are overcome by use of ratios of immobile elements. In order for these ratios to remain undisturbed during fractional crystallization, these elements must also be equally incompatible in the fractionating phases. We have used Ce/Y and Zr/Nb (both ratios of immobile, incompatible elements) in a simple equilibrium partial melting model. The results of modelling suggest that the Forties alkali basalts reflect less than 0.5 per cent partial melting of a depleted MORB source. We consider an asthenospheric source most likely, and note that it must have been LILE enriched by some mechanism. If an enriched (streaky) asthenopshere source has melted to give the Forties alkali basalts, then the maximum extent of melting will have been greater than 0.5 per cent and is probably closer to 2 per cent. Some of the other Mesozoic rocks may be produced by smaller amounts of melting of the same streaky asthenosphere source. However, the ultra-potassic rocks in the Central Graben and parts of the Netherlands require a source in the MBL, because nothing similar is observed in OIB. The full range of Mesozoic rocks can be explained by input from two sources, streaky asthenosphere, and enriched lithosphere (MBL).

6.5. Extension and magmatism

We have so far considered, independently, the extents of melting from igneous rock composition, and the stretching estimates (given in the first section). We can now put these two parameters together. The maximum extent of melting seen anywhere in the rift is in the Forties province, where we concluded that the most plausible model for the alkali basalts involves less than 2 per cent partial melting of a streaky MORB source. It must be conceded that a volatile-induced lowering of the alkali basalt liquidus in the lithosphere would invalidate much of the subsequent discussion except that which limits the amounts of extension and the potential temperature upwards; for reasons already explained we consider such a source less likely, at least so far as the alkali basalts are concerned. The other Mesozoic rocks must all have been produced by melting in the presence of volatiles and some (the ultra-potassic varieties) can originate in the lithosphere.

Our initial intention was to exploit the theory relating upwelling to melt production derived by McKenzie and Bickle (1988), and to use the stretching estimates to generate extent of melting values with which to compare those derived independently. There are two parts to this. To convert stretching to upwelling we need to know the initial thickness of the lithosphere (and hence the starting geotherm); to convert upwelling to extent of melting we need to know both the positions of the solidus and the adiabat for the asthenosphere (i.e. the Tp).

6.5.1. *Initial lithosphere thickness*

The extent to which the temperature of the TBL (thermal boundary layer) is taken towards and over the solidus during upwelling depends upon its initial value. For a given asthenosphere, Tp, this depends on the depth at which the conductive geotherm intersects the adiabat, i.e. the initial thickness of the lithosphere. The degree of melting, and hence the amount of melt produced for any value of β and Tp, is thus highly sensitive to the initial temperature profile in the lithosphere, a function of the initial thickness (TL). Colder (thicker) lithosphere requires larger values of β (at the same Tp) than hotter (thinner) lithosphere to produce the same amount of melt.

Barton and Wood (1984) suggested that the thermal time constant for the North Sea lithosphere, derived from the thermal subsidence of the basin, is the same as that obtained from ocean ridge subsidence, and therefore requires the same lithosphere thickness. For the North Sea, TL would therefore be about 120 to 125 km (McKenzie and Bickle 1988), corresponding to an MBL (mechanical boundary layer) thickness of approximately 100 km. The Barton and Wood value appears to be the one most commonly used in subsidence calculations (Wood 1982; N. J. White 1988).

6.5.2. *Relating upwelling to extent of melting*

Once the depth to the top of the TBL (i.e. the thickness of the MBL) is fixed, upwelling at a given point is directly related to the amount of stretching for any instantaneous McKenzie-type process. The TBL follows the adiabat for the appropriate distance down pressure (Fig. 6.13). The problem reduces to knowing the location of the adiabat relative to the solidus. As these are independent they must be treated separately. At this stage we will ignore finite rates of extension which can add downward, conduction-driven components to the motion of the TBL.

6.5.3. *Asthenospheric potential temperature and uplift*

The simplest assumption is that the Tp is 1280 °C, the 'normal' value which gives the best fit between MORB melt volume, composition, experimental melting data, and the upwelling necessary to replace separating oceanic lithosphere. However, the value of this parameter is so crucial in determining melt volume and composition (Fig. 6.2, Fig. 6.13.) for a given β value that some discussion of its possible variation is required. Small variations in oceanic crustal thickness are inferred by R. S. White and McKenzie (1989) to represent a range for the 'normal' Tp of 1280 ± 30 °C. However, the only apparent instances of significantly anomalous Tp are above plumes or localized uprising convection currents (R. S. White *et al.* 1987; R. S. White and McKenzie 1989). In the Pacific (McKenzie *et al.* 1980) these have a clear long-wavelength gravity and bathymetric signature which can only be successfully modelled if they are hotter than the surrounding asthenosphere, and also contribute dynamic support to the overlying lithosphere. Not all plumes recognized in this way have volcanoes or even sea mounts above them. This may imply that they exist on a range of scales of excess temperature or size. Cooler down-going flows of asthenosphere are also implied by the gravity data. Elongate rising ridges may also exist, perhaps, for example, under the Cameroon line (West Africa), but in general plumes appear to be centri-symmetric (e.g. Hawaii, Cape Verde, Iceland). They will give rise to a broad mushroom-like anomaly at the base of the lithosphere as they spread out (R. S. White and McKenzie 1989) and they can move relative to one another only slowly in plate terms (<1 cm/year). Plumes may predispose the lithosphere to rupture above them when it is put in tension, by inducing local stress through uplift, elevating the heat flow, and so causing mechanical softening; they may cause thermal thinning by conductive heat transfer, but no mechanism is known whereby stretching the lithosphere can initiate a plume or influence the location of one in the short term.

If a plume existed beneath the North Sea in the Middle Jurassic prior to rifting it should have created a radial signature of uplift and excess Tp. The presence of a plume will continue to distort uplift and subsidence relationships after rifting (R. S. White and McKenzie 1989, Fig. 7). Furthermore for it to be centred on the axis of a basin which was already there requires a major coincidence and invites the questions: Where was the plume prior to rifting? and Where did it go afterwards? A pre-existing basin, produced by extension during the Triassic, does appear to be required to explain the sediment accumulation (Ziegler 1982*a*), and the discrepancy between β values deduced from well subsidence studies and those implied by present crustal thickness from seismic refraction modelling (Barton and Wood 1984). Even an off-centred plume developing in the North Sea between extensional events would be a considerable coincidence. A further possibility, providing that there was little relative movement of the Eurasian plate, is to postulate a long-lived plume which was present from the Triassic onwards.

Even if there is evidence for pre-faulting regional

uplift during the Middle Jurassic, this does not necessarily require the presence of a plume. Calculations (R. S. White and McKenzie 1989, Fig. 7) suggest up to 300 m of excess elevation for every 100 °C excess in Tp above a plume, due to a combination of lowered density and dynamic support. For plumes which are hot enough to induce melting without extension a further component of elevation is caused by melt separation. However, low-temperature (±30 °C) broad-scale thermal anomalies which are not convectively driven plumes may also occur within the ambient thermal structure of the asthenosphere (D. P. McKenzie, pers. comm. 1989). Simple isostatic calculations, using the equation below, for an instantaneous temperature increase in the asthenosphere suggest that between 17 and 34 metres of uplift will occur per 10 °C rise in temperature, depending upon the depth of compensation (50 or 100 km below the base of the lithosphere respectively).

$$U = a(kT_2 - kT_1)/(1 - kT_2)$$

U is the change in elevation in km from some reference level; a is the depth in km to compensation from the base of the lithosphere; k is the thermal coefficient of expansion (3.28×10^{-5}); T_1 is the initial or steady state temperature, and T_2 is the new temperature.

If such an anomaly is present for a sufficiently long period it will change the thermal structure of the lithosphere, and the amounts of uplift will be greater. The regional nature of any temperature anomaly, and therefore its isostatic effects, needs to be stressed. Uplift due to thermal effects is likely to occur over a radius of 500–1000 km at least.

Perhaps the best way of determining the Tp parameter while ignoring the magmatic evidence is to consider the question of whether or not there was regional uplift prior to the Mesozoic extensional phase in the North Sea. Our present reading of the informal opinions offered by various oil companies and the literature, is that some pre-rift uplift did occur, and that this is possibly centred on the triple-junction area, but is not on the scale favoured by Ziegler and Van Hoorn (1989). This must be explained, and its magnitude is likely to be crucially important in determining the Tp conditions of the asthenosphere at the time. If the uplift is no more than say 100 m, then only a slight excess in Tp may be required (perhaps less than 30 °C).

Once rifting begins, the relationship between uplift and subsidence of extending lithosphere is essentially governed by four factors; the potential temperature and dynamic support, β, the amount of melt generated, and the density of the residue (R. S. White and McKenzie 1989). In areas of 'normal' (1280 °C) Tp conditions, the initial subsidence is purely an isostatic response to the extension, providing that the extension is insufficient to generate any melt. In cases where melting occurs, the effect of the low density melt and a reduction in the density of the residual mantle from which the melt has been extracted (usually rather small) can be calculated, and causes uplift which thus reduces the net subsidence of the column. When the Tp is elevated by the presence of a mantle plume then, in addition to the effects of melt and residue, dynamic stresses and the thermal effects of the plume operate to reduce subsidence.

Figure 7 in R. S. White and McKenzie's paper (1989) shows the total subsidence at the time of rifting under different Tp conditions (assuming MBL thickness = 100 km) as a function of the four governing factors. For a β factor of 1.6 (our maximum value for the North Sea) positive relief only occurs when Tp = 1480 °C. In this case the amount of uplift predicted varies from zero to 600 m depending on the depth of compensation used in the calculation. For Tps of 1280 °C and 1380 °C the melt produced at large amounts of extension causes relative uplift but is never sufficient to cause positive relief. The implication is that, if pre-rift uplift indicates the presence of a thermal anomaly, then the amount of regional syn-rift uplift (other than the flank and footwall types) may help to discriminate between low-temperature effects and plumes of different temperature. Positive relief during the syn-rift stage, on the scale required by Ziegler and Van Hoorn (1989) requires a high temperature plume (Tp > 1380 °C) to be located below the North Sea at that time.

Syn-rift uplift of a regional type is difficult to discern because of rift-flank and fault-footwall effects which tend to dominate the well data. Until we have seen convincing evidence for pre-rift uplift in the form of onlapping relationships and/or erosional unconformities in the seismic records prior to the onset of faulting, then we have no justification for adopting an anomalously high Tp; as we explain below, a Tp as high as 1480 °C is considered very unlikely on the basis of the magmatic and estimated-extension evidence alone. A further point is the good agreement shown between estimates of extension from subsidence calculations with those made from the faults (N. J. White this volume). Any large-scale thermal anomaly would cause heterogeneous extension, with far greater subcrustal thinning than crustal thinning, and this would affect a very large area—probably the whole North Sea. There is no compelling evidence for such a large discrepancy in the extension estimates from the Northern North Sea and so, on these grounds, a plume seems unlikely. Without positive evidence for a plume, we prefer to assume a Tp close to 1280°C; this requires no special pleading, models oceanic data well (McKenzie and Bickle 1988), and seemingly agrees with evidence from cratonic areas as well (McKenzie 1989).

6.5.4. *Potential temperature: implications from β and per cent melt estimates*

We will now consider the amount of melting which McKenzie and Bickle (1988) would predict (Fig. 6.13, from unpublished diagrams provided by D. P. McKenzie) for a value of 1.6 (the value in the Forties), with an initial MBL thickness of 100 km for three different Tp examples (1480°C, 1380°C, and 1280°C). We will not consider other areas where the β factors are smaller (e.g. the Egersund Basin) because here the melting clearly involves the presence of volatiles and a quite different solidus.

Figure 6.13a shows that when Tp = 1480°C with a β value of 1.6 the amount of partial melting will be close to 15 per cent. Such a large degree of partial melting in the Forties area would produce magmas of similar composition to that of primary MORB, with Zr/Nb around 15–20 and Ce/Y less than 1! The amount of melt produced would correspond to a thickness of between 3 and 5 km, which is approaching the melt production at some ocean ridges! Because the alkali basalts erupted relatively soon after the onset of rifting (in the Lower Jurassic) then we cannot appeal to conductive heat loss to reconcile the observations with such a temperature. Clearly, if our 2 per cent value for the maximum extent of partial melting is correct and the stretching estimates are accurate, then the 1480°C model cannot be correct. This conclusion appears to be inescapable whether or not the solidus temperatures are revised downwards for the lithosphere or the asthenosphere.

Now consider a case where the stretching takes place over asthenosphere with a Tp of 1380°C (Fig. 6.13b), at the edge of a large 1480°C plume which can be up to 2000 km or so in diameter with only the central part very hot (R. S. White and McKenzie 1989), or over the centre of a lower temperature plume. In this case when the MBL = 100 km and $\beta = 1.6$ there is close to 0 per cent melting which would be about right for the alkali basalts.

Simple passive stretching (McKenzie 1978) implies 'normal' (Tp = 1280°C) asthenosphere temperatures, and is a model that has been strongly supported for the North Sea. However, Fig. 6.13c immediately shows that melt generation is much more difficult to achieve under these Tp conditions. The solidus is not intersected until β values go above 2.5. The apparent conclusion to draw is that melting would not have occurred at all in the Forties area of the North Sea if the Tp was 1280°C, or, that if it had, the composition would have been more extreme than alkali basalt. Clearly melting did occur and alkali basalts were produced, therefore if we wish to retain the simple passive stretching model, with 'normal' temperatures, for the North Sea we must either revise the position of the solidus used by McKenzie and Bickle, or, alternatively propose that the extension estimates are too small. Even if we increase stretching estimates to take account of a residual thermal effect from the Triassic event it is very unlikely that a value of 2.5 could be justified and, as has already been discussed there are a number of authors who feel that the estimates of extension that we are using are already too high.

6.5.5. *Solidus position, mantle composition, the melting process, and small melt fractions*

Drawing from the discussion above, we conclude that the model constructed by McKenzie and Bickle (1988) cannot simultaneously meet all of the melt composition against β-value constraints imposed by the North Sea data. The model is clearly highly sensitive to the position of the solidus, which itself is dependent on mantle composition. We are now in a position to summarize a best fit set of values for the solidi, and for the potential temperature of the asthenosphere.

Our conclusion from the modelling section was that an exclusively lithospheric source was unlikely for the Forties alkali basalts. We are therefore forced

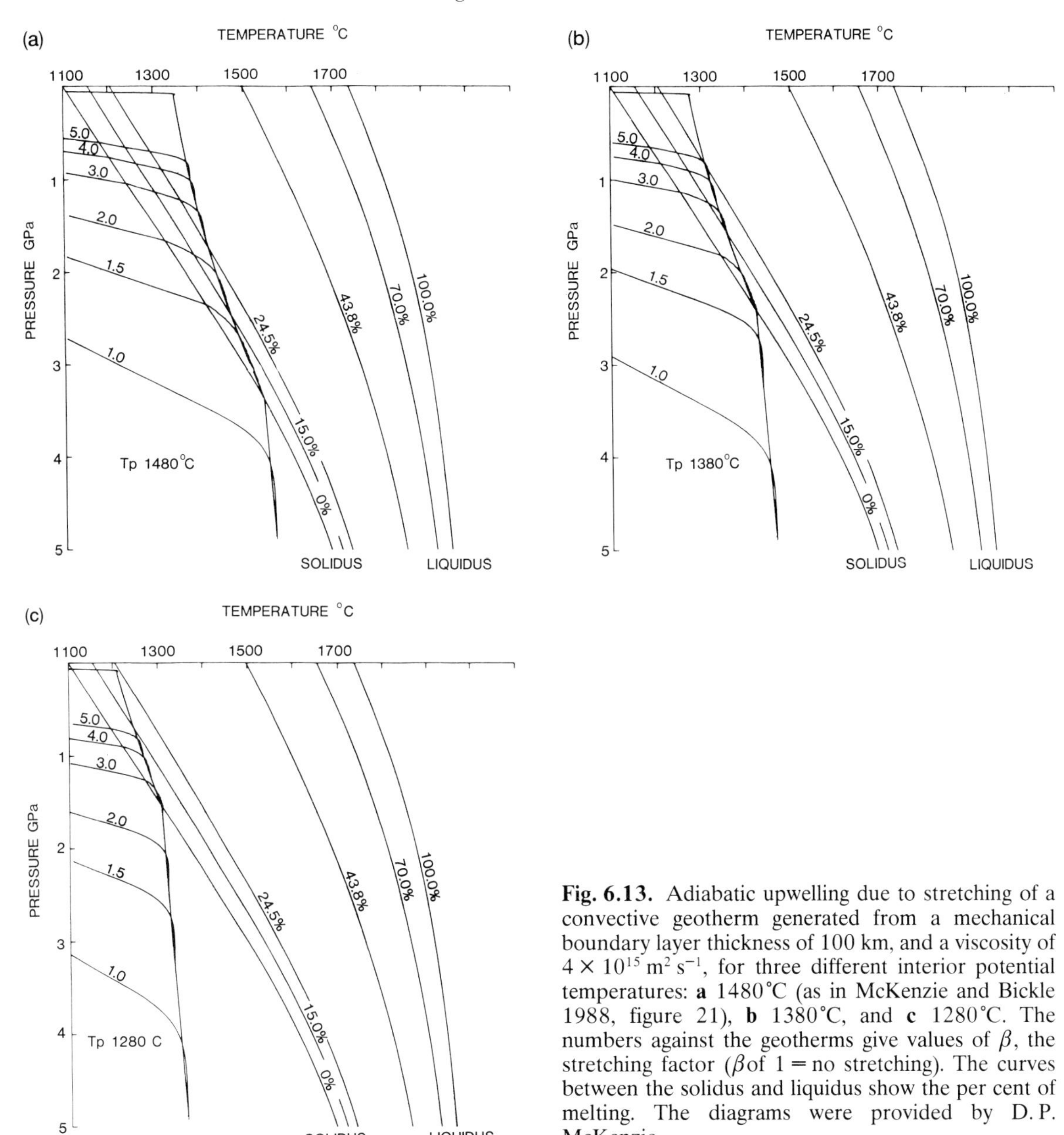

Fig. 6.13. Adiabatic upwelling due to stretching of a convective geotherm generated from a mechanical boundary layer thickness of 100 km, and a viscosity of $4 \times 10^{15}\ m^2\ s^{-1}$, for three different interior potential temperatures: **a** 1480°C (as in McKenzie and Bickle 1988, figure 21), **b** 1380°C, and **c** 1280°C. The numbers against the geotherms give values of β, the stretching factor (β of 1 = no stretching). The curves between the solidus and liquidus show the per cent of melting. The diagrams were provided by D. P. McKenzie.

to reach at least the alkali basalt liquidus for a depleted dry MORB-source asthenosphere (<0.5 per cent melting) and to go beyond it if there is a contribution from either streaks or enriched lithosphere incorporated in the Forties magmas (<2 per cent melting). The K_2O content of the Forties basalts at 0.5–0.8 wt% (Table 6.3) is significantly higher than the predicted value of 0.26 wt% for the dry peridotite (depleted MORB source) at the lowest extents of melting generated by the parameterization (McKenzie and Bickle 1988, table A1). The Forties basalts are not perfectly dry melts but contain ocelli—implying late separation of an immiscible, probably volatile-rich melt—and small amounts of primary hydrous phases, such as biotite and hornblende. This component of K_2O and volatiles could be from the lithosphere, requiring an appropriate overstep of the true dry alkali basalt liquidus, or it could be from streaks in the asthenosphere, requiring an overstep compositionally (i.e. more melting), but the appropriate liquidus might be displaced down pressure because of the volatile component. Isotopic data may help resolve the component sources of the Forties basalts, but our feeling in the absence of firm data is that the Forties liquidus temperatures might turn out to be lower than in the dry experiments despite increased melting of enriched sources. In effect we cannot confirm or quantify three probable real deviations from the idealized model of McKenzie and Bickle, each of which could produce a low temperature ($T' < -0.5$) 'tail' to the parameterized dry solidus:

(1) the effect of fusible streaks;
(2) the effect of volatiles in the asthenosphere arising from indigenous amphiboles, micas, or carbonate phases, or from passing small melt fractions from deeper down; and
(3) the effect of experimental uncertainty in determination of the true dry solidus.

None of these three uncertainties will affect the results of the model when applied to high degree melts but they are crucial at small degrees of melting.

Streaks (1) will have most effect on the amount of small melt fraction generated at a given temperature, and may also lower the solidus itself so that, for example, the Egersund nephelinites might be simultaneously generated at $\beta = 1.2$ with Forties basalt at 1.6. Volatiles (2) will have the most marked effect on the solidus temperature. Wyllie (1987) suggests a lowering of the solidus temperature by 500°C at 3 GPa (~100 km) for peridotite saturated in H_2O. Peridotite saturated in CO_2 could begin to melt at temperatures some 200–300°C lower than in the dry case at the same pressures (Eggler 1976, 1978; Wyllie 1987; Edgar 1987; Green and Wallace 1988; Wallace and Green 1988; Hunter and McKenzie 1989). The degree of liquidus lowering for alkali basalt is difficult to assess. Experimental uncertainties (3) introduce the possibility of a lower-temperature, but essentially undetectable, true dry solidus; however, this will not alter the fact that the dry alkali basalt liquidus must be at a higher temperature than the inferred solidus of McKenzie and Bickle (i.e. at $T' > -0.5$). How much higher, and and exactly what melt fraction alkali basalt actually represents, are still not clear. The trace-element controls are probably more precise in giving 0.5 per cent.

Our conclusion is that an alkali basalt liquidus representing 1–2 per cent melting, but lowered by some 100°C through the effect of perhaps 1 wt% volatiles, is attainable for Tp = 1280°C and $\beta = 1.6$. The experimental data, as well as recent theoretical work by McKenzie (1989), suggests that a true volatile-dependent solidus might be intersected by the steady-state geotherm, and that small amounts of melt produced under steady-state conditions might move at melt fractions as low as 10^{-3} per cent. These continuously forming melts in the upper part of the convecting asthenosphere would migrate upwards along the geotherm, as they contain insufficient heat to affect the temperature of their environment. They would fractionally crystallize from the 'knee' in the geotherm to the point where they again intersect their solidus. Such small melt fractions provide a means of enriching the lithosphere in volatiles and incompatible elements. As they are present in the asthenosphere in regions moving upwards but not downwards (such regions have already lost the small melt fraction) the question of whether a continuum exists between these initial melts and those produced during upwelling is a difficult one to answer. A continuum in melting behaviour may exist under some rifts, but not necessarily lead to eruption. Under others, a real gap in the melting regime may arise if all of the volatiles in the asthenosphere have already been melted out and there are none being actively melted below the rifted region. The same divergence may occur beneath oceanic ridges, which may show the small melt fraction component, or not, according to the ridge's proximity to upwelling or downgoing asthenosphere.

This model generates a steady-state picture (Fig. 6.14) with the following features:

(1) an MBL with the fractionally crystallized products of steady-state melt fractions spread through a finite thickness (i.e. temperature range) near its base as metasomatized and veined (previously-depleted) mantle; event-generated small melt fractions may be present still higher;
(2) a TBL from which these melts have been extracted with steady-state melt fractions migrating upwards through it, and possibly with

some enrichment from recently crystalllized largest-fraction (the first to solidify) steady-state melts;

(3) an asthenosphere which is releasing into the TBL a range of low-melt fractions, their compositions dependent on the exact pattern of convective circulation below the sampling point.

The response of these three layers to stretching perturbations is the key to understanding rift magmatism. The only layer which is guaranteed to produce more melt (than in the steady state), to whatever extent it is perturbed, is the enriched MBL (and the TBL if the metasomatism extends that far), because the low melt fractions are on their respective solidi there. The volumes of these melts will depend on the enrichment history (which decides their distribution and concentration in the MBL) and not purely on the size of the perturbation. The position and shape of the solidus curve for these melts at lower pressures is also crucially important in determining how much further melting ensues with extensional thinning. Whether they are erupted is a plumbing problem. It is possible that if small melt fractions and host mantle are both involved in the remelting process, magmas compositionally similar to single-stage low melt fractions could be duplicated, possibly even before asthenosphere melting is initiated.

The asthenosphere itself may contribute more melt immediately it is subjected to upwelling (or heating) if the melting is truly a continuum from lowest fraction onward. If the steady-state melting during convection leaves residual K- and volatile-bearing phases in the asthenosphere, adiabatic decompression may well create more melt, particularly if a volatile-bearing ultra-low melt fraction is passing through the asthenosphere from below. If equilibrium is established between percolating melt and host asthenosphere even without K-bearing phases appearing, some increase in melt volume may occur. If however steady-state convection effectively melts out all the low-melt fraction components from the upper parts of the asthenosphere (and TBL), no additions to the melt will occur from here until the 'dry' solidus is crossed, even though steady-state melts from below might be present in passing.

Until the melting behaviour of realistic asthenosphere mantle compositions is characterized for small melt fractions, and the behaviour of metasomatized and veined lithospheric mantle is similarly well-

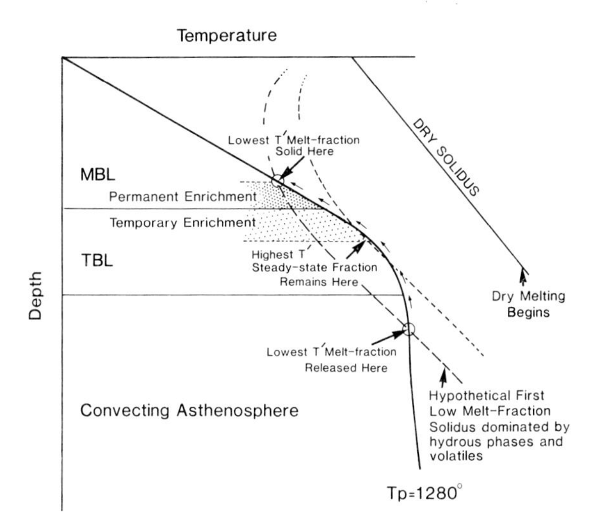

Fig. 6.14. The steady-state production of ultra-low melt fractions from a hypothetical mantle with a solidus (bold dashed line) dominated by hydrous phases and volatiles. The first melts are produced when this solidus intersects the geotherm. After upward migration, these smallest melt fractions freeze at the lowest temperatures as an enriched layer in the mechanical boundary layer (MBL) where the geotherm re-intersects their solidus. The largest steady-state melt fractions are those produced by the largest amount of upwelling over the hypothetical solidus; they will crystallize at the highest temperatures where their respective solidus (the fine dashed line) intersects the geotherm.

T′ is the temperature above the solidus at a given pressure. Arrows show the path taken (in P-T space) by the migration melt fractions whose thermal Peclet number is less than 1 (McKenzie 1989) and which therefore carry no heat. Any melts which solidify in the thermal boundary layer (TBL) will only reside there temporarily because of this layer's periodic convective overturn. Melts reaching the MBL will reside permanently in steady-state conditions, and will be remobilized if the geotherm is perturbed. Also marked on the diagram is the dry peridotite solidus (*see* Figs 6.1, 6.2, and 6.13) which must be intersected in order for large-scale melt production to occur. **Note**: The hypothetical solidi (dashed lines) are highly schematic; they may not have this shape in reality, they may not be parallel to each other or to the dry solidus, and they will not necessarily intersect the geotherm at the depths indicated.

known, no quantifiable magma composition *versus* stretching model can be created for compositions, outside the present experimental range, or for volatile-dependent melting. If the generation of small melt fractions of any sort is entirely 'event'-driven, then the problem is still complex for cratonic situations which have accumulated these enriched melts. The event under investigation is still liable to generate magmas from all three layers and mix them. In either situation it may prove possible to identify isotopically melts which have a definite asthenospheric signature and so, with a model of asthenosphere low-melt fraction behaviour, characterize the response of the asthenosphere to upwelling.

There remains the persistent close similarity of ocean-island nephelinites and alkali basalts to their rift counterparts. If ocean islands are produced by melting of enriched lithosphere rather than streaky asthenosphere then there should be enrichment/age relationships. More enriched OIBs with greater isotopic differences from MORB should be found on older oceanic lithosphere; whether or not this is the case has not yet been determined.

6.5.6. *Timing and location of magmatism: rift propagation and rates of upwelling*

A number of questions concerning the timing, location and relative extents of magmatic activity in the Mesozoic North Sea have to be addressed. First is the noticeable age progression of magmatic activity from north to south in the rift system (Table 6.1 from top to bottom, and Fig. 6.15) which was commented on, though not explained, by Dixon and colleagues (1981). The second question arises from the observed lack of magmatic activity in many parts of the system, most notably in the Viking Graben and much of the Central Graben, in marked contrast to the large outpourings of alkali basalts in the Forties area. Why should it appear that only some and not all on- and off-axis areas are active in terms of melt generation? Why is most of the off-axis activity seemingly confined to long-lived structural highs? We do not know the answers to all of these questions, but we suggest that some may lie in the nature of the rifting process, and are possibly also related to fundamental differences in the compositional and rheological properties of the lithosphere between different areas.

6.5.7. *Timing of magmatic activity*

Figure 6.15 shows radiometric dates, given as error bars, for igneous activity from north to south in the North Sea rift system during the Mesozoic. Note that different symbols are used for the different dating techniques. For reasons discussed earlier we place more reliability on $^{40}Ar/^{39}Ar$ than K/Ar dates which are likely to be disturbed in altered samples. The general southerly progression of magmatic activity from Mid Jurassic in the north to Late Jurassic on the flanks of the Central Graben, to Early Cretaceous in the south, may be a function of a slow southwards migration of the rift as it formed. Propagation of the rift system, which began in the north in Mid-Jurassic times, is supported by the Upper-Jurassic (late Oxfordian to early Kimmeridgian) onset of well defined block-faulting accompanied by differential fault-controlled subsidence and uplift in the Dutch Sector of the North Sea. In the Dutch Sector regional (thermal) subsidence did not get under way until the Late Cretaceous (Van Wijhe 1987), whereas the thermal subsidence sedimentation started some tens of Ma earlier, during the Early Cretaceous period, in northern parts of the rift system (Wood 1982; N. J. White 1988). These differences are probably sufficient to explain the observed age progression of the magmatism, especially when one considers the somewhat doubtful validity of many of the K/Ar dates used. We hope to constrain the picture better with further $^{40}Ar/^{39}Ar$ dating to be carried out in the near future. It should be noted that if the timing of igneous activity corresponds directly to propagation of the rift system, then melts formed at any locality during the extensional phase will escape and express themselves at the surface at that time.

6.5.8. *Location of magmatic activity*

The first and most basic observation concerning locations of Mesozoic igneous rocks in the North Sea is the compositional on-axis/off-axis relationship noted by Dixon and colleagues (1981). The largest degrees of melting, represented by the Forties alkali basalts, are found within the main graben structure, and are located at the rift triple junction. All of the other rocks are chemically more exotic, represent much smaller degrees of melting, occur in much smaller quantities, and are only found on the rift flanks or in sub-basins away from the rift proper. This simplistic view is immediately supported by estimates for the degrees of extension, which are much smaller in the off-axis areas; some of these off-axis areas have undergone no crustal extension at all but have been thinned in the sub-crust by the downward widening of the stretching envelope of the main rift, resulting in rift phase uplift and

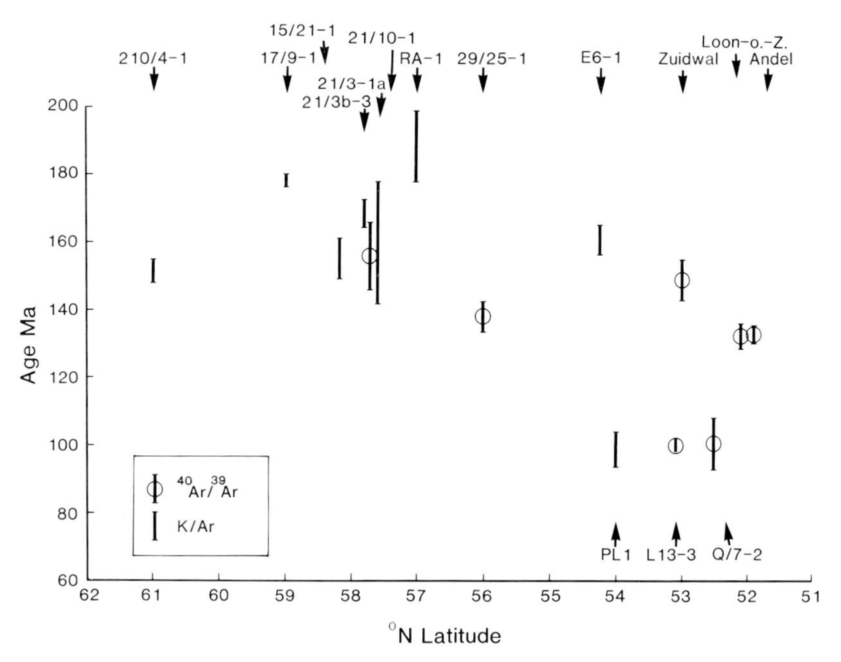

Fig. 6.15. Radiometric age data (given as error bars) for the Mesozoic North Sea igneous rocks (*see* Table 6.2) plotted against distance (given as degrees of latitude) down the rift system from north to south. Circles indicate samples for which $^{40}Ar/^{39}Ar$ dates have been determined; all other dates are by K/Ar techniques. Well numbers relate to individual samples indicated by the arrows. Please note (with reference to Table 6.2) that, in some cases, several age determinations for one sample are included within the single error bar (e.g. 21/3b-3, Zuidwal).

erosion, followed later by thermal subsidence. However, the relationship between extension and location is clearly not as simple as current models would predict. The relationships are more subtle. We must try to explain why the bulk of on-axis activity centres on the triple junction, why only certain flank areas are magmatically active, and why in general the magmatism appears to be so fickle in its choice of location.

According to estimates for the degree of extension in the Viking Graben (N. J. White 1988), the Central Graben, and in the triple junction (Wood 1982; Barton and Wood 1984), there is very little difference in lithospheric attenuation between these areas. If anything the amount of attenuation may reach its maximum values in the two graben structures some distance away from the triple junction. In terms of the model framework developed so far, the occurrence of magmatic activity only in the triple junction is somewhat disconcerting. Earlier we suggested that all of the North Sea rocks could be explained by differing degrees of melting of asthenosphere combined with melting of an enriched source (either streaks in the asthenosphere or an enriched layer, formed by small melt fractions, in the lithosphere). This means that even in the less stretched areas such as the Egersund Basin and the flanks of the Central Graben some solidus has been intersected and melting has occurred. Why then is no melt seen in the more highly attenuated Viking and Central Grabens?

Perhaps the answer lies in a combination of three separate arguments;

(1) the way in which the geometry of the triple junction is related to the amount of extension;
(2) the ways in which long-lived crustal lineaments control magma plumbing;
(3) the enrichment history of the lithosphere and how it may control the production of magma from the MBL.

Stretching at the triple junction

Our first observation is that, in terms of its geometry, the triple junction may in fact be an area where the amount of extension is larger than anywhere else.

For example, if the surrounding rifts (Moray Firth Basin, Viking Graben, Central Graben) all had similar β values of 1.5, then geometrical considerations require the triple junction to stretch by a factor of β^2, in this example by 2.25. The amount of stretching in the Moray Firth arm is, however, likely to be rather less than in the other two rifts, and this would reduce the β value in the triple junction while still keeping it above that of the Viking and Central Grabens. If this is the case, then it would explain why the largest extents of melting occur in the Forties area. However, the fact that the observed subsidence in the triple junction does not appear to fit with larger β values requires some explanation, as does the absence of melts within the surrounding rifts.

If the amounts of extension in the area of the triple junction are indeed larger than in the surrounding rifts, then there are four possible explanations as to why this has not already been reported. First the subsidence calculations of Wood (1982) and Barton and Wood (1984) assume that the top of the volcanics in the Forties province (the Rattray Formation) represents already-compacted basement, and only consider the thickness and compaction of sediments deposited after this period. If the calculations could be performed from the base of the volcanic sequence, when the rift phase really began, then stretching estimates might be more realistic, and would be larger. Second, the subsidence calculations above do not take into account the marked reduction in subsidence for a given value of β that can be caused by the production of melts during rifting which decrease the overall density of the lithosphere, and the reduced density of the residual asthenosphere from which the melt has been extracted. Third, the three-dimensional nature of the triple junction in terms of heat conduction may have had a profound effect on the subsidence history. Finally, flow of the lower crust during rifting and accompanying magmatic activity could lead to much reduced initial subsidence, and even uplift. The work of producing revised subsidence calculations and extension estimates is already under way.

Larger degrees of extension in the triple junction still do not explain the paucity of magmatism elsewhere within the main rifts. This problem might simply be due to sampling (there are fewer wells drilled in the centres of the main rifts) but we think this unlikely and that the absence is real. Another explanation is given by Holliger and Klemperer (this volume), who suggest that melts are 'intraplated' into the base of the crust in parts of the Viking Graben on the basis of an observed discrepancy between the Moho depth as determined from gravity models and seismic reflection.

Magma plumbing considerations

The coincidence between magmatic activity and major lines of weakness within rift zones has long been known (Cartwright, this volume). The triple junction area in the North Sea appears to coincide with a north-east extrapolation of the Highland Boundary fault and may also coincide with the line of the Iapetus Suture, although new evidence from deep seismic reflection profiling suggests that this may not be the case (Klemperer and Hurich, this volume). The Central Graben province lies on the north western extension of the Tornquist line where it intersects the Mid North Sea High. Such major zones of weakness in the lithosphere may allow easy passage for magma bodies where elsewhere they are largely intruded at base-crust levels.

Enrichment history considerations

The explanation which we will now put forward relies on the small-melt fraction model for lithospheric enrichment rather than the 'streaky-mantle' model (as we are reluctant to postulate very large differences in streak-population between different areas), and also on the observation that, apart from the Forties, the magmatism is largely confined to areas which were formerly stable highs.

We suggest that melting of the asthenosphere occurred below the Viking and Central Grabens but that, unlike the more attenuated Forties area, the melt fractions were not able to escape to the surface. The reason that melts are observed in central North Sea (flanks of the Central Graben), and the southern North Sea may be due to the fact that in these areas the graben intersects pre-existing and long-lived stable highs, the central North Sea High, the Texel-Isselmeer High, and the London-Brabant Massif (Ziegler 1982*a*). If the enriched-lithosphere model is correct, then long-lived and stable areas such as these may be expected to have accumulated more in the way of enriched veins than the surrounding more mobile areas. If this is true then perhaps below these areas small amounts of melt from the asthenosphere could combine with enough remobilized enriched lithosphere to create larger volumes of melt able to reach the surface. In the vicinity of the main graben the extension during Carboniferous and Permian may have removed much of the enriched-lithosphere component in areas where there was magmatism, and Triassic extension may have remobilized and moved any remaining small melt fractions to higher levels in the lithosphere, levels which could not be reached by migrating melts in the Mesozoic before they reached their freezing temperature, or which, if they could be reached, did not provide enough enriched melt to constitute a magma body. This explanation also explains the lack of magma of intermediate composition between the alkali basalts and the more exotic types (lamprophyres and nephelinites). The existence of nephelinites and alnöites in the Egersund Basin may indicate a similar process.

One of the most notable features of the North Sea rift system in general is the very minor amount of magmatic activity. This feature was brought out by Dixon and colleagues (1981) in their comparison with the East African rift system. The North Sea shows substantially less magmatism, even though the degrees of extension appear to be similar. The Eastern Rift of East Africa is developed in a stable Proterozoic cratonic lithosphere, whereas the Western Rift is in Archean lithosphere and shows more extreme (and potassic) rock types. The North Sea lithosphere is, in comparison, only recently stabilized.

6.5.9. *Summary*

Magmatic activity in the North Sea rift system shows an approximate 60 Ma time progression from north to south, which may be related to southwards migration of the rift. The location of magmatic activity is controlled by the relative degrees of extension coupled with the location of zones of weakness in the lithosphere and the degree of enrichment of the lithosphere. The degree of extension is greatest in the triple junction, which may also coincide with major lineaments, hence the extruded alkali basalts of the Forties province. Elsewhere the asthenosphere probably melted to a small degree in most places, but the majority of these melts either could not escape or were frozen during their passage to the surface except where they could amalgamate with remobilized small-melt fractions in the lithosphere. The amounts of small-melt fraction veining/metasomatizing the lithosphere is likely to be higher in formerly-stable blocks of lithosphere, and it is in these areas that most other igneous rocks are observed. The past history of the lithosphere may therefore play a crucial role in locating areas of magmatic activity and in determining the overall preponderance of magmatic activity in any rift system for a given amount of extension.

6.6. Conclusions

The primary aims of this paper were threefold. First, we intended to review both the existing theoretical framework concerning melt generation in rifts and also the character of Mesozoic rift-related magmatism in the North Sea Basin. Secondly, we wished to say something about degrees of partial melting and upwelling implied by the observed magma chemistries. Finally, and as our primary objective, we set out to test the quantitative model of McKenzie and Bickle (1988) using the observations on melt chemistry, degrees of extension, pre-rift lithosphere thickness, and the scale of pre-rift uplift in the North Sea.

The amounts and compositions of melts generated by the decompression of upwelling asthenosphere below rifts are controlled by a number of interrelated and interdependent parameters: degree of extension; potential temperature; pre-rift lithosphere thickness; rate of extension; mantle source composition; lithospheric interaction; and the nature of the extraction process. In the North Sea the first four parameters are well constrained, the mantle source composition and the nature of lithospheric interaction are problematic, and extraction, in theory poses no problems.

The rift-related magmatism varies from mildly to highly undersaturated and alkaline in character, it varies in age with postulated rift propagation from north to south, and is restricted to specific locations in the system. Modelling using trace-element ratios suggests that, if the magmas are produced from dry depleted MORB-type asthenosphere, then they result from very small degrees of melting (less than 0.5 per cent) and that there must be some input from an enriched mantle source/sources either in the lithosphere or the asthenosphere or both, probably requiring greater extents of melting (<2 per cent).

The McKenzie and Bickle (1988) model is unable to explain all the observations in the North Sea. The model, which predicts no melting given the most likely set of parameters applicable for the North Sea, breaks down because of its poor resolution on the scale of small-melt fractions. However, it still provides an inescapable constraint in that the temperature of the alkali basalt liquidus in the asthenosphere must have been attained below the Forties province. On balance we feel that 1–2 per cent melting of the real asthenosphere (rather than the dry experimental asthenosphere) may well be accessible with a Tp of 1280°C and $\beta = 1.6$ corresponding to a 100°C lowering of the dry solidus due to the presence of volatiles and other minor components. The dry solidus model will, however, still apply in areas where the degree of melting is large. The fact that melting does occur and that the melt compositions are generally highly enriched compared to those which can be produced from a single-stage melting event of a depleted MORB-source asthenospheric mantle (Forties excepted) leads us to envisage a more complex model involving migration of ultra-small melt fractions into the lithosphere, their crystallization, and their subsequent remobilization during 'events'. Enriched low melting-point lithosphere may operate in conjunction with a streaky asthenosphere, or with asthenosphere which may show a continuum in melting between ultra-low melt fractions dominated by CO_2/H_2O and the beginning of melting of 'dry' peridotite. As yet we have no way of choosing between these different mantle sources, nor do we know exactly where in T-P space melting first takes place.

We cannot yet provide quantitative estimates of the degrees of extension from melt chemistries when the degrees of melting are small and melt sources are enriched. However, when such melts are seen, the degrees of extension are not likely to be as great as 2.5 for Tp = 1280°C, and may be considerably less. In order to proceed, careful analysis of experimental data concerning the melting points and compositions of such small melt fractions must be undertaken, and the chemical and isotopic nature of the lithosphere must be characterized.

To conclude on a more optimistic note, it seems likely that the McKenzie and Bickle (1988) model will be applicable in areas where considerably larger degrees of melting have occurred, and could be used to provide first order estimates of extension and, from that, heat flow and generalized subsidence history.

Acknowledgements

We are indebted to the following oil companies for provision of the igneous material used in this study as well as additional information and discussion: Amerada Hess Ltd, Amoco UK Exploration Co., Britoil, BP Exploration Co. Ltd, Chevron Exploration North Sea Ltd, Elf (UK), Esso (Norway), Nederlandse Aardolie M.B.V., Occidental Petroleum Ltd, Petroland (Elf), Phillips Petroleum Co. UK Ltd, Premier Oil Exploration Ltd, Ranger Oil, Shell International M.B.V., Shell UK Exploration Ltd, Texaco Ltd, Total Oil-Marine Plc. We are grateful to Dan McKenzie for much useful and enlightening discussion, and for providing preprints of his work. Martin Menzies made many useful comments when reviewing the original manuscript. Peter Ziegler is thanked for his help in initiating this study, and for provision of preprints. Nicky White kindly introduced D.M.L. to the arguments about extension estimates and subsidence *vs* uplift. We thank Paul Beattie and Scott Babcock who provided useful discussion on earlier drafts of this paper. I. Brown, P. Condon, D. James, S. Klemperer, H. Nicholson, and M. Spiegelman are thanked for helpful discussion. We are grateful to Dodie James for assistance with XRF analyses and to Yvonne Cooper for photographic services beyond the call of duty. D.M.L. gratefully acknowledges the support of a N.E.R.C. studentship during the course of this work.

References

Allègre, C. J. (1982). Chemical geodynamics. *Tectonophysics*, **81**, 109–32.

Allègre, C. J. (1987). Isotope geodynamics. *Earth and Planetary Science Letters*, **86**, 175–203.

Ahern, J. L. and Turcotte, D. L. (1979). Magma migration beneath the ocean ridge. *Earth and Planetary Science Letters*, **45**, 115–22.

Baker, B. H. (1987). Outline of the petrology of the Kenya rift alkaline province. In *Alkaline igneous rocks* (ed. J. G. Fitton and B. G. J. Upton), *Geol. Soc. Special Publication*, **30**, 293–312. London, Blackwell Scientific Publications.

Barberi, F., Santacroce, R., and Varet, J. (1982). Chemical aspects of rift magmatism. In *Continental and oceanic rifts* (ed. G. Palmason), 223–58. AGU Geodynamics Series, Vol. 8.

Barton, P. and Wood, R. (1984). Tectonic evolution of

the North Sea Basin: crustal stretching and subsidence. *Geophys. J. Roy. Astron. Soc.*, **79**, 987–1022.

Bosworth, W. (1987). Off-axis volcanism in the Gregory Rift, East Africa: Implications for models of continental rifting. *Geology*, **15**, 391–400.

Brooks, C., James, D. E., and Hart, S. R. (1976). Ancient lithosphere: Its role in young continental volcanism. *Science*, **193**, 1086–94.

Burke, K. and Dewey, J. F. (1973). Plume generated triple junctions: Key indicators in applying plate tectonics to old rocks. *J. Geol.*, **81**, 387–405.

Christie, P. A. F. and Sclater, J. G. (1980). An extensional origin for the Buchan and Witchground Graben in the North Sea. *Nature*, **283**, 729–32.

Cottençon, A., Parant, B. and Flacelière, G. (1975). Lower Cretaceous gas fields in Holland. In *Petroleum and the continental shelf of north west Europe*, Vol. 1; *Geology* (ed. A. W. Woodland), 403–12, London, Applied Science Publishers.

Cox, K. G. (1983). The Karoo province of southern Africa: origin of trace element enrichment patterns. In *Continental basalts and mantle xenoliths* (ed. C. J. Hawkesworth and M. J. Norry), 139–57. Nantwich, Cheshire, Shiva Publishing.

Dewey, J. F. (1982). Plate tectonics and the evolution of the British Isles. *J. Geol. Soc. of London*, **139**, 371–412.

Dixon J. E., Fitton, J. G. and Frost, R. T. C. (1981). The tectonic significance of post-Carboniferous igneous activity in the North Sea Basin. In *Petroleum geology of the continental shelf of north west Europe* (ed. L. V. Illing and G. D. Hobson), 121–37. London, Heyden and Son.

Edgar, A. D. (1987). The genesis of alkaline magmas with emphasis on their source regions: inferences from experimental studies. In *Alkaline igneous rocks* (ed. J. G. Fitton and B. G. J. Upton), Geological Society Special Publication **30**, 29–52. Oxford, Blackwell Scientific Publications.

Eggler, D. H. (1976). Does CO_2 cause partial melting in the low-velocity layer of the mantle? *Geology*, **4**, 69–72.

Eggler, D. H. (1978). Stability of dolomite in a hydrous mantle, with implications for the mantle solidus. *Geology*, **6**, 397–400.

Erlank, A. J., Waters, F. G., Hawkesworth, C. J., Haggerty, S. E., Allsopp, H. L., Rickards, R. S., and Menzies, M. A. (1987). Evidence for mantle metasomatism in peridotite nodules from the Kimberley Pipes, South Africa. In *Mantle metasomatism* (ed. M. A. Menzies and C. J. Hawkesworth), 221–311. London, Academic Press.

Eynon, G. (1981). Basin development and sedimentation in the Middle Jurassic of the northern North Sea. In *Petroleum geology of the continental shelf of northwest Europe* (ed. L. V. Illing and G. D. Hobson), 196–204. London, Heyden and Son.

Faerseth, R. B., Macintyre, R. M., and Naterstad, J. (1976). Mesozoic alkaline dykes in the Sunnhørdland region, Western Norway: ages, geochemistry and regional significance. *Lithos*, **9**, 331–48.

Fall, H. G. (1980). The petrology and geochemistry of Jurassic igneous rocks from the northern North Sea. Unpublished PhD thesis, University of Sheffield.

Fall, H. G., Gibb, F. G. F., and Kanaris-Sotiriou, R. (1982). Jurassic volcanic rocks of the northern North Sea. *J. Geol. Soc.*, **139**, 277–92.

Fitton, J. G. (1987). The Cameroon line, West Africa: a comparison between oceanic and continental alkaline volcanism. In *Alkaline igneous rocks* (ed. J. G. Fitton and B. G. J. Upton). *Geol. Soc. Special Publication*, **30**, 273–92. Oxford, Blackwell Scientific Publications.

Fitton, J. G. and Dunlop, H. M. (1985). The Cameroon Line, West Africa, and its bearing on the origin of oceanic and continental alkali basalt. *Earth and Planetary Science Letters*, **72**, 23–38.

Fitton, J. G. and James, D. (1986). Basic volcanism associated with intraplate linear features. *Phil. Trans. Roy. Soc. of London*, **A317**, 253–66.

Fleck, R. J., Sutter, J. F., and Elliot, D. H. (1977). Interpretation of discordant $^{40}Ar/^{39}Ar$ age spectra of Mesozoic tholeiites Antarctica. *Geochimica et Cosmochimica Acta.*, **41**, 15–32.

Foucher, J. P., Le Pichon, X., and Sibuet, J. C. (1982). The ocean-continent transition in the uniform lithosphere stretching model: role of partial melting in the mantle. *Phil. Trans. Roy. Soc. of London*, **A305**, 27–43.

Furnes, H., Elvsborg, A. and Malm, O. A. (1982). Lower and Middle Jurassic alkaline magmatism in the Egersund sub-basin, North Sea. *Marine Geology*, **46**, 53–69.

Gast, P. W. (1968). Trace element fractionation and the origin of tholeiitic and alkaline magma types. *Geochimica et Cosmochimica Acta*, **32**, 1057–86.

Gibb, F. G. F. and Kanaris-Sotiriou, R. (1976). Jurassic igneous rocks of the Forties Field. *Nature*, **260**, 23–5.

Gibbs, A. D. (1987*a*). Linked tectonics of the northern North Sea basins. In *Sedimentary basins and basin forming mechanisms* (ed. C. Beaumont and A. J. Tankard) *Can. Soc. Petrol. Geologists*, *Memoir* **12** 163–71.

Gibbs, A. D. (1987*b*). Deep seismic profiles in the northern North Sea. In *Petroleum geology of north west Europe* (ed. J. Brooks and K. Glennie), 1025–8. London, Graham and Trotman.

Gibbs, A. D. (1989*a*). Structural styles in basin formation. In *Extensional tectonics and stratigraphy of the North Atlantic margins* (ed. A. J. Tankard and H. R.

Balkwill). *Am. Assoc. Petrol. Geologists Memoir* **46**, 81–93.

Gibbs, A. D. (1989*b*). A model for linked basin development around the British Isles. In *Extensional tectonics and stratigraphy of the North Atlantic margins* (ed. A. J. Tankard and H. R. Balkwill). *Am. Assoc. Petrol. Geologists Memoir* **46**, 501–509.

Green, D. H. and Ringwood, A. E. (1967). The genesis of basaltic magmas. *Contributions to Mineralogy and Petrology*, **15**, 103–90.

Green, D. H. and Wallace, M. E. (1988). Mantle metasomatism by ephemeral carbonatite melts. *Nature*, **336**, 459–62.

Hallam, A., Hancock, J. M., La Breque, J. L., Lowrie, W., and Channell, J. E. T. (1985). Jurassic to Palaeogene: Part 1. Jurassic and Cretaceous geochronology and Jurassic to Palaeogene magnetostratigraphy. In *The chronology of the geological record* (ed. N. J. Snelling). *Geological Society Memoir* **10**, 118–40. London, Blackwell.

Harland, W. B., Cox, A. V., Llewellyn, P. B., Pickton, C. A., Smith, A. G., and Walters, R. (1982). *A Geological time scale*. Cambridge University Press.

Harland, W. B., Armstrong, R. C., Craig, L., Smith, A. G., and Smith, D. G. (in press). *A geologic time scale*. Cambridge University Press.

Harrison, R. K., Jeans, C. V., and Merriman, R. J. (1979). Mesozoic igneous rocks, hydrothermal mineralisation and volcanogenic sediments in Britain and adjacent regions. *Bull. Geol. Survey of Great Britain*, **70**, 57–69.

Hart, R. A. (1973). A model for chemical exchange in the basalt-sea water system of oceanic layer II. *Can. J. of Earth Sciences*, **10**, 799–816.

Harte, B. (1983). Mantle peridotites and processes—the kimberlite sample. In *Continental basalts and mantle xenoliths* (ed. C. J. Hawkesworth and M. J. Norry), 46–91. Nantwich, Cheshire, Shiva Publishing.

Hawkesworth, C. J., Van Calsteren, P., Rogers, N. W., and Menzies, M. A. (1987). Isotope variations in recent volcanics: a trace-element perspective. In *Mantle metasomatism* (ed. M. A. Menzies and C. J. Hawkesworth), 365–89. London, Academic Press Geology Series.

Hellinger, S. J., and Sclater, J. G. (1983). Some comments on two-layer extensional models for the evolution of sedimentary basins. *J. Geophys. Research*, **88**, 8251–69.

Hofmann, A. W. and White, W. M. (1982). Mantle plumes from ancient oceanic crust. *Earth and Planetary Science Letters*, **57**, 421–36.

Holliger, K. and Klemperer, S. L. (1989). A comparison of the Moho interpreted from gravity data and from deep seismic reflection data in the northern North Sea. *Geophys. J. Roy. Astron. Soc.*, **97**, 247–58.

Howitt, F., Aston, E. R., and Jacque, M. (1975). The occurrence of Jurassic volcanics in the North Sea. In *Petroleum and the continental shelf of north west Europe, Vol. 1, Geology*, (ed. A. W. Woodland), 379–88. Barking, Essex, Applied Science Publishers Ltd.

Hunter, R. H. and McKenzie, D. P. (1989). The equilibrium geometry of carbonate melts in rocks of mantle composition. *Earth and Planetary Science Letters*, **92**, 347–56.

James, S. D. (1987). Volcanism in sedimentary basins and its implications for mineralization. Unpublished PhD thesis, University of Newcastle-upon-Tyne.

Jaques, A. L. and Green, D. H. (1980). Anhydrous melting of peridotite at 0–15 kb pressure and the genesis of tholeiitic basalts. *Contributions to Mineralogy and Petrology*, **73**, 287–310.

Jarvis, G. T. and McKenzie, D. P. (1980). Sedimentary basin formation with finite extension rates. *Earth and Planetary Science Letters*, **48**, 42–52.

Kent, P. E. (1975). Review of North Sea basin development. *J. Geol. Soc. of London*. **131**, 435–68.

Klemperer, S. L. (1988). Crustal thinning and nature of extension in the northern North Sea from deep seismic reflection profiling. *Tectonics*, **7**, 803–21.

Klemperer, S. L. and White, N. J. (1989). Coaxial stretching or lithospheric simple shear in the North Sea? Evidence from deep seismic profiling and subsidence. In *Extensional tectonics and stratigraphy of the North Atlantic margins* (ed. A. J. Tankard and H . R. Balkwill). *Am. Assoc. Petrol. Geologists*, Memoir **46**, 511–22.

Klemperer, S. L. and Hurich, C. A. (this volume). Lithospheric structure of the North Sea from deep seismic reflection profiling.

Kostopoulos, D. and James, S. D. (submitted). The nature, composition and melting regime of the upper mantle and the influence of variably stretched lithosphere on magma production and trace element geochemistry. *J. Petrology*.

Latin, D. M. and White, N. J. (1990). Generating melt during lithospheric extension: pure shear *vs* simple shear. *Geology*, **18**, 327–31.

Leeder, M. R. (1983). Lithospheric stretching and North Sea Jurassic sourcelands. *Nature*, **305**, 510–14.

Le Pichon, X. and Sibuet, J. C. (1981). Passive margins: A model of formation. *J. Geophys. Research*, **86**, 3708–20.

McKenzie, D. P. (1967). Some remarks on heat flow and gravity anomalies. *J. Geophys. Research*, **72**, 6261–73.

McKenzie, D. P. (1978). Some remarks on the development of sedimentary basins. *Earth and Planetary Science Letters*, **40**, 25–32.

McKenzie, D. P. (1984). The generation and compaction of partially molten rock. *J. Petrology*, **25**, 713–65.

McKenzie, D. P. (1989). Some remarks on the movement of small melt fractions in the mantle. *Earth and Planetary Science Letters*, **95**, 53–72.

McKenzie, D. P. and Bickle, M. J. (1988). The volume and composition of melt generated by extension of the lithosphere. *J. Petrology*, **29**, 625–79.

McKenzie, D. P., Watts, A., Parsons, B., and Roufosse, M. (1980). Planform of mantle convection beneath the Pacific Ocean. *Nature*, **303**, 602–3.

Mohr, P. (1982). Musings on continental rifts. In *Continental and oceanic rifts* (ed. G. Palmason), 293–309. AGU Geodynamics Series, Vol. 8.

Parsons, B. and McKenzie, D. P. (1978). Mantle convection and the thermal structure of the plates. *J. Geophys. Research*. **83**, 4485–96.

Perrot, J. and Van der Poel, A. B. (1987). Zuidwal—a Neocomian gasfield. In *Petroleum geology of north west Europe* (ed. J. Brooks and K. Glennie), 325–35. London, Graham and Trotman.

Ribe, N. M. (1985). The generation and composition of partial melts in the earth's mantle. *Earth and Planetary Science Letters*, **73**, 361–76.

Ribe, N. M. (1987). Theory of melt segregation—a review. *J. Volcanology and Geothermal Research*, **33**, 241–53.

Richter, F. M. and McKenzie, D. P. (1984). Dynamical models for melt segregation from a deformable matrix. *J. Geology*, **92**, 729–40.

Ritchie, J. D., Swallow, J. L., Mitchell, J. G., and Morton, A. C. (1988). Jurassic ages from intrusives and extrusives within the Forties igneous province. *Scot. J. Geology*, **24**, 81–8.

Royden, L. and Keen, C. E. (1980). Rifting process and thermal evolution of the continental margin of eastern Canada determined from subsidence curves. *Earth and Planetary Science Letters*, **51**, 345–61.

Sclater, J. G. and Christie, P. A. F. (1980). Continental stretching: An explanation of the post mid-Cretaceous subsidence of the Central North Sea Basin. *J. Geophys. Research*, **85**, 3711–39.

Sclater, J. G., Royden, L., Horvath, F., Burchfiel, B. C., Semken, S., and Stegena, L. (1980). The formation of the Intra Carpathian Basins as determined from subsidence data. *Earth and Planetary Science Letters*, **51**, 139–62.

Scott, D. R. and Stevenson, D. J. (1986). Magma ascent by porous flow. *J. Geophys. Research*, **91**, 9283–96.

Shaw, D. M. (1970). Trace element fractionation during anatexis. *Geochimica et Cosmochimica Acta*, **34**, 237–43.

Sleep, N. H. (1971). The thermal effects of the formation of Atlantic continental margins by continental breakup. *Geophys. J. Roy. Astron. Soc*., **24**, 325–50.

Sleep, N. H. (1984). Tapping of magmas from ubiquitious mantle heterogeneities: an alternative to mantle plumes? *J. Geophys. Research*, **89**, 10029–41.

Spiegelman, M. and McKenzie, D. P. (1987). Simple 2-D models for melt extraction at mid-ocean ridges and island arcs. *Earth and Planetary Science Letters*, **83**, 137–52.

Steiger, R. H. and Jäger, E. (1977). Subcommission on geochronology: convention on the use of decay constants in geochronology and cosmochronology. *Earth and Planetary Science Letters*, **36**, 359–62.

Stolper, E., Walker, D., Hager, B. H., and Hays, J. F. (1981). Melt segregation from partially molten source regions: the importance of melt density and source region size. *J. Geophys. Research*, **86**, 6261–71.

Takahashi, E. and Kushiro, I. (1983). Melting of a dry peridotite at high pressure and basalt magma genesis. *Am. Mineralogist*, **68**, 859–79.

Van Hinte, J. E. (1976). A Jurassic time scale. *Am. Assoc. Petrol. Geologists Bulletin*, **60**, 489–97.

Van Wijhe, D. H. (1987). The structural evolution of the Broad Fourteens Basin. In *Petroleum geology of north west Europe* (ed. J. Brooks and K. Glennie), Vol. 1, 315–24. London, Graham and Trotman.

Verhoogen, J. (1954). Petrological evidence on temperature distribution in the mantle of the earth. *Trans. Am. Geophys. Union*, **35**, 85–92.

Wallace, M. E. and Green, D. H. (1988). An experimental determination of primary carbonatite magma composition. *Nature*, **335**, 343–6.

Wernicke, B. (1985). Uniform sense simple shear of the continental lithosphere. *Can. J. Earth Sciences*, **22**, 108–25.

White, N. J. (1988). Extension and subsidence of the continental lithosphere. Unpublished PhD thesis. Cambridge University.

White, N. J. (1989). The nature of lithospheric extension in the North Sea. *Geology*, **17**, 111–14.

White, N. J., Jackson, J. A., and McKenzie, D. P. (1986). The relationship between the geometry of normal faults and that of sedimentary layers in their hanging walls. *J. Struct. Geol*., **8**, 897–909.

White, N. J. and McKenzie, D. P. (1988). Formation of the 'steer's head' geometry of sedimentary basins by differential stretching of the crust and mantle. *Geology*, **16**, 250–3.

White, R. S. (1988*a*). The Earth's crust and lithosphere. In *Oceanic and continental lithosphere: similarities and differences* (ed. M. A. Menzies and K. G. Cox), 1–10. *J. Petrology*, special volume 1988. Oxford University Press.

White, R. S. (1988*b*). A hot-spot model for Early Tertiary volcanism in the North Atlantic. In *Early Tertiary volcanism and the opening of the north east Atlantic* (ed. A. C. Morton, and L. M. Parson), *Geol. Soc, Special Publication*, **39**, 3–13.

White, R. S. and McKenzie, D. P. (1989). Magmatism at rift zones: The generation of volcanic continental margins and flood basalts. *J. Geophys. Research*, **94**, 7685–729.

White, R. S., Fowler, S. R., McKenzie, D. P., Westbrook, G. K., and Bowen, A. N. (1987). Magmatism at rifted continental margins. *Nature*, **330**, 439–44.

Whiteman, A., Naylor, D., Pegrum, R., and Rees, G. (1975). North Sea troughs and plate tectonics. *Tectonophysics*, **26**, 39–54.

Wood, R. J. (1982). Subsidence in the North Sea. Unpublished PhD thesis, Cambridge University.

Woodhall, D. and Knox, R. W. (1979). Mesozoic volcanism in the northern North Sea and adjacent areas. *Bull. Geol. Survey of Great Britain*, **70**, 34–56.

Wyllie, P. J. (1987). Transfer of subcratonic carbon into kimberlites and rare earth carbonatites. In *Magmatic processes: physiochemical principles* (ed. B. O. Mysen). *Geochem. Soc. Special Publication*, **1**, 107–19.

Yoder, H. S. (1976). *Generation of basaltic magma*. Washington, DC, National Academy of Science.

Ziegler, P. A. (1975). North Sea history in the tectonic framework of north-western Europe. In *Petroleum geology and continental shelf of north west Europe*, *Vol. 1, Geology* (ed. A. W. Woodland), 131–50. Barking, Essex, Applied Science Publishers.

Ziegler, P. A. (1981). Evolution of sedimentary basins in north-west Europe. In *The petroleum geology of the continental shelf of north-west Europe* (ed. L. V. Illing and G. D. Hobson), 3–39. London, Heyden and Son.

Ziegler, P. A. (1982*a*). Faulting and graben formation in Western and Central Europe. *Phil. Trans. Roy. Soc. London*, **A305**, 113–43.

Ziegler, P. A. (1982*b*). *Geological atlas of western and central Europe*. The Hague, Shell Internationale Maatschappij B.V.

Ziegler, P. A. (1983). Crustal thinning and subsidence in the North Sea: matters arising. *Nature*, **304**, 561.

Ziegler, P. A. and Van Hoorn, B. (1989). Evolution of the North Sea rift system. In *Extensional tectonics and stratigraphy of the North Atlantic margins* (ed. A. J. Tankard and H. R. Balkwill), *Am. Assoc. Petrol. Geologists Memoir*, **46**, 471–500.

7 Linked fault tectonics of the North Sea

A. D. Gibbs

Abstract

The North Sea, comprising the Viking and Central Grabens, is predominantly a Jurassic extensional rift with a superimposed thermal subsidence basin of Cretaceous and Tertiary age. The architecture of the northern and southern parts of the rift is dominated by tilted fault blocks forming terraced margins. The detailed geometry of these terraces is discussed and the role of second-order crestal faults and linking transfer fault components is emphasized.

Transfer faults and fault zones are critical in the structural development of the North Sea basin, and are a key factor in the prospectivity of the area. The variety of resulting structures is summarized, and the degree to which the extensional basement and cover faults form a linked system is discussed.

At a regional scale, the Viking Graben and much of the Central Graben are asymmetric in cross-section. This results in differing distributions of hydrocarbon traps on either side of the rift. Possible models for this observed asymmetry include linked crustal simple shear and partially decoupled simple and pure shear delamination of the lower crust, as well as the traditional symmetric pure shear system.

The key factor in each of these end member models is that they can be tested by reference to regional stratigraphic and structural data bases. These tests involve cross-sectional area balance with appropriate allowances for compaction of the cover, ductile deformation, and isostatic and thermal effects. The approach combining modelling and geometric tests at each stage of interpretation used in the North Sea has proved to be a powerful tool in the analysis of rift basins.

7.1. Introduction

Regional maps of the North Sea rift systems clearly show that the Viking and Central Grabens are bounded by continuous fault zones along both margins. These form part of the major tectonic lineament system of north-western Europe (Fig. 7.1). The North Sea rift is contiguous with marginal faulted basins such as the Møre and Mid Norwegian basins to the north (Nelson and Lamy 1987; Gage and Dore 1986), the Witch Ground, Moray, Aberdeen, and Southern Gas basins to the west (Glennie 1986; Ziegler 1982), the Stord and Tornquist systems to the east (Pegrum 1984*a*, *b*) and, to the south, the Dutch Central Graben and eventually the Rhine Graben (Clark-Lowes *et al.* 1987; Olsen 1987).

Each of these basins has its own complex history of ancestral basin growth prior to rifting in the Mesozoic. Regional isopach maps such as those by Ziegler (1982, 1989), Pegrum and Spencer (1990) and more detailed work on single formations (Brown *et al.* 1987) demonstrate that the North Sea rift formed a more or less continuously subsiding basin throughout the Triassic and Jurassic. Additional depocentres lie on the terraced margins.

Along the margins of the rift, local uplift and tilting of individual horst blocks led to erosional truncation or non-deposition on the crests, and along parts of the rim of both the Viking and Central Grabens. This

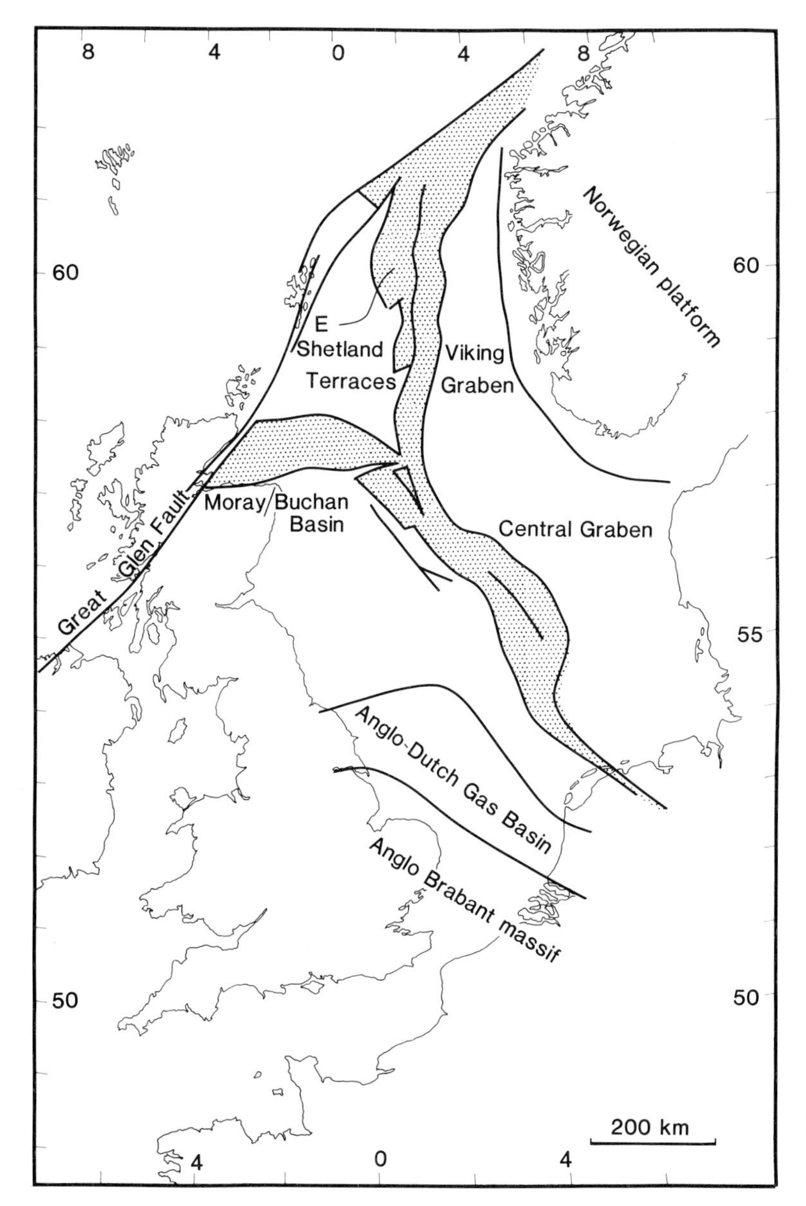

Fig. 7.1. Generalized tectonic elements of the North Sea basin. Stippled area indicate main graben.

combination of extensional faulting, uplift, and subsidence has led to the formation of the prolific oil and gas fields of the North Sea (Fig. 7.2). The fields in the Viking and Central Grabens are located largely on the tilted Jurassic horst blocks and along reactivated transfer faults which have controlled later sedimentation. This pattern becomes modified in the Central Graben by associated inversion and salt movement on pre-existing structural trends. In the Anglo-Dutch Gas Basin inversion of earlier structures and salt movement are more important than rift forming structures in trap development.

Towards the end of the Jurassic, marked rift-controlled sedimentation ceased and the basin broadened markedly as thermal subsidence took over as the dominant process, producing the characteristic 'steer's head' geometry of an extensional and superimposed thermal basin (White and McKenzie

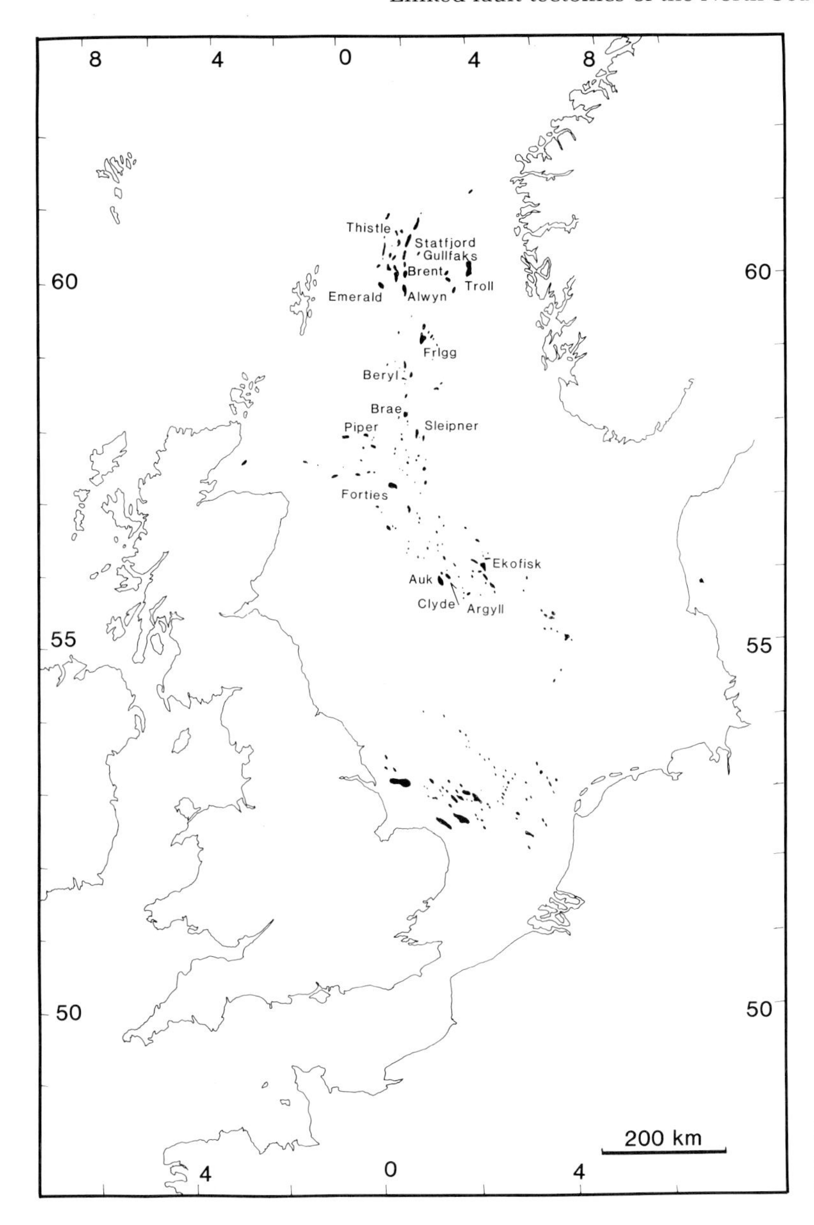

Fig. 7.2. Distribution of principal oil and gas fields in the North Sea rift and the southern gas basin.

1988). At this stage, fault-controlled subsidence became essentially confined to one or two of the major axial fault zones. In the north Viking Graben, for example, substantial subsidence continued throughout the lower to mid Cretaceous along the western boundary fault of the Viking Graben rather than on the Shetland Terraces. Hence faulting activity became more concentrated in the axial zone of the graben despite a general broadening of the depositional basin as a whole. This observation led to the interpretation of the graben forming essentially by hangingwall collapse (Gibbs 1984*a*, *b*), with the major western boundary faults of the Viking Graben forming the short-cut or break-away fault (cf. Wernicke 1985, 1986).

McKenzie's (1978) model for rift basin formation placed these observations on a sound theoretical basis where the extension of the upper crust in the rift phase is related to extension of the entire lithosphere. In turn, the thermal history, subsidence, and hence hydrocarbon maturation in the basin, depend on the amount of extension. Thermal equilibriation of the asthenosphere eventually leads to the formation of the post-rift subsidence basin whose width is controlled by the wavelength of the extensional thinning of the lower lithosphere. McKenzie thus showed that tectonic extension, basin subsidence, timing, and thermal history are all related to the amount of lithospheric extension given as the 'beta factor' measured across the visible basin.

McKenzie's pure shear model is inherently volume balanced within the lithosphere in the sense of Dahlstrom (1969), when allowances are made for water load and compaction of the sediment fill, and the P/T volume effect on the lower crust and mantle lithosphere is taken into account. Within the limits of observation, predictions of crustal thinning above the Moho from a pure shear model (Barton and Wood 1984; Barton 1985) and development of the subsidence basin (Sclater and Christie 1980; Thorne and Watts 1989) fits the observed data well enough for the approach to have wide and general applications not only to the North Sea but many other predominantly extensional rifts.

7.2. Regional extension

A number of authors have now pointed out that a major anomaly exists in the North Sea between the observed extension on faults mapped at the pre-rift and early syn-rift markers and the predicted values for extension derived by formal backstripping the basin of its thermal component to derive a value for beta (extension). Ziegler (1989) for example, quotes extensional beta values of 1.15 and 1.02 for the Viking and Central Grabens respectively. These values were measured from fault offsets on industry seismic interpretations, and are consistently low compared with values computed from subsidence rates. Ziegler implies that 'intra plate' processes other than extension must operate, for example subcrustal erosion or 'basification' of the lower crust by plutonic processes related to thermal processes in the mantle. There is, however, no direct evidence for these other processes on a wide enough scale in the North Sea rift.

Other authors suggest that the success of the McKenzie model is a consequence of 'aliasing', i.e. that the lithospheric extension is achieved by low angle faults spread over broad regions, such that the rift basin of one detachment system approximately overlies the region of lithospheric mantle pure shear and associated thermal subsidence of another (Coward 1986; Kusznir *et al.* 1987; Moretti and Pinet 1987). Thorne and Watts (1989) further point out that the superposition of more than one stretching event leads to the inability to adequately discriminate between different models for stretch, or to separate the subsidence component relating to each event. In particular they suggest that a late-stage Tertiary stretching event may be represented relating to reactivation of the North Sea rift system during opening of the North Atlantic. Local extension as well as inversion, particularly along transfer fault trends, is widespread during the Tertiary (Cornforth *et al.* 1986; Gibbs, 1989 *a* and *b*).

Further discrepancies arise in estimating tectonic extension of the pre-rift sequence from interpretations of seismic data. Gibbs (1983, 1984*a*) pointed out the need to produce structurally balanced interpretations from seismic data, first in two dimensions and then using maps in three dimensions. He commented on the frequent tendency for published interpretations to show a preponderance of steep normal faults where the same data could be interpreted with much shallower dipping faults. While many major faults in the North Sea rift zone are undoubtedly steep, a significant number are probably much shallower dipping than previously supposed, and extension estimates will vary accordingly.

Although this is itself may be insufficient to account for the observed discrepancies, Wernicke and Axen (1988) recently pointed out that in general it may be difficult to measure the position of both foot and hangingwall cut-offs correctly, due to isostatic rebound of the footwall in response to unloading. This process has been previously discussed for some North Sea basins (Barr 1987), but if Wernicke and Axen's model is generally correct, it would be exceedingly difficult, if not impossible, to identify the pre-rift footwall cut-offs on seismic data. For example, in the Viking Graben the base Triassic does not form a prominent seismic marker, and in the Central Graben salt tectonics frequently obscures the geometry of the rift basin (Shorey and Sclater 1988).

Even without such a radical reinterpretation as that implied by Wernicke and Axen, estimates of extension may vary by a factor of two or three depending on which model for hangingwall deformation is chosen by the interpreter (White *et al.* 1986). For example, it is common to assume that extension on a geological section is equivalent to the sum of the horizontal offsets (heave) on each of the faults represented. This is the assumption used, for example, by Ziegler in his work. Where extension is equal to heave, this implies that the hangingwall has behaved as if by vertical simple shear (Gibbs 1983). If some other shear angle is chosen the extension can be much greater; for example, an antithetic shear angle of 60° increases the value of extension by 300 per cent. Area balancing a known hangingwall geometry using different shear angles implies very different detachment depths but, in general, available reflection data does not image detachment surfaces so it is difficult to use the seismic data to choose the appropriate shear angle. In estimates of extension across the Viking and Central Grabens, this implies that extension measured across the Viking Graben could lie anywhere between Ziegler's value of 16 km and 48 km, depending on hangingwall deformation style alone without changing the interpreted fault cut-offs.

With these caveats and different approaches in mind, it is clear that in a basin such as the North Sea, although there may be general agreement that it is predominantly an extensional rift related to lithospheric extension, it is not yet clear which of a range of extensional models is appropriate, nor even what data can be used to discriminate between different lithospheric stretching models. Even in the North Sea, much interpretational work remains to be done, in particular to integrate regional and the highly detailed local mapping of prospects.

7.3. Regional linkages

Following the realization that crustal and lithospheric processes which gave rise to the North Sea rift are regionally linked, and in some way 'balanced' (in that they preserve crustal and possibly lithospheric section), it was suggested by Gibbs (1984*a*) that the rift system might be linked on an essentially mechanical system of detachment and sidewall faults. This model arose from the realization that extensional fault systems could be mapped in a way analogous to that currently used for contractional (thrust) faults following the work of Dahlstrom (1969) and more recently the synthesis of Boyer and Elliott (1982). Gibbs proposed that extension faults formed as emergent ramps to a general ramp/flat extensional detachment, and hence the geometry of the emergent faults, and of the stratigraphic syn-rift sequence, could be used to deduce the likely geometry of the whole system. Along strike, transfer faults (Gibbs 1984*a*, 1989*a*) compartmentalize the extensional dip-slip system into strain-compatible domains.

At the largest scale Gibbs (1987*a*, *b*) proposed that the whole crust might be involved to generate an essentially asymmetric graben detaching on a single fault or shear zone. Industry data was interpreted in this manner to build a model for the Central Graben (Fig. 7.3). This model linked on a common detachment system the structural development of the graben boundary faults and the mid-graben highs along parts of the Central Graben system.

Following the successful experiments on several lines to the north of Scotland to image the deep crust and the crust mantle boundary by the British Institutes Reflection Profiling Syndicate (BIRPS; Brewer and Smythe 1984; Matthews 1983; Warner 1985), a commercial oil company, Britoil, acquired a deep seismic profile across the Viking Graben to test, amongst other things, the applicability of a linked asymmetric model for graben formation during the rift stage. The results of the interpretation of this profile were published showing that the data was consistent with an asymmetric simple shear of the crust (Beach 1986; Beach *et al.* 1987). These interpretations were influenced by the concurrent work of Wernicke (1986) in the Basin and Ranges, with the publication of his simple shear detachment model for extension.

Using a more extensive grid of deep seismic profile, Gibbs (1987*a*, *b*, 1989*b*) proposed that the North Sea could be subdivided at a regional scale into major crustal domains separated by detachment faults and by major strike slip systems which form sidewalls and act as transfers to the detachment system. Major European lineaments such as the Great Glen fault zone may therefore form not only domain boundaries and penetrate crust (McGeary 1989), but form the linkage between the detachment sets in the third dimension. Gibbs' (1989*b*) model proposes that intra-plate deformation is accomplished by selective activation or reactivation of one or more of these linked systems between crustal domains. The strong dipping reflectors seen on some

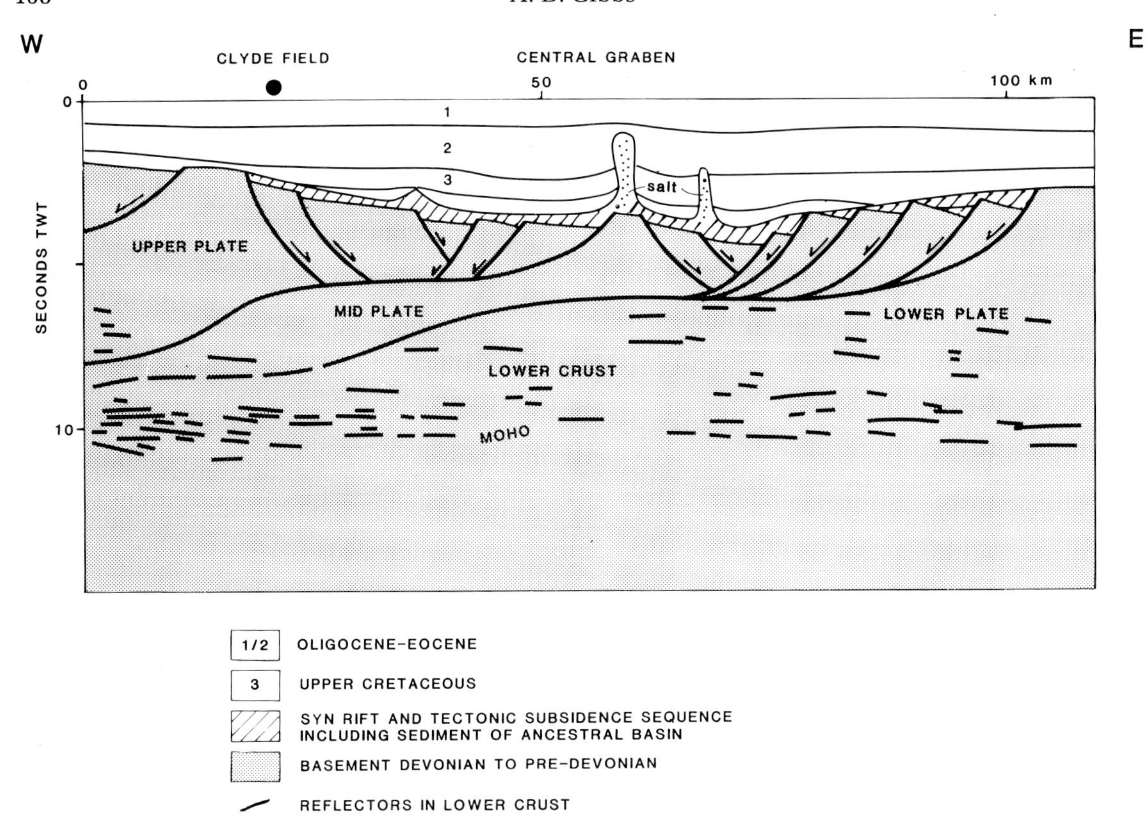

Fig. 7.3. Cross section model of Central Graben using a variety of data based on Gibbs (1984a), and Barton and Wood (1984). Note that the basin is regionally asymmetric, and the mid-graben high is developed on a detachment flat.

of the North Sea data, and in particular on the BIRPS data off the north coast of Scotland (Brewer and Smythe 1984), are taken as strong evidence that a Wernicke-type model is appropriate for the tectonic development of the area. Indeed there now seems to be general agreement that the mantle reflectors on the BIRPS data represent a discrete deformation or shear zone within the mantle (Klemperer 1988; White, this volume). In effect, long range linked deformations comprising basin growth or inversion can occur well within, but linked to deformation at, the plate margins (Beach 1987).

Individual data sets can be interpreted in different ways. Klemperer (1988), for example, uses some of the deep profiles across the Viking Graben to support the view that the mantle lithosphere (and possibly lower crust) deform by pure shear beneath a simple shear in the mid to upper crust. At its logical conclusion, Klemperer's model suggests that extensional or compressional deformation is thin skinned, and is entirely within the elastic lid of the crust and decoupled from, but coincident with, pure shear deformation of the lower crust and mantle. The implications of these two end-member models for tectonics and understanding of structural development on a local scale are far reaching, but not yet fully explored.

7.4. Extensional linkages at field scale

While still controversial at a regional scale, the linked model for extension has been widely adopted in oil industry interpretations of both the Viking and Central Grabens. Several well-documented analyses of producing oil fields have recently been published demonstrating the usefulness of linked dip slip and transfer systems.

Earlier published maps and sections of the North Sea extensional fault blocks show a preponderance of simple, large, steep, normal faults in cross-section which die out rather rapidly along strike on the map. Figure 7.4 shows such an interpretation of the typical Thistle Field area of the Shetland Terrace system using the 2D seismic data available at that time. On Fig. 7.4, major faults are seen to die out rapidly along strike (Hay 1977; Hallett 1981), essentially in the distance beteen adjacent seismic control lines. Transfer faults as proposed by Gibbs (1984) may be used as an alternative in such areas to the high tip strains implied in the Hallett interpretation. The geometry of such faults (Gibbs 1989) is such that they are difficult to interpret on seismic data, as their vertical component of displacement varies markedly along their length.

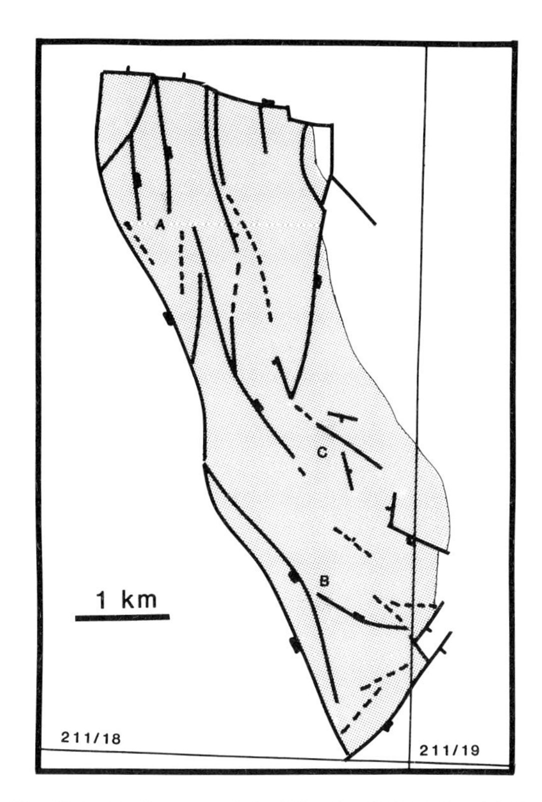

Fig. 7.4. Fault map of the Thistle Field (after Hallett 1981). Solid lines represent faults with their continuation shown dashed where uncertain. Note the termination of major faults at A, and the unlinked short fault segments at B and C. Interpretations such as this on the older 2D seismic data led to the proposal that transfer faults could account for mapping problems where there was some evidence from pressure testing of the reservoir that some of these faults linked in some way to form structural discontinuities. More recent seismic data suggests that both linked transfer faults and accommodation zones are present.

For this general area it was originally assumed (Gibbs 1984*a*) that most of the transfer faults would be orthogonal or only slightly oblique to the transport direction. Their mapped geometry can therefore be used to determine the direction of tectonic transport (Figs 7.5 and 7.6). Beach (1985) used this approach to derive a model for opening of the Central and Viking Grabens with both dip and oblique slip components implied. Moreover, using this approach, it can be shown that the sections along which extension has been frequently measured (Thorne and Watts 1989) do not lie in the extension direction, and will therefore give misleading estimates of fault extension. Analysis of detailed local

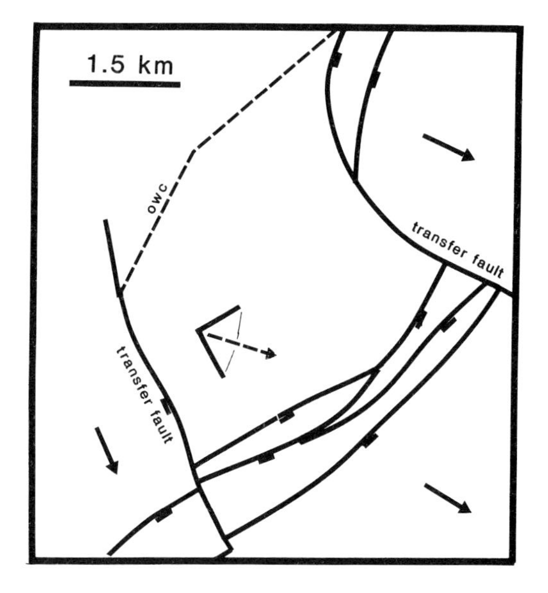

Fig. 7.5. Sketch map of Murchison Field (Engelstad 1987) showing transfer fault boundaries. Solid arrows denote extension direction for pure strike-slip transfer faults. Oil-water contact marked as OWC. A common extension direction, say parallel to the dashed arrow, would result in dip slip components on the transfer faults and oblique slip on the field boundary faults. The regional context of the field is important in resolving this issue.

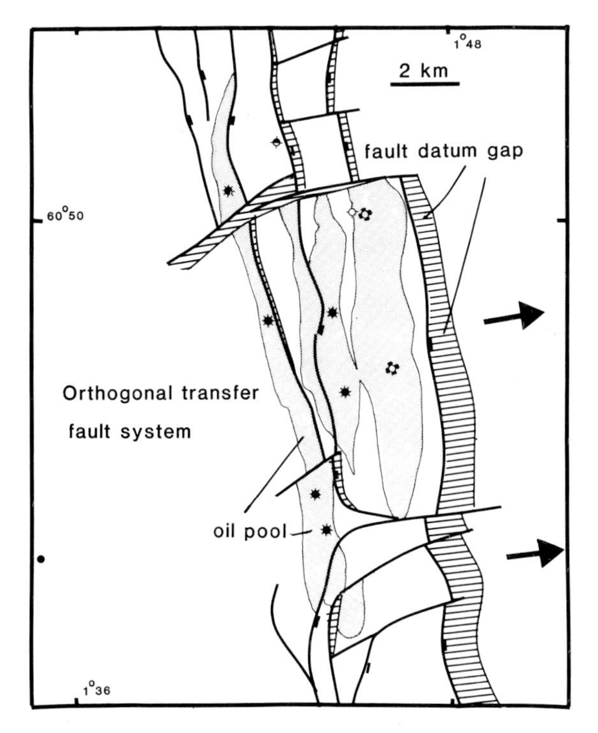

Fig. 7.6. The Alwyn Field (after Johnson and Eyssautier 1987) comprises a transfer-faulted crustal area. Displacements on an almost orthogonal fault set indicate that the transfer faults lie parallel to the transport or extension direction. (Parallel ruling indicates separation-heave on the top reservoir horizon.)

maps is therefore vital in building up a reliable picture of regional extension. Figure 7.5, however, demonstrates that extension direction can vary locally as major hangingwall elements rotate. It is therefore necessary to integrate the mapping of large areas with the detailed field maps before regional extension directions can be deduced.

From the same general area as the Thistle Field, recently published maps of producing fields show examples of transfer fault geometry with the major transfer faults separating the oil accumulations along strike (Fig. 7.7, after Engelstad, 1987; Johnson and Eyssautier, 1987). On the Shetland Terrace region and elsewhere in the North Sea rift system, the transfer faults either compartmentalize the oil accumulations by providing lateral fault seal or the dip changes (monoclinal folds) associated with the transfer fault provide an additional element of dip closure.

On a smaller scale, i.e. within the transfer compartment, further subdivision may sometimes be developed. Commonly in the Shetland Terrace region it is apparent from analysis of 3D seismic data that the compartments are subdivided by linking branching faults and by accommodation zone arrays in the sense of Rosendahl (1987) and Larsen (1988). Roberts and co-workers (1987) show these patterns for Statfjord (Fig. 7.8). With many of the North Sea Fields now well into their production life, it is becoming clear that the ability to differentiate between completely linked transfer and detachment

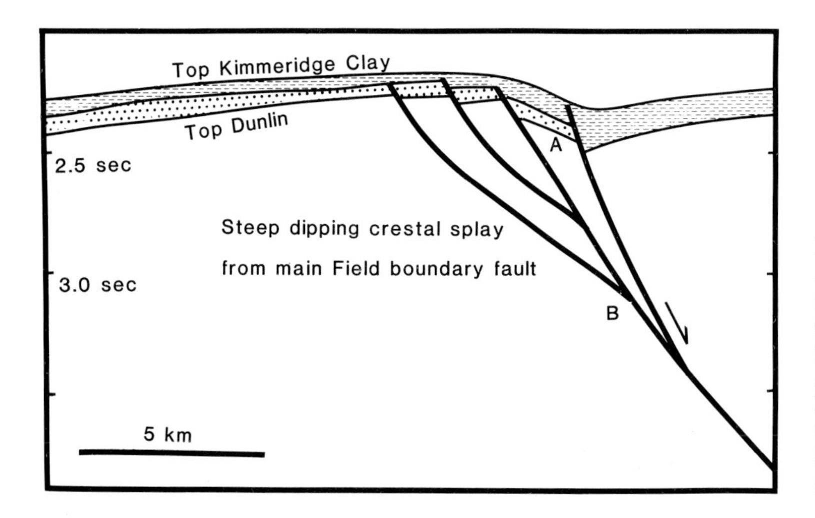

Fig. 7.7. Cross-section of Murchison Field (after Engelstad 1987) showing typical synthetic splay from main extension fault. The strong dip to the right (A) may be caused by the concave downward fault (B), or may be an interpretation artefact. These crestal splays are frequently bounded by transfer faults at the scale of Figs 7.5 and 7.6.

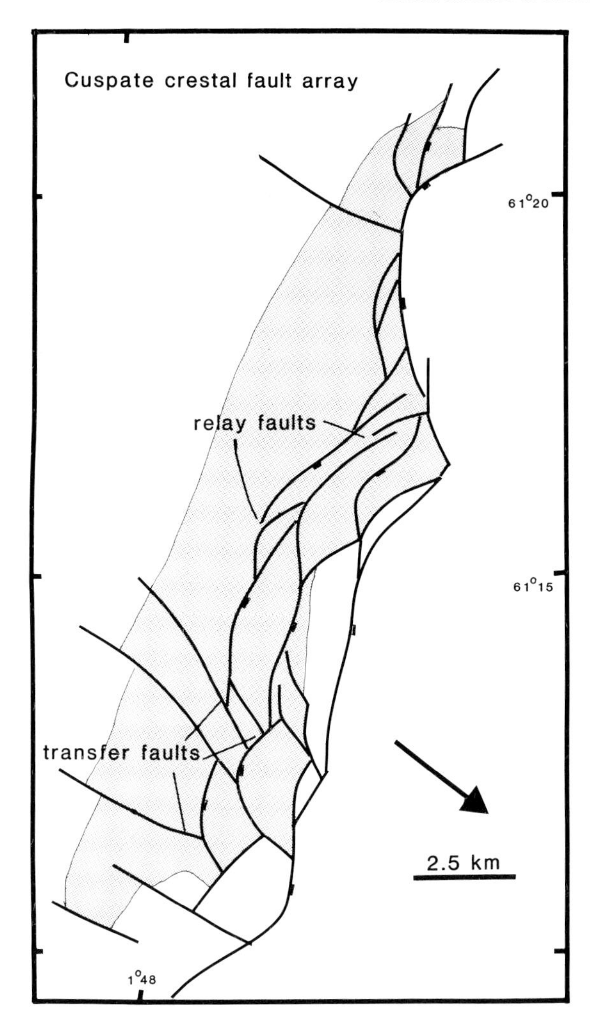

Fig. 7.8. Map of the Statfjord Field at top Brent (after Roberts *et al.* 1987). Note cuspate arrays of faults on crest in contrast to Figs 7.6 and 7.7. Also several of these faults have *en echelon* terminations, and are linked by relay faults, e.g. at A (Larsen 1988). Transfer faults T are also present.

sets and accommodation arrays is vital. Figures 7.6 and 7.8 summarize the basic geometric elements of each of these end-member structural types in map view. In many cases both transfer and accommodation components will be present, and the problem then becomes one of deciding which, if any, is significant to the economic development of the field.

Mapping fault cut-offs and producing hanging and footwall cut-off maps can help to distinguish these different types of linkage, particularly where good data exists. Barnett and co-workers (1987) and Walsh and Watterson (1987) have pioneered this technique for extensional faults. Basically extension must be maintained or vary smoothly along strike, otherwise cleavage or breccia belts will develop, even in near-surface rocks. Sudden jumps in displacement across a fault zone mean either that displacement has been transferred and a transfer fault should be mapped, or that an extension fault which should be present in the accommodation array has been missed from the interpretation. Figures 7.9 and 7.10 show cut-off diagrams for both of these types of system. These diagrams can therefore be used at any stage in the interpretative process, with section balancing techniques, to highlight problem areas in the interpretation and likely solutions.

In cross section, interpretations which were previously dominated by single, steep planar faults now also show more complex linked dip slip arrays. Gibbs (1984*b*) demonstrated that the Clyde Field in the Central Graben could be interpreted as resulting from a linked basement fault with an overlying cover detachment. The upper detachment coincides with the Zechstein evaporites; throughout the southern part of the rift system complex, upper detachments frequently partition extension of the Jurassic sediments from synchronous extension or reactivation of the pre-salt basement. Interpretation of such areas relies being able to discriminate effectively between extensional detachment and the sliding associated with halokinesis, where there is no overall change in the regional length of the section.

To the north of the Viking Graben no salt is present, but upper detachments (Gibbs 1987*c*) are now being identified in several areas. The Jurassic Dunlin shales frequently serve as the secondary detachment surface, and interpretations by Gabrielsen (1986) and Fossen 1989 are two recent examples from the Shetland Terrace region. In the former example, Gabrielsen interprets the reservoir section as having deformed on a shallow detachment with strongly rotated listric fault blocks, across which the base-Cretaceous unconformity cuts, structurally attenuating both the upper and lower reservoir units (Fig. 7.11). This listric array sits above steeper planar faults linked to the main bounding fault of the terrace.

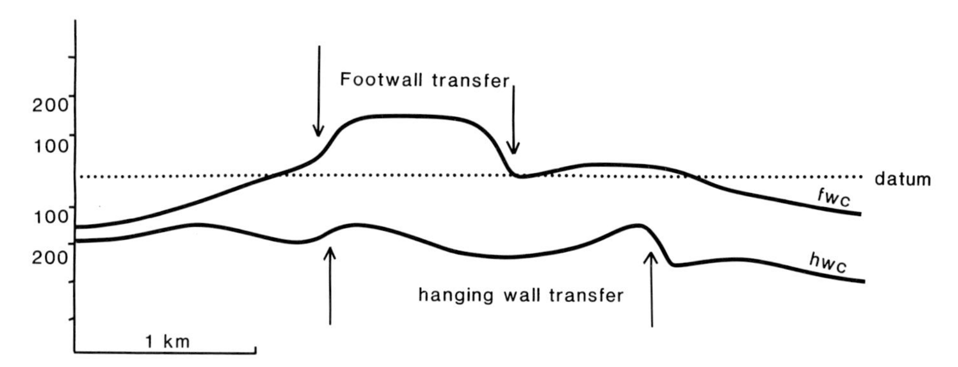

Fig. 7.9. Diagram for separation of hangingwall cut-offs (hwc) and footwall cut-offs (fwc) on a horizon on a major fault zone in the Shetland Terrace area. Rapid changes in fault separation are indicative of position of transfer faults mapped in both foot and hangingwalls.

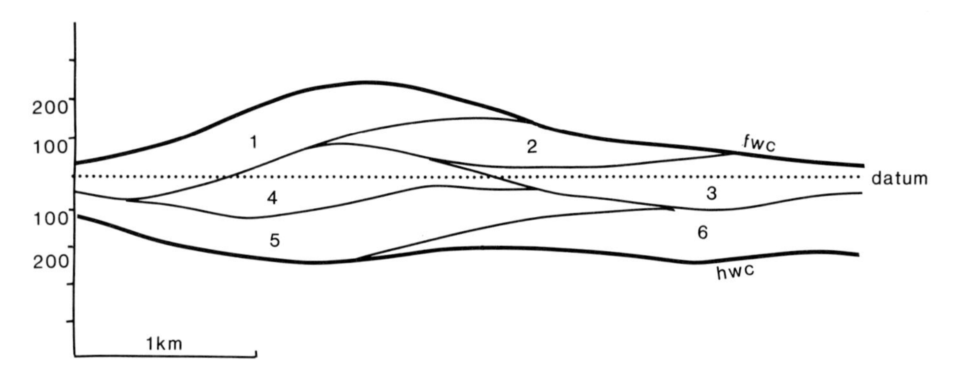

Fig. 7.10. Cut-off diagrams for an *en echelon* accommodation array of faults 1 to 6 without transfer faults in the Shetland Terrace region. Note cumulative footwall and hangingwall separation should have smooth envelope.

7.5. Conclusions

Linked tectonic models can be applied to all scales in the North Sea rift basin to assist understanding of the tectonic and structural development of the basin. Regionally a purely structural approach suggests conclusions which are at variance with some of the implications of purely theoretical basin models. In these cases, structural analysis provides a powerful constraint on a more theoretical approach, as it incorporates a large body of consistent seismic and interpretational data. However, the structural models themselves are sensitive to changes in assumptions, for example, on models for hanging-wall deformation; it is thus important to discriminate between strict geometric observations of the structure and its interpretation.

At a more local scale, conventional balanced-section analysis coupled with map analysis provides an essential tool in understanding the structural development of the region and of structures within it, particularly where fault arrays and accommodation zones are present. The majority of the mapping problems facing either the regional or development geologist can be resolved without consideration of crustal structure. The differences in, say, thermal or subsidence history predicted by different models may be well within the errors of observation. The principal problem is to map the fault and rotational geometry in the area adequately, and to determine the extent to which transfer faults and accommodation zones jointly contribute to the structure along strike.

An integrated approach must eventually lead to

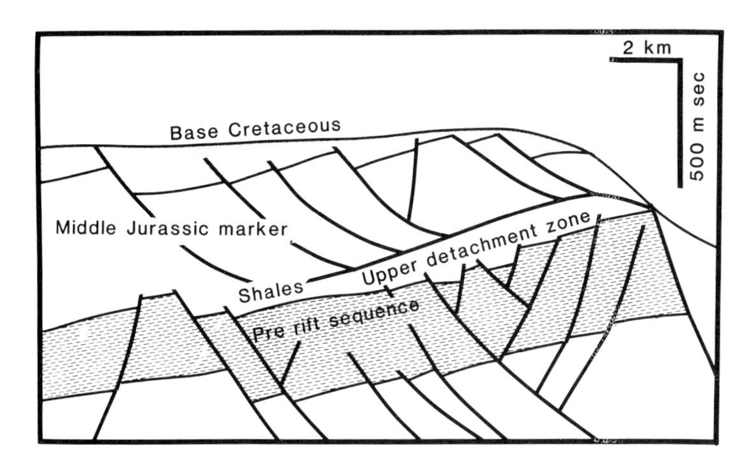

Fig. 7.11. Cross-section of linked upper detachment array (after Gabrielsen 1986). In map view, the upper detachment array will be similar to that shown in Fig. 7.8. Note that this is a geoseismic interpretation, and balanced section tests should be made on a depth-converted section. Gabrielsen implies that the faults in the pre-rift sequence join the upper detachment faults in a zone of complex strain in a major shale horizon.

an improved understanding of the processes of rift evolution and prediction of prospectivity within the basin, but must be based on thorough structural mapping at all scales.

References

Barnett, J. A. M., Mortimer, J., Rippon, J. H., Walsh, J. J., and Watterson, J. (1987). Displacement geometry in the volume containing a single normal fault. *Am. Assoc. Petrol. Geol.*, **71**, 925–37.

Barr, D. (1987). Lithospheric stretching, detached normal faulting and footwall uplift. In *Continentlal extension tectonics* (ed. M. P. Coward, J. F. Dewey and P. L. Hancock). *Geol. Soc. Special Publication*, **28**, 75–94.

Barton, P. J. and Wood, R. (1984). Tectonic evolution of the North Sea Basin; crustal stretching and subsidence. *Geophys. J. Roy. Astron. Soc.*, **79**, 987–1022.

Barton, P. J. (1985). Comparison of deep reflection and refraction structures in the North Sea. In *Reflection seismology: a global perspective* (ed. M. Barazangi and L. Brown). *Am. Geophys. Union*, Geodynamics Series, **13**, 297–300.

Beach, A. (1985). Some comments on sedimentary basin development in the northern North Sea. *Scot. J. Geol.*, **21**, 493–512.

Beach, A. (1986). A deep seismic reflection profile across the northern North Sea. *Nature*, **323**, 53–5.

Beach, A. (1987). A regional model for linked tectonics in NW Europe. In *Petroleum geology of north west Europe* (ed. J. Brooks and K. Glennie), 43–8. London, Graham and Trotman.

Beach, A., Bird, T., and Gibbs, A. D. (1987). Extensional tectonics and crustal structures; deep seismic reflection data from the northern North Sea Viking Graben. In *Continental extension tectonics* (ed. M. P. Coward, J. Dewey and P. L. Hancock). *Geol. Soc. Special Publication*, **28**, 467–76.

Boyer, S. M. and Elliott, D. W. (1982). Thrust systems. *Am. Assoc. Petrol. Geol. Bull.*, **66**, 1196–230.

Brewer, J. A. and Smythe, D. K. (1984). The Moine and Outer Isles Seismic Traverse (MOIST). In *Seismic expression of structural styles* **3** (ed. A. W. Bally). Tulsa, USA, *Am. Assoc. Petrol. Geol*.

Brown, S., Richards, P. C., and Thompson, A. R. (1987). Patterns in the deposition of the Brent Group (Middle Jurassic) UK, North Sea. In *Petroleum geology of north west Europe* (ed. J. Brooks and K. Glennie), 889–915. London, Graham and Trotman.

Clark-Lowes, D. D., Kuzemko, N., and Scott, D. A. (1987). Structure and petroleum prospectivity of the Dutch Central Graben and neighbouring platform areas. In *Petroleum geology of north west Europe* (ed. J. Brooks and K. Glennie), 337–56. London, Graham and Trotman.

Coward, M. P. (1986). Heterogeneous stretching, simple shear and basin development. *Earth and Planetary Science Letters*, **80**, 325–36.

Dahlstrom, C. D. A. (1969). Balanced cross sections. *Can. J. Earth Sci*., **6**, 743–57.

Engelstad, N. (1986). Murchison. In *Geology of the Norwegian oil and gas fields* (ed. A. M. Spencer *et al*.), 295–308. London, Graham and Trotman.

Fossen, H. (1989). Indication of transpressional tectonics in the Gullfaks oil-field, northern North Sea. *Marine and Petroleum Geol.*, **6**, 22–30.

Gabrielsen, R. H. (1986). Structural elements in graben systems and their influence on hydrocarbon trap and types. In *Habitat of hydrocarbons on the Norwegian continental shelf* (ed. A. M. Spencer *et al*.), 55–60. London, Graham and Trotman.

Gibbs, A. D. (1983). Balanced cross-section construction from seismic sections in areas of extensional tectonics. *J. Struct. Geol.*, **5**, 153–60.

Gibbs, A. D. (1984*a*). Structural evolution of extensional basin margins, *J. Geol. Soc.*, **141**, 609–20.

Gibbs, A. D. (1984*b*). Clyde Field growth fault secondary detachment above basement faults in the North Sea. *Am. Assoc. Petrol. Geol. Bull.*, **68**, 1029–39.

Gibbs, A. D. (1987*a*). Linked tectonics of the northern North Sea basin. pp. 163–172. In *Sedimentary basins and basin-forming mechanisms* (ed. C. Beaumont, C., and A. J. Tankard). *Can. Soc. Petrol. Geol. Mem.* **12.**

Gibbs, A. D. (1987*b*). Deep seismic profiles in the northern North Sea. In *Petroleum geology of north west Europe* (ed. J. Brooks and K. Glennie), 1025–8. London, Graham and Trotman.

Gibbs, A. D. (1987*c*). Development of extensional and mixed mode sedimentary basins. In *Continental extensional tectonics* (ed. M. P. Coward, J. F. Dewey, and P. L. Hancock). *Geol. Soc. Special Publication*, **28**, 19–33. Oxford, Blackwell Scientific Publications.

Gibbs, A. D. (1989*a*). Structural styles in basin evolution. In *Extensional tectonics and stratigraphy of the North Atlantic margins* (ed. A. J. Tankard and H. R. Balkwill). *Am. Assoc. Petrol. Geol. Mem*, **46**, 81–93.

Gibbs, A. D. (1989*b*). Model for linked basin development around the British Isles. In *Extensional tectonics and stratigraphy of the North Atlantic margins* (ed. A. J. Tankard and H. R. Balkwill). *Am. Assoc. Petrol. Geol. Mem*, **46**, 501–9.

Glennie, K. W. (1986). The structural framework and the pre-Permian history of the North Sea area. In *Introduction to the petroleum geology of the North Sea*, 2nd edn. (ed. K. Glennie), 25–62. Oxford, Blackwell Scientific Publications.

Hallett, D. (1981). Refinement of the geological model of the Thistle Field. In *Petroleum geology of the continental shelf of north west Europe* (ed. L. V. Illing and G. D. Hobson), 315–25. London, Institute of Petroleum, Heyden and Son.

Hay, J. C. (1977). *The Thistle Oilfield in NPF.* Mesozoic Northern North Sea Symposium, Section II, 1–20. Oslo, Norwegian Petroleum Society.

Johnson, A. and Eyssautier, M. (1987). Alwyn North Field and its regional context. In *Petroleum geology of north west Europe* (ed. J. Brooks and K. Glennie), 963–78. London, Graham and Trotman.

Klemperer, S. L. (1988). Crustal thinning and nature of extension in the northern North Sea from deep seismic reflection profiling. *Tectonics*, **7**, 803–21.

Kusznir, N. S., Karner, G. D. and Egan, S. (1987). Geometric, thermal and isostatic consequences of detachments in continental lithosphere extension and basin formation. In *Sedimentry basins and basin-forming mechanisms* (ed. C. Beaumont and A. J. Tankard). *Can. Soc. Petrol. Geol. Mem.*, **12**, 185–203.

Larsen, P. H. (1988). Relay structures in a Lower Permian basement-involved extension system, East Greenland. *J. Struct. Geol.*, **10**, 3–8.

McGeary, S. (1989). Reflection seismic evidence for a Moho offset beneath the Walls Boundary strike-slip fault. *J. Geol. Soc. of London*, **146**, 261–9.

McKenzie, D. P. (1978). Some remarks on the development of sedimentary basins. *Earth and Planetary Science Letters*, **40**, 25–32.

Matthews, D. H. (1983). Deep seismic profiling around Britain. *Terra Cognita*, **3**, 7–11.

Moretti, I. and Pinet, B. (1987). Discrepancy between Lower and Upper crustal thinning. In *Sedimentary basins and basin-forming mechanisms* (ed. C. Beaumont and A. J. Tankard). *Can. Soc. Petrol. Geol. Mem.*, **12**, 223–40.

Nelson, P. H. H. and Lamy, J. M. (1987). The More/West Shetland area: a review. In *Petroleum geology of north west Europe* (ed. J. Brooks and K. Glennie), 775–84. London, Graham and Trotman.

Olsen, J. C. (1987). Tectonic evolution of the North Sea region. In *Petroleum geology of north west Europe* (ed. J. Brooks and K. Glennie), 389–402. London, Graham and Trotman.

Pegrum, R. M. (1984*a*). Structural development of the south western margin of the Russian-Fennoscandian Platform. In *Petroleum geology of the north European margin* (ed. A. M. Spencer, *et al.*), 359–70. London, Graham and Trotman.

Pegrum, R. M. (1984*b*). The extension of the Tornquist Zone in the Norwegian North Sea. *Norsk. Geologisk. Tidsskrift*, **64**.

Pegrum, R. M. and Spencer, A. M. (1990). Hydrocarbon plays in the northern North Sea. In *Classic petroleum provinces* (ed. J. Brooks), *Geol. Soc. Special Publication*, **50**, 441–70.

Roberts, J. D., Mathieson, A. S., and Hampern, J. M. (1987). Statfjord. In *Geology of the Norwegian Oil and Gas Fields* (ed. A. M. Spencer *et al.*), 319–40. London, Graham and Trotman.

Rosendahl, B. R. (1987). Architecture of continental rifts with special reference to East Africa. *Earth and Planetary Science Letters*, **15**, 445–503.

Sclater, J. G. and Christie, P. A. F. (1980). Continental stretching: an explanation of the post mid-Cretaceous subsidence of the Central North Sea Basin. *J. Geophys. Res.*, **85**, 3711–39.

Shorey, M. D. and Sclater, J. G. (1988). Mid-Jurassic through mid-Cretaceous extension in the Central Graben of the North Sea, part 2. Estimates from faulting observed on a seismic reflection line. *Basin Research*, **1**, 201–16.

Thorne, J. A. and Watts, A. B. (1989). Quantitative analysis of North Sea subsidence. *Am. Assoc. Petrol. Geol. Bull.*, **73**, 88–116.

Walsh, J. J. and Watterson, J. (1987). Distributions of cummulative displacement and seismic slip on a single normal fault surface. *J. Struct. Geol.*, **9**, 1039–46.

Warner, M. R. (1985). Deep seismic reflection profiling of the continental crust at sea. In *Reflection seismology: a global perspective* (ed. M. Barazangi and L. Brown), 281–6. *Am. Geophys. Union. Geodynamics Series*, **13**, 281–6.

Wernicke, B. (1985). Uniform sense of normal simple shear of the continental lithosphere. *Can. J. Earth Sci.*, **22**, 108–25.

Wernicke, B. (1986). Whole lithosphere normal simple shear: an interpretation of deep reflection profiles in Great Britain. In *Reflection seismology: the continental crust* (ed. M. Barazangi and L. Brown). *Am. Geophy. Union Geodynamics Series*, **14**, 331–9.

Wernicke, B. and Axen, G. J. (1988). On the role of isostacy in the evolution of normal fault systems. *Geology*, **16**, 848–51.

White, N. J. (this volume). Does the uniform stretching model work in the North Sea?

White, N. J., Jackson, J. A., and McKenzie, D. P. (1986). The relationship between the geometry of normal faults and that of the sedimentary layers in their hanging walls. *J. Struct. Geol*., **8**, 897–910.

White, N. J. and McKenzie, D. (1988). Formation of the 'steer's head' geometry of sedimentary basins by differential stretching of the crust and mantle. *Geology*, **16**, 250–3.

Ziegler, P. A. (1982). *Geological atlas of western and central Europe*. The Hague, Shell Internationale Petroleum Mastschappij B. V.

Ziegler, P. A. (1989). Evolution of the Arctic-North Atlantic rift system. *Am. Assoc. Petrol. Geol. Mem.*, **43**, 198.

8 Architectural styles of basin fill in the northern Viking Graben

R. H. Gabrielsen, R. B. Færseth, R. J. Steel, S. Idil, and O. S. Kløvjan

Abstract

The northern Viking Graben area has been affected by two major extensional events of Permo-Triassic and Jurassic-Cretaceous age. Both these events are subdivided into pre-rift, active stretching, and thermal cooling stages, each of which has characteristic basin configurations and sedimentary patterns.

This idealized picture is modified by several effects:

(1) thermal equilibrium was not reached between the rift episodes; this affects not only the basin development in the interlude between the rift episodes, but also the subsidence pattern after termination of the final rift episode;
(2) there are indications of local inversion; and
(3) additional minor extensional events may have interfered with the major extensional pulses.

In general it is found that, during active stretching stages, sand was more locally deposited because of topographic trapping, and the distribution of sand bodies was dependent on the local drainage pattern. During stages of thermal cooling, the actual basin was more extensive and basin floor topography less pronounced. This tended to cause more widespread sand distribution, except during periods of relatively rapid thermal subsidence, when the basin was mud-dominated, with sands restricted to the basin margin areas and to lensoid tracts in the basin.

8.1. Introduction

Due to intense hydrocarbon exploration since the discovery of the Ekofisk Field in 1969, the North Sea is now one of the best explored continental shelves in the world. In spite of this, the amounts of data available vary considerably within the area; this is partly because of differences in data acquisition, as a consequence of the area being cut by several national border lines, and partly a consequence of exploration history and strategy.

After the early hydrocarbon discoveries in the Norwegian North Sea, which were Cretaceous chalk plays in the Central Graben, and Palaeocene–Eocene sands in the southern Viking Graben, the exploration activity to a large extent has been concentrated upon Jurassic rotated fault blocks along the graben margins. This situation is particularly pronounced in the Viking Graben area, where the lack of wells away from the graben margins, as well as lack of data from the deeper levels, now hampers the construction of detailed tectono-sedimentological models.

Following the discovery of the Troll Field in 1979, a larger number of wells were drilled on the eastern flank of the Viking Graben (Fig. 8.1) in the Horda Platform area (Rønnevik and Johnsen 1984; Hellem *et al.* 1986; Gray 1987). Some of these wells have penetrated pre-Jurassic strata (e.g. wells 31/2-4, 31/4-3), and one reached its total depth in the basement (well 31/6-1). Accordingly, in establishing a tectono-sedimentological model for the Viking Graben system, it is convenient to use the Horda

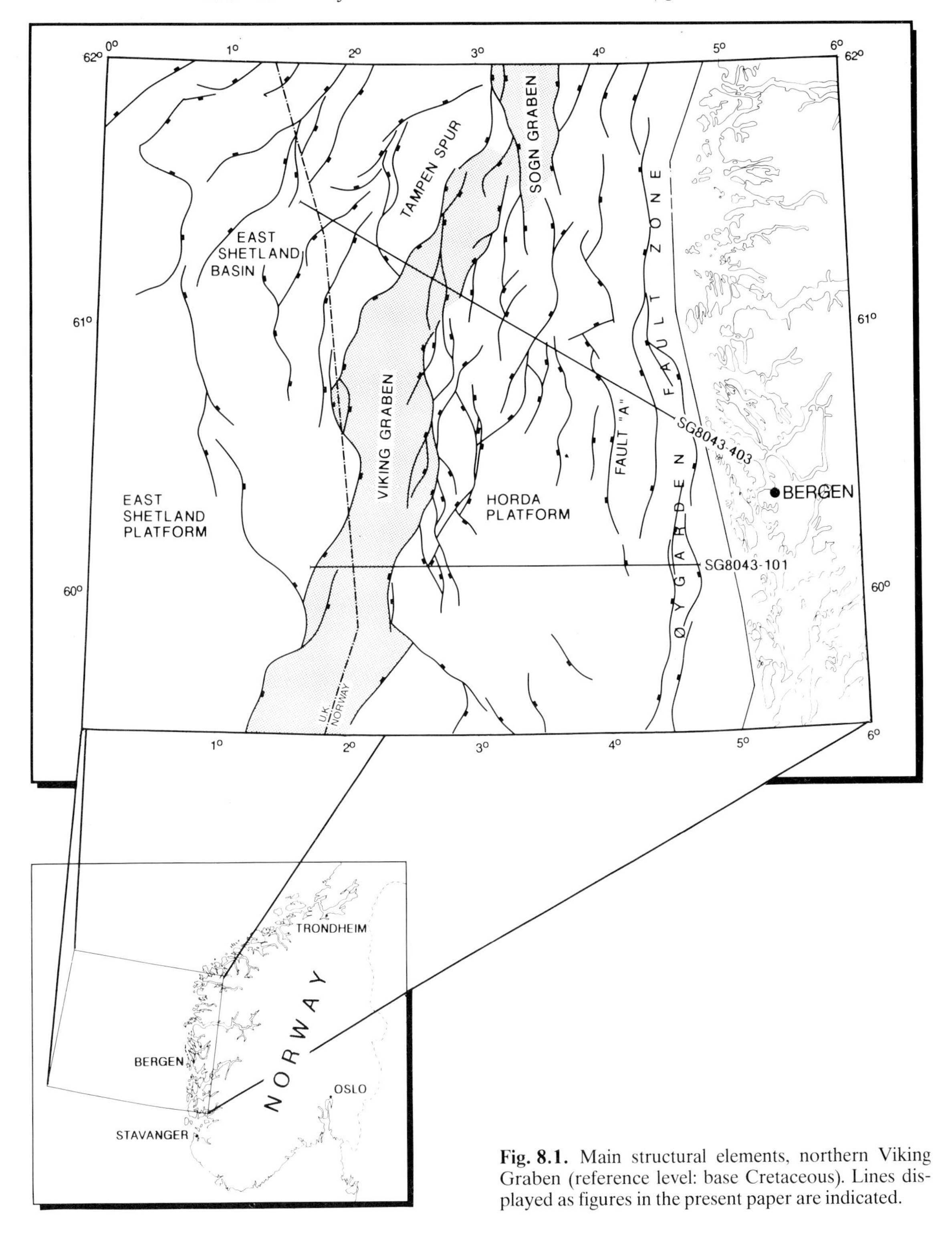

Fig. 8.1. Main structural elements, northern Viking Graben (reference level: base Cretaceous). Lines displayed as figures in the present paper are indicated.

Platform as a reference for a correlation across the graben area.

The present study, based upon the regional seismic survey SG8043, deep seismic lines NSDP-84, and a series of adjacent wells, aims at outlining the structural evolution of the central segment of the Viking Graben, and evaluating how its Triassic to Cretaceous infill is influenced by the structural style and rates of subsidence. To accomplish this the paper is built up as follows.

First, a summary of tectonic models in use for the study area is presented, with an idealized model for the influence of the tectonic development on the architectural style of the graben fill through the different stages. Thereafter, an evaluation of the complications due to repeated extensional events is given, followed by structural and sedimentological descriptions of the study area. Finally, the structural styles and the associated architectural styles of the graben fill are compared to the model. The stratigraphic nomenclature used is in accordance with Vollset and Doré (1985).

8.2. Tectonic and sedimentological models

The North Sea has attracted considerable interest from basin modellers. In fact, McKenzie (1978), in his classical and widely cited paper on basin subsidence, used the North Sea as one of his type examples. This was followed by a number of related works, which, to a large extent concentrated upon the Central Graben (Sclater and Christie 1980; Wood 1981; Dewey 1982; Leeder 1983; Wood and Barton 1983; Ziegler 1983; Barton and Wood 1984).

The deep seismic reflection data acquired in 1983 by Western Geophysical, and similar data acquired in 1984 by the Geophysical Company of Norway (North Sea Deep Profiles, NSDP-84), have brought a new dimension into this discussion (Beach 1985, 1986; Beach *et al.* 1987; Gibbs 1987; Klemperer 1988). These data have gradually moved the interest in basin modelling to the Viking Graben area (Giltner 1987; Scott and Rosendahl 1989; Thorne and Watts 1989), and have focused interest on the deep structure of the graben system.

In the early models for the deep structure and the development of the North Sea graben system, triple junctions in a plate tectonic sense (Whiteman *et al.* 1975) and asthenospheric doming (see summary, Ziegler 1982) were proposed. It is now, however, realized that, even if mid-Jurassic doming were related to the triple junction in central North Sea, this effect is not recorded in the northern Viking Graben (Giltner 1987; Badley *et al.* 1988). Even though workers in the northern North Sea today agree upon a model involving multiple active stretching, each event being followed by thermal cooling, there are disagreements about the deeper geometry of the graben structure as well as the amount, timing, and nature of the stretching.

The models which have been used for the northern Viking Graben can be summarized as follows.

1. Symmetrical pure shear models. These are congruent with the model of McKenzie (1978), and are favoured for the Viking Graben by Giltner (1987) and Badley and colleagues (1988).
2. Inhomogeneous shear-models, with a sub-horizontal mid-crustal decoupling horizon. This class of models, which involves a sub-horizontal crustal detachment, has been proposed by Klemperer (1988) and has much in common with the 'sub-horizontal mid-crustal decoupling model' advocated by some authors for the Basin and Range (Stewart 1971; Eaton *et al.* 1978; Miller *et al.* 1983; Allmendinger *et al.* 1987).
3. Asymmetrical simple shear model (Wernicke 1985). This model has been used to explain the asymmetry of the graben structure (Beach 1986; Beach *et al.* 1987; Scott and Rosendahl 1989), and the distribution of active seismic zones (Gabrielsen 1989). The flats introduced in the shear zone by Beach and co-workers makes this model an analogue to the one proposed by Lister and colleagues (1986).

A close relation between basin characteristics and the main stages in the graben formation should be expected. This includes parameters such as subsidence rates, styles of faulting, and basin topography (relief and basin gradients). At the next level, this is reflected in the architecture of the basin fill.

The models in use for the Viking Graben area all have in common a two-stage development, namely the crustal stretching and the thermal subsidence stages (McKenzie 1978; Royden and Keen 1980; Sclater and Christie 1980; Hellinger and Sclater 1983). In addition, pre-rift subsidence is well known from many rifts (Harding 1983, 1984; Scott and

Rosendahl 1989). Within this framework, three main tectono-sedimentological stages can be sketched (Gabrielsen 1986, 1989). It should be emphasized, however, that the development of the Viking Graben demonstrates smaller and larger deviations from this generalized pattern.

During the pre-rift subsidence stage, the basin covers a wider area than the later graben *sensu stricto*. Movements are located on steep crustal fractures with only limited vertical throws. The basin gradients would be moderate; the sedimentary environments (other than continental) and sediment transport directions would not be easily predictable. Both axial (for example, through-flowing fluvial) and transverse sedimentary systems can be variably present as feeder systems. Nevertheless, the basin would probably be an elongated feature (Fig. 8.2a). Deposition would commonly be in a continental environment, though not necessarily so.

In the crustal stretching stage, thinning of the upper crust takes place by low-angle faulting. This process gives rise to a pronounced relief with valleys and ranges parallel to the graben margins (Fig. 8.2b). The gradient across and along the rifted area may still be moderate. However, due to the developing relief, any direct transverse sediment transport system is likely to be disrupted. Towards the end of this stage, high stand of sea level can be expected if the rise of subsidence-controlled sea-level outpaces sediment supply.

The graben floor gradient, on a regional scale, would increase dramatically in the thermal cooling/sediment loading stage, whereas the relief would be smoothed (Fig. 8.2c). Fault activity would again be restricted to steep planar faults with capability to take up vertical movements. During times of rapid subsidence, transverse sediment transport would dominate, and margin-attached sand bodies would be produced. In non-marine environments, sand tracts would tend to form isolated and lensoid, elongate sand bodies. During times of less rapid subsidence, particularly if sediment input has been significant, transverse sediment systems tend to amalgamate to produce axial systems. This, in turn,

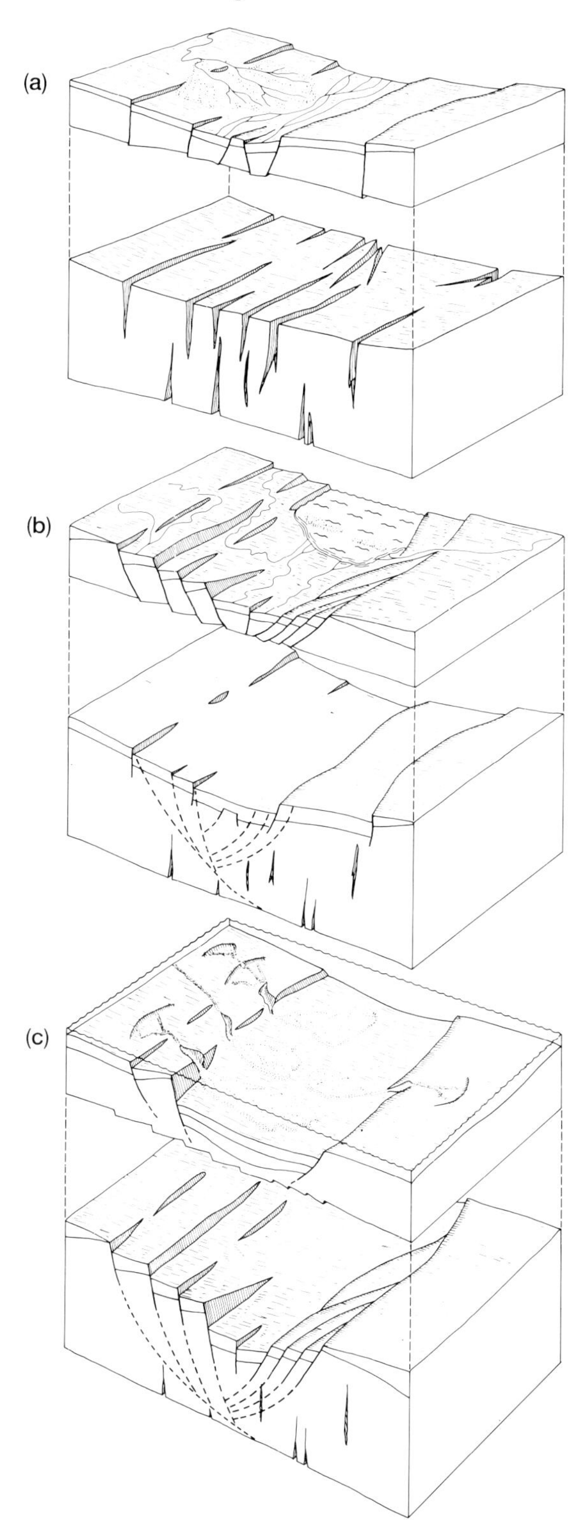

Fig. 8.2. Idealized sketch showing basin configurations and sediment transport systems related to **a.** pre-rift stage, **b.** active stretching stage, and **c.** thermal cooling stage.

results in sand being spread out over most of the basinal area (Steel and Gjelberg 1989).

8.3. Timing of tectonic events in the northern North Sea

This simplistic picture is obscured in systems where multiple rifting takes place, and where thermal equilibrium is not attained between the separate extensional episodes. This situation would affect not only the style and extent of subsidence in the interlude between two rift episodes, but also the subsidence history succeeding the last of the episodes. As late Palaeozoic to Cainozoic tectonism in the North Sea has been subdivided into several discrete phases, namely Devonian–Carboniferous, Permo-Triassic, middle–late Jurassic, and late Cretaceous–early Tertiary these complications should be expected (Halstead 1975; De'Ath and Schuyleman 1981; Eynon 1981; Gray and Barnes 1981; Hallett 1981; Harding 1983; Whiteman *et al.* 1975; Ziegler 1975, 1978, 1982; Beach *et al.* 1987; Giltner 1987).

The Devonian–Carboniferous part of the structural history is not very well understood for the study area, mainly due to lack of data. The north-western margin of Europe is believed to have been affected by pre-Viséan sinistral movements and post-Viséan stretching (Ziegler 1982). Several Devonian basins are found along the western coast of Norway (Siedlecka and Siedlecki 1972; Steel 1976). These have been set in connection with extension (Bryhni and Skjerlie 1975) and shear (Steel and Gloppen 1980), and with extensional (Hossack 1984; Norton *et al.* 1987; Steel 1988) and compressional (Torsvik *et al.* 1987) reactivation of Caledonian thrusts.

Although not well dated, a Permo-Triassic rift event has been postulated by several workers in the North Sea (Ziegler 1978, 1982; Eynon 1981; Badley *et al.* 1984, 1985, 1988; Beach *et al.* 1987; Frost 1987; Giltner 1987; Scott and Rosendahl 1989; Thorne and Watts 1989). The sediment transport systems and the resultant sequence-signatures from these basins are best known from the Hebridean Province off north-west Scotland (Steel 1974, 1977). In the literature there is considerable uncertainty as to the dating of the crustal thinning and thermal subsidence stages related to this event.

The timing of the Jurassic–Cretaceous rift episode is well constrained (Ziegler 1982; Brown 1984; Giltner 1987; Beach *et al.* 1987; Badley *et al.* 1988; Thorne and Watts 1989), even though the tectonic setting for this event is debated (Beach *et al.* 1987; Badley *et al.* 1988).

Regarding the late Cretaceous–early Tertiary subsidence (Hamar *et al.* 1980; Beach *et al.* 1987; Frost 1987) it is highly questionable if it should be regarded as a separate extensional event (Badley *et al.* 1988; Donato and Tully 1981).

The timing and duration of the main tectono-sedimentological events in the northern North Sea as proposed by the different workers in the area are summarized in Fig. 8.3. In the present work in the northernmost North Sea we focus on the Permo-Triassic event and its related subsequent thermal subsidence, and the late Jurassic–Cretaceous event.

8.4. Description of key cross-sections

The structural and sedimentological evolution of the northern North Sea Basin has been considered to be relatively simple and consistent from section to section along strike (Giltner 1987). The present study, however, reveals significant inhomogeneities in this simple picture. This effect is well illustrated in the two transects (lines SG8043-101 and SG8043-403 for locations, see Fig. 8.1).

Several regional and semi-regional unconformities have been identified. The most prominent are represented by the transition between basement and the Triassic (U1 in Fig. 8.4) and the base of the Draupne Formation (U3). In addition there are steeply-dipping growth sequences within the Lower Triassic, but these seem to flatten out progressively into Middle and Upper Triassic strata which are broadly conformable with the overlying Lower Jurassic sequence. Also, rotated fault blocks in places may have produced local unconformities near the base of the Middle Triassic (U2).

The following description of the graben's structural elements refers to the graben as developed from Late Jurassic time onwards. As will be demonstrated later, the structural picture prior to this may have been very different.

8.4.1. *Line SG8043-403; structural description*

This line (Fig. 8.4), which reveals a typical section across the northern Viking Graben, strikes NW–SE and crosses the graben axis at about 61°N; it is referred to as 'the northern area' in the following text. To its extreme east a faulted shallow basement is identified. The faults in this segment are characteristically normal syn- and antithetic faults, locally with minor sediment-filled halfgraben-like basins.

The Øygarden Fault Zone is an extensional fault with westerly vergence separating the shallow basement area from the Horda Platform. A small horst is commonly developed in its footwall, whereas its hangingwall geometry varies along strike. Some sections display a narrow roll-over anticline in the hangingwall, whereas normal drag is seen in others. In line SG8043-403 the Jurassic and Cretaceous parts of the sequence situated in the hangingwalls of the faults of the Øygarden Fault Zone dip regionally towards the west, i.e. towards the basin. On the contrary, the late Palaeozoic units below the unconformity U1 dip in the opposite direction, i.e. towards the Øygarden Fault Zone. These differences may be indicative of reactivation of the faults of this zone.

Both the Mesozoic and late Palaeozoic mega-units can be followed westwards across the faulted Horda Platform where an abrupt change in structural dip occurs at one of the major faults (Fault A, Fig. 8.4). Crossing this fault going from the Norwegian mainland towards the Viking Graben, the dip in the Jurassic sequence changes from westerly to easterly. On the other hand, the uppermost Jurassic and the Cretaceous units still dip towards the west, down-lapping the previous sequence. It is also noted that some faults, which are extensional structures, have geometries indicative of reactivation.

The Horda Platform area is bordered to the west by the eastern graben margin fault system. This is seen as a series of normal faults at the Mesozoic levels. The eastern faults in this system border easterly-tilted fault blocks, whereas the faults closer to the graben axis are associated with westerly-tilted blocks. Here the number of antithetic faults also increases. The overlying Cretaceous sequence laps onto the basin margins.

The eastern margin of the Viking Graben *sensu stricto* may be defined at the point where the tilt of the fault blocks shifts from easterly to westerly. This point coincides with onlap of the Lower Cretaceous sequence in line SG8043-403, and west of this the Mesozoic sequence dips westward. This is consistent across the graben structure all the way down to the deepest point of the graben as defined at the base Cretaceous level.

Even though the general westerly dip is almost constant in this part of the graben, considerable relief is seen at the Mesozoic levels, as emphasized by the horst on the eastern side of the graben axis, and a fault-bounded depression with a large hangingwall anticline occurring in the deepest area of the graben. The depression is filled by the oldest Cretaceous sediments in the system (?pre Valanginian). The western shoulder of the depression coincides with a subplatform (Gabrielsen 1986) which in turn is bounded by a series of westerly rotated fault blocks defining the western graben margin fault system.

The Tampen Spur consists of several large, rotated fault blocks with internal listric fault systems. The rotated fault blocks, which contain several of the giant oil fields of the Viking Graben province like Gullfaks, Snorre and Statfjord, are delineated by large faults with deep roots.

8.4.2. *Line SG8043-101; structural description*

This line, offering an E–W-section across the Horda Platform immediately north of 60°N (referred to as 'the southern area' in the following description), displays a very different geometry from that seen in line SG8043–403.

The master fault of the Øygarden Fault Zone in this section is also an extensional fault, but in contrast to the previously described section, it dips east; it does not show the same signs of reactivation as recorded further north. On its eastern side it has a parallel fault with a westerly throw; together these structures define a narrow graben with a strong internal westerly tilt. Both these faults are best identified on the Triassic and older levels. In the shallower (Jurassic) parts of this section only minor off-sets are seen, and these faults cut backwards into the footwalls relative to their deeper counterparts, possibly indicating a level of detachment in the Upper Triassic part of the sequence.

Although dipping relatively steeply to the west, the Jurassic and Cretaceous strata are practically undeformed in the Horda Platform area. At deeper levels, however, stratigraphic units with strong westerly tilts are identified. The rotation seems to be associated with large easterly-dipping growth faults, of which two are seen in this line. Towards the western border of the platform the westerly tilt of the Mesozoic strata becomes less pronounced, and the

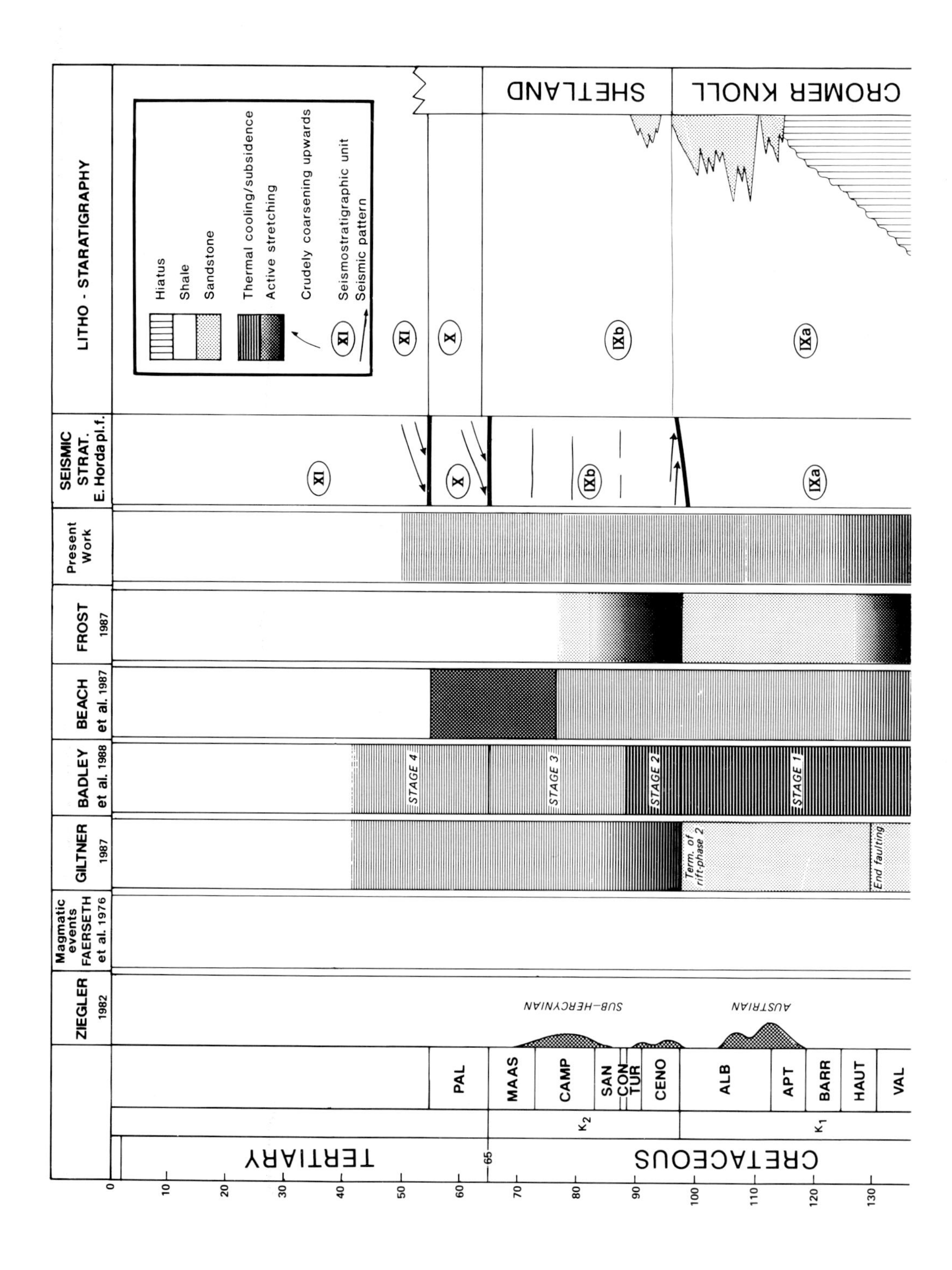
LITHO - STARATIGRAPHY
SEISMIC STRAT. E. Horda pl.f.
Present Work
FROST 1987
BEACH et al. 1987
BADLEY et al. 1988
GILTNER 1987
Magmatic events FAERSETH et al. 1976
ZIEGLER 1982
Hiatus
Shale
Sandstone
Thermal cooling/subsidence
Active stretching
Crudely coarsening upwards
Seismostratigraphic unit
Seismic pattern
XI
X
IXb
IXa
SHETLAND
CROMER KNOLL
STAGE 4
STAGE 3
STAGE 2
STAGE 1
Term. of rift-phase 2
End faulting
SUB-HERCYNIAN
AUSTRIAN
PAL
MAAS
CAMP
SAN
CON
TUR
CENO
ALB
APT
BARR
HAUT
VAL
K_2
K_1
TERTIARY
CRETACEOUS
65
0
10
20
30
40
50
60
70
80
90
100
110
120
130

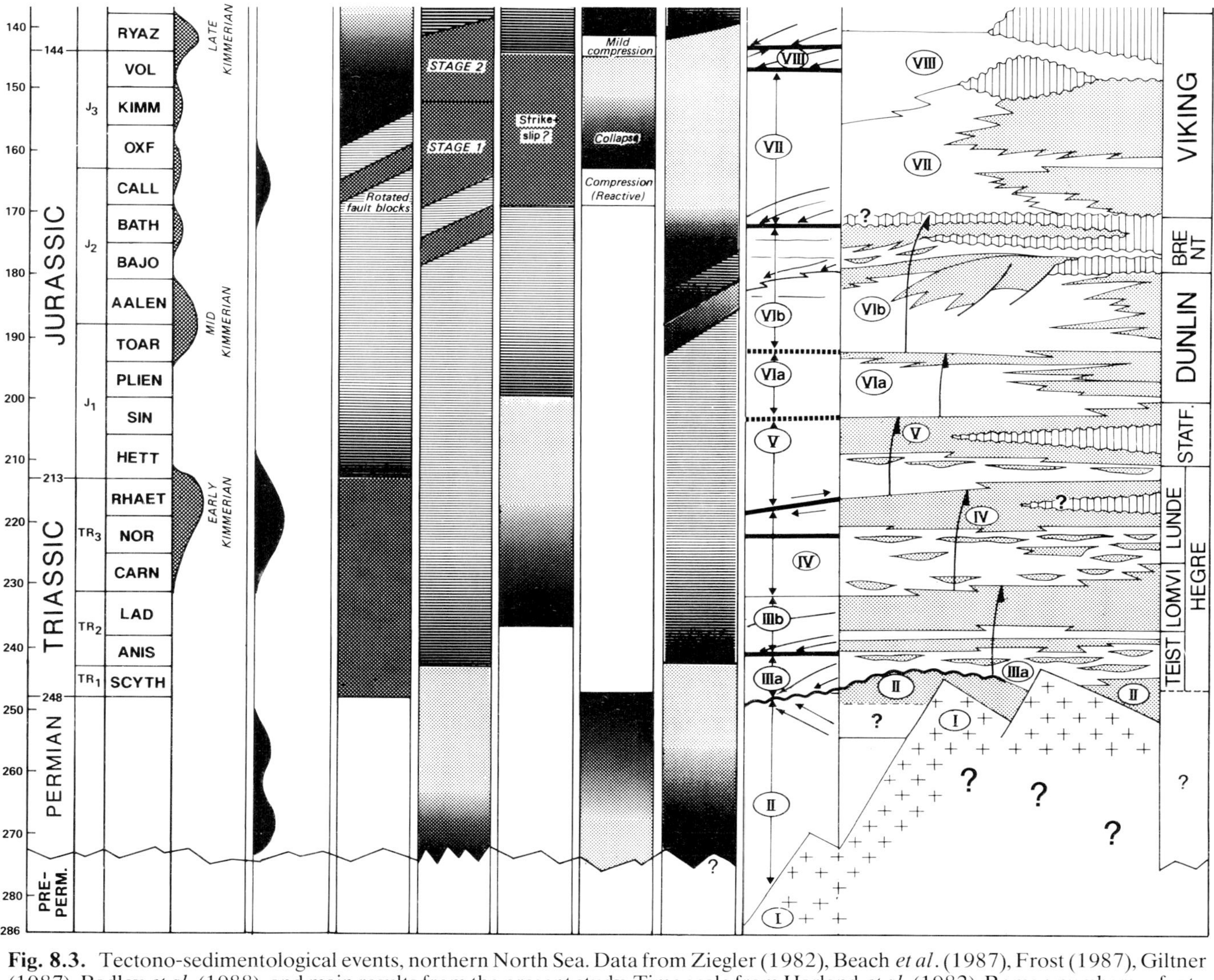

Fig. 8.3. Tectono-sedimentological events, northern North Sea. Data from Ziegler (1982), Beach *et al.* (1987), Frost (1987), Giltner (1987), Badley *et al.* (1988), and main results from the present study. Time scale from Harland *et al.* (1982). Roman numbers refer to sedimentary sequences described in text.

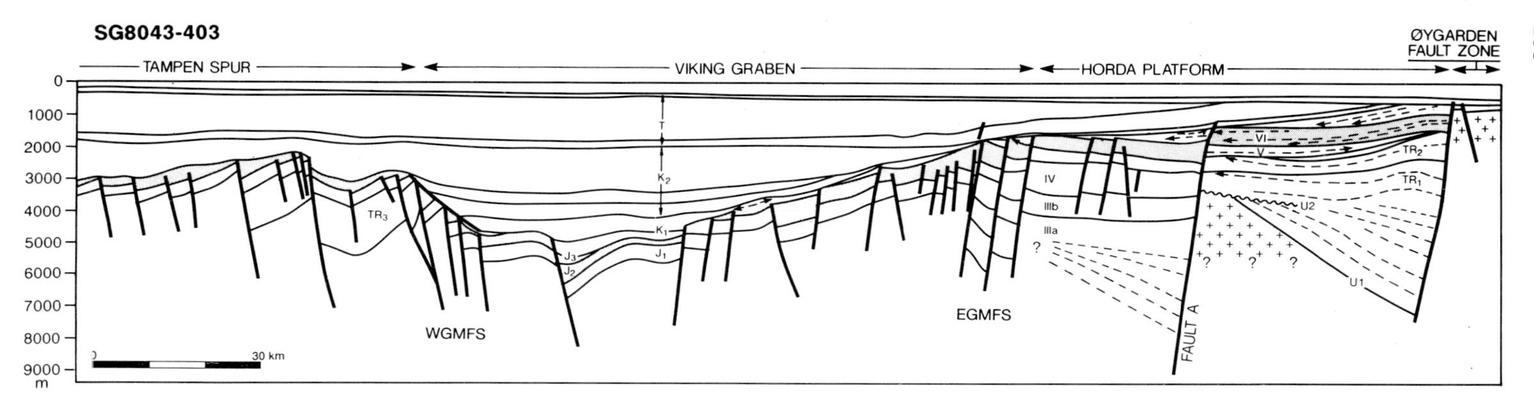

Fig. 8.4. Geoseismic section, line SG8043-403. For location of line, *see* Fig. 8.1. U1, U2 and U3 are major unconformities. U3, which is not marked specifically in the figure, is set at the base of the Cretaceous sequence. WGMFS and EGMFS are western and eastern graben margin fault systems respectively. Roman numbers refer to seismostratigraphic units described in text; *see also* Fig. 8.3. The Lower/Middle Jurassic (sequence VI) is shaded for clarity. Note vertical exaggeration which implies that fault geometries are not real.

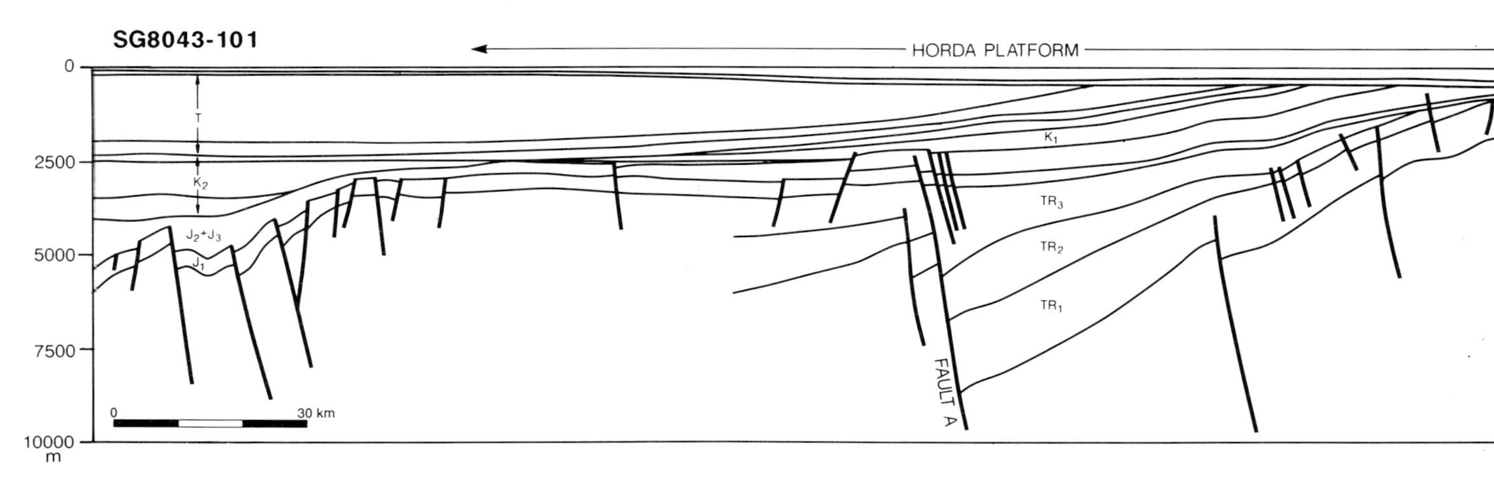

Fig. 8.5. Geoseismic section, line SG8043-101. For location of line, *see* Fig. 8.1. Roman numbers refer to seismostratigraphic units described in text; *see also* Fig. 8.3. Note vertical exaggeration which implies that fault geometries are not real.

few faults identified are mainly westerly-dipping planar normal faults.

The eastern margin fault system as seen on base Cretaceous level is not well developed in this area, and the margin of the graben has the characteristics of a flexure. At the intra Jurassic level, however, the sequence is heavily faulted. As in line SG8304-403, the eastern margin of the Viking Graben *sensu stricto* is set at the flexural line where the Mesozoic strata reveal strong westerly tilts. Even here these units are heavily structured.

8.4.3. *Comments on the general structure of the Horda Platform and the Viking Graben*

In previous descriptions and cross-sections, the northern Viking Graben is described as a symmetric structure with a well defined graben floor, which is delineated by systems of normal rotational or non-rotational normal faults (Ziegler 1981, 1982; Giltner 1987; Klemperer 1988). Alternatively, an asymmetric simple-shear model has been depicted, with a major detachment dipping east below the Horda Platform and associated antithetic faults in the same area (Beach 1986; Beach *et al.* 1987; Gabrielsen 1988).

However, detailed mapping of the interior graben (Færseth 1984) reveals that the graben floor is heavily segmented, and that the geometry may vary along the graben axis. The above descriptions seem to confirm the latter picture, and we would like to emphasize the following points:

(1) the eastern and western margin fault systems of the Viking Graben *sensu stricto* have highly variable geometries along strike, and the eastern and western graben margins show contrasting geometries;
(2) even though the larger part of the Horda Platform is dominated by westerly-dipping, north–south-striking normal faults, an area around 60°N has the opposite configuration; this anomalous structuring probably mirrors a deep structural grain and accordingly also the depositional pattern developed in late Palaeozoic times.

8.4.4. *Seismo-stratigraphic/stratigraphic subdivision*

The seismic sections described above have been depth-converted and 'back-stripped' with the use of program ECHO/PAL (Welldrill 1988; Figs 8.6 and 8.7). The sections have not been decompacted. Due to poor data quality in the central part of the Viking Graben, the following seismo-stratigraphic descriptions lean heavily on observations made on the Horda Platform.

Several seismic stratigraphic sequences separated by regional or local unconformities (sequences I–XI) have been identified, and are described in the following paragraphs, with general reference to lines SG8043-403 and SG8043-101 (Figs 8.4 and 8.5). These observations are in turn compared to, and integrated with, independent data on the basin's stratigraphic sequences (Fig. 8.3).

Pre Permo-Triassic sequences

Sequence Ia (basement) is seen only locally in a few sections (e.g. lines SG8043-203) and is not identified with certainty in lines SG8043-403 or 101. In line SG8043-203, it is seen as an uneven surface which may be correlated to top basement as drilled in Well 31/6-1. At this site high-grade gneisses were encountered.

Sequence Ib. The basement reflection has a stratified unit on top of it. In general, it is difficult to separate the basement reflection from the reflections of Sequence II in the seismic sections, though in places the latter can be seen to onlap presumed basement by a slight angular unconformity (U1). This unit, which can be identified only locally, is also truncated by unconformities U1 and U2, and may be eroded along the leading edges of the rotated fault blocks (Fig. 8.4). Sequence Ib was accordingly not penetrated in Well 31/6-1, and its age may only tentatively be given to be late Palaeozoic.

Permo-Triassic sequences

A Permo-Triassic rift phase is frequently referred to in the literature, but details from this event are sparse. Where well penetration is deep enough on the Horda Platform and on the Tampen Spur, the data suggest that this phase terminated in Scythian times, as judged by the overlap and drowning of the tilt-block topography (Badley *et al.* 1988). The initiation of this phase is not datable in the study area but, based upon regional information, it is generally

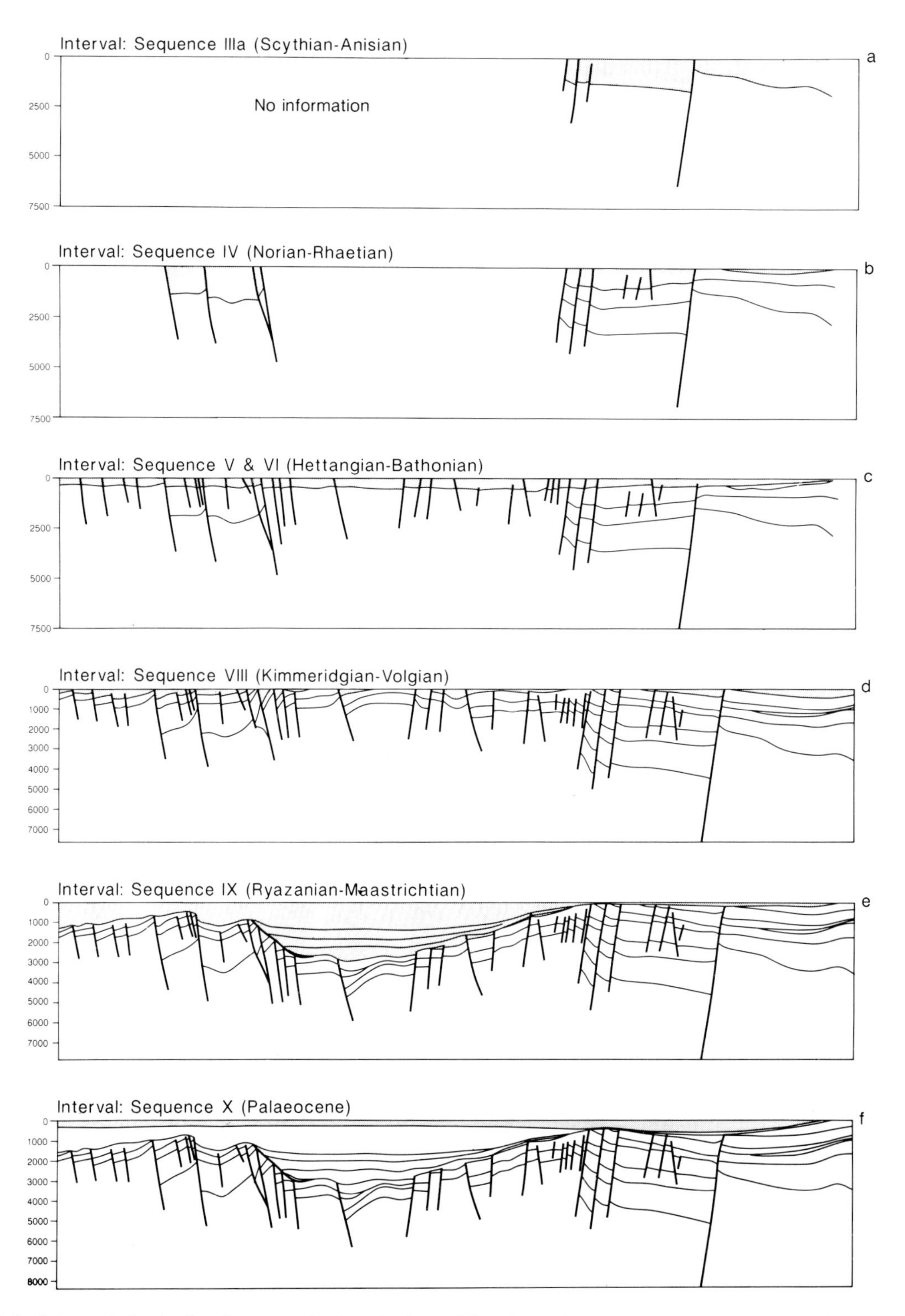

Fig. 8.6. Schematic basin development; 'back-stripping' of line SG8043-403 (depth converted, vertical scale in metres). Note vertical exaggeration. For location of line, *see* Fig. 8.1.

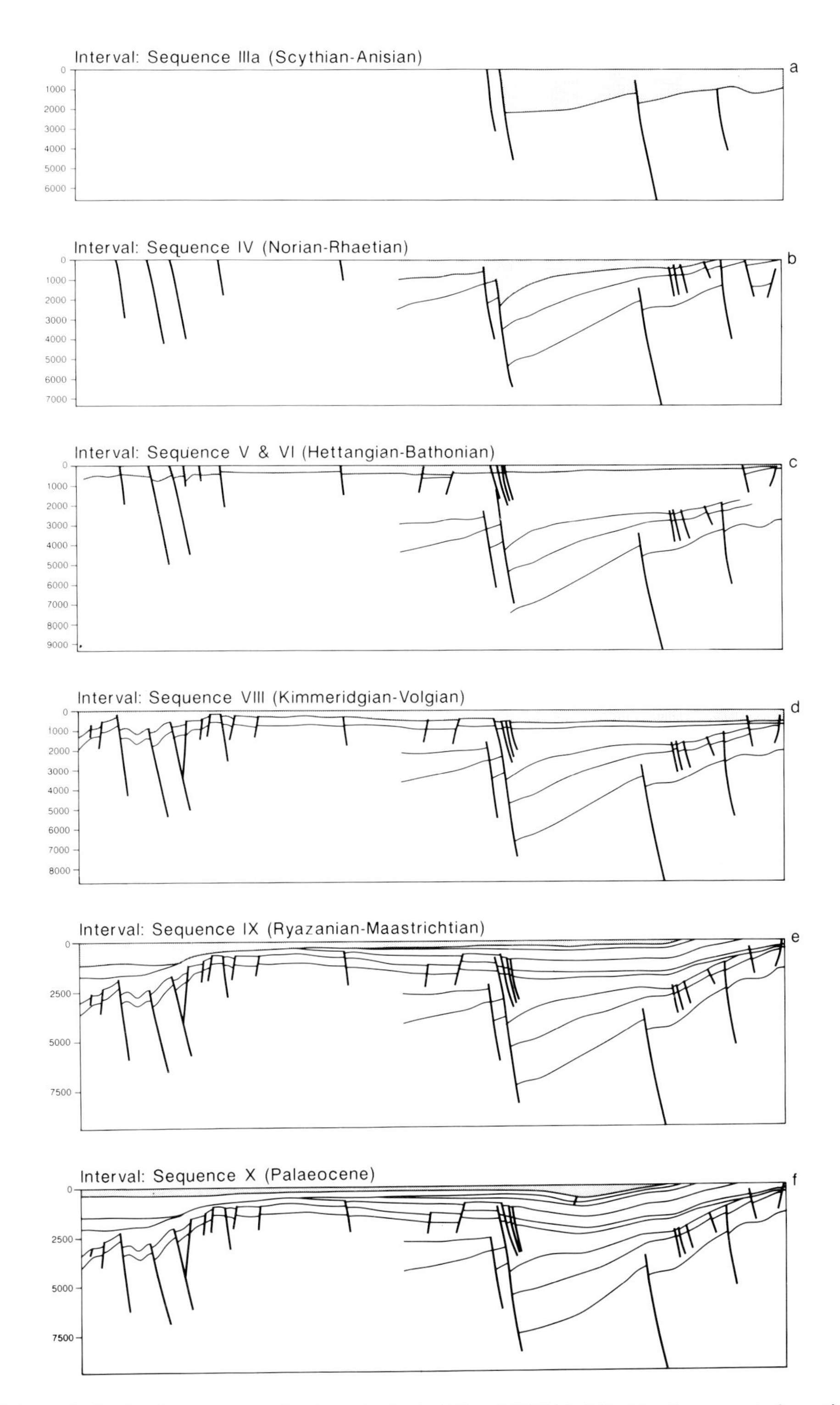

Fig. 8.7. Schematic basin development; 'back-stripping' of line SG8043-101 (depth converted, vertical scale in metres). Note vertical exaggeration. For location of line, see Fig. 8.1.

assumed to have been during the Permian. The information which can be extracted about this rift event from the seismic lines used in the present study is very limited, but tilted fault-blocks with much greater throws than are seen at Jurassic levels are sporadically distinguished, suggesting extensive faulting and rotation.

Sequence II (?late Palaeozoic–early Triassic) is so far identified only in the easternmost parts of the sections. It is developed by the infilling of low areas between strongly tilted fault blocks (easterly tilt in line SG8043-403, Fig. 8.4, and westerly tilt in line SG8043-101, Fig. 8.5). This sequence seems to lap onto sequences Ia and Ib as seen on the leading edges of some of the major rotated fault blocks; in places it displays a divergent seismic reflection pattern (e.g. line SG8043–101).

The unconformity defining the top of sequence II possibly corresponds in places to the 'Hardegsen unconformity' as defined by W. H. Ziegler (1975). The uneven topography, which is seen locally at the unconformity, is believed to reflect differences in lithology and hence differences in resistance to erosion of the strata of sequence II.

The deposits of sequence II were laid down directly on basement (sequence Ia) and sequence Ib in a late Palaeozoic basin of uncertain extent and geometry, where synsedimentary faulting seems to have been an important factor in the development. Sparse well data suggest that at least 7–800 metres of earliest Triassic sediments infilled the various half-graben structures. These deposits are often dominantly sandy, but also contain thick shaly units in places. It is likely that the shaly sections occur preferentially adjacent to the active fault scarps, where lakes or terminal floodbasins developed. The extent to which this part of sequence II is truly 'syn-rift' (i.e. condensing upflank within the halfgrabens), or is passively onlapping an already created topography is still uncertain.

Sequence III (Scythian–Ladinian). In the north-eastern part of the Horda Platform area sequence III appears in places to lap on to the early Triassic erosional surface, filling into the relief associated with the unconformity. In the south (line SG8043-403), the sequence thickens towards faults with easterly vergence. There is no evidence of rotation of fault blocks in association with synsedimentary faulting in this area. This is regarded as an indication that the movements took place along steep planar faults.

Well data and correlations show sequence III to be about 1000 metres thick on the central Horda Platform; there is a slight thinning to the east of the region. In detail, however, thickening at the hanging-walls of the major fault blocks is recorded. Sequence III encompasses the upper part of the Teist Formation and the Lomvi Formation (Fig. 8.3). There is a crude overall increase in the sand/shale ratio upwards. Sequence III is of great lateral extent, and can be recognized from the Horda Platform, across the Tampen Spur, and up to 62°N. The unit may, from seismo-stratigraphic evidence, be divided into two sub-sequences.

Sequence IIIa (Scythian–Anisian) oversteps the Triassic tilted fault blocks, and is overlain by sequence IIIb (Figs. 8.4 and 8.5). The strata of this sub-sequence laps on to the crests of the rotated fault blocks previously described. It displays abrupt thickness variations across major faults (Fig. 8.6a), and downlap configuration is locally recorded (line SG8043-403). The sequence, which is correlated to the Teist Formation, consists of fine-grained lacustrine and floodbasin deposits with occasional belts of sandy, ephemeral stream channels and sheetflood deposits. At higher levels the stream channels merge to create laterally extensive sand bodies (Fig. 8.8a).

Sequence IIIb (intra Anisian–Ladinian) follows the same patterns as sequence IIIa. Its base is limited by a slight unconformity below low-angle clinoforms, and a seismic reflection pattern characterized by downlap from east towards west. The unit varies in thickness, but shows less regional thickness variations than sequence IIIa. Locally, however, there are clear indications of synsedimentary fault activity, with abrupt thickness changes across major faults, particularly in the southernmost area. Sequence IIIb is dominantly sandy, and is interpreted as a fluvial deposit. It can be recognized over much of the northernmost North Sea (Lomvi Formation). The laterally extensive nature of the sands suggests low rates of change of base level during this period.

Sequence IV (Carnian–Rhaetian) also follows the same general depositional pattern. It thins and disappears at the easternmost graben margin, lapping on to the top of sequence IIIb; it has an internal concordant seismic configuration. The effects of an acceleration of subsidence along some of the major faults is recorded in the southernmost areas (Honian-Rhaetian, Fig. 8.7b). Eastern areas and crests of some fault blocks were extensively eroded towards the end of Triassic times, as seen from the local erosional truncations at the top of the sequence.

Well data indicate that sequence IV is some 550 metres thick, but thins on the eastern reaches of the Horda Platform. It consists of the sandy/shaly lower part of the middle member and the overlying sand-rich parts of the upper members of the Lunde Formation. The shalier portions of sequence IV originate from lacustrine and floodbasin deposition, whereas the sandy parts represent the deposits of ephemeral streamflow in broad, shallow channels. The fluvial channels occur both on alluvial fans and as floodplain river systems.

Sequence V (Rhaetian–Sinemurian) is some 300–600 metres thick, and includes a shaly/silty unit below Statfjord Formation sandstones (included in the uppermost Lunde Formation in some classifications, or within the Statfjord Formation as the Raude Member in others; Deegan and Scull 1977, Vollset and Doré 1984), and the Statfjord Formation itself. Both on the Horda Platform and the Tampen Spur, sequence V is thus of broadly upward-coarsening character, but shows greater lateral variability in lithology and in thickness than is the case in sequences III and IV. There are clear signs of local fault movement controlling lateral thickness and lithology changes in this unit, and the activity seems to have spread out to incorporate several minor faults rather than a few master faults, as in the case of the underlying sequence.

The lower part of the sequence represents lacustrine, floodbasin, and, in places, brackish lagoonal deposits (Fig. 8.8b). The sandstones were deposited largely by low-sinuosity and braided streams, and in the uppermost part of the sequence, by shoreline processes.

Sequence VI (Sinemurian–Bathonian) is divided into two sub-units, both of which follow the general outline of sequence V, having an internal parallel pattern and concordant tops and bases at a regional scale. Also, the basin configuration was similar to that of deposition of sequence V.

Sequence VIa (Sinemurian–Toarcian) includes the Nansen member of the Statfjord Formation, the Amundsen, Johansen, Burton, and Cook Formations, and is some 200–300 metres in thickness on the eastern edges of the Viking Graben and the Horda Platform. The succession consists of alternating shaly and sandy units. The shale-dominated Amundsen, Johansen, and Burton Formations represent mainly off-shore and prograding shelf conditions formed during periods of significant basinal subsidence, or of eustatic sea-level rise. The Cook Formation shelf sandstones extend farthest out from the basin margin (Fig. 8.8c and 8.8d) and represent minimum rates of subsidence.

Sequence VIb (Toarcian–Bathonian) consists of Drake Formation shales and the overlying sand-rich Brent Group, generally exceeding 300 metres in thickness in the Viking Graben, thinning considerably on to the Horda Platform. In the shale-dominated lower part of the sequence there occur a number of basin margin-attached sandstone units (Drake Formation sand units and Oseberg/Broom Formations) which again represent basin-subsidence 'minima', or eustatic falls of sea level. The overlying Brent Group also represents sediment supply from the basin margins, but abundance of sediment supply or slow subsidence rates caused extensive regression and the development of axial sediment transport, so that the Brent delta system (Fig. 8.8e) prograded generally northwards in the northern Viking Graben. Maximum spreading of sand occurred by mid Bajocian times, and reached the northern Tampen Spur area. Increasing rates of relative sea level rise from then on (argued to have been caused by increasing subsidence rates by Graue *et al.* 1987) led to a southwards and basin marginwards retreat of the deltaic system. This led to deposition of the Heather Formation shales and the initiation of sequence VII.

The late Jurassic was dominated by increasing destabilization of the basin area. It has been generally accepted that the Viking Graben started to develop as a major topographic feature in Callovian times (Ziegler 1982; Leeder 1983; Giltner 1987; Badley *et al.* 1988). However, according to the present reconstructions (Figs 8.6d and 8.7d), the basin gradient was moderate into latest Jurassic times even though increasing fault activity and extensive fault block rotation may have contributed to local strong relief. The existence of the Viking Graben as a continuous regional graben at this time may therefore be questioned. It is also important to note that the extension was unevenly distributed, and that rotation of fault blocks was more significant

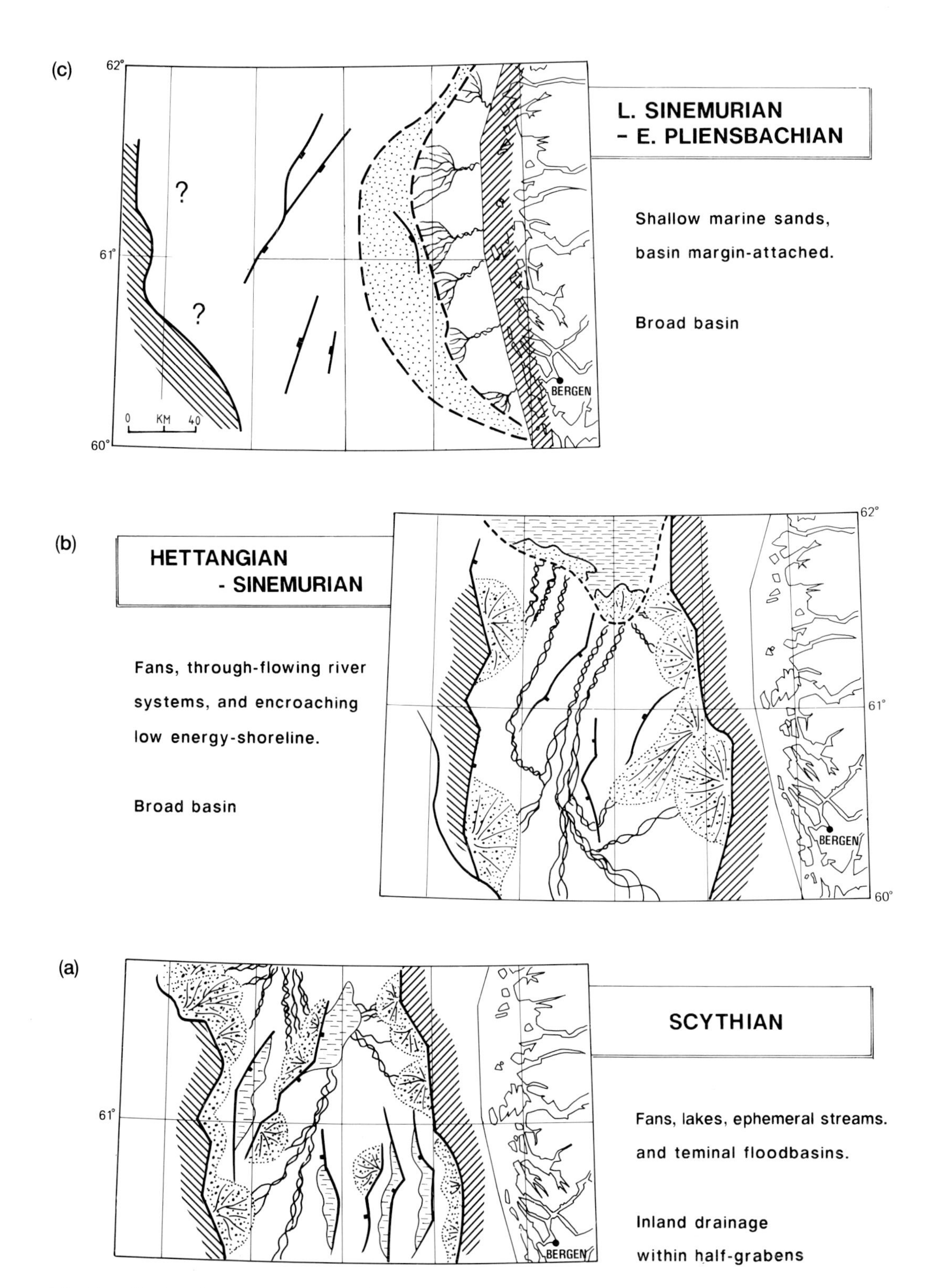

Fig. 8.8. Simplified palinspastic reconstructions. **a, b, c** Scythian-Pliensbachian. **d, e, f,** Pliensbachian-Oxfordian.

(f)

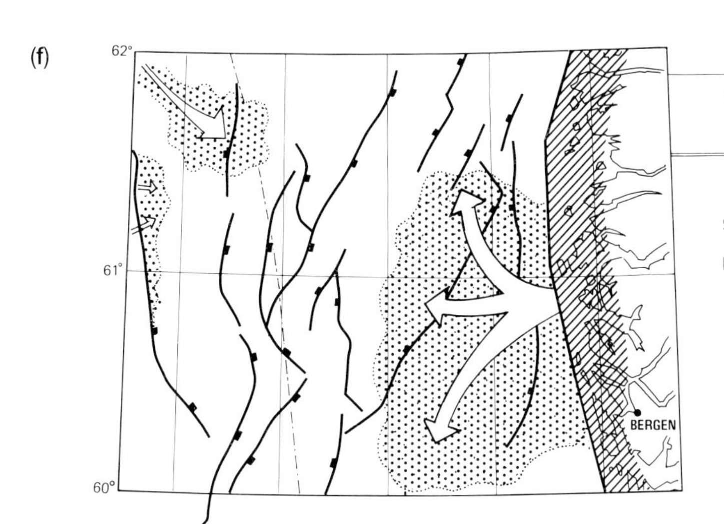

CALLOVIAN
- OXFORDIAN

Shallow marine sands,
basin margin-attached

(e)

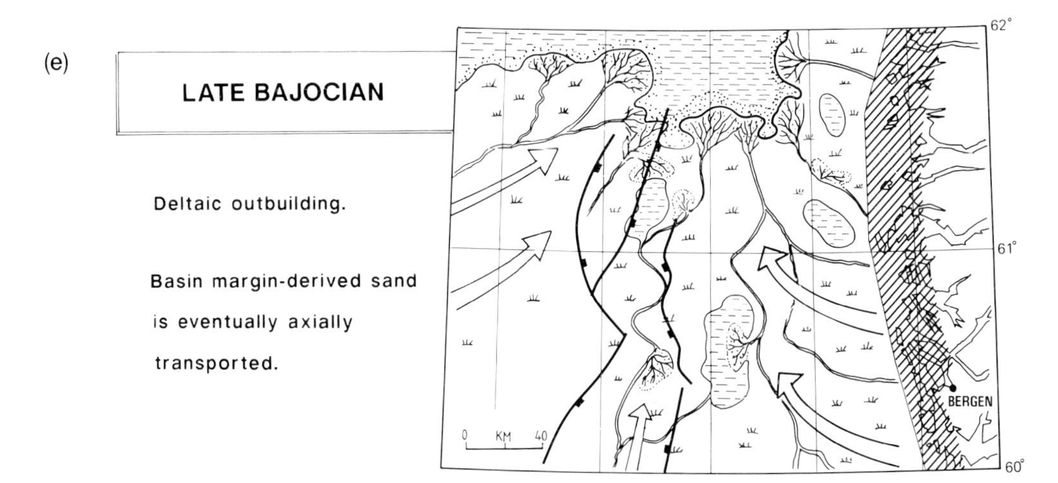

(d)

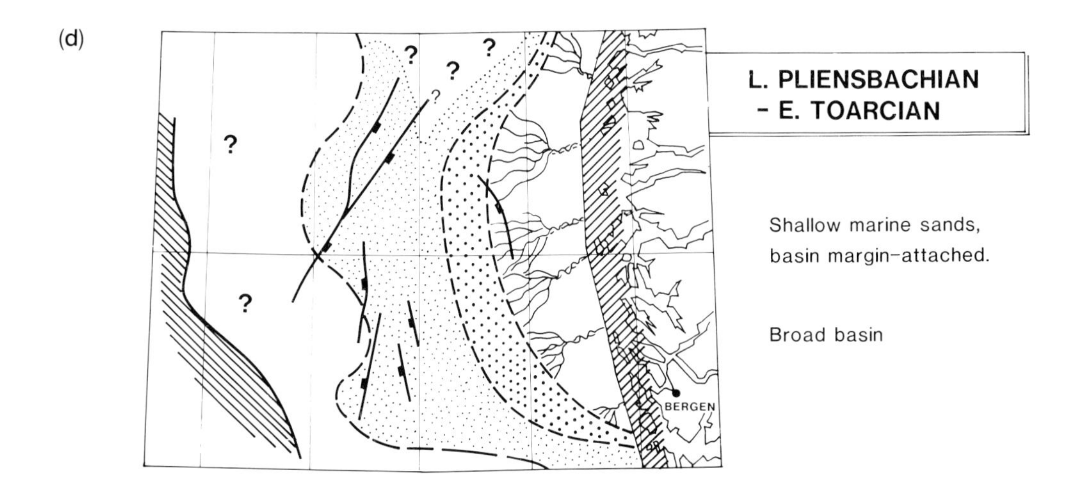

closer to the present graben centre than within the platforms.

The Late Jurassic rift sequence (Bathonian–Ryazanian) contains the shaly Heather and organic-rich Draupne Formations as well as their sandy equivalents (Krossfjord, Fensfjord, and Sognefjord Formations) on the Horda Platform (see also Hellem *et al.* 1986). The latter sandy units are again basin-margin attached, and generally do not prograde into or along the axial reaches of the basin because of the high stand of relative sea level, caused by increasing rates of subsidence and/or eustatic sea-level rise. General aspects of sands within this interval have been considered by Færseth and Pederstad (1988).

During deposition of *Sequence VII* (Callovian–Kimmeridgian) and *Sequence VIII* (Kimmeridgian–Ryazian), the northern and southern areas again show indications of differences in structuring. In both areas the influence of faulting and fault-block rotation increased but, in the south, the graben axis was more clearly developed at this time than to the north, where local uplift and erosion of the axial part of the graben area took place (Figs 8.6 and 8.7).

Sequences VII and VIII are characterized by well developed clinoforms over the Horda Platform, indicative of westerly sediment transport (Fig. 8.8f); the sequences are locally separated by an unconformity at the top of the Viking Group.

The Cretaceous and Tertiary sequences

The closing of the Jurassic period announced the most dramatic shift in basin development of the study area. The latest Jurassic–earliest Cretaceous periods are characterized by development of several regional and local unconformities, caused partly by eustatic changes (Vail and Todd 1981; Rawson and Riley 1982), and partly by tectonic events (Rawson and Riley 1982).

Sequence IX, which incorporates the entire Cretaceous sequence, is thin or lacking on the graben flanks. In the platform areas, however, the thickness of the Cretaceous sequence may exceed 100 metres. The sequence thickens to more than 3000 metres in the centre of the graben where the early Cretaceous strata lap on to the graben margin. The only significant sand accumulations within the area are the Aptian–Albian turbidite sands, which are thickly developed (200–300 metres), but only extend 10 kilometres or so out from the basin margin north of the Horda Platform. These again are clearly of the margin-attached type, with their geometry being controlled by the balance between sediment supply and the high relative sea level stand in Early Cretaceous times.

The Tertiary and younger sedimentary units (*sequences X and XI*) are again characterized by westward-prograding clinoforms. In contrast to the upper Cretaceous sequence, however, the lower Tertiary exhibits a pronounced landward thickening, although the easternmost part of the sequence was eroded later (Figs 8.6f and 8.7f). In the southern area (Fig. 8.7f) it is noteworthy that a thickening of these sequences in the hangingwall of one of the major easterly-throwing faults may suggest a reactivation of this deep feature. In detail there are also indications of mild reactivation (inversion?) of other major faults of the Øygarden Fault Zone.

8.5. Discussion

Assuming that the sedimentary infilling of the northernmost North Sea basin from late Palaeozoic times can be considered in terms of sedimentary response during phases of active stretching (during the intervals Late Permian/Early Triassic, and late Mid-Jurassic/earliest Cretaceous in the considered case), and of sedimentary response during the intervening phases of thermal subsidence, then it is possible to evaluate whether the models suggested (8.2, 8.3) are in accordance with the data presented.

According to Gabrielsen (1986, 1988), the pre-rift stage is accompanied by initiation of deep and steep planar faults. On land in west Norway, a fracture system of regional scale parallels the coastline (Ramberg *et al.* 1977; Gabrielsen and Ramberg 1979). This is partly filled with alkaline dykes emplaced in three episodes from Early Permian to Mid-Jurassic (Fig. 8.3; Naterstad *et al.* 1975; Færseth *et al.* 1976). The dykes carry mantle-derived lherzolite nodules, and augite and spinel megacrysts, indicative of derivation from depths of approximately 60 km (Færseth 1976). Similar dykes of early Jurassic age have been encountered in the Egersund Basin off southern Norway (Furnes *et al.* 1982). Accordingly, fracturing and dyke emplacement may be taken as heralding initial stretching during both the Permo-Triassic and the Jurassic–Cretaceous events.

8.5.1. *The Permo-Triassic rifting*

Using this criterion, it is reasonable to assume the initiation of the Permo–Triassic event to have taken place in early Permian times. The subsequent crustal extension, which is visualized by synsedimentary rotation of major fault blocks away from the assumed basin centre, seems to have lasted well into Scythian times (end of deposition of sequence II).

Sequences IIIb–VIb (Lower Triassic–late Middle Jurassic) were laid down in a broad basin with a moderate gradient across the marginal areas. There is good evidence for synsedimentary fault activity in this period, but lack of rotation of single fault blocks are rare, and instead a regional rotation towards the assumed basin centre is observed. This is taken as an indication that the active stretching stage had terminated when deposition of sequence IIIb commenced.

From the pattern of sequence IV it is evident that rotation slowed down in most of the north-eastern part of the Viking Graben in Norian to Rhaetian times, probably indicating that the subsidence was an effect of thermal contraction. Even though there are exceptions from this pattern (see Fig. 7.7b), the transition from crustal stretching to thermal cooling is accordingly proposed to be a pre-Anisian event. Thus, we agree with Badley and colleagues (1988), who suggested that the Triassic–Mid-Jurassic sequence above the Early Triassic unconformity was deposited during the thermal cooling stage of the Permo-Triassic event.

In the opinion of the authors, the subsidence induced by the thermal contraction associated with the Permo-Triassic extensional event tended to slow down periodically from the Mid-Triassic.

Giltner (1987), based upon geoseismic sections across the Viking Graben taken from Ziegler (1982), estimated the Permo-Triassic stretching factor in the northern Viking Graben to be a maximum of 1.6 in the central part of the rift as presently seen, diminishing to 1.2 at the flanks. Using a different set of parameters, and based upon the model of Jarvis and McKenzie (1980), Beach and co-workers (1987) arrived at a similar stretching factor (1.5). Both Giltner (1987) and Beach and co-workers estimated the stretching to have taken place over approximately 40 Ma from Triassic to earliest Jurassic. Bearing these results in mind and comparing with the present datings, we would prefer to await better constrained data on the sediment thicknesses in the deeper graben area before the calculation of a reliable stretching factor is made. For the same reason, we question the basin configuration of the northern Viking Graben in Permian and Triassic times as recently proposed by Thorne and Watts (1989) and Scott and Rosendahl (1989).

8.5.2. *Mild late Triassic stretching?*

To disturb this simplistic picture, it is noted that a strong magmatic episode is also recorded in the late Triassic between *c*.230–210 Ma, i.e. Carnian–Rhaetian, (Færseth *et al*. 1976), and it may be speculated that the Permo-Triassic rift episode was superimposed by a separate extensional event. This tectonic episode coincides with initiation of a large number of faults (Figs 8.6c and 8.7c) which, although subordinate in size to the master faults of the area, may be said to represent a significant tectonic event. The structuring was accompanied by deposition of the particularly coarse-grained parts of the Statfjord Formation (Røe and Steel 1985), and may justify the 'early Cimmerian phase' frequently mentioned in the literature.

Intrabasinal highs (possibly exposing basement) and delineated by major faults were present in the basin. In this connection, the system of faults with easterly throws identified around 60°N (line SG8043-101) is of particular interest, and lends support to the assumption that transfer faults may have existed (Scott and Rosendahl 1989).

Considering the aftermath of the thermal cooling stage, and initiation of a new stretching event dated to early Jurassic by analogy to the magmatic activity reported by Færseth and co-workers (1976), the North Sea basin was a hybrid during the earliest Jurassic times, experiencing the effect of thermal cooling from the Permo-Triassic event followed by the initial extension of the Jurassic–Cretaceous event. This may be the explanation of relatively high synsedimentary fault activity as recorded during deposition of sequence V (Rhaetian–Hettangian), diminishing during deposition of sequence VI (Sinemurian–Bathonian).

From the Bajocian the effect of increased fault activity again is noticeable in the basin. High-standing crests of rotated fault blocks influenced the transportation system within the basin area, acting both as local sand sources and as barriers to transverse sediment transport. Following the regional

Oxfordian–Kimmeridgian transgression, these effects diminished before fault activity peaked, and a regression took place in Volgian times. This complicated interaction between tectonic and regional eustatic processes led to establishment of the 'late Cimmerian unconformity' (Rawson and Riley 1982), and deposition of sand bodies in a number of different positions relative to the basin margins (Færseth and Pederstad 1988).

From the reconstructions (Figs 8.4 and 8.5) it is particularly noted that there are considerable differences in the basin configurations when flattened to the Lower Cretaceous levels. On line SG8043-101, a considerable and rugged relief is evident, whereas, on the contrary, line SG8043-403 indicates local domal uplift and erosion of the Late Jurassic sequence. We are not yet prepared to explain these differences, but notice that modifications of the Viking Graben system by Late Jurassic strike-slip movements have been proposed recently (Beach 1985; Fossen 1989).

It is generally considered that low-angle listric fault processes halted in Ryazanian times (Ziegler 1982; Badley *et al*. 1988) marking the transition to the thermal subsidence stage; from this point on in the development, the axial transportation systems lost much of their significance.

8.5.3. *Influence of eustatic sea level changes*

An alternative control on the extensive sand distribution at the specific levels listed above could have been low rates of eustatic sea level rise, or intervals of eustatic sea level fall. This argument could be specially relevant for the widespread sand distribution in the upper parts of sequences III and VIb, since these levels correspond well with short-term eustatic sea level 'lows' (Haq *et al*. 1987). However, in general we suggest that there is likely to have been variation in subsidence rates within the post-rift interval, and that such variations in base level may well have triggered the types of lithological changes demonstrated above, not least because of the large scale of the sequences.

8.6. Conclusions

As discussed above and summarized in Fig. 8.6, we suggest that sand development and distribution was controlled by tectonics (and eustacy) in two different ways.

1. During stages of rifting and half-graben development, sand was usually 'locally' distributed (albeit in great thickness in places) because of topographic trapping and/or synsedimentary tilt-trapping. The relative distribution of sand and mud was dependent on sand availability and drainage patterns, and sand was commonly laterally derived from erosion areas. Fine-grained deposits commonly accumulate to great thicknesses in actively subsiding areas because standing water and/or sea level react quickly to base level change, and because fine-grained deposits are best preserved in such situations. Sand accumulation was less of a hallmark in the Jurassic–Cretaceous rift episode than in the Permo-Triassic episode, largely because of higher sea level stand and restricted sand availability during the former.

2. During stages of thermal cooling the area of basin development was more extensive, subsidence was often greater and basin floor topography less prominent than in the preceding stage, although synsedimentary faulting (on steep, planar faults) could be prominent in places. This caused sand to be distributed more widely and, in general, it is easier to correlate sequences across the basin. Nevertheless, we suggest that the main control on sand distribution in the thermal cooling stages was rate of subsidence on the broad scale. We have divided the thermal cooling stage following the Permo-Triassic rift event into five mega-sequences (sequences III, IV, V, VIa and VIb of Fig. 8.3), where we suggest there was a tendency to decreasing rates of subsidence in the late stages of each interval. We suggest that sand was more restricted in its distribution (either margin-attached, or trapped in channel belts) during periods of relatively rapid subsidence, and was more widely distributed (reached farther out in the basin) during intervals of less rapid subsidence and greater basin floor stability. This notion is supported by hiatuses preferentially developed in the upper part of sequences III to VIb.

Acknowledgements

The authors are indebted to Norsk Hydro a.s. for permission to publish this paper. Thanks also to colleagues in Norsk Hydro, to two anonymous referees whose comments and suggestions have been very helpful, and to Masaoki Adachi, Paul Gnanalingam, Mette Hole, and Hal Thomas, who drafted the figures.

References

Allmendinger, R. W., Hauge, T. A., Hauser, E. C., Potter, C. J., and Oliver, J. (1987). Tectonic heredity and the layered lower crust in the Basin and Range Province, western United States. In *Continental extensional tectonics* (ed. M. P. Coward, J. F. Dewey, and P. L. Hancock). *Geol. Soc. Special Publication*, **28**, 223–46.

Badley, M. E., Egeberg, T., and Nipen, O. (1984). Development of rift basins illustrated by the structural evolution of the Oseberg structure, Block 30/6, offshore Norway. *J. Geol. Soc. London*, **141**, 639–49.

Badley, M. E., Egeberg, T., and Nipen, O. (1985). Discussion on development of rift basins, illustrated by the structural evolution of the Oseberg structure, Block 30/6, offshore Norway. *J. Geol. Soc. London*, **142**, 933–4.

Badley, M. E., Price, J. D., Rambech Dahl, C., and Abdestein, T. (1988). The structural evolution of the northern Viking Graben and its bearing upon extensional modes of graben formation. *J. Geol. Soc. London*, **145**, 455–72.

Barton, P. and Wood, R. (1984). Tectonic evolution of the North Sea basin: crustal stretching and subsidence. *Geophys. J. Roy. Astron. Soc.*, **2(6)**, 987–1022.

Beach, A. (1985). Some comments on sedimentary basin development in the northern North Sea. *Scot. J. Geol.*, **21, 4,** 493–512.

Beach, A. (1986). A deep seismic reflection profile across the northern North Sea. *Nature*, **323**, 53–5.

Beach, A., Bird, T., and Gibbs, A. (1987). Extensional tectonics and crustal structure: deep seismic reflection data from the northern North Sea Viking Graben. In *Continental extensional tectonics*. (ed. M. P. Coward, J. F. Dewey, and P. L. Hancock). *Geol. Soc. Special Publication*, **28**, 467–76.

Brown, G. (1984). Jurassic. In *Introduction to the petroleum geology of the North Sea*. (ed. K. W. Glennie). Oxford, Blackwell Sci. Publ. 103–31.

Bryhni, I. and Skjerlie, J. F. (1975). Syndepositional tectonism in the Kvamshesten district (Old Red Sandstone), western Norway *Geol. Mag.* **112**, 593–600.

Campbell, C. J. and Ormaasen, E. (1987). The discovery of oil and gas in Norway: an historical synopsis. In *Geology of the Norwegian oil and gas fields* (ed. A. M. Spencer *et al.*), 1–37. London, Graham and Trotman.

Dé'Ath, N. G. and Schuyleman, S. F. (1981). The geology of the Magnus Field. In *Petroleum geology of the continental shelf of north-west Europe* (ed. L. W. Illing and G. D. Hobson), 342–51. London, Institute of Petroleum, Heyden.

Deegan, C. E. and Scull, B. J. (1977). A standard lithostratigraphic nomenclature for the central and northern North Sea. *Inst. of Geol. Sci. Report*, 77/25, Norw. Petrol. Directorate, Bull, **1**.

Dewey, J. F. (1982). Plate tectonics and the evolution of the British Isles. *J. Geol. Soc. London*, **139**, 317–412.

Donato, J. A. and Tully, M. C. (1981). A regional interpretation of North Sea gravity data. In *Petroleum geology of the continental shelf of north-west Europe* (ed. L. W. Illing and G. D. Hobson), 65–75. London, Institute of Petroleum, Heyden.

Eaton, G. P., Wahl, R. R., Prostka, H. J., Mabey, D. R., and Kleinkopf, M.-D. (1978). Regional gravity and tectonic patterns: Their relation to late Cenozoic epeirogeny and lateral spreading in the western Cordillera. In *Cenozoic tectonics and regional geophysics of the western Cordillera* (Ed. R. B. Smith and G. P. Eaton). *Geol. Soc. Amer.*, **152**, 93–106.

Eynon, G. (1981). Basin development and sedimentation in the Middle Jurassic of the northern North Sea. In *Petroleum geology of the continental shelf of north-west Europe* (ed. L. W. Illing and G. D. Hobson), 196–204. London, Institute of Petroleum, Heyden.

Færseth, R. B. (1978). Mantle-derived lherzolite nodules and megacrysts from Permo-Triassic dykes, Sunnhordland, western Norway. *Lithos*, **11**, 23–35.

Færseth, R. B. (1984). *Tectonic map of the northeast Atlantic*. Norsk Hydro a.s.

Færseth, R. B. and Pederstad, K. (1988). Regional sedimentology and petroleum geology of the marine, late Bathonian-Valanginian sandstone in the North Sea. *Marine Petrol. Geol.*, **5**, 17–33.

Færseth, R. B., McIntyre, R. M., and Naterstad, J. (1976). Mesozoic alkaline dykes in the Sunnhordland region, western Norway: ages, geochemistry and regional significance. *Lithos*, **9**, 332–45.

Fossen, H. (1989). Indication of transpressional tectonics in the Gullfaks oilfield, northern North Sea. *Marine Petrol. Geol.*, **6**, 22–30.

Frost, R. E. (1987). The evolution of the Viking Graben tilted fault block structures: a compressional origin. In *Petroleum geology of north-west Europe* (ed. J. Brooks and K. W. Glennie), 1009–24. London, Graham and Trotman.

Furnæs, H., Elvsborg, A. and Malm, O. A. (1982). Lower and Middle Jurassic alkaline magmatism in the Egersund sub-basin, North Sea. *Marine Geol.*, **46**, 53–69.

Gabrielsen, R. H. (1986). Structural elements in graben systems and their influence on hydrocarbon trap types. In *Habitat of hydrocarbons on the Norwegian continental shelf* (ed. A. M. Spencer *et al*), 55–60. Norw. Petrol. Soc., Graham and Trotman.

Gabrielsen, R. H. (1989). Reactivation of faults on the Norwegian continental shelf and its implications for earthquake occurrence. In *Causes and effects of earthquakes at passive margins and in areas with post-glacial rebound on both sides of the North Atlantic* (ed. S. Gregersen and P. Basham), 69–92. Amsterdam, Elsevier.

Gabrielsen, R. H. and Ramberg, I. B. (1979). Fracture patterns in Norway from LANDSAT imagery: results and potential use. *Proceedings, Norwegian Sea Symposium*, Tromsø 1979, Norwegian Petroleum Society, NSS/23, 1-28.

Gabrielsen, R. H., Ekern, O. F., and Edvardsen, A. (1986). Structural development of hydrocarbon traps, Block 2/2, Norway. In *Habitat of hydrocarbons on the Norwegian continental shelf* (ed. A. M. Spencer *et al*.), 129–41. Norw. Petrol. Soc. Graham and Trotman.

Gibbs, A. D. (1983). Balanced cross-section construction from seismic sections in areas of extensional tectonics. *J. Struct. Geol.*, **5**, 153–60.

Gibbs, A. D. (1987). Deep seismic profiles in the northern North Sea. In *Petroleum geology of north west Europe* (ed. J. Brooks and K. Glennie), 1025–8. London, Graham and Trotman.

Giltner, J. P. (1987). Application of extensional models to the northern Viking Graben. *Norsk geol. tidsskr.*, **67**, 339–52.

Graue, E., Helland-Hansen, W., Johnson, J., Lomo, L., Nøttverdt, A., Rønning, K., Ryseth, A., and Steel, R. (1987). Advance and retreat of the Brent Delta system, Norwegian North Sea. In *Petroleum geology of north west Europe* (ed. J. Brooks and K. Glennie), 915–38. London, Graham and Trotman.

Gray, D. I. (1987). Troll. In *Geology of the Norwegian oil and gas fields* (ed. A. M. Spencer *et al*.) 389–401. London, Graham and Trotman

Gray, W. D. T. and Barnes, G. (1981). The Heather Oil Field. In *Petroleum geology of the continental shelf of north-west Europe* (ed. L. W. Illing and G. D. Hobson), 335–41. London, Institute of Petroleum, Heyden.

Hallet, D. (1981). Refinement of the geological model of the Thistle Field. In *Petroleum geology of the continental shelf of north-west Europe* (ed. L. W. Illing and G. D. Hobson), 315–25. London, Institute of Petroleum, Heyden.

Halstead, P. H. (1975). Northern North Sea faulting. In *Jurassic North Sea Symposium* (ed. K. G. Finstad and R. C. Selley). Norwegian Petroleum Society, 10/1–38.

Hamar, G. P., Jacobsen, K. H., Ormaasen, D. E. and Skarpnes, O. (1980). Tectonic development of the North Sea north of the Central Highs. The sedimentation of the North Sea reservoir rocks. Norwegian Petroleum Society, 23 pp.

Haq, B. U., Harbendol, J., and Vail, P. R. (1987). Chronology of fluctuating sea levels since the Triassic. *Science*, **235**, 1156–67.

Harding, T. P. (1983). Graben hydrocarbon plays and structural styles. *Geol. en Mijnbouw,* **62**, 3–23.

Harding, T. P. (1984). Graben hydrocarbon occurrences and structural styles. *Amer. Assoc. Petrol. Geol. Bull.*, **68**, 33–362.

Hellem, T., Kjemperud, A., and Øvrebø. O. K. (1986). The Troll Field: a geological/geophysical model established by the LP085 Group. In *Habitat of hydrocarbons on the Norwegian continental shelf* (ed. A. M. Spencer *et al*.), 217–38. Norw. Petrol. Soc., Graham and Trotman.

Hellinger, S. J. and Sclater, J. G. (1983). Some comments on two-layer extensional models for evolution of sedimentary basins. *J. Geophys. Res.*, **88(B10)**, 8251–70.

Hossack, J. R. (1984). The geometry of listric growth faults in the Devonian basins of Sunnfjord, W. Norway. *J. Geol. Soc. London*, **141**, 629–37.

Jarvis, G. T. and McKenzie, D. P. (1980). Sedimentary basin formation with finite extension rates. *Earth Planet Sci. Lett*., **48**, 42–52.

Klemperer, S. L. (1988). Crustal thinning and nature of extension in the northern North Sea from deep seismic reflection profiling. *Tectonics*, **7**, 803–21.

Leeder, M. R. (1983). Lithospheric stretching and North Sea Jurassic clastic source lands. *Nature*, **305**, 510–4.

Lister, G. S., Etheridge, M. A., and Symonds, P. A. (1986). Detachment faulting and the evolution of passive continental margins. *Geology*, **14**, 246–50.

McKenzie, D. P. (1978). Some remarks on the development of sedimentary basins. *Earth and Planetary Science Letters*, **40**, 25–32.

Miller, E. L., Gans, P. B., and Garing, J. (1983). The Snake Range décollement: An exhumed mid-Tertiary ductile–brittle transition. *Tectonics*, **2**, 239–63.

Naterstad, J., Færseth, R. B., and McIntyre, R. M. (1975). Mesozoic alkaline dykes in the Sunnhordland region. In *Jurassic North Sea Symposium* (ed. K. G. Finstad and R. C. Selley). Norwegian Petroleum Society, JNNSS/8, 1-2.

Norton, M. G., McClay, K. R., and Way, N. A. (1987). Tectonic evolution of Devonian basins in northern Scotland and southern Norway. *Norsk Geol. Tidsskr.* **67**, 323–38.

Ramberg, I. B., Gabrielsen, R. H., Larsen, B. T., and Solli, A. (1977). Analysis of fracture pattern in southern Norway. *Geol. en Mijnbouw*, **56**, 295–310.

Rawson, P. F. and Riley, L. A. (1982). Latest Jurassic–early Cretaceous events and the 'Late Cimmerian Unconformity' in North Sea area. *Amer. Assoc. Petrol. Geol., Bull.*, **66**, 2628–48.

Røe, S.-L. and Steel, R. J. (1985). Sedimentation, sea-level rise and tectonics at the Triassic-Jurassic boundary (Statfjord Formation), Tampen spur, northern North Sea. *J. Petrol. Geol.*, **8**, 163–86.

Rønnevik, H. and Johnsen, S. (1984). Geology of the Greater Troll Field. *Oil and Gas J.*, **82**, **4**, 100–6.

Royden, L. and Keen, C. W. (1980). Rifting processes and thermal evolution of the continental margin of eastern Canada determined from subsidence curves. *Earth and Planetary Science Letters*, **51**, 343–61.

Sclater, J. G. and Christie, P. A. F. (1980). Continental stretching: an explanation of the post mid-Cretaceous subsidence of the Central North Sea Basin. *J. Geophys. Res.*, **85**, 3711–39.

Scott, D. L. and Rosendahl, B. R. (1989). North Viking Graben: An East African perspective. *Am. Assoc. Petrol. Geol. Bull.* **73**, 155–65.

Siedlecka, A. and Siedlecki, S., 1972. A contribution to the geology of the Downtonian sedimentary rocks of Hitra. *Norges Geol. Unders.*, **275**, 1–28.

Steel, R. J. (1974). New Red Sandstone piedmont and floodplain sedimentation in the Hebridean Province, Scotland. *J. Sedim. Petrol.*, **44**, 336–57.

Steel, R. J. (1976). Devonian basins of western Norway: sedimentary response to tectonism and varying tectonic context. *Tectonophysics*, **36**, 207–24.

Steel, R. J. (1977). Triassic rift basins of westernmost Scotland: their configuration, infilling and development. In *Proceedings, Mesozoic Northern North Sea Symposium* (ed. K. G. Finstad and R. C. Selley), Norwegian Petroleum Society, MNNSS.

Steel, R. J. (1980). Some aspects of Triassic sedimentation and basin development: East Greenland, North Scotland and North Sea. In *The sedimentation of the North Sea reservoir rocks*. Norwegian Petroleum Society, XVII.

Steel, R. J. (1988). Coarsening-upward and skewed fan bodies: symptoms of strike-slip and transfer fault movement in sediment basins. In: *Fan Deltas: Sedimentology and Tectonic Settings* (ed. W. Nemec and R. J. Steel), Blackie & Sons, Glasgow, 75–83.

Steel, R. J. and Gjelberg, J. (1989). Reservoir sand sequence development on the Norwegian Shelf (abstract). Abstracts. *Europ. Assoc. Petrol. Geol.*, Berlin 1989.

Steel, R. J. and Gloppen, T. G. (1980). Late Caledonian (Devonian) basin formation, western Norway: signs of strike-slip tectonics during infilling. In *Sedimentation in oblique-slip zones* (ed. P. F. Ballance and H. G. Reading) Spec. Publ. Int. Assoc. Sediment. **4**, 79–103.

Stewart, J. H. (1971). Basin and range structure—a system of horsts and grabens produced by deep-seated extension. *Geol. Soc. Amer. Bull.*, **82**, 1019–44.

Thorne, J. A. and Watts, A. B. (1989). Quantitative analysis of North Sea subsidence. *Amer. Assoc. Petrol. Geol. Bull.*, **73**, 88–116.

Torsvik, T. H., Sturt, B. ., Ramsay, D. M., and Vetti, V. (1987). The tecto-magmatic signature of the Old Red Sandstone and pre-Devonian strata in the Håsteinen area, western Norway, and implications for the later stages of the Caledonian Orogeny. *Tectonics*, **6**, 305–22.

Vail, P. R. and Todd, R. G. (1981). Northern North Sea Jurassic unconformities, chronostratigraphy and sea-level changes from seismic stratigraphy. In *Petroleum geology of the continental shelf of north-west Europe* (ed. L. W. Illing and G. D. Hobson), 216–35. London, Institute of Petroleum, Heyden.

Vollset, J. and Doré, A. (eds) (1984). A revised Triassic and Jurassic lithostratigraphic nomenclature for the North Sea. *NPD. Bull*, **3**; Stavanger, Oljedirektoratet.

Wernicke, B. (1985). Uniform-sense normal simple-shear of the continental lithosphere. *Can. J. Earth Sci.*, **22**, 108–25.

Whiteman, A., Naylor, D., Pegrum, R., and Rees, G. (1975). North Sea troughs and plate tectonics. *Tectonophysics*, **26**, 39–54.

Wood, R. J. (1981). The subsidence history of Conoco well 15/30-1, central North Sea. *Earth and Planetary Science Letters*, **54**, 306–12.

Wood, R. and Barton, P. (1983). Crustal thinning and subsidence in the North Sea. *Nature*, **302**, 134–6.

Ziegler, P. A. (1975). Geologic evolution of the North Sea and its tectonic framework. *Amer. Assoc. Petrol. Geol. Bull.*, **59**, 1073–97.

Ziegler, P. A. (1978). North-western Europe: tectonics and basin development. *Geol. en. Mijnbouw*, **57**, 589–626.

Ziegler, P. A. (1981). Evolution of sedimentary basins in north-west Europe. In *Petroleum geology of the continental shelf of north-west Europe* (ed. L. W. Illing and G. D. Hobson), 3–39. London, Institute of Petroleum, Heyden.

Ziegler, P. A. (1982). *Geological atlas of western and central Europe*. Shell International Petrol., Maatschappij B.V.

Ziegler, P. A. (1983). Crustal thinning and subsidence in the North Sea. *Nature*, **304**, 561.

Ziegler, W. H. (1975). Outline of the geological history of the North Sea. In *Petroleum and the continental shelf of north-west Europe. Vol. 1. Geology* (ed. A. W. Woodland), 131–50. London, Institute of Petroleum.

9 A kinematic model for the orthogonal opening of the late Jurassic North Sea rift system, Denmark–Mid Norway

A. M. Roberts, G. Yielding, and M. E. Badley

Abstract

Selective reading of recent literature on north-west European tectonics could readily leave the impression that the North Sea rift, from the Danish Central Graben to the mid-Norwegian Haltenbanken area, developed during the late Jurassic–Early Cretaceous as a large 'pull-apart' basin, within which all the major faults have a significant strike-slip component. Indeed it is not uncommon to find Late Jurassic structures locally interpreted with a compressional geometry in a purportedly-transtensional, regional strain-field.

This paper aims to present a simpler model in which each of the major basins of the rift, with the exception of the Inner Moray Firth Basin, opened by essentially orthogonal movement on the main intra-basinal and basin margin faults. Thus the Central Graben, Outer Moray Firth Basin, Viking Graben, Møre Basin, West Shetland Basin, and Mid-Norwegian margin are all interpreted as dip-slip basins, not pull-apart basins. The exception to this model, the Inner Moray Firth Basin, opened as a result of dextral strike-slip on the Great Glen fault system. This is the only major strike-slip fault system, as opposed to local-accommodation transfer faults, involved in the opening of the North Sea rift.

The north-west European margin can essentially be viewed as four crustal blocks, the Northern Highland, Grampian Highland, Norwegian/Danish, and Møre Plateau blocks, which moved apart from each other in the Late Jurassic. Vector triangles drawn to illustrate the relative motion of these blocks, and estimates of the minimum extension magnitude, support the conclusion that the main basins all opened orthogonally.

Outside the Inner Moray Firth, Late Jurassic structures previously interpreted as the product of strike-slip or compressional deformation can be better interpreted within an extensional framework. In particular it is likely that a number of uplifted and eroded structural highs within and flanking the rift were elevated not by compression but by syn-extensional footwall uplift.

9.1. Introduction

The last ten years or so have seen a complete revolution in our understanding of the sub-surface geology of the North Sea rift system. This has largely come about by integrating data obtained during hydrocarbon exploration with increasingly sophisticated basin-modelling techniques. Thus, in general terms, the idea is now accepted that the North Sea has undergone two significant periods of lithospheric extension (Permo–Triassic and Late Jurassic), followed by a period of lithospheric cooling (Cretaceous–present day, Eynon 1981; Barton and Wood 1984; Sclater *et al.* 1986; Giltner 1987; Badley *et al.* 1988). There is a less uniform agreement, however, on the kinematics of the rift development. While some authors still argue for a dominantly-extensional, dip-slip origin for the major graben systems within the rift (e.g. Viking Graben, Badley *et al.* 1988; Central Graben, Cartwright 1987; Roberts *et al.* in press; Mid-Norway, Bukovics *et al.* 1984), there has been an increasing tendency to invoke a significant strike-slip component within the rift system kinematics (e.g. Viking Graben, Beach *et al.* 1987; Larsen 1987; Fossen 1989; Central Graben, Gowers and Sæbøe 1985; Olsen 1987; Mid-Norway, Gabrielsen and Robinson 1984; Price and Rattey 1984; Caselli 1987).

One factor in the increased advocation of strike-slip movements within the North Sea rift has been the application of purely-geometric section-balancing techniques to basin-modelling problems (Williams and Vann 1987, has a summary of these techniques). It has been argued, for example (Gibbs 1985), that deep-seismic reflection geometries from the Central Graben cannot be adequately modelled by the application of the Chevron construction, and that this discrepancy largely results from a significant component of strike-slip displacement across the basin. The implicit assumption made here is not that the particular balancing technique might be inapplicable, but that the geology must be more complex than is readily apparent. The problem with such an argument is that balancing techniques are inherently non-unique, as Williams and Vann have pointed out. Different assumptions about the mode of deformation will lead to quite different conclusions about fault geometry.

Our own work throughout the North Sea rift system (performed largely on a confidential commercial basis) has led to us interpreting most areas within the rift as characterized by extensional dip-slip structures (Badley *et al.* 1984, 1988; Roberts *et al.* in press; Roberts and Yielding in press), with the exception of the Inner Moray Firth (Roberts *et al.* 1990). Thus our interpretations were clearly at variance with much, although not all, of the recently published literature. In particular many interpretations based on commercial reflection profiles favour the strike-slip models, while those interpretations based on geophysical basin-modelling generally assume a largely-extensional model. We therefore set out to investigate whether a kinematic model on the scale of the rift system (Danish Central Graben–Mid Norway) could be erected which would be consistent with our interpretations from disparate areas. At the same time we realized that, alongside such a kinematic model, new interpretations of structures attributed in the literature to 'strike-slip deformation' would be necessary. This paper presents the results of this investigation.

9.2. Interpretative methodology

Timing of extension

While we readily recognize the existence of two periods of extensive lithospheric stretching within the North Sea rift, of probable Triassic and Late Jurassic age, and significantly-more-minor extension during the Early–Middle Jurassic, it is in general the geometric effects of the Late Jurassic extensional episode which are most apparent on seismic reflection data. It is structures of this age which have been interpreted in the literature as the product of strike-slip deformation, and it is therefore on the kinematics of Late Jurassic lithospheric stretching that our model concentrates.

Assessment of extension vectors

Any realistic regional kinematic assessment must attempt to quantify extension vectors, i.e. displacement direction and magnitude, throughout the various components of the rift system. In our attempt to do so for the late Jurassic North Sea rift we have relied on a number of data sources and attempted to integrate them into a single model. These data sources fall into three categories;

(1) our own measurements of fault-controlled, upper-crustal extension, explained in detail in the section on the Viking Graben;
(2) published estimates of upper-crustal extension;
(3) published estimates of whole-crust extension from subsidence and refraction data.

While it could be argued that these data sources may not be directly comparable, we believe that an attempt must be made to integrate all available information.

Components of the North Sea rift system

We consider the Late Jurassic North Sea rift system (Denmark–Mid Norway) to comprise five main kinematic components, each a sedimentary basin in its own right (Fig. 9.1):

the Viking Graben
the Moray Firth Basin
the Central Graben

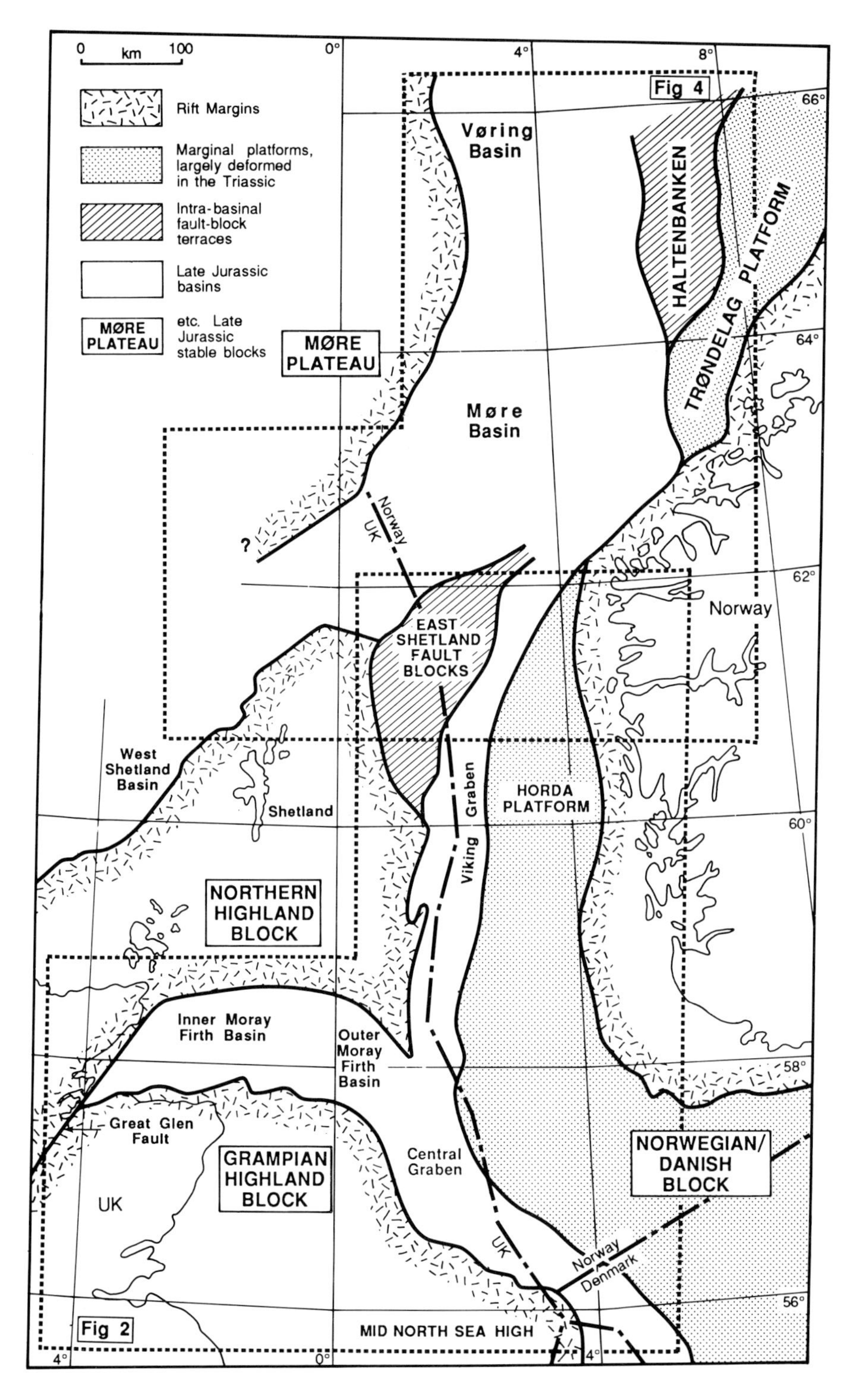

Fig. 9.1. Regional location map showing the main tectonic elements of the North Sea rift system. The areas covered in more detail by Figs 9.2 and 9.4 are also shown.

the Møre Basin, including Haltenbanken (Mid-Norway)
the West Shetland Basin.

The kinematics of each basin are distinct, but together they form an interlinked system and must ultimately be assessed as such.

We therefore look at the evidence for defining a Late Jurassic extension vector in each of the five basins, although in the case of the West Shetland Basin displacement magnitude is only inferred very indirectly. We then integrate this information into a vector diagram and assess the requirements for any non-dip-slip movement within the rift system. Strictly, we should describe the relative motion between each pair of stable blocks by a pole of rotation rather than a single displacement vector. However, such an approach is not warranted here, due to the short lengths of the individual rift basins (*c*.300 km). Insofar as deformation patterns are relatively consistent within each basin, it is sufficient to assume that the pole of rotation is at a large distance, and therefore a single vector describes each basin. It is unlikely that seismic reflection data can ever provide a more detailed picture of slip vector distribution within each basin.

Although this discussion considers the internal kinematics and geometry of the rift, the resulting vector diagram is best expressed in terms of relative movement between the stable platform areas flanking the rift. We consider the rift margins to have comprised four stable blocks during the Late Jurassic (Fig. 9.1):

the Northern Highland (and Shetland) Block
the Grampian Highland Block
the Norwegian/Danish Block
the Møre Plateau (attached to Greenland before the Tertiary opening of the northernmost Atlantic).

It is the relative movement between these blocks for which we ultimately suggest a solution.

A previous kinematic assessment of the Viking and Central Grabens (Beach 1985) has suggested that our Norwegian/Danish Block in fact comprises two distinct kinematic blocks, 'Norway' and 'Denmark', separated by the strike-slip Tornquist Zone (Pegrum 1984). We accept that the Tornquist Zone, extending south-east towards the Polish 'Caledonides', may exist in the basement to the Norwegian/Danish Block. There is, however, a lack of convincing evidence that Tornquist-related structures deform the Mesozoic fill of the rift. Certainly no single structural element comparable with, for example, the strike-slip Great Glen fault can be mapped from commercial seismic data, while Klemperer and Hurich (this volume) have stated that no indication of Tornquist zone structurs has yet been identified on deep seismic profiles across the North Sea. Pegrum and Ljones (1984) have suggested that faults transverse to the main rift, at *c*. 58°N, may be controlled by strike-slip movement on the Tornquist trend. Similar structures within the Central Graben have, however been interpreted by Roberts and co-workers (in review) as discontinuous transfer faults, linking offset extensional faults without any necessary basement control. Likewise, although it is much argued in the literature that Cretaceous inversion in the Norwegian/Danish Central Graben was controlled by movement on transverse strike-slip structures (Gowers and Sæbøe 1985), both Cartwright (1989) and Roberts and colleagues have demonstrated that simple dip-slip reactivation (in a reverse sense) of the main extensional faults could have produced the observed inversion geometries. We thus conclude that evidence for Mesozoic strike-slip movement on the Tornquist 'trend' in the central North Sea is poor, perhaps non-existent, and therefore we do not include it in our kinematic assessment.

9.3. The Viking Graben

The Viking Graben is chosen as the starting point in our assessment for a number of reasons. First, it is the central component of our five-basin rift system. Secondly, its structure is extremely well-documented in the literature. Thirdly, lying as it does north of the Zechstein salt basin, its fault block topography is very clearly defined. We ourselves have, in addition, worked extensively within this basin (Badley *et al*. 1984, 1988, 1989).

The extension vector

Both on the UK and Norwegian sides of the Viking Graben the internal structure of the basin comprises a series of N–S-trending fault blocks (Fig. 9.2). This structural pattern is only disrupted in the NW corner of the basin; in the Magnus/Tampen Spur area, NE–SW-trending faults become dominant. This fault

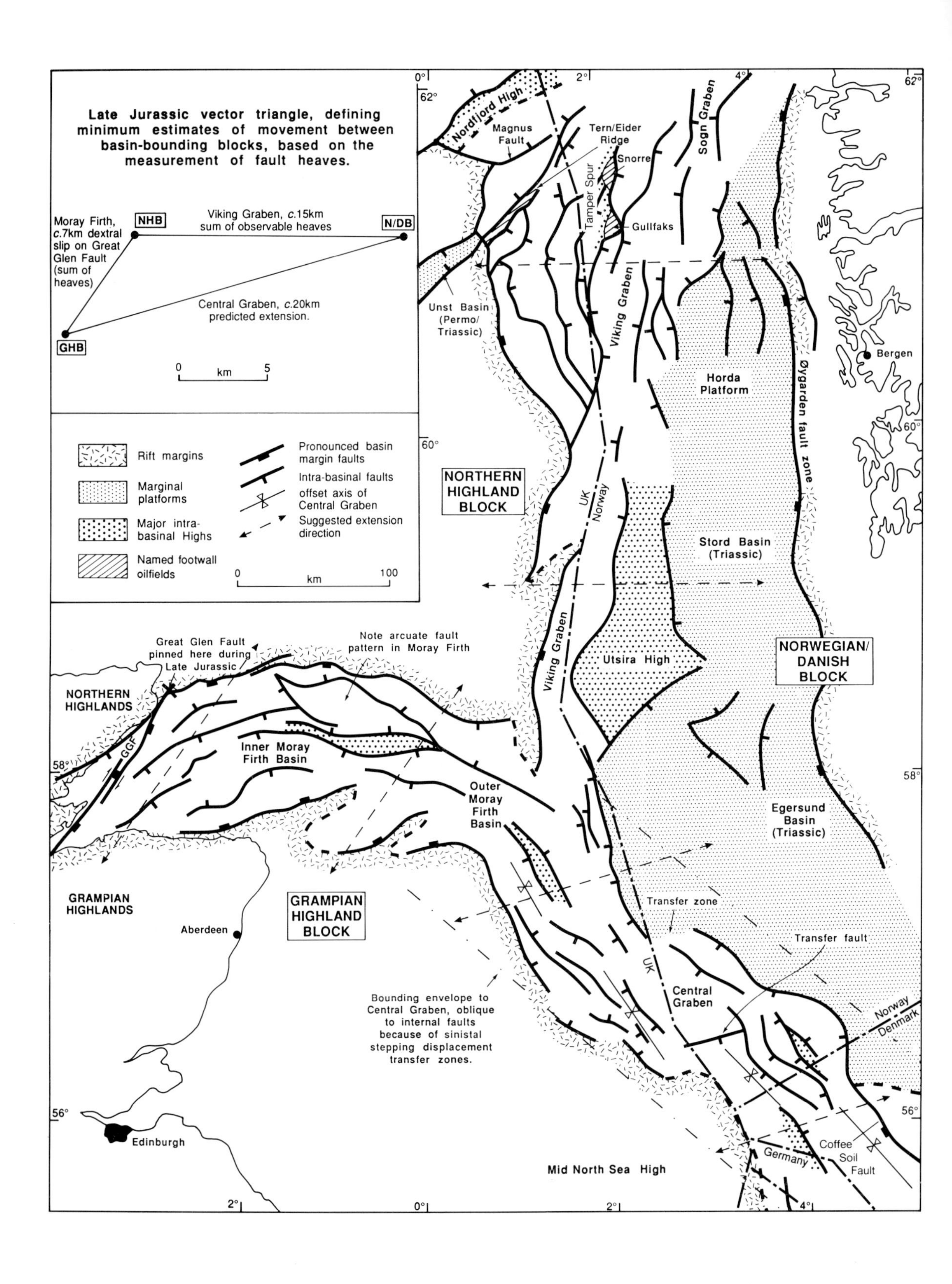

Late Jurassic vector triangle, defining minimum estimates of movement between basin-bounding blocks, based on the measurement of fault heaves.
NHB
N/DB
GHB
Moray Firth, c.7km dextral slip on Great Glen Fault (sum of heaves)
Viking Graben, c.15km sum of observable heaves
Central Graben, c.20km predicted extension.
0 km 5
Rift margins
Marginal platforms
Major intra-basinal Highs
Named footwall oilfields
Pronounced basin margin faults
Intra-basinal faults
offset axis of Central Graben
Suggested extension direction
0 km 100
Nordfjord High
Magnus Fault
Tern/Eider Ridge
Snorre
Tamper Spur
Gullfaks
Sogn Graben
Unst Basin (Permo/Triassic)
Viking Graben
Horda Platform
Bergen
Øygarden fault zone
NORTHERN HIGHLAND BLOCK
UK Norway
Stord Basin (Triassic)
Utsira High
NORWEGIAN/ DANISH BLOCK
Great Glen Fault pinned here during Late Jurassic
Note arcuate fault pattern in Moray Firth
NORTHERN HIGHLANDS
GGF
Inner Moray Firth Basin
Outer Moray Firth Basin
Egersund Basin (Triassic)
GRAMPIAN HIGHLANDS
GRAMPIAN HIGHLAND BLOCK
Aberdeen
Transfer zone
Transfer fault
Bounding envelope to Central Graben, oblique to internal faults because of sinistal stepping displacement transfer zones.
Central Graben
Norway Denmark
Germany
Coffee Soil Fault
Edinburgh
Mid North Sea High
62°
60°
58°
56°
0°
2°
4°

system is considered in more detail in our assessment of the West Shetland Basin, and in particular the Magnus Basin/Tampen Spur transfer system.

The N–S-trending fault blocks of the Viking Graben are all defined by major extensional faults (Badley *et al.* 1988; Yielding 1990); on first inspection there appears to be little need to assume anything other than orthogonal E–W extension on these faults. With this supposition in mind we have constructed, from reflection seismic data, five approximately E–W cross-sections across the entire graben, between 60°N and 61°30′N (see Marsden *et al.*, this volume). Summation of seismically-observable fault heaves (at top Middle Jurassic) on these sections is consistently *c.*15 km (range 14–18 km).

If we assume a 'rigid-domino' model (Barr 1987*a*, 1987*b*; Jackson *et al.* 1988; Yielding 1990) these measurements of heave can be directly related to extension, and imply a Late Jurassic extension of *c.*10 per cent. On the other hand, applying a 'soft-domino' model (Gibson *et al.* 1989; Jackson and White 1989; Childs *et al.* in press; Walsh and Watterson in press) would lead to an estimate of extension significantly greater than the measurable heaves, perhaps by a maximum factor of about two. A similar disparity exists between various listric fault models. For example, applying the Chevron construction (Gibbs 1983) would give an extension estimate equal to the sum of heaves, whereas assuming inclined simple shear in the hangingwall (White *et al.* 1986) would give an extension greater than the heaves. The approach we have taken here is to use the heave measurements directly, because they are observable on seismic data and are the nearest to objective data. Any other approach is more model-dependent. We recognize that the heave measurements may well be an underestimate of the true extension, but as long as they represent a constant proportion of the true extension within the rift our subsequent kinematic arguments remain unaffected.

As stated above, summing the Late Jurassic fault heaves on E–W sections consistently gives an observable extension of *c.* 15 km. While such consistency in the magnitude of E–W extension is not in itself irrefutable evidence that this was the true extension direction, it is consistent with such a hypothesis. It is to be expected that stretching measurements in the extension direction will be more systematically consistent than those made oblique to it. Oblique measurements are likely to incorporate or cross transfer zones.

Thus, we suggest that the simplest kinematic assessment of the Viking Graben involves seismically-observable, Late Jurassic E–W extension (between the Northern Highland Block and Norway) of *c.*15 km, which in reality may equate to an extension maximum of *c.*30 km (soft-domino model). Our observable figure of 15 km is similar to previous estimates by Beach *et al.* (1987), Giltner (1987), and Badley and co-workers (1988). An E–W extension direction is compatible with the sub-regional kinematic assessment of Speksnijder (1987).

It was acknowledged earlier that the Late Jurassic extension within the North Sea probably comprised the second of two significant stretching events which formed the component basins of the rift, the first stretching episode perhaps being of Triassic age. Thus the suggested *c.* 15 km, or more, of Late Jurassic E–W extension in the Viking Graben is not a measurement of total crustal extension within the basin. The integration of Giltner's (1987) total extension curves across the Viking Graben suggests a value of *c.* 60 km, although these curves do not quite extend to the basin margins. Likewise Hamar and Hjelle's (1984) gravity modelling of the Viking Graben suggests a total extension of *c.*60 km, while Klemperer's (1988) interpretation of crustal thickness from deep seismic data suggests a similar figure. Solli's (1976) refraction line, as quoted by Ziegler (1982), suggests a total Viking Graben extension of 75–100 km; this now rather old refraction line, however, appears to overestimate extension in the basin axis by comparison with more recent

Fig. 9.2. Tectonic elements, including the pattern of major Late Jurassic faults, of the Viking Graben, Central Graben and Moray Firth areas. Inset, the suggested kinematic relationship between the stable bounding blocks, with the resulting extension directions superimposed on the main fault pattern. Note that the Viking Graben, Central Graben, and Outer Moray Firth are suggested to have opened orthogonal to the major intra-basinal faults. The Inner Moray Firth opened as a result of dextral slip on the Great Glen fault. (Compiled largely from mapping by Badley, Ashton and Associates, with infill from published maps of the Outer Moray Firth.)

estimates. Thus a consensus estimate of total extension in the Viking Graben may be c. 60 km, or possibly slightly more. Of this we believe that at least 15 km occurred during the Late Jurassic, and perhaps as much as 30–40 km in the Triassic.

Previously-interpreted 'strike-slip' structures

Three lines of reasoning have been used to invoke significant transpression within the kinematics of the Viking Graben fault blocks:

(1) the thickness of the Cretaceous–Tertiary basin above the fault blocks (Beach *et al.* 1987);
(2) the elevation and erosion of the Snorre (V. B. Larsen 1987) and Gullfaks (Fossen 1989) structures;
(3) the interpreted geometry of individual fault blocks (Beach 1985; Frost 1987; Speksnijder 1987).

We believe that each of these arguments is readily discounted.

Beach and colleagues (1987) suggested that the considerable thickness of Cretaceous and Tertiary sediments, up to c. 3 km in the rift axis, deposited after Late Jurassic extension, was incompatible with their observed E–W fault-controlled extension of 5 per cent, i.e. application of a McKenzie-type model to a basin in which $\beta = 1.05$ would not produce the observed Cretaceous–Tertiary subsidence. They therefore argued that crustal extension during the Late Jurassic had occurred along NE–SW-trending strike-slip 'through-crust shears', dipping at c. 60°. The need to invoke such 'extra' extension has been negated by the work of Giltner (1987) and Marsden and co-workers (this volume), who have shown that application of a two-stage (Triassic and Late Jurassic), time-dependent rift model to the Viking Graben accounts very well for the post-rift thermal subsidence basin, when dip-slip E–W extension is assumed. We conclude therefore that sediment thicknesses cannot be used to argue for significant strike-slip deformation within the Viking Graben.

Some authors have suggested that the clearly-eroded profiles of the Snorre/Tampen Spur structure (V. B. Larsen 1987) and the Gullfaks structure (Fossen 1989) argue for their uplift above both sea-level and a 'regional datum' by transpressional deformation. Application of Barr's (1987*a*, *b*) model for fault block footwall uplift, and in particular his curves of footwall-uplift/hangingwall-subsidence *versus* extension, show, however, that Snorre and Gullfaks can be considered as footwall blocks uplifted during lithospheric extension. Both the Snorre and Gullfaks structures sit at the crest of large fault blocks, bounded by normal faults with large displacements. Considerable footwall uplift is the anticipated norm in such circumstances.

We conclude therefore that the clear uplift of the Gullfaks and Snorre structures is not evidence of transpressional or compressional deformation (Badley *et al.* 1989; Yielding 1990).

Beach (1985, Fig. 3) has illustrated a number of major faults in the northern Viking Graben which he considered to show a compressional geometry. In addition, Speksnijder (1987) has illustrated a number of structures, within an overall E–W extensional setting, which he believed show a local transpressional geometry. Notable in both these interpretations is the compressional nature of the Tern-Eider Ridge. We present an alternative interpretation of the Tern–Eider Ridge (Fig. 9.3) which shows it as a horst with high basement, flanked by normal faults. Likewise the other compressional structures identified by Beach and Speksnijder are more readily interpreted as structures bounded by normal faults. We therefore do not believe that any significant compression has occurred within the Viking Graben, and base our kinematic model on such an argument.

To conclude our discussion of the Viking Graben, we believe it to be a two-stage rift, with a total c. E–W extension of c. 60 km, of which at least 15 km (and possibly as much as 30 km) occurred in the Late Jurassic. Previously published 'evidence' of significant strike-slip deformation can, at the least, be better interpreted within an extensional framework, and in some instances discounted altogether. Thus we believe the major N–S trending fault sets which define the structure of the Viking Graben to be dip-slip normal faults.

9.4. Moray Firth Basin

Extension vector

As discussed in detail by Roberts and co-workers (1990), we largely follow the kinematic model of McQuillin and colleagues (1982) for the opening of the Moray Firth Basin. Incorporation of the heaves measurable from Barr's (1985) palinspastic map restorations into this model suggests that, during the

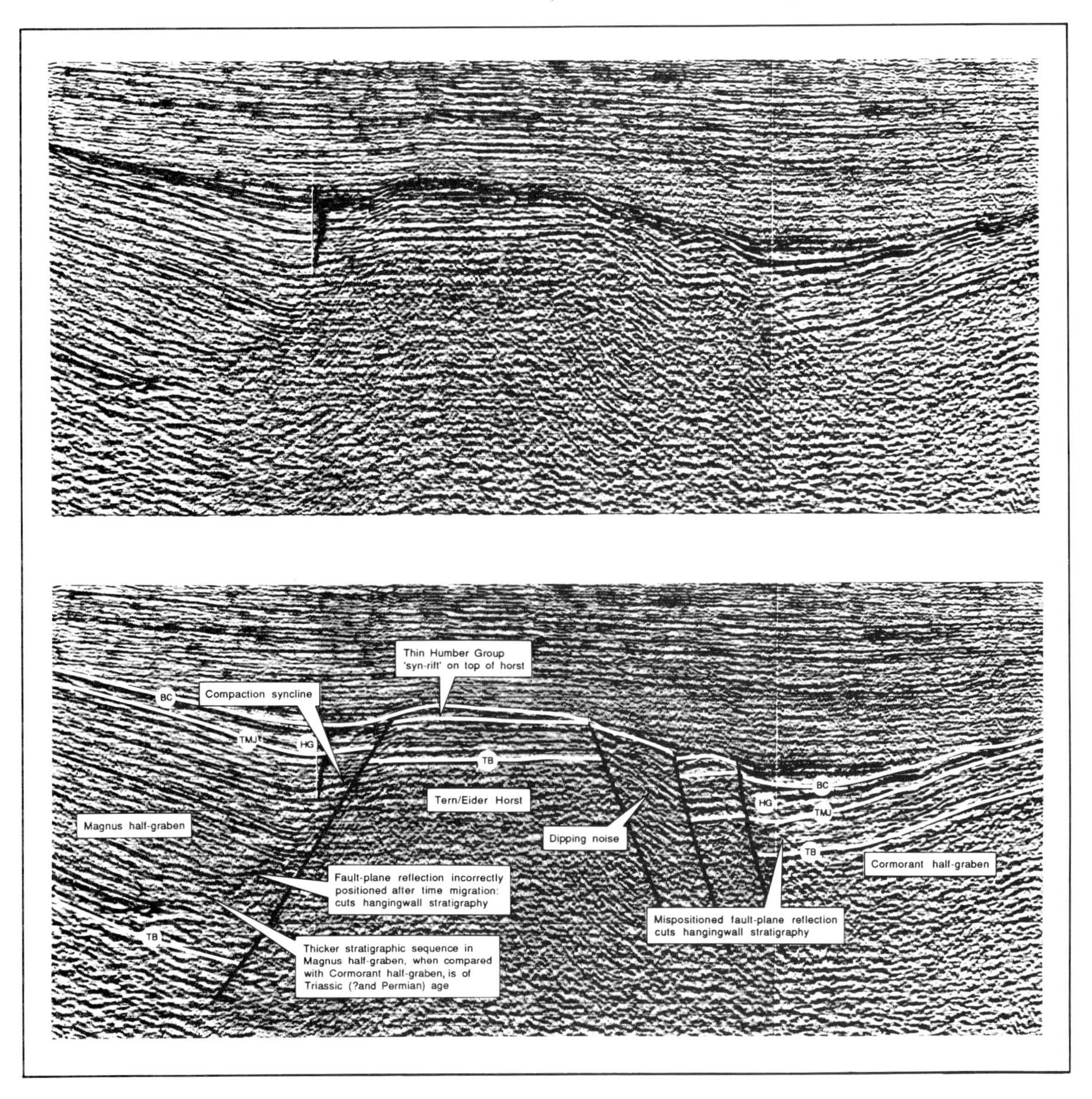

Fig. 9.3. Uninterpreted and interpreted seismic line across the Tern/Eider Ridge. The interpretation put forward here shows it as a horst with high basement, flanked by major normal faults and rotated half-graben. Previous interpretations have suggested that the Ridge is bounded by reverse faults. TB = near top basement, TMJ = near top Middle Jurassic, HG = Humber Group 'syn-rift' (Callovian-Ryazanian), BC = near base Cretaceous. (*See* Fig. 9.5 for location.)

Late Jurassic, the Moray Firth Basin opened as a result of a *c.* 7 km dextral movement on the Great Glen fault (Figs 9.1 and 9.2). This is a figure similar to our own more crude measurements of post-Triassic fault heave within the Inner Basin, and to the base Jurassic extension measured by Beach (1984, Fig. 2) in the Outer Basin. This displacement caused the Northern Highland and Grampian Highland blocks to slide past each other, but, as a result of a 'pinned' point on the Great Glen fault close to the north Scottish coast, also caused the Moray Firth Basin to open as a 'pull-apart' structure.

Again we add the caveat that Barr's, Beach's, and our own extension measurements are all the sum of observable fault heaves, and may be an underestimate of total Late Jurassic extension. The methodology applied, however, is consistent with our approach to the Viking Graben and thus we anticipate that the ratio of 7 km:15 km (Moray: Viking) will reflect the true Jurassic stretching ratio between these basins.

Thus, following both our own and others' earlier work, we suggest that the extension vector for the Late Jurassic Moray Firth Basin trends NNE–SSW, parallel to the Great Glen fault, and has a magnitude of at least 7 km.

Some authors, such as Chesher and Bacon (1975) and Frostick and co-workers (1988) have suggested that the Great Glen fault has accommodated fault-normal extension in addition to strike-slip displacement. As discussed, however, by Roberts and co-workers (1989) such movement is ruled out by the vertical nature of the Great Glen fault on seismic reflection profiles; vertical faults cannot accommodate any fault-normal extension.

The arcuate fault pattern of the Moray Firth Basin

The Moray Firth Basin comprises two distinct structural provinces; an Inner (western) Basin, in which faults trend sub-parallel to the Great Glen fault (NNE–SSW) in the west, fanning round to an E–W trend in the east; and an Outer (eastern) Basin in which faults trend NNW–SSE, i.e. perpendicular to the Great Glen fault (Fig. 9.2).

A consideration of the kinematic model for the basin makes this 90° swing in fault-strike understandable. The NNW–SSE-trending faults of the Outer Basin are the true extensional, dip-slip faults to the Moray Firth Basin; they lie perpendicular to the extension vector. As the strike-slip sidewall of the Inner Basin is approached, however, the major faults within the basin become near-synthetic with the controlling Great Glen fault, and strike close to NNE–SSW. Between the truly exstensional Outer Basin and the strike-slip sidewall to the Inner Basin lies a zone of arcuate-trending, oblique-slip faults. This is the main part of the Inner Basin, forming a zone of distributed strain between the two orthogonal end-member fault sets (Fig. 9.2).

Thus, in conclusion, we suggest that the Outer Moray Firth Basin is a dip-slip extensional basin (Beach 1984), which opened orthogonally on its main NNW–SSE fault set. The Inner Moray Firth Basin is, however, distinct. It is bounded to the west by the strike-slip Great Glen fault and shows fault trends and internal geometries consistent with extensional oblique-slip on the main internal fault sets (Roberts *et al.* 1990).

Lithospheric stretching below the Moray Firth Basin

In the Viking Graben (Giltner 1987) and Central Graben (Sclater *et al.* 1986) it has proved possible to attempt an integration of lithospheric stretching estimates with fault-controlled kinematic models for the basin, and we have indeed already suggested here that such an approach has proved worthwhile in the Viking Graben. Such an approach cannot, however, be applied with such confidence to the Moray Firth Basin. Although Christie and Sclater (1980) have demonstrated considerable crustal thinning and isostatic compensation below the Outer Moray Firth, Donato and Tully (1981) have shown the Inner Moray Firth to be underlain by unthinned crust, with the result that the basin is not in isostatic equilibrium. In essence the Outer Basin confirms to an origin by lithospheric stretching, whereas the Inner Basin is a mechanically-created, upper-crustal hole, now filled with sediment. The western limit of Tertiary sediments within the Moray Firth corresponds closely with the western limit of crustal thinning below the basin (Barr 1985). The change from thinned to unthinned crust is transitional (Donato and Tully 1981), it is not marked by a major, deep-crustal wrench fault.

Barr (1985) has pointed out that, in order to balance upper crustal extension of *c*.7 km in the Inner Moray Firth, a similar amount of lower crustal and mantle lithosphere attenuation must have occurred elsewhere. It is not at present known, nor indeed are there any clear indications of, the site of such attenuation. Although perhaps unlikely because of the NNE–SSW extension direction within the basin, it is possible that some additional thinning may have occurred as a compensating measure below the Outer Basin. While this possibility remains, it would be imprudent to assume implicitly that Christie and Sclater's (1980) measurements of crustal- (and initial lithospheric) thinning below the Outer Basin (which we estimate to yield a total Mesozoic thinning of 40–45 km) necessarily truly reflect the fault-controlled upper-crustal extension in this part of the basin. We are constrained therefore in the Moray Firth to use measurements of fault-controlled extension in our attempt to quantify the kinematic history of the basin.

9.5. Central Graben

The extension vector

As discussed above, in both the Viking Graben and Moray Firth Basin seismic reflection data readily allow the measurement of fault heaves at Jurassic levels. Both these basins lie beyond the limit of the Zechstein salt basin, and thus their internal structural geometry is unaffected by halokinesis. In the Central Graben, however, mobile Zechstein salt is almost ubiquitous. Mobile salt has severely disrupted the internal geometry of Mesozoic sequences on the major fault blocks. This means that, on a regional scale, it is impossible to measure the magnitude of Late Jurassic extension reliably across active normal faults.

In order to measure fault-controlled extension without a halokinetic overprint we must investigate a marker horizon below the salt. The only such marker imaged with any consistency is the intra-Permian top Rotliegend reflection. Even measuring extension at this level, however, is fraught with difficulty. First, the top Rotliegend is never imaged without interruption across the whole basin. This is partly because of its depth (up to 8 km in the hangingwall of major faults) and partly because of the ray-path distortion and dispersion achieved by overlying salt diapirs banked against such major faults. For this reason we regard Ziegler's (1983) fault-controlled extension estimate at top Rotliegend level (*c*.30 km) as extremely unreliable, and likely to include almost-unquantifiable errors. In addition, the top Rotliegend marker does not solely record extension imposed during the Late Jurassic, it also records Triassic extension, which in the Central Graben may have been quite considerable (Barton and Wood 1984; Sclater *et al*. 1986). By fitting a time-dependent rift-model to the subsidence history of the Central Graben, Sclater and co-workers (1986) concluded that the late Jurassic component of extension in the basin was 30–40 km, approximately normal to the graben margins.

We make our own assessment of the Late Jurassic extension by closing the Viking Graben/Moray Firth/Central Graben vector triangle (Fig. 9.2). By defining *c*. 15 km (minimum) of E–W extension across the Viking Graben and *c*. 7 km (minimum) of NNE–SSW extension across the Moray Firth, we have in fact defined the seismically-observable relative motion between the Northern Highland, Grampian Highland, and Norwegian/Danish blocks. Extension across the Central Graben corresponds to the relative motion between the Grampian and Norwegian/Danish blocks (Figs 9.1 and 9.2). Our constraints from the Viking Graben and Moray Firth suggest this motion to have been *c*.20 km (minimum) is the direction 075°–255°, obtained by closing the vector triangle (Fig. 9.2). This estimate for Late Jurassic extension across the Central Graben is thus rather lower than obtained by Sclater and colleagues (1986). However, application of a soft-domino model to the quantities defined by our vector triangle would put an upper bound on Late Jurassic extension in the Central Graben of *c*. 40 km, more in line with the subsidence modelling of Sclater and his colleagues.

Superposition of the resolved Central Graben extension directly onto a fault map for the UK/Norwegian/Danish Central Graben (Fig. 9.2) shows the major fault sets within the basin to be dominantly extensional, and to have accommodated movement almost orthogonal to their strike. Deviation in strike of *c*.10° from the true orthogonal trend is apparent for some fault sets, but this is too small an obliquity to manifest oblique-slip geometries locally on these structures (McCoss 1986). Thus, like the Viking Graben and Outer Moray Firth, we conclude that the Late Jurassic Central Graben opened orthogonally to its main fault trend (*c*.NNW–SSE) as a dip-slip basin.

Previous authors, such as Gowers and Sæbøe (1985), Vejbæk and Andersen (1987), and Olsen (1987), have suggested that varying amounts of strike-slip deformation controlled the opening of the Central Graben. Note, however, that none of these models are completely internally consistent with each other. Such strike-slip control has generally been invoked in order to explain observed inversion geometries resulting from a later (Cretaceous) reversal of motion on the main fault sets. It has, however, been argued by one of us elsewhere (Roberts *et al*. in press) that inversion geometries in the Norwegian/Danish Central Graben are readily explained by reversal of motion on the main dip-slip fault sets, coupled with synchronous halokinesis in some cases (Cartwright 1989). Thus there is no need to appeal to the presence of either extensional or contractional strike-slip 'flower structures' within the Central Graben.

Transfer faults and transfer zones

Inspection of a regional map of the Central Graben (Fig. 9.2) shows that, although the main fault trend, and consequently that of the main sub-basins, within the Graben lies approximately orthogonal to the resolved (075°–255°) extension direction, the bounding envelope to the Late Jurassic Central Graben as a whole trends more-nearly NW–SE. This is because the graben margins, unlike those of the Viking Graben, are not defined by continuous major faults, with the exception of the Coffee Soil Fault in the Danish sector. Rather the graben margins are marked by discontinuous faults which progressively transfer motion between them in a south-eastwards direction. This south-eastwards transfer of the main displacements is responsible for stepping the graben axis eastwards from the UK to the Norwegian sector and then eastwards again into the Danish sector.

Displacement transfer between dip-slip fault sets in the Central Graben is accomplished both on discrete transfer faults and across *en échelon* transfer zones ('relay structures' of P. H. Larsen, 1988). It is important to realize that such transfer faults are of only local extent and accommodate lateral displacements linked into dip-slip faults (Roberts *et al.* in press). They did not drive the extension of the graben system in the manner suggested by Pegrum and Ljones (1984) and Gowers and Sæbøe (1985).

9.6. The Viking Graben/Moray Firth/ Central Graben triangle

It is suggested here that during the Late Jurassic the three basins of the North Sea rift system defined a kinematically-closed system (Fig. 9.2), which need not have accommodated any extension from outside the rift system (Beach 1985). Our kinematic model has the additional appealing feature of suggesting that all three grabens opened approximately orthogonally to their dominant fault set and can thus be viewed as 'simple' extensional basins. We believe such an interpretation to be consistent with fault geometries as seen on seismic reflection data. The only exception to this dominantly extensional setting is the western sidewall to the Inner Moray Firth. This is marked by the Great Glen fault, the only strike-slip fault within the rift 'triangle' to exert a regional kinematic control.

This southern rift 'triangle' does not, however, complete the interpretation of North Sea rift kinematics. The Viking Graben opens northwards into the less well-known Møre Basin, and from there to the mid-Norwegian shelf. It is the relationship of the Viking Graben to these more northerly basins which we explore next.

9.7. The Møre Basin and Haltenbanken (Mid-Norway)

Very few publications have addressed the relationship of the mid-Norway continental margin (Møre Basin and Haltenbanken) to the better-known graben system of the North Sea. During the course of work on the Haltenbanken area, however, we were struck by the many broad similarities between the timing and geometry of structures in this area when compared with the more familiar Viking Graben to the south. We felt that a temporal and kinematic link was suggested. Clearly if there is such a link between Haltenbanken and the Viking Graben it must continue through the large and poorly-known Møre Basin, which lies between the two petroleum provinces (Fig. 9.1).

For a number of reasons, namely water-depth, basin-depth, and political policy, the Møre Basin remains unexplored as a hydrocarbon province. Thus our knowledge of it is poor by comparison with its flanking petroleum provinces. Probably the most comprehensive regional account of the Møre Basin published to date is that by Hamar and Hjelle (1984). We have drawn most of our data and inferences about the Møre Basin from there, as there is little additional readily available data to be obtained.

The Møre Basin stands distinct from the Viking Graben, Central Graben, and Haltenbanken areas because of its greath depth. The base of the Cretaceous in the basin centre attains a depth of 10 km (Hamar and Hjelle 1984), twice that of the deepest parts of the Viking and Central Grabens. Because of the great thickness of Cretaceous sediments within the basin it is sometimes referred to as a 'Cretaceous basin', implying initiation of subsidence at this time.

Indeed, Hamar and Hjelle have suggested that the Cretaceous was the time of fastest subsidence. There is, however, apparently little Cretaceous fault control on the basin; rather it has been suggested that buried, pre-Cretaceous structural highs, detectable at the limit of seismic resolution, are in fact buried Late Jurassic fault blocks. Thus the basin is likely to be a Jurassic, or possibly older, rift basin, filled by Cretaceous and Tertiary thermal subsidence sediments.

The great depth to the base Cretaceous in the basin implies considerable crustal (intially lithospheric) attenuation below it. Gravity modelling by Hamar and Hjelle shows that, on average, the pre-rift crust is stretched by a factor of $\beta = 2$, while in the centre of the basin the pre-rift crust is possibly thinned to less than 10 km thickness, i.e. $\beta \geqslant 3$. The axis of the Møre Basin must have been close to continental break-up in the pre-Cretaceous. The accelerated Cretaceous 'subsidence' in the basin, inferred by Hamar and Hjelle from sediment thickness data only, may in fact in part be the product of the filling-in of a deep-water basin at the end of such extreme rifting, rather than true Cretaceous thermal subsidence in its entirety.

The extension vector

The Møre Basin is too deep to allow any analysis of its internal fault-block geometry from seismic data; thus at present we can establish no kinematic indicators within the basin. Instead we must turn to the better-known and structurally-shallower Haltenbanken area, adjacent to the northern margin of the Møre Basin, for indications of the Late Jurassic extension direction in the Mid-Norway area.

The Haltenbanken area (Figs 9.1 and 9.4) consists of a 'terrace' of tilted fault blocks, at a similar depth to the fault blocks of the Viking Graben, flanked to the west by the much deeper Møre/Vøring Basin and to the east by the shallower Trøndelag Platform, an area similar in many respects to the Horda Platform of the Viking Graben.

The tilted fault blocks of this area appear to be extensional structures, possibly locally internally-disrupted by movement of Triassic salt (Jackson and Hastings 1986). We have made detailed measurements of seismically-observable extension across the Njord structure in southern Haltenbanken (Fig. 9.4), normal to the dominant NNE–SSW fault trend in this area. Measurements in this WNW–ESE direction show a consistent extension of *c*.6 km across this internally-faulted tilted fault block, suggesting that such measurements do not incorporate the displacement on any transfer faults. Similar measurements of fault-normal extension (i.e. WNW–ESE) have been made by Wheeler (*pers. comm.*) across the Midgard structure in central Haltenbanken. Again a consistent estimate of extension is achieved in this direction. While such measurements do not prove a WNW–ESE extension direction across Haltenbanken, they provide strong evidence to support such an argument.

Detailed analysis of faults at right-angles to the main NNE–SSW trend shows some to possess the classic features of transfer faults within a predominantly extensional fault set; notably, their sense of throw and inferred sense of lateral displacement vary along strike. The NNE–SSW faults always maintain their sense of throw along strike, even when the heave varies. Such observations again provide supporting evidence for a fault-normal WNW–ESE extension direction across Haltenbanken.

The Trøndelag Platform and Nordland Ridge, both platform areas adjacent to Haltenbanken, show evidence in places of more than 1 km of erosion adjacent to major NNE–SSW-trending faults. Roberts and Yielding (in press) have shown that this erosion can be quantitatively modelled as the result of isostatic uplift in the footwalls of dip-slip, planar, normal faults.

Thus, we favour orthogonal (WNW–ESE) opening of the Late Jurassic Haltenbanken rift margin, and by inference also therefore a similar extension direction within the related Møre Basin.

Assessment of the magnitude of extension across the Møre/Haltenbanken Graben system is more difficult. The clearly-imaged fault blocks of Haltenbanken define only the one-sided margin to the basin, and much of the same graben system probably lies to the west in the Møre/Vøring Basin. To our knowledge the only assessment of extension across the entire Møre Basin is the result of gravity modelling by Hamar and Hjelle (1984, Fig. 10). Their gravity modelling shows an approximately symmetric basement structure within the Møre Basin, which, perhaps surprisingly, was not exploited by the younger Atlantic opening. The Møre Basin thus appears to lie in its entirety within the Norwegian shelf, and has not been partitioned between Norway and Greenland during subsequent sea-floor spreading.

Hamar and Hjelle's gravity modelling shows the

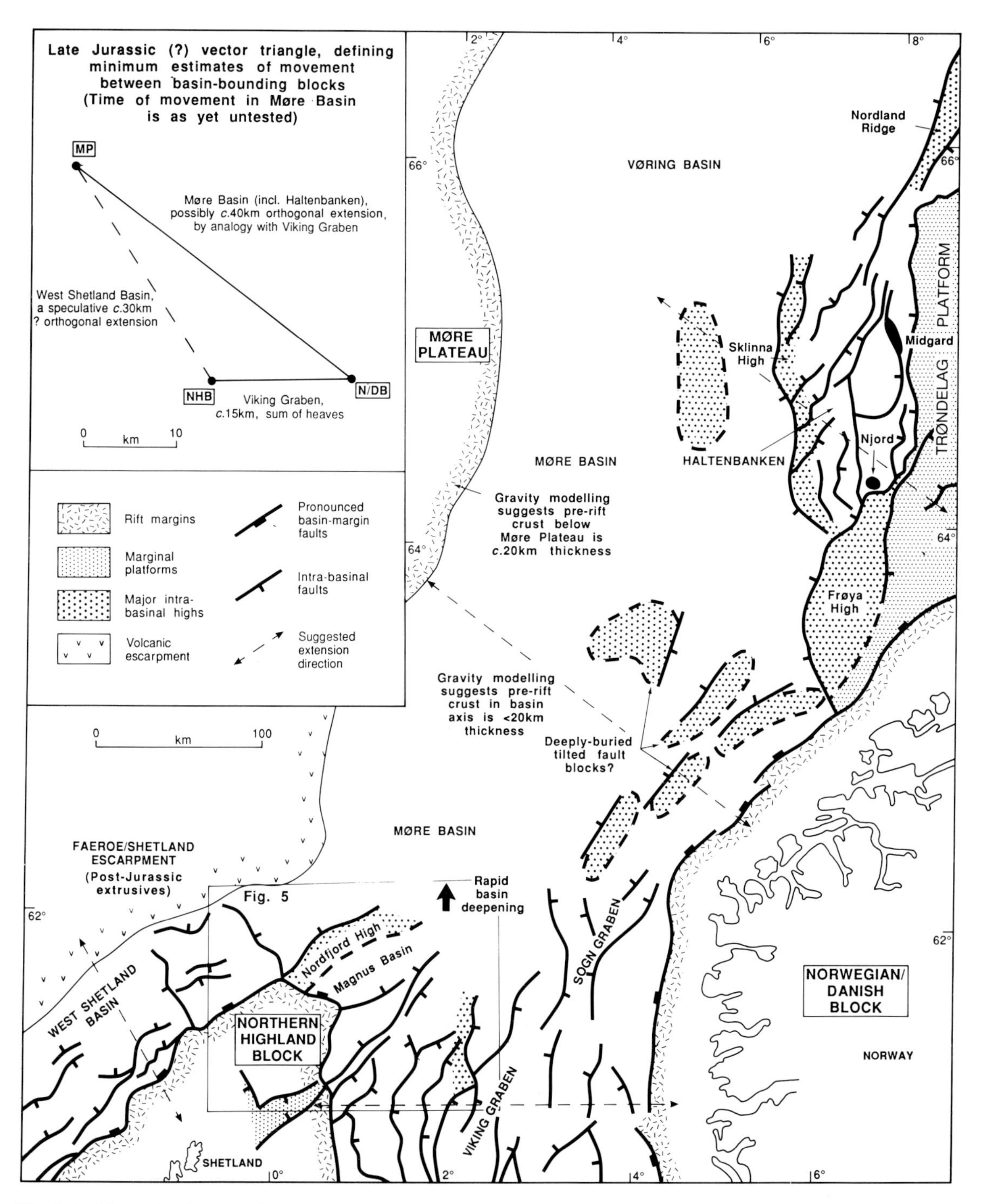

Fig. 9.4. Tectonic elements, including the pattern of major Late Jurassic faults, of the Viking Graben, Møre Basin (including Haltenbanken), and West Shetland areas. Inset, the suggested kinematic relationship between the stable bounding blocks, with the resulting extension directions superimposed on the main fault pattern. Note that all the main fault systems are suggested to have opened approximately orthogonally. (Compiled from mapping by Badley Ashton and Associates, Hamar and Hjelle 1984, and Duindam and van Hoorn (1987).)

pre-rift crust to the Møre Basin to be extended on average by $\beta = 2$. The present width of their modelled basin is *c.* 300 km, thus the Møre Basin has probably extended in total by *c.* 150 km. Both in the flanking Haltenbanken and Viking Graben areas, however, there is evidence for considerable pre-Late Jurassic extension. Thus the 150 km of extension across the Møre Basin is probably not the product of extension during the Late Jurassic alone.

It was earlier argued, on the basis of published geophysical modelling, that total extension across the Viking Graben was *c.*60 km. Of this we believe that *c.* 15 km (minimum) occurred during the Late Jurassic. The remarkable structural and stratigraphic continuity along the whole North Sea/mid-Norway rift suggests that all the basins have shared a similar history. By analogy therefore with the contiguous Viking Graben this would suggest that *c.* 40 km (minimum) of the total *c.* 150 km extension in the Møre Basin occurred in the Late Jurassic.

Thus our best estimate of Late Jurassic extension across the Møre Basin is *c.* 40 km (minimum) in a WNW–ESE direction. This marks the separation between the Møre Plateau and Norwegian blocks. Clearly the magnitude of this vector is prone to considerable uncertainties, but again we suggest orthogonal opening.

Previously-interpreted 'strike-slip' structures

The Møre Basin is too deep and too poorly-known for anyone previously to have speculated within the literature on its detailed internal kinematics. The kinematics of Haltenbanken have, however, been addressed. While some authors, such as Bukovics and colleagues (1984), have interpreted the structure of Haltenbanken in terms of an orthogonal rift basin during the Mesozoic, other authors, notably Gabrielsen and Robinson (1984) and Caselli (1987) have interpreted Haltenbanken as a dextral strike-slip fault system, with the main NNE–SSW fault set active as the controlling strike-slip fault system. The main evidence that these authors cite in favour of such a kinematic model is the recognition of a number of uplifted and eroded structural highs, the Frøya, Sklinna, and Nordland highs (Fig. 9.4), within the Haltenbanken fault system. These highs, elevated both above sea-level and their regional datum during the Late Jurassic–Early Cretaceous, could, it has been argued, only be the product of local transpression within a dominant strike-slip fault set.

We would argue, however, that these structures, like the Snorre and Gullfaks structures in the Viking Graben, are more adequately interpreted as the product of isostatic uplift, in the footwalls of major normal faults (Vening Meinesz 1950, Jackson and McKenzie 1983; Barr 1987*a*, *b*; Kusznir *et al.* 1988; Buck 1988; King *et al.* 1988; Stein *et al.* 1988), and indeed we have quantitatively modelled them as such (Roberts and Yielding in press). Such a model is consistent with our own kinematic deductions about the Haltenbanken area, while our own structural analyses have shown that in detail the strike-slip model is not supported by structural geometries on a prospect scale.

9.8. The West Shetland Basin

The previous section has suggested that the Møre Basin has probably been extended by more than twice the amount of the Viking Graben, with possibly a total extension of *c.* 150 km in the former and *c.* 60 km in the latter. Thus although the Møre Basin and Viking Graben are contiguous parts of the North Sea rift system, the Møre Basin clearly does not transfer all of its displacement into the Viking Graben. Thus the discrepant extension, in total *c.* 90 km, must either have been transferred out of the rift system by a major transcurrent fault or have passed southwestwards into the West Shetland Basin. Given that we know Mesozoic faulting to have occurred in the West Shetland Basin (Nelson and Lamy 1987; Duindam and van Hoorn 1987), we favour the interpretation which kinematically links it with the North Sea rift system.

We are not the first authors to suggest an element of structural continuity between the Møre and West Shetland basins, Nelson and Lamy in particular adopting this theme. Their kinematic solution to the Viking Graben/Møre/West Shetland 'triangle' was, however, different from ours. They suggested that the West Shetland Basin accommodates all of the crustal extension shown within the Møre Basin, none being transferred into the Viking Graben. Extension of the Viking Graben, they suggested, terminates at a major dextral transcurrent fault, which transferred the Viking Graben extension out of the rift system to the north-west. Such a major transcurrent fault has never been identified at the transition of the Viking Graben into the Møre Basin. Given the clear structural continuity, evidenced on seismic data, between the two basins, it seems extremely unlikely that such a structure exists. We favour a kinematic model whereby the temporally-similar Mesozoic basins of the North Sea area are all linked structures. We do not believe it is necessary either to transfer extension out of or into the system along major transcurrent faults (Beach 1985; Nelson and Lamy 1987).

The extension vector

We have access to very little data from the West Shetland area itself. That which we have seen, however, together with the published literature on the area, gives no reason to suspect that, it is anything other than a dominantly-westwards-facing extensional basin, like the mid-Norway margin. To the best of our knowledge this basin, alone among those of the North Sea rift, has not yet been interpreted as a strike-slip basin.

In order to define a putative Late Jurassic extension vector for the basin we can apply a similar technique to that used in the Central Graben, whereby estimating the displacement across the Møre Basin and Viking Graben allows us to 'close the triangle' and resolve a vector for the West Shetland Basin (Fig. 9.4). This vector turns out to be *c*.30 km (minimum) NW–SE, close to orthogonal to the dominant fault trend in the basin. This vector defines the relative movement, in our linked rift scheme, between the Northern Highland and Møre Plateau blocks. Such an extension estimate, as part of *c*. 90 km total extension, has yet to be tested in the West Shetland area by refraction, subsidence, or gravity modelling of crustal thinning; we would thus stress that our minimum estimate is, at present, very speculative.

9.9. The Magnus Basin/Tampen Spur transfer system

As previously highlighted by Nelson and Lamy (1987), the West Shetland fault system, unlike the Viking Graben, is not structurally contiguous with the Møre Basin. Clearly, therefore, if our kinematic model is correct, a zone of displacement transfer must link the two basins. It is at this point that we return to the possible significance of the NE–SW-trending faults which disrupt the N–S structure of the north-western corner of the Viking Graben (Figs 9.2, 9.4 and 9.5). These faults, such as the Magnus fault, the East Magnus fault, and the West Tern/Eider fault, are predominantly west-facing structures of similar trend to the West Shetland and Mid-Norway fault systems. They show considerable displacement variation along strike, defining a very narrow Tampen Spur 'high' north of the Snorre field, but a much wider Magnus-Statfjord 'high' further south. One fault in particular, the Magnus fault, is an enormous structure, on which we estimate the observable Late Jurassic heave alone to be *c*. 4 km. This fault, with a similarly large fault *c*. 50 km to the north-west, has caused substantial rotation of the intervening fault block, the crest of which has therefore suffered massive footwall uplift and erosion, modelled by Marsden and colleagues (this volume). The effects of this erosion are now seen in the tabular structure of the Nordfjord High (Nelson and Lamy, Figs 7 and 8).

Thus, at its south-western extremity, this fault set, normally considered as part of the Viking Graben, is contiguous, around the Nordfjord High, with the West Shetland fault system; while at its north-western extremity, north of Snorre, it is contiguous with the southern limits of the Møre Basin. We therefore suggest that it is this NE–SW-trending set of faults which transfers displacement from the Møre Basin into the West Shetland fault system. The Magnus fault apparently links southwards into a NW–SE-trending transfer fault, defining the southern margin of the Magnus Basin. Other faults of the system, with smaller displacement, may, however, transfer displacement in an *en échelon* 'relay structure' manner, as we have similarly identified in the Central Graben.

Our suggested zone of displacement transfer can be seen on one of the regional maps of Hamar and Hjelle (1984 Fig. 4), where it is marked by a pronounced northwards plunge of the base Cretaceous from the Tampen Spur area into the Møre Basin. This steep plunge of the base Cretaceous, considerably oblique to all *c*. E–W cross-sections drawn across the structure of this area, demonstrates the importance of investigating how the rift system links together in three dimensions.

Our final point of discussion concerning the transfer system concerns the role of two similar horst structures marginal to the NE–SW-trending faults. The Tern/Eider Ridge and East Magnus Ridge have both previously been interpreted as transpressional highs (Beach 1985). We believe, however, that they

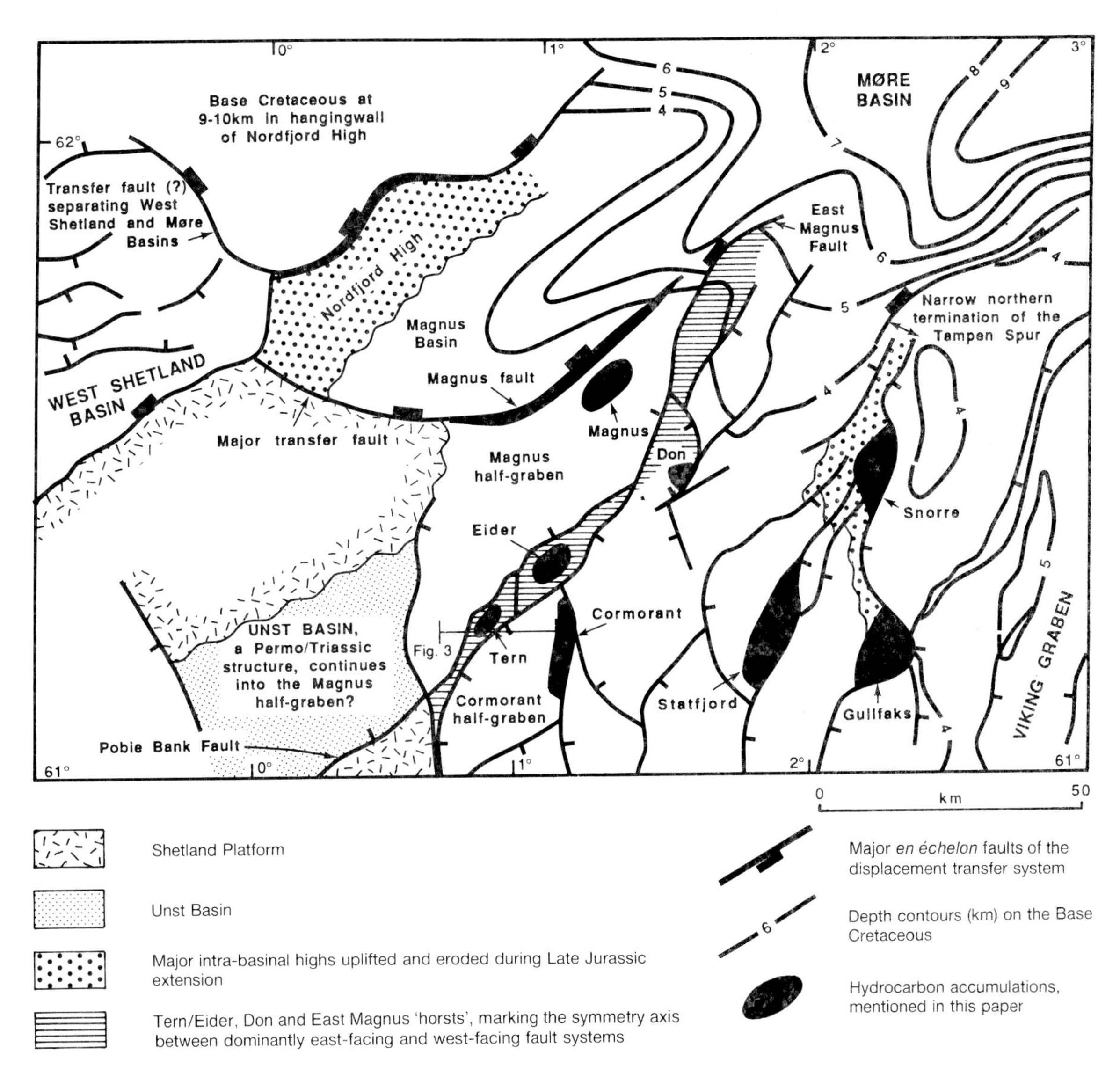

Fig. 9.5. Tectonic elements of the northwestern Viking Graben and southern Møre Basin. The major *en échelon* faults of the Magnus/Tampen Spur transfer system, which kinematically link the Møre and West Shetland Basins, are highlighted (refer to Fig. 9.4 for location). Base Cretaceous contours below 4 km are shown, in order to illustrate the rapid northwards deepening into the Møre Basin. The base of the Cretaceous over the East Shetland fault blocks lies above 4 km. Also shown is the Tern/Eider, East Magnus horst system, marking the symmetry axis between dominantly east-facing faults of the Viking Graben trend and dominantly west-facing faults of the Møre/West Shetland trend (*see* Marsden *et al.* this volume for regional cross-sections through the symmetry axis). West of the Tern/Eider Ridge, structural contiguity between the Permo/Triassic Unst Basin and the Magnus half-graben, now separated by the Late Jurassic basin-margin fault, is suggested. (Compiled from mapping by Badley Ashton and Associates, with infill from Hamar and Hjelle 1984, and Duindam and van Hoorn 1987).

are more simple residual horsts, flanked by normal faults, which mark an axis of symmetry in the rift system between the east-facing structures of the Viking Graben and the west-facing structures of the mid-Norway/West Shetland system. This axis of symmetry continues south into the eastern margin of the Unst Basin (Johns and Andrews 1985), defined by the Pobie Bank Fault.

9.10. Conclusions

By integrating minimum extension estimates with kinematic indicators and a knowledge of internal basin geometry, we have attempted to erect a simple kinematic model for the Late Jurassic opening of the North Sea rift system. We believe that this rift system comprises two kinematically closed triangles, linked through the Viking Graben (Fig. 9.6).

Our main conclusions from the exercise are as follows.

1. The dominant fault sets in the Viking Graben, Outer Moray Firth, Central Graben, Møre Basin, and West Shetland Basin can all be interpreted as normal fault sets which opened almost orthogonal to their strike in a dip-slip manner.
2. The Great Glen fault, forming the sidewall to the Inner Moray Firth, is the only strike-slip fault involved in the regional kinematics of the rift. Other smaller structures with a lateral displacement are, in general, local transfer faults linked into the dominant extensional structures.
3. The most extended basin within the rift system is the Møre Basin. Displacement within this basin is resolved southwards into two components passing both east of the UK (Viking and Central Graben) and west of the UK (West Shetland Basin).
4. Many structures interpreted in the literature as the product of transpressional deformation can be better interpreted within an extensional framework. In particular there is a common misconception that structural elevation above a regional datum equates with compression. Models, such as those of Jackson and McKenzie (1983), Barr (1987*a*, *b*), and Kusznir and colleagues (1988), show that structural elevation may be achieved as the result of footwall uplift during extensional faulting. We believe such models to be widely applicable within the North Sea rift (Marsden *et al.* this volume; Yielding 1990; Roberts and Yielding in press).
5. Basin-modelling, together with seismic reflection data, suggests a considerable pre-Late Jurassic, possibly Triassic, extensional history within the

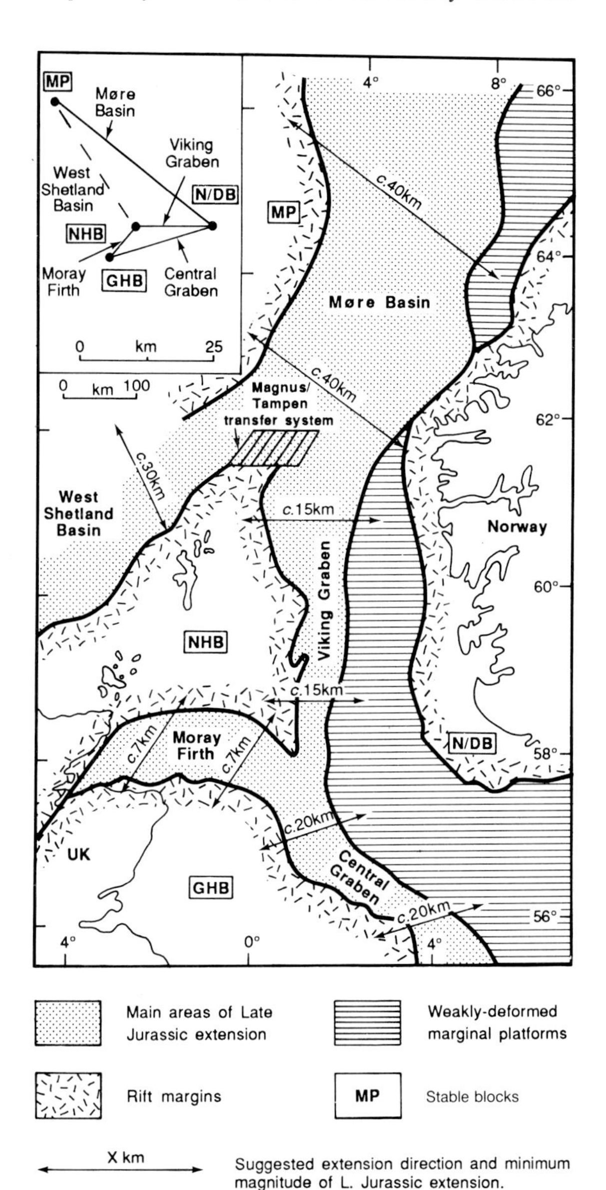

Fig. 9.6. Summary diagram of suggested Late Jurassic kinematics throughout the North Sea rift. Inset shows the composite vector diagram of minimum Late Jurassic stretching estimates, which relates the four basin-bounding blocks to each other.

rift system. The kinematics of this earlier extension are difficult to quantify, but we see no evidence to suggest that stretching directions need have been significantly different from those proposed here for the Late Jurassic.

We recognize that our kinematic solution for the opening of the North Sea rift, based on two closed and overlapping vector triangles, is not a unique solution to this problem. Work is currently in progress with the aim of quantifying the effects of 'soft-domino' deformation on our extension estimates. We believe, however, that our current model, based to a large extent on the measurement of fault heaves, provides an elegantly simple model, devoid of the need to appeal to major, *ad hoc* strike-slip movements within the basin.

Acknowledgements

We wish to thank our Badley Ashton colleagues, John Price and Brett Freeman, for the many discussions we have had on all aspects of North Sea tectonics. We also thank Terkel Olsen of Statoil who has acted as the main external 'sounding-board' for our wilder flights of fantasy. We have benefited greatly from discussions with Dave Barr, Juan Watterson, John Walsh, Nick Kusznir, Gary Marsden, and James Jackson, all of whom have convinced us of the importance of footwall uplift as a deformation mechanism within extending sedimentary basins. Finally, we thank David Kemp for draughting the figures.

References

Badley, M. E., Egeberg, T., and Nipen, O. (1984). Development of rift basins illustrated by the structural evolution of the Oseberg feature, Block 30/6, offshore Norway. *J. Geol. Soc.*, **41**, 639–49.

Badley, M. E., Price, J. D., Rambech Dahl, C., and Agdestein, T. (1988). The structural evolution of the northern Viking Graben and its bearing upon extensional modes of basin formation. *J. Geol. Soc.*, **145**, 455–72.

Badley, M. E., Price, J. D., Rambech Dahl, C. and Agdestein, T. (1989). Discussion on the structural evolution of the northern Viking Graben and its bearing upon extensional modes of basin formation. *J. Geol. Soc.*, **146**, 1038–40.

Barr, D. (1985). 3-D palinspastic restoration of normal faults in the Inner Moray Firth: implications for extensional basin development. *Earth and Planetary Science Letters*, **75**, 191–203.

Barr, D. (1987*a*). Lithospheric stretching, detached normal faulting and footwall uplift. In *Continental extensional tectonics* (ed. M. P. Coward, J. F. Dewey, and P. L. Hancock), 75–94. *Geol. Soc. Special Publication*, **28**.

Barr, D. (1987*b*). Structural/stratigraphic models for extensional basins of half-graben type. *J. Struct. Geol.*, **9**, 491–500.

Barton, P. and Wood, R. (1984). Tectonic evolution of the North Sea basin: crustal stretching and subsidence. *Geophys. J. Roy. Astron. Soc.*, **79**, 987–1022.

Beach, A. (1984). Structural evolution of the Witch Ground Graben. *J. Geol. Soc.*, **141**, 621–8.

Beach, A. (1985). Some comments on sedimentary basin development in the northern North Sea. *Scot. J. Geol.*, **21**, 493–512.

Beach, A., Bird, T., and Gibbs, A. D. (1987). Extensional tectonics and crustal structure: deep seismic reflection data from the northern North Sea Viking Graben. In *Continental extensional tectonics* (ed. M. P. Coward, J. F. Dewey, and P. L. Hancock). *Geol. Soc. Special Publication*, **28**, 467–76.

Buck, W. R. (1988). Flexural rotation of normal faults. *Tectonics*, **7**, 959–73.

Bukovics, C., Shaw, N. D., Cartier, E. G., and Ziegler, P. A. (1984). Structure and development of the mid-Norway continental margin. In *Petroleum geology of the north European margin* (ed. A. M. Spencer *et al.*), 407–26. London, Graham and Trotman.

Cartwright, J. A. (1987). Transverse structural zones in continental rifts—an example from the Danish Sector of the North Sea. In *Petroleum geology of north west Europe* (ed. J. Brooks and K. Glennie), 441–452, London, Graham and Trotman.

Cartwright, J. A. (1989). The kinematics of inversion in the Danish Central Graben. In *Inversion tectonics* (ed. M. A. Cooper and G. D. Williams), *Geol. Soc. Special Publication*, **44**, 153–175.

Caselli, F. (1987). Oblique-slip tectonics, Mid-Norway shelf. In *Petroleum geology of north west Europe* (ed. J. Brooks and K. Glennie), 1049–64. London, Graham and Trotman.

Chesher, J. A. and Bacon, M. (1975). A deep seismic survey in the Moray Firth. *Inst. Geol. Sciences*, Report No. 75/11.

Childs, C., Walsh, J. J., and Watterson, J. (in press). A method for estimation of the density of fault displacements below the limit of seismic resolution in reservoir formations. In *North Sea oil and gas reservoirs*. Norwegian Institute of Technology (Trondheim), Graham and Trotman.

Christie, P. A. F. and Sclater, J. G. (1980). An extensional origin for the Buchan and Witchground Graben in the North Sea. *Nature*, **283**, 729–32.

Donato, J. A. and Tully, M. C. (1981). A regional interpretation of North Sea gravity data. In *Petroleum geology of the continental shelf of north west Europe* (ed. L. V. Illing and G. D. Hobson), 65–75. London, Heyden and Son.

Duindam, P. and van Hoorn, B. (1987). Structural evolution of the West Shetland continental margin. In *Petroleum geology of north west Europe* (ed. J. Brooks and K. Glennie), 765–74. London, Graham and Trotman.

Eynon, G. (1981). Basin development and sedimentation in the Middle Jurassic of the northern North Sea. In *Petroleum geology of the continental shelf of north west Europe* (ed. L. V. Illing and G. D. Hobson), 196–204. London, Heyden and Son.

Fossen, H. (1989). Indication of transpressional tectonics in the Gullfaks oil-field, northern North Sea. *Marine and Petroleum Geology*, **6**, 22–30.

Frost, R. E. (1987). The evolution of the Viking Graben tilted fault-block structures: a compressional origin. In *Petroleum geology of north west Europe* (ed. J. Brooks and K. Glennie), 1009–24. London, Graham and Trotman.

Frostick, L., Reid, I., Jarvis, J., and Eardley, H. (1988). Triassic sediments of the Inner Moray Firth, Scotland: early rift deposits. *J. Geol. Soc.*, **145**, 235–48.

Gabrielsen, R. H. and Robinson, C. (1984). Tectonic inhomogeneities of the Kristiansund-Bodø fault complex, offshore Mid-Norway. In *Petroleum geology of the north European margin* (ed. A. M. Spencer *et al.*), 397–406. London, Graham and Trotman.

Gibbs, A. D. (1983). Balanced cross-section construction from seismic lines in areas of extensional tectonics. *J. Struct. Geol.*, **5**, 153–60.

Gibbs, A. D. (1985). Discussion on the structural evolution of extensional basin margins. *J. Geol. Soc.*, **142**, 941–2.

Gibson, J. R., Walsh, J. J., and Watterson, J. (1989). Modelling of bed contours and cross-sections adjacent to planar normal faults. *J. Struct. Geol.*, **11**, 317–28.

Giltner, J. P. (1987). Application of extensional models to the Northern Viking Graben. *Norsk Geologisk Tidsskrift*, **67**, 339–52.

Gowers, M. B. and Sæbøe, A. (1985). On the structural evolution of the Central Trough in the Norwegian and Danish sectors of the North Sea. *Marine and Petroleum Geology*, **2**, 298–318.

Hamar, G. P. and Hjelle, K. (1984). Tectonic framework of the Møre Basin and the northern North Sea. In *Petroleum geology of the north European margin* (ed. A. M. Spencer *et al.*), 349–59. London, Graham and Trotman.

Jackson, J. A. and McKenzie, D. (1983). The geometric evolution of normal fault systems. *J. Struct. Geol.*, **5**, 471–82.

Jackson, J. A. and White, N. J. (1989). Normal faulting in the upper continental crust: observations from regions of active extension. *J. Struct. Geol.*, **11**, 15–36.

Jackson, J. A., White, N. J., Garfunkel, Z., and Anderson, H. (1988). Relations between normal-fault geometry, tilting and vertical motions in extensional terrains, an example from the southern Gulf of Suez. *J. Struct. Geol.*, **10**, 155–70.

Jackson, J. S. and Hastings, D. S. (1986). The role of salt movement in the tectonic history of Haltenbanken and Trænabanken and its relationships to structural style. In *Habitat of hydrocarbons on the Norwegian continental shelf* (ed. A. M. Spencer *et al.*), 241–57. London, Graham and Trotman.

Johns, C. and Andrews, I. J. (1985). The petroleum geology of the Unst Basin, North Sea. *Marine and Petroleum Geology*, **2**, 361–72.

King, G. C. P., Stein, R. S., and Rundle, J. B. (1988). The growth of geological structures by repeated earthquakes, 1, Conceptual framework. *J. Geophys. Research*, **93**, 13307–19.

Klemperer, S. (1988). Crustal thinning and nature of extension in the northern North Sea from deep seismic reflection profiling. *Tectonics*, **7**, 803–22.

Klemperer, S. L. and Hurich, C. A. (this volume). Lithospheric structure of the North Sea area from deep seismic reflection profiling.

Kusznir, N. J., Marsden, G., and Egan, S. (1988). Fault block rotation during continental lithosphere extension: a flexural cantilever model. *Geophys. J.*, **92**, 546.

Larsen, P. H. (1988). Relay structures in a Lower Permian basement-involved extension system, East Greenland. *J. Struct. Geol.*, **10**, 3–8.

Larsen, V. B. (1987). A synthesis of tectonically-related stratigraphy in the North Atlantic–Arctic region from Aalenian-Cenomanion time. *Norsk Geologisk Tidsskrift*, **67**, 281–94.

Marsden, G., Yielding, G., Roberts, A. M., and Kusznir, N. J. (this volume). Application of a flexural cantilever simple-shear/pure-shear model of continental lithosphere extension to the formation of the northern North Sea basin.

McCoss, A. M. (1986). Simple constructions for deformation in transpression/transtension zones. *J. Struct. Geol.*, **8**, 715–9.

McQuillin, R., Donato, J. A., and Tulstrop, J. (1982). Development of basins in the Inner Moray Firth and the North Sea by crustal extension and dextral displacement of the Great Glen Fault. *Earth and Planetary Science Letters*, **60**, 127–39.

Nelson, P. H. H. and Lamy, J. M. (1987). The Møre/ West Shetland area: a review. In *Petroleum geology of north west Europe* (ed. J. Brooks and K. Glennie), 775–84, London, Graham and Trotman.

Olsen, J. C. (1987). Tectonic evolution of the North Sea region. In *Petroleum geology of north west Europe* (ed. J. Brooks and K. Glennie), 398–402. London, Graham and Trotman.

Pegrum, R. M. (1984). Structural development of the southwestern margin of the Russian–Fennoscandian Platform. In *Petroleum geology of the north European margin* (ed. A. M. Spencer *et al.*), 359–69. London, Graham and Trotman.

Pegrum, R. M. and Ljones, T. E. (1984). 15/9 gamma gas field offshore Norway, new trap type for North Sea Basin with regional structural implications. *Am. Assoc. Petrol. Geol. Bulletin*, **68**, 874–902.

Price, I. and Rattey, R. P. (1984). Cretaceous tectonics off Mid-Norway: Implication for the Rockall and Færoe-Shetland troughs. *J. Geol. Soc.*, **141**, 985–92.

Roberts, A. M. and Yielding, G. (in press). Deformation around basin-margin faults in the North Sea/Mid-Norway rift. In *The geometry of normal faults* (ed. A. M. Roberts, G. Yielding, and B. Freeman). *Geol. Soc. Special Publication*.

Roberts, A. M., Price, J. D., and Badley, M. E. (1989). Discussion of Triassic sediments of the Inner Moray Firth, Scotland: early rift deposits. *J. Geol. Soc.*, **146**, 361–2.

Roberts, A. M., Badley, M. E., Price, J. D., and Huck, I. W. (1990). The structural history of a transtensional basin, Inner Moray Firth, NE Scotland. *J. Geol. Soc.*, **147**, 87–103.

Roberts, A. M. Price, J. D., and Olsen, T. S. (in press). Late Jurassic half-graben control on the siting and structure of hydrocarbon accumulations: UK/ Norwegian Central Graben. In *Tectonic events controlling Britain's oil and gas reserves* (ed. R. Hardman and J. Brooks). *Geol. Soc., Special Publication*.

Sclater, J. G., Hellinger, S. J., and Shorey, M. (1986). An analysis of the importance of extension in accounting for the post-Carboniferous subsidence of the North Sea basin. *University of Texas Instit. for Geophysics*, Internal Report.

Solli, M. (1976). En Seismisk skorpeunderskelse Norge-Shetland. Unpublished Ph.D. thesis, University of Bergen.

Speksnijder, A. (1987). The structural configuration of Cormorant Block IV in context of the northern Viking Graben structural framework. *Geologie en Mijnbouw*, **65**, 357–79.

Stein, R. S., King, G. C. P., and Rundle, J. B. (1988). The growth of geological structures by repeated earthquakes, 2, field examples of continental dip-slip faults. *J. Geophys. Research*, **9**, 13319–31.

Vejbæk, O. V. and Anderson, C. (1987). Cretaceous-Early Tertiary inversion tectonism in the Danish Central Trough. *Tectonophysics*, **137**, 221–8.

Vening Meinesz, F. A. (1950). Les grabens africains, resultat de compression ou de tension dans le croute terrestre? *Inst. R. Colonial Belge Bull.*, **21**, 539–52.

Walsh, J. J. and Watterson, J. (in press) Geometric and kinematic coherence and scale effects in normal fault systems. In *The geometry of normal faults* (ed. A. M. Roberts, G. Yielding, and B. Freeman). *Geol. Soc. Special Publication*.

White, N. J., Jackson, J. A., and McKenzie, D. P. (1986). The relationship between the geometry of normal faults and that of the sedimentary layers in their hanging walls. *J. Struct. Geology*, **8**, 879–909.

Williams, G. and Vann, I. (1987). The geometry of listric normal faults and deformation in their hangingwalls. *J. Struct. Geol.*, **9**, 789–95.

Yielding, G. (1990). Footwall uplift associated with Late Jurassic normal faulting in the northern North Sea. *J. Geol. Soc.*, **147**, 219–22.

Ziegler, P. A. (1982). Faulting and graben formation in western and central Europe. *Phil. Trans. Roy. Soc., London*, **A305**, 113–43.

Ziegler, P. A. (1983). Discussion on crustal thinning and subsidence in the North Sea. *Nature*, **304**, 561.

10 The structural evolution of the Ringkøbing-Fyn High

J. Cartwright

Abstract

The Ringkøbing-Fyn High (RFH) is a major E–W trending positive structural element that crosses central Denmark and continues westwards to the border of the Central Graben of the North Sea. It was formed during a regional tectono-magmatic event in late Carboniferous–Early Permian. Interpretation of a regional grid of seismic data in the Danish Sector shows that the RFH is composed of a series of shallow basement horsts transected by several roughly N–S trending grabens. The RFH is bounded on its northern and southern margins by WNW–ESE trending fracture zones. These fracture zones are postulated to have been active in a transform sense during Late Caledonian orogenesis, and were subsequently reactivated to varying degrees in later deformational episodes. This postulate stems from the recognition of westward-dipping basement fabrics on the seismic data immediately east of the border faults to the Central Graben. These basement structures are thought most likely to represent the eroded remnants of the southward continuation of the Norwegian Caledonide deformation front.

10.1. Introduction

The Ringkøbing-Fyn High (RFH) is a WNW–ESE-trending series of rhombohedral horst massifs with a generally thin cover of Mesozoic and Tertiary sediments. The horsts are separated by several N–S- or NNE–SSW-trending grabens that contain thick sequences of Late Palaeozoic to Mesozoic sediments. With the Mid North Sea High, the RFH forms a barrier between the northern and southern Permian Basin (Fig. 10.1).

This positive feature was formed during a major phase of regional tectonic activity in the Late Carboniferous–Early Permian (Ziegler 1982). Although several hypotheses for the origins of the highs have been proposed (Ziegler 1982; Dewey 1982; Donato *et al*. 1983), little has been published on the structure of these positive elements. This paucity of descriptive material stems in part from the fact that the highs have been considered to be largely unprospective for hydrocarbons, and as such the density of well and seismic coverage is limited in comparison with neighbouring regions of the North Sea. However, several regional seismic surveys of the offshore sector of the RFH were carried out in the early 1980s in response to changes in licensing arrangements in Denmark. The main aims of this paper are to present the results of an interpretation of one of these surveys, to document the principal structures of the RFH, and to discuss the implications of the interpretation for the regional structural evolution of the central North Sea. This study of the offshore tract of the RFH is complementary to the work being carried out on the eastern part of the RFH as part of the European Geotraverse (EGT), which is centred on mainland Denmark (EUGENO-S Working Group 1988).

10.2. Regional stratigraphy and structure

10.2.1. *Seismic data base*

The seismic grid used for this mapping comprised some 7000 line kilometres of 7 s record multi-channel reflection data supplied by Merlin Geophysical Ltd. The data coverage amounted to an

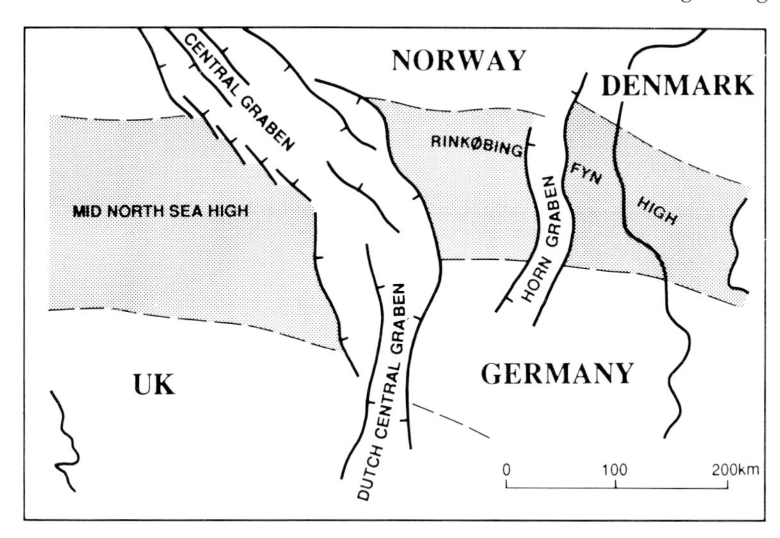

Fig. 10.1. Principal tectonic elements of the Central North Sea (after Ziegler 1982).

average line spacing of 10 km, which proved adequate for the mapping of all the major structures.

10.2.2. *Regional stratigraphy*

The stratigraphy offshore in the North Sea sector of the RFH is known from only six wells, none of which calibrate the Palaeozoic succession. Constraints on the deeper foundations of the RFH come from three wells which penetrate through to the crystalline basement, Per-1 on the extreme west, immediately adjacent to the Tail End Graben, and Glamsbjerg and Grinsted, located on basement horsts onshore in Denmark (Fig. 10.2). Results of radiometric dating of the basement cores from these wells have been summarized by Frost and co-workers (1981), and by Ziegler (1982). The basement in Per-1 shows evidence of a major Caledonian overprint, whereas both Glamsbjerg and Grinsted have ages in the range 800–900 Ma, and have been correlated with the Dalslandide Province of southern Norway (Watson 1976; EUGENO-S Working Group 1988). These limited samples of the crystalline basement indicate that a major crustal province boundary must be present on the RFH, somewhere between the onshore and offshore calibration points.

The crystalline basement is overlain by a parallel-bedded sequence which attains a maximum thickness of some 5 km in areas protected from erosion by major block faulting. This sequence has not been directly calibrated by any wells drilled in the offshore sector. A representative seismic section showing the typical seismic response of this sequence is illustrated in Fig. 10.3. The sequence is bound at its base by a high-amplitude low-frequency event that may represent the basement/cover interface. The overlying sequence is remarkably parallel and concordant with the basal sequence boundary over the entire extent of the offshore extent of the RFH. The internal reflection character is uniform and easily correlatable over large distances, suggesting that it was deposited in an epeirogenic setting with minimal basin floor relief. The upper part of the sequence is characterized by a high frequency–high amplitude seismic facies, which is locally discontinuous, but regionally extensive. This seismic character is highly reminiscent of Coal Measures successions in the Southern Gas Basin of the North Sea. Indeed, there is a striking overall similarity of gross thickness and bulk seismic character between the platform cover of the basement of the RFH, and that of the cover to the basement on the Mid North Sea High. The pre-Permian succession on the Mid North Sea High is known, from drilling, to range from Middle Devonian to Late Carboniferous in age and, on the basis of the similarity in reflection character, it is suggested that the undrilled platform cover of the RFH is also of a similar age range. This sequence has been referred to as Lower Palaeozoic by Ziegler (1982) and, until there is a positive calibration of this sequence in a borehole, the age relations must remain an open question.

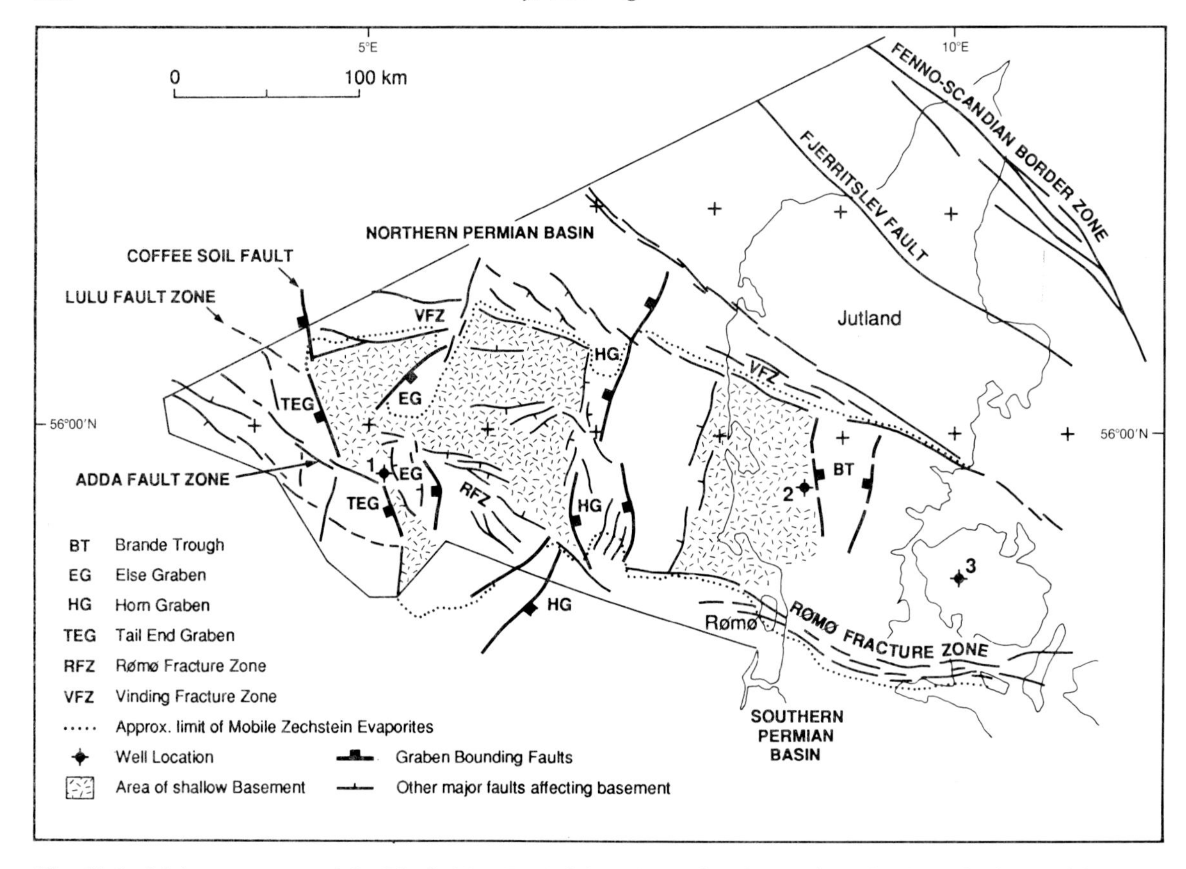

Fig. 10.2. Major structures of the Ringkøbing-Fyn High and associated areas in Jutland and in the Danish North Sea. Wells penetrating crystalline basement indicated by numbers: 1- Per-1; 2- Grinsted; 3- Glamsbjerg.

The upper boundary of the postulated Upper Palaeozoic sequence is everywhere found to be a major unconformity, with often considerable angularity and evidence of pronounced erosional truncation. The large normal faults that comprise the two marginal fracture zones and bound the transecting grabens preserve large thicknesses of the Upper Palaeozoic sequence in their hangingwalls, but over large areas of the basement horsts there has been almost complete erosional denudation of the platform cover. The major angular unconformity is referred to as the Saalian Unconformity (Ziegler 1982) but, owing to the absence of direct well calibration, the duration of the erosional event cannot be defined with any precision. Coeval with the development of the unconformity there was a period of intense block faulting, tilting of fault blocks, uplift, and erosion. From wells drilled in the vicinity of the Horn Graben, the Saalian Unconformity can be seen to be onlapped by Lower Rotliegend (Early Permian) sediments, suggesting that an approximate upper limit for the tectonic event is Early Permian (Olsen 1983). Published stratigraphic syntheses are in agreement that the Saalian Unconformity developed in response to a regional tectonic phase that took place in the Late Carboniferous/Early Permian (Ziegler 1982; Arthaud and Matte 1977).

The Late Carboniferous–Early Permian tectonic activity was accompanied by widespread intrusive and extrusive magmatism (Dixon *et al.* 1981; Latin *et al.* this volume). Several of the wells drilled on the RFH penetrated volcanic rocks or volcaniclastic sediments referred to the Lower Rotliegend. Radiometric dating of these lavas is inconclusive owing to their severe degree of alteration, but detailed geochemical analysis has established their affinity with other volcanic rocks of the Late Carboniferous/Early Permian volcanic province of north-west Europe (Dixon *et al*. 1981).

From the Permian to the early part of the Cretaceous, the RFH acted as a stable, rigid, non-subsiding massif. It was flanked to the north and south by the major Permo-Triassic depocentres. There is no evidence that the RFH was at any stage in the Late Permian or Mesozoic an area of elevated topography, and it is probable that the RFH acted only as a minor source of sediment for the adjacent Permo-Triassic depocentres. The Upper Permian and Triassic sequences thin by stratal condensing, with neither basal onlap nor suprastratal erosional truncation; their depositional extent was controlled by activity on the marginal fracture zones. The close spatial relationship between the fundamental structure of the margins of the RFH and the depositional limits of the Permo-Triassic can be seen for example in the mapped limits of mobile Zechstein evaporites (Fig. 10.2). The RFH remained an area of non-deposition until it was completely overstepped in the Late Cretaceous.

10.2.3. *Regional structural elements*

The major faults and structural elements of the offshore tract of the RFH are illustrated in Fig. 10.2. The RFH is delimited on its northern margin by a generally WNW–ESE-trending zone of intense faulting at Mesozoic and Palaeozoic levels referred to in this paper as the Vinding Fracture Zone (VFZ). This linear alignment of cover and basement faults can be traced from Jutland to the Tail End Graben, a distance of over 200 km. The VFZ formed the southern limit of deposition of mobile Zechstein evaporites in the northern Permian Basin. The southern margin of the RFH is also marked by a WNW–ESE-trending zone of intense cover and basement faulting referred to as the Rømø Fracture Zone (RFZ). This too forms a limit for Zechstein evaporite deposition over much of its strike extent. West of the Horn Graben, the zone of fracturing is more dispersed, with one suite of faults appearing to splay and die out in the central area of the Else Graben, whilst the other fault suite is probably offset to the south, and continues as a major WNW-trending fracture zone that transects the Tail End Graben (Cartwright 1987). The other principal structural elements in the offshore region are the Else and Horn Grabens. Both these grabens have NNE–SSW trends, although their principal boundary faults deviate substantially from this orientation in detail.

The Vinding Fracture Zone (VFZ)

The VFZ is a broad, linear zone of intense faulting at basement and cover levels that marks the northern boundary of the Ringkøbing-Fyn High. On Jutland, it coincides with a prominent WNW–ESE-trending aeromagnetic anomaly known as the Vinding Line (Dikkers 1977), from which the more general term used here is derived. The VFZ is mapped offshore to the west of Jutland as a 10–15 km wide zone of intense faulting; it is further characterized by the tilting, uplift and truncation of the Upper Palaeozoic sequence (Fig. 10.3). The VFZ intersects and offsets the boundary fault of the northern Horn Graben, and continues along a WNW trend as far as the boundary fault of the northern Else Graben (Fig. 10.2). At this point, the VFZ changes orientation to a WSW trend. The VFZ continues along this new trend towards the uplifted shoulder of the Tail End Graben but, as a result of the severe attenuation of the Upper Palaeozoic strata on the shoulder, it is difficult to identify intra-basement faulting with the same confidence as when some overlying sedimentary cover is preserved. There is a suggestion of disruption within the crystalline basement of the footwall immediately adjacent to the Coffee Soil fault, and it is considered probable that the VFZ continues as far west as the Coffee Soil fault, where a projected intersection would be close to the point of strike reorientation of the Coffee Soil fault at the Lulu Transverse Zone.

In detail, the geometry of the VFZ varies considerably along its strike. Most of the individual faults comprising the fault zone have normal components of displacement, ranging from 100–2000 m. Minor thrust components have been observed on some faults. Most of the faulting is concentrated in the pre-Permian sequence, which is invariably found to be tilted and severely truncated at the Saalian Unconformity. The larger faults are often observed to propagate upwards into the Mesozoic section, and salt withdrawal collapse structures tend to nucleate above the traces of the more significant deeper faults.

It is difficult to draw any direct conclusions as to the kinematic evolution of the VFZ simply on the basis of the seismic expression and the mapped

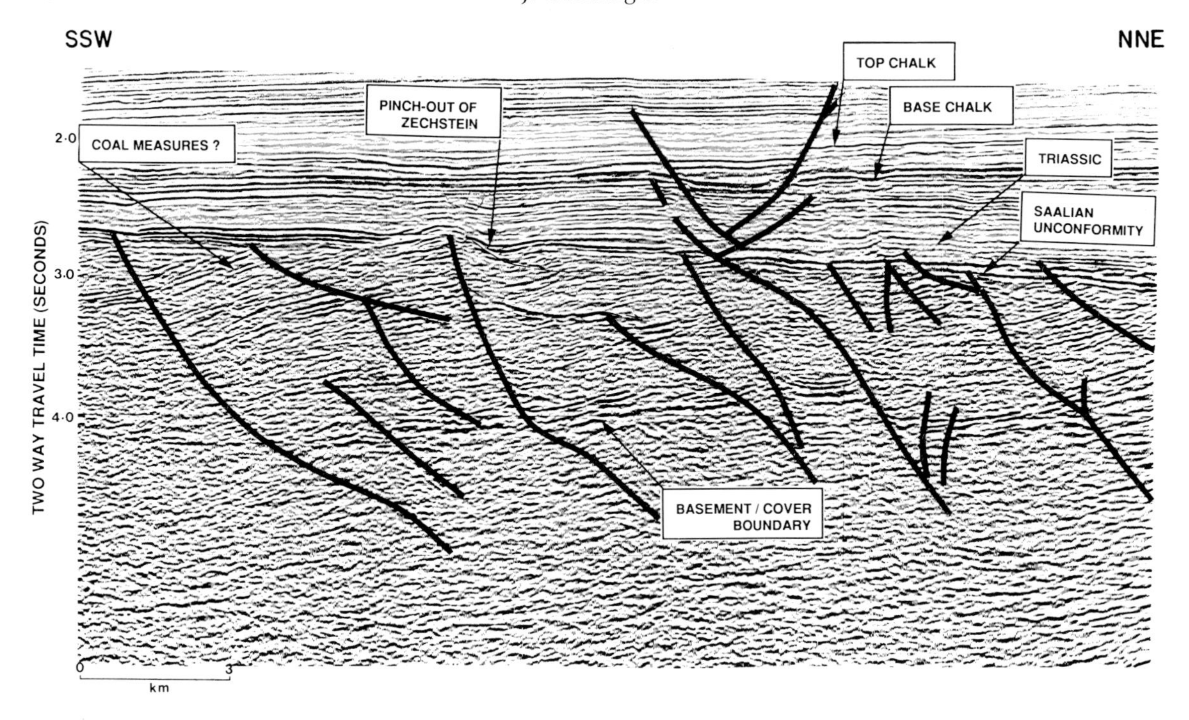

Fig. 10.3. Representative seismic section across the Vinding Fracture Zone, showing the highly faulted Upper Palazozoic succession, the pinch-out of the Zechstein, the severe truncation at the Saalian Unconformity, and the localized reactivation of deeper faults in the Mesozoic.

structural configuration. There is undoubtedly good evidence for a major structural mobilization in the Late Carboniferous–Early Permian which involved extremely localized block rotations and erosional truncation, but whether this involved dominantly dip-slip deformation or strike-slip deformation is a matter for speculation. The extreme localization of the intense deformation along the markedly linear extent of the VFZ does, however, argue for some form of deep-seated basement control. Thereafter, the VFZ acted as a structurally-controlled depositional hinge, separating the steadily subsiding basin in the north from the stable RFH to the south. The minor accommodational deformation required of this hinge behaviour was expressed in limited extensional reactivation of some of the constituent faults of the VFZ.

The Rømø Fracture Zone (RFZ)

The RFZ is most clearly defined as a structural entity in South Jutland where it is found to be associated with a major, linear, WNW-trending magnetic anomaly superimposed over a linear, WNW-trending steep gravity gradient (EUGENO-S Working Group 1988). From well penetrations in southern Denmark and northern Germany, it is evident that this lineament is related to a major crustal province boundary or suture between the North German Caledonides (Ziegler 1982) and the Dalslandian crystalline basement, termed the Trans-European fault, but the exact nature of this relationship is not clear.

The upper crustal structure of the RFZ is similar to that described for the VFZ, and the deeper basement faults that form the foundations to the structure exerted a similar control on the depositional limits of the northern margin of the southern Permian Basin as those of the VFZ exerted on the southern margin of the northern Permian Basin.

The extension of the RFZ into the North Sea is suggested by alignments of major basement faults that follow a prominent WNW trend along almost the entire southern margin of the offshore part of the RFH. A linear network of generally WNW-trending basement faults can be traced from the island of Rømø to the hinged eastern margin of the southern Horn Graben (Fig. 10.2). This alignment of basement faults appears to terminate in the axis of the graben along a boundary fault system separating the two southern segments of the Horn Graben. Thereafter, the WNW trend is taken up by a major splaying and bifurcating basement fault zone which clearly offsets the large boundary faults of the two southern segments of the Horn Graben. Across this intersection, there is an abrupt change in orientation of the boundary faults from a NE–SW to a N–S trend. The WNW-trending fault zone continues for 60 kilometres along strike further to the west with a large combined northward heave at basement level but, still further to the west a series of bifurcating basement faults throwing down to the south takes up the identical WNW–ESE trend.

Further to the west, the WNW-trending fault zone appears to terminate at the margins of the Else Graben. The projected point of intersection coincides directly with an intensely faulted zone in the axis of the graben which is thought to mark the boundary between two oppositely verging segments of the Else Graben (Figs 10.2 and 10.4). From Fig. 10.4, it can be observed that the westward prolongation of the WNW-oriented fault trend is collinear with the prominent segment boundary in the Tail End Graben, the Adda Fault Zone. Unfortunately, the seismic grid in the area of shallow crystalline basement between the Else and Tail End Grabens is too coarse to allow any definitive correlation of faults identified in the crystalline basement, but it is nevertheless considered that the similarity in the alignments of the RFZ and the Adda Fault Zone is too striking to be a coincidence. The intervening basement horst forming the peneplained eastern shoulder of the Tail End Graben is permeated by fault plane reflections and, whilst much of the basement deformation is expressed as low-angle thrusts or shear zones (*see* later section on Else Graben), there are some indications that an obliquely intersecting set of basement faults connects the RFZ and the Adda Fault Zone. The confirmation of this connection must await a tighter grid in this western sector of the RFH, but it is interesting to consider that both the VFZ and RFZ are mapped with confidence as striking alignments of major faulted zones virtually to the borders of the Tail End Graben, and their projected intersections are both at points of transverse segmentation.

The Horn Graben

The Horn Graben was formed during the Late Carboniferous to Early Permian tectonic phase, and continued to act as a major depocentre until the end of the Triassic (Ziegler 1982; Olsen 1983). The Horn Graben is segmented, and the major basement border faults display a range of orientations about the mean NNE–SSW trending rift axis (Fig. 10.5).

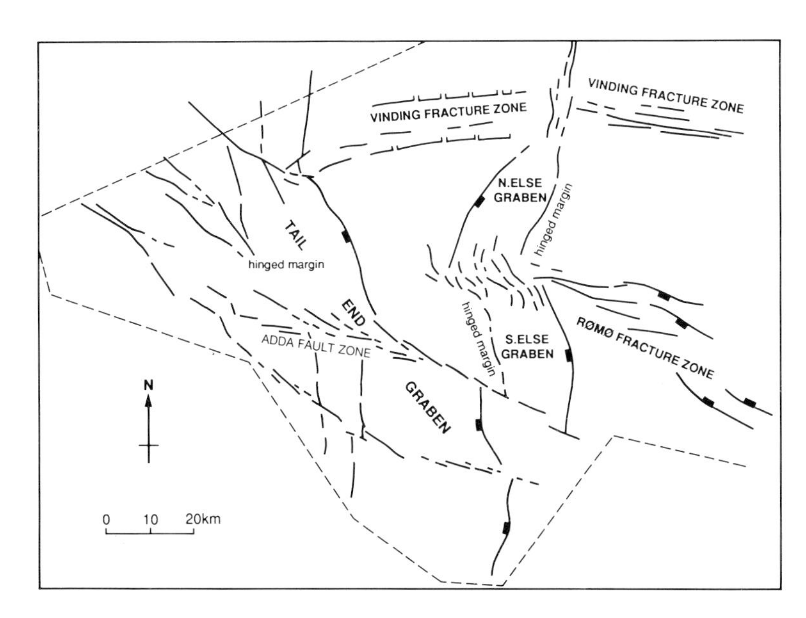

Fig. 10.4. Structural links between the Vinding and Rømø Fracture Zones, the Else Graben, and the Tail End Graben.

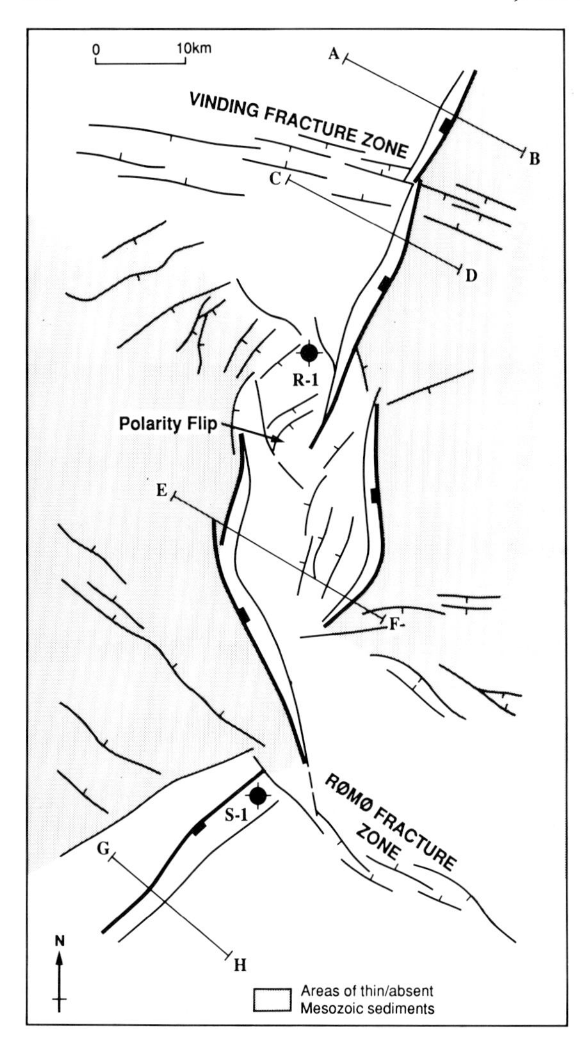

Fig. 10.5. Major and minor faults of the Horn Graben. Figure 10.6 shows the cross-sections marked.

The segmentation of the graben is directly related to the intersections of the Rømø Fracture Zone (RFZ) and the Vinding Fracture Zone (VFZ) with the graben. The segment boundaries are intensely faulted zones several kilometres wide, which have a WNW trend; individual constituent faults comprising the segment boundaries often follow more closely the trends of the faults bounding the graben margins in the respective segments. The outer segment boundaries appear to be almost direct continuations of the RFZ and the VFZ, although the precise offset relations between the transverse and the rift-parallel fault systems are far from clear, and must await a much tighter seismic grids for their definition

Each segment has a distinctive rift fill, and the differences in the fill characteristics are attributable to changes in the geometry and the displacement patterns of the major border structures. These differences are illustrated in type cross sections across each of the four segments (Fig. 10.6). The southern segment is strongly asymmetric with divergent fill towards the west (Fig. 10.6, line G-H.

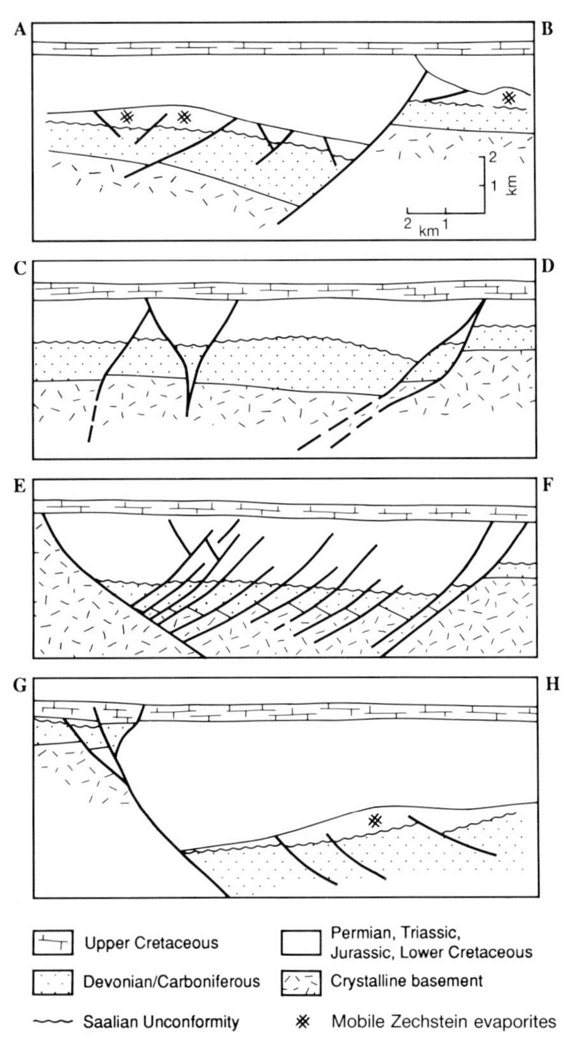

Fig. 10.6. Representative geological sections across the four segments of the Horn Graben. The sections are located on Figure 10.5.

Across the intersecting RFZ, the form becomes more symmetric, albeit with a fine-scale asymmetry imposed by a series of closely spaced domino-style fault blocks (Fig. 10.6, line E-F). This highly faulted segment is bound by NNW-trending basement faults, in contrast with the NE-trending basement faults of the western margin further south. The syn-rift displacement on the NNW-trending western border fault is a maximum at the intersection with the RFZ, decreasing rapidly northwards towards the zone of polarity reversal. The border fault of the two northern segments strikes NNE in both cases (Fig. 10.5), but the stratigraphy of each of the two segments is quite distinct (Fig. 10.6 sections A-B, C-D). However, in both segments, there is a pronounced thickening of the Mesozoic section towards the border fault on the eastern margin, in marked contrast to the situation in the southernmost segment.

The boundary between the two central graben segments is particularly interesting in that it constitutes a zone of interchange of polarity in the inclination of the major rift border faults (Fig. 10.5). This type of polarity 'flip' is increasingly being recognized in continental rifts (Chapin *et al.* 1978; Rosendahl *et al.* 1986), but no satisfactory mechanical explanations of the extraordinary bi-polar geometry have been proposed. In the central two segments of the Horn Graben, the displacement on the two oppositely verging border faults decreases towards the segment boundary. This loss of net displacement (and extension) is partly taken up by the development of splaying networks of minor faults branching off the major border faults in the immediate vicinity of the segment boundary, but may also partly reflect a genuine decrease of bulk extension in this central axis of the graben. This observation may have implications for the mechanism responsible for the adoption of the bi-polar form.

Strike seismic lines running along the graben axis show that the structure of this central 'torsional axis' is an intensely faulted dome which was severely eroded at the pre-rift Saalian Unconformity. The splay faults curve away from their primary rift-parallel orientation to form the crestal collapse faults of the domal structure (Fig. 10.5). This fracture pattern is possibly analogous to coalescing caldera ring systems along an analogous segment boundary in the Oslo Graben (Ramberg and Spjeldnaes 1978).

The geometry and truncation relationships of the dome are indicative of a net relative uplift of this axial part of the Horn Graben of 2–3 kilometres. In addition, the axial dome appears on aeromagnetic maps as an area of high magnetic intensity. Late Carboniferous–Early Permian volcanic rocks were encountered close to the Saalian Unconformity in the R-1 well (Fig. 10.5), but it is not clear whether they were erupted prior to, during, or immediately after the strong phase of uplift and erosion. Nevertheless, it is probable that the intense magnetic signature is related to the presence of a considerable thickness of extrusive magmatic rocks whose eruptive locus was the axial dome of the central Horn Graben.

The close spatial and temporal association of doming, magmatism, and nucleation of oppositely inclined rift fractures suggests that the development of the bi-polar graben form is causally linked to the formation of the domal structure. Furthermore, the location of the dome exactly mid-way between the bounding fracture zones of the RFH seems too symmetrical to be a coincidence, and suggests that the WNW-trending fracture zones are also intimately connected with the doming and initial rift propagation. This notion gains further credence when it is recalled that the main fracturing and tilting within the two fracture zones also took place during the Late Carboniferous–Early Permian tectonic phase.

The Else Graben

The Else Graben is composed of two segments arranged in a bi-polar form equivalent to that described for the two central segments of the Horn Graben (Figs 10.2 and 10.4). The oppositely inclined basement faults controlling the half-graben geometries of the two segments also meet in an intensely faulted axial domal structure. The southern segment is bound to the east by a basement fault that is steeply inclined along its upper portion, but flattens at depth into a low amplitude fault plane reflection dipping approximately 10 degrees to the west (Fig. 10.7). Beneath the floor of the graben and the adjacent footwall, the basement is permeated by similarly oriented, sub-parallel fault plane reflections, which occasionally anastamose to produce lensoid areas of zero reflectivity. The reflection amplitude of this suite of dipping events is highly variable, and is suggestive, therefore, of lateral heterogeneity in the lithological contrasts from which the reflections emanate. The deepest fault plane reflection event can be traced to a depth of approximately

10 kilometres into the crystalline basement but, beyond this depth, the high noise/signal ratio on the record sections does not permit the discrimination between genuine reflections and migration artefacts.

The seismic expression of the fault plane reflections is similar to that described for many shallow inclination basement faults/shear zones which have been imaged on deep reflection profiles acquired by the BIRPS group (Cheadle *et al.* 1987). The origin of the reflectivity of the inclined surfaces is a matter of

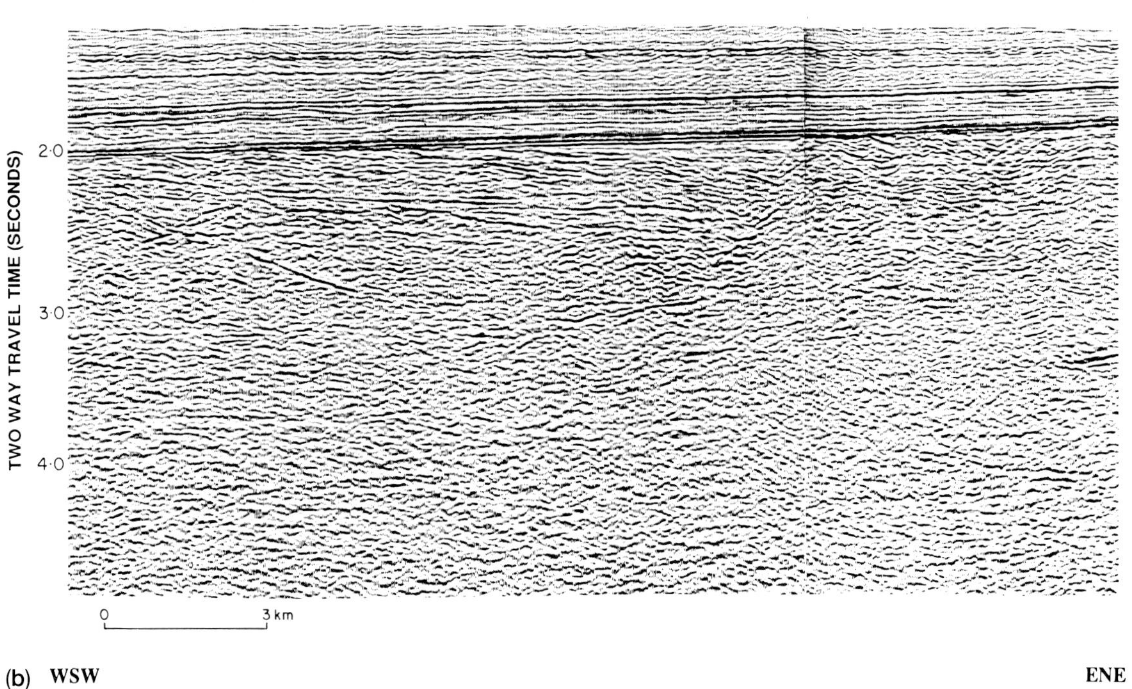

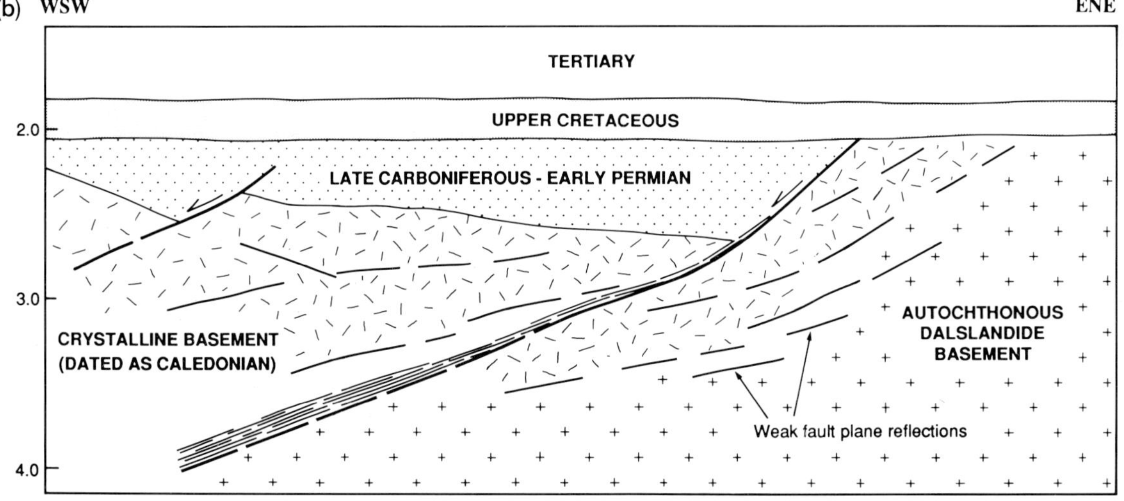

Fig. 10.7. Representative **a** seismic section and **b** line interpretation across the southern segment of the Else Graben. Weak to moderately reflective fault plane reflections can be seen dipping to the west from within the crystalline basement. These are interpreted as emanating from mylonitic shear zones.

debate, but modelling of the Outer Isles thrust, for example, has shown that mylonitic shear zones are capable of producing the observed acoustic signatures (Peddy 1984). The fact that the 'fault planes' indicated on Fig. 10.7 give rise to discrete reflections implies that either the faults juxtapose basement lithologies of greatly contrasting density and velocity, or that the 'faults' themselves represent a lithological banding. The observed lateral variability in amplitude of the reflections could be accounted for in either of these two cases but, considering the relatively minor evidence of brittle deformation seen in the cover sequences of the Else Graben, it is considered more probable that the fault plane reflections emanate from mylonitic shear zones which formed under the mid- or lower crustal P/T conditions usually associated with the development of such ductile fabrics.

The west-dipping 'shear zones' are observed on the seismic data throughout the shallow basement horst separating the Else and Tail End Grabens, and also from beneath the Else Graben itself, but beyond the bounding fault of the eastern margin of the graben, the basement is devoid of reflections. The relationship of the brittle faults forming the main graben structure to the underlying shear zones depicted on Fig. 10.7 implies that the graben developed through the selective reactivation of certain elements of the pre-existing shear zone network (Fig. 10.8). Similar processes of brittle extensional reactivation of pre-existing ductile inclined shear zones are now established for a number of rift basins in north-west Europe (Cheadle *et al*. 1987), one of the best documented of which is the Minches Basin in relation to the Outer Isles thrust (Stein 1989).

No direct relationship between basement fabrics and extensional faults can be established, however, for the northern segment of the Else Graben. The major bounding fault is on the western margin and the thick Late Carboniferous–Early Permian rift fill sequence is strongly divergent towards this boundary fault (Fig. 10.9). The eastern margin is a structural and depositional hinge which acted as a fulcrum for rotation of the graben floor. The basement beneath the floor of the graben is buried beyond the limits of seismic resolution and there is no clear indication in the shallower basement horst west of the graben of any comparable intra-basement structure to that seen beneath the southern segment. Weak inclined reflections dipping to the east are recognized from the basement horst west of the north Else Graben, but they cannot be interpreted with anything approaching the confidence applied to the basement reflections under the south Else Graben. It remains to be seen, therefore, whether the polarity flip between the two segments of the Else Graben was itself pre-conditioned by a bi-polar segmentation of a pre-existing basement fabric, but this must be considered as a distinct possibility.

One of the most striking aspects of the Else

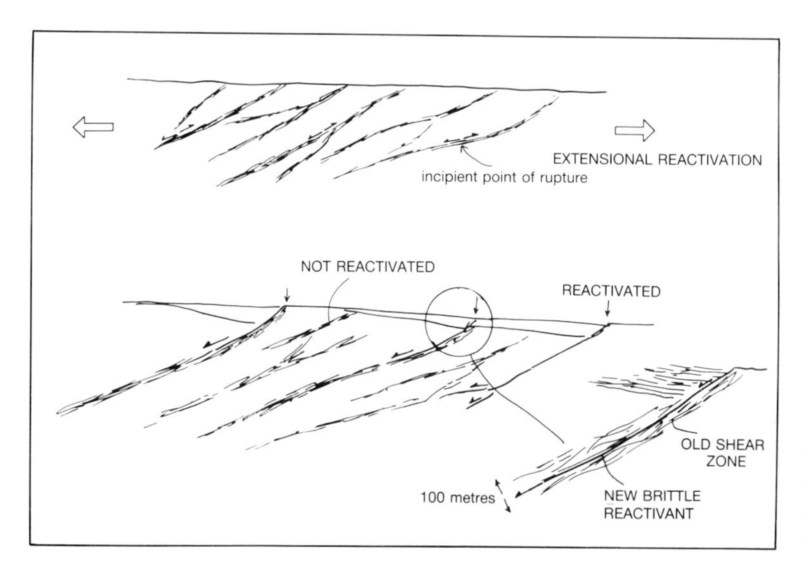

Fig. 10.8. Selective brittle reactivation of shear zones of the Else 'Duplex' to form the South Else Graben.

Graben is the great contrast in the thickness of the Late Carboniferous to Permian fill of the northern and southern segments (Figs 10.7 and 10.9). There is no obvious explanation for this contrast, although the proximity of the northern segment to the northern margin of the RFH may be significant. Presumably, the answer lies in the detailed relative kinematics of the RFZ and VFZ and in the spatial evolution of the opening along the axis of the Tail End Graben.

10.3. The early origins of the basement structure of the RFH

The most important results of the seismic mapping of the Ringkøbing-Fyn High are summarized below.

1. The northern and southern margins of the Ringkøbing-Fyn High are defined by two major WNW–ESE-trending fault lineaments, the Rømø Fracture Zone (RFZ) and the Vinding Fracture Zone (VFZ).

2. The earliest activity on these fracture zones recorded in the deformation of the sedimentary cover took place in the Late Carboniferous–Early Permian, and this resulted in the intense fracturing, tilting, and erosional truncation of the platform sequence of presumed Upper Palaeozoic age.

3. The marginal fracture zones are co-linear with large gravity and magnetic anomalies, particularly along the VFZ in Central Jutland. The amplitude and wavelength of these linear anomalies is suggestive of major contrasts at depth in crustal density and composition. In the case of the RFZ, the lineament is related to the crustal province boundary between the Mid-European Caledonides to the south and the Sveco-Norwegian crystalline basement to the north.

4. The linear trends of the two fracture zones are interrupted at intervals along their strike by their intersection with NNE–SSW-trending bounding faults of the Horn and Else Grabens.

5. The segmentation of the Horn and Else Grabens is spatially related to the trace of the fracture zones, such that the positions at which the bounding faults change strike or sense are in almost all cases the points of intersection of grabens and the fracture zones.

6. The central two segments of the Horn Graben and the two segments comprising the Else Graben

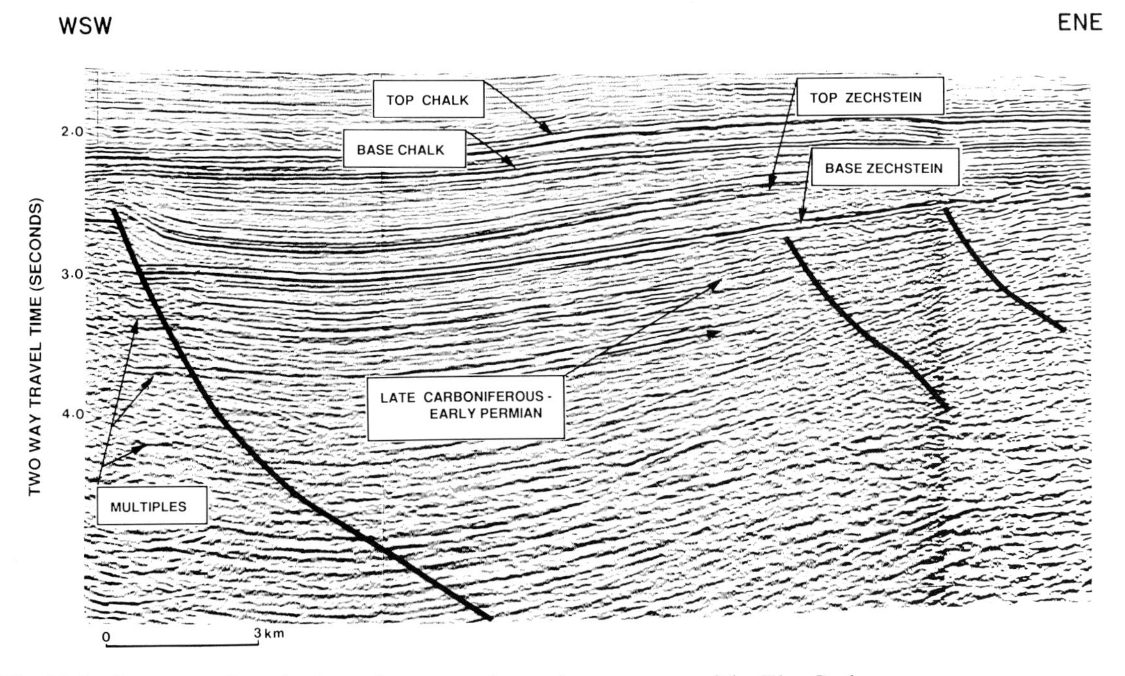

Fig. 10.9. Representative seismic section across the northern segment of the Else Graben.

are arranged in a bi-polar form. In both grabens, the oppositely inclined basement faults interchange over areas of intense faulting of the basement and sedimentary cover. The gross structure of these areas of polarity reversal is domal, and the pre-rift sequences are strongly truncated over the crest of the uplift at the Saalian Unconformity. Furthermore, there is a suggestion that considerable eruptive magmatic activity was centred on the domal structures. The doming, erosion, fracturing, magmatism, and initial opening of the grabens all occurred during the Late Carboniferous–Early Permian tectonic phase, but the precise temporal relationship between the magmatism and the fracture propagation is not known.

Several important questions arise from the observations summarized above. For example, does the continuity of shallow structural expression of the fracture zones from Jutland westwards into the North Sea necessarily imply that the crustal province boundaries interpreted from well and potential field data also continue as the deep crustal roots to the structural margins of the RFH offshore? How and when did these crustal boundaries originate, during the Silurian to Early Devonian, or even earlier?

The question of the early origins of the WNW–ESE-trending fracture zones has been addressed recently by Pegrum (1984) in connection with an analysis of the structural evolution of the Tornquist Zone. Both Dikkers (1977) and Pegrum (1984) have postulated that the margins of the RFH are fault controlled but, whereas Dikkers considered that this tectonic lineation was subordinate to the main NE–SW Caledonide trend in the central North Sea, Pegrum has argued for a much greater influence of the WNW–ESE-trending structural elements in the tectonic evolution of the area than hitherto conceived.

The solution of the tectonic problems mentioned above hinges predominantly on the question of the provinciality of the crystalline basement of the Danish Sector of the North Sea. Existing reconstructions of the early evolution of the Central North Sea can be divided into two main categories. The first group considers that the contact between the Mid-European Caledonides and the Precambrian basement of the Baltic Shield is a suture between the microcontinental fragments of Central Europe, which rifted away from a southerly Gondwana and which, after the northerly subduction of the Rheic Ocean, collided with the cratonic shield of Baltica (Ziegler 19892; Cocks and Fortey 1982). The second group regards the contact between these two basement provinces as a transform plate boundary running along the line of the Tornquist Zone through Poland and mainland Denmark, and then either continuing on a north-westerly trajectory to the Orkney Islands (Pegrum 1984) or curving on a WNW to E–W trending arc to link with the major Caledonian fault zones of southern Scotland (Brochwicz-Lewinski *et al.* 1984). It is clear from the critical location of the Danish Sector at the confluence of the 'Mid-European' and Scottish–Norwegian Caledonides that any possible means for discriminating between these contrasting reconstructions must lie in a clearer appreciation of the basement provinciality and structure of the Ringkøbing-Fyn High.

10.4. Radiometric dating of basement provinces

Direct evidence concerning the nature of the crystalline substrate of the RFH is restricted to only three wells. Two of these (Glamsbjerg and Grinsted) are located in onshore Denmark, whilst the third (Per-1) is located on the shoulder of the Tail End Graben near the western limit of the RFH (Fig. 10.2). The onshore wells reached T.D. in medium-grade metamorphic rocks dated by the K/Ar method at 825 ± 15 Ma (Glamsbjerg) and 880 ± 15 Ma (Grinsted). These dates (Ziegler 1982) are broadly comparable to those obtained from samples taken from the Sveco-Norwegian terrane of southern Norway (Watson 1976) and the Precambrian basement of southern Jutland is therefore usually correlated with that of southern Norway and south-western Sweden, and regarded as part of Baltica (Watson 1976; Ziegler 1982; Pegrum 1984; Brochwicz-Lewinski *et al.* 1984).

The basement composition of the offshore sector of the RFH is even less well constrained than the onshore sector. The sole direct sample of the crystalline basement is the Per-1 well, which reached T.D. in a microgranite dated by the K/Ar method at 435 Ma (Ziegler 1982). This well has only recently been released into the public domain, and it is not known whether this 'Caledonian' age refers to a cooling age for the granite or whether it is a metamorphic re-setting of an older basement unit. Ziegler notes that an Ar/Ar dating of this core yielded a 'Scourian' age of 2400 Ma but, in the absence of accompanying detils on the petrography of the sample and the level of alteration, these results should be regarded with caution.

10.5. Regional tectonic setting of basement provinces and Caledonian structure of the central North Sea

Frost and co-workers (1981) compiled the well data in samples of crystalline basement in the North Sea and adjacent areas and, from a combination of published radiometric age determinations and original petrographic analysis, constructed a basement province map of the North Sea (Fig. 10.10). This construction is comparable to Ziegler's (1982) Caledonian framework, and portrays the Caledonian deformation front curving on an arcuate trajectory through the RFH so that the Mid-European 'branch' of the Caledonides link with their Scottish–Norwegian counterparts. If the radiometric determination of a 'Caledonian' age for the Per-1 sample is accepted as genuine, then it suggests that the westernmost block of the RFH was probably entrained in the collisional zone. This further implies that the deformation front must lie somewhere to the east of the well location on the shoulder of the Tail End Graben, and must separate the Precambrian basement onshore, sampled in Grinsted (well 2) and Glamsbjerg (well 3), from the Caledonian basement of the westernmost block. There is no indication in this present seismic mapping of the basement structure of the RFH of any arcuate zone of intense basement deformation such as might be expected as a manifestation of the contact between the allochthonous Caledonide and the autochthonous Dalslandian (or Moravian) crustal domains. Instead, the structural framework of the RFH is dominated by an orthogonal set of major basement faults oriented generally WNW–ESE and NNE–SSW (Fig. 10.2). There is, however, circumstantial evidence to suggest that such a deformation front indeed exists within the RFH west of Jutland but that, far from arcuate in form, it is linear and is oriented NNE–SSW. It is proposed here that the low angle, west-dipping shear zones, loosely termed the Else Duplex, is the structure most likely to represent the postulated 'missing' deformation front. As argued previously, the fault plane reflections are probably indicative of mylonitic shear zones, and their geometry is consistent with an origin or early history as the mid-crustal roots to an imbricated suite of basement thrust nappes. Although highly speculative, given the single calibration point, it seems reasonable to link the imbricated structure of the

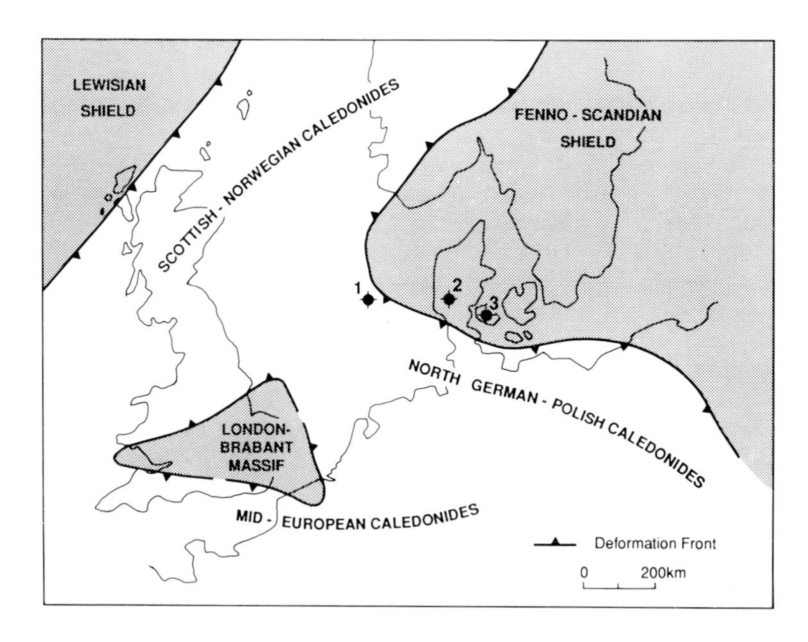

Fig. 10.10. Crystalline basement provinces of the North Sea region, modified from Frost *et al.* (1981). Numbered well locations in the Danish Sector are equivalent to those ilustrated in Figure 10.2. Note the arcuate form of the link between the Norwegian and German-Polish Caledonides passing across the Danish Sector.

Else 'Duplex' with collisional orogenesis of Late Caledonian age. The reactivation of this basement fabric in a brittle extensional mode in the Late Carboniferous–Early Permian is entirely consistent with current thinking on the extensional collapse of orogenic belts, and with the simple observation that major extensional fault propagation is often the hand-maiden of pre-existing basement fabrics. This interpretation of the Else 'Duplex' as the Caledonian Deformation Front is depicted schematically in Fig. 10.11.

The mapped position of the 'Else Duplex' is illustrated in Fig. 10.11. Even allowing for the coarseness of the seismic grid over the area, the west-dipping structures can be correlated with confidence along a NNE–SSW strike for some 80 km. The strike, vergence, and disposition of the 'Else Duplex' immediately brings to mind the possibility that this east-facing structure is somehow related to the similarity oriented and similarly east-vergent nappes of the southern Norwegian Caledonides. The frontal thrust complexes of the Scandinavian Caledonides are remarkably linear on a gross scale over the 2000 km of their NNE strike length, and it does not seem unreasonable that they should continue, albeit with a necessary sinistral offset across the south Norwegian Sea and into the Danish Sector. Unfortunately, across the intervening area of the north Permian Basin, the crystalline basement is buried too deeply to allow any direct correlation of basement structure from conventional seismic data; in addition, the magnetics and gravity data are too low in resolution to be of any assistance. What is clear, however, is that several major basement fracture zones transect the intervening area with predominantly WNW–ESE trends; any or all of these are capable of having acted as lateral ramps, rooted in basement, for the necessary sinistral offset from the Norwegian coast to the Danish Sector (Fig. 10.12a). If such a correlation is not accepted as the most plausible given the limited number of observations currently at our disposal, then the only real alternative is to regard the 'Else Duplex' as an allochthonous fragment of an in-situ portion of the Caledonides transported into the current position (Fig. 10.12b) by major strike-slip activity along the RFZ and Tornquist Zone (Pegrum 1984).

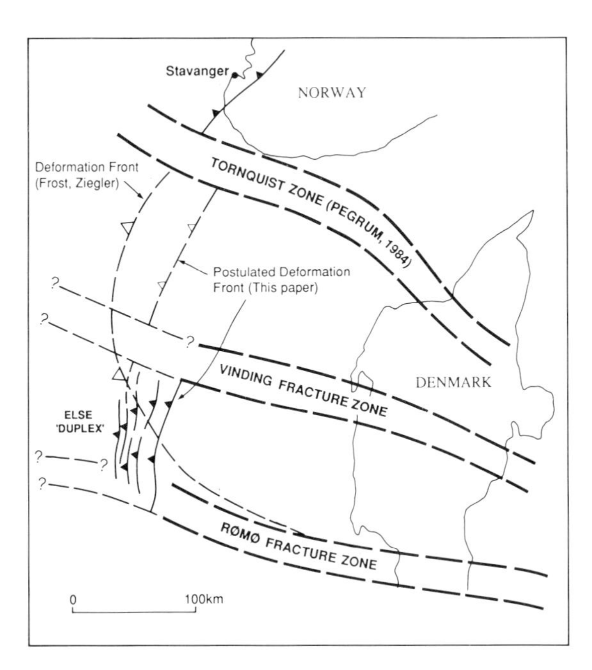

Fig. 10.11. An alternative interpretation of the position of the Caledonian deformation front in the Danish and Norwegian sectors of the North Sea. It is postulated that the Else 'Duplex' represents the extreme southward limit of the Norwegian Caledonides, and that the connection is through a series of minor offsets across transecting crustal-scale fracture zones along an otherwise linear strike alignment.

In either of these interpretations, however, the Rømø Fracture Zone RFZ, (and its sister lineament, the Trans-European Fault) assumes a highly significant role. In that of Fig. 10.12a, it must represent the effective southern margin of the Precambrian crust of the Fenno-scandian Shield Province at least as far west as the Else Graben, with the corollary that the Caledonian Deformation Front exhibits a rhombohedral rather than an arcuate form. Whereas in that of Fig. 10.12b, the RFZ assumes the role of a boundary structure to a strike-slip allochthon. In both interpretations, the RFZ and the VFZ may have been active as transform-type tectonic features during the Caledonian consolidation of the Central North Sea.

The initial rupture pattern that developed in the Late Carboniferous–Early Permian tectonic phase can be understood in the context of a lithospheric structure which was inherited from Late Caledonian collisional orogenesis, and which was dominated by transform splays of the ancestral Tornquist Zone centred on the margins of the RFH and of the northern Permian Basin (Fig. 10.12). Building on the original thesis of Arthaud and Matte (1977) in

(a)

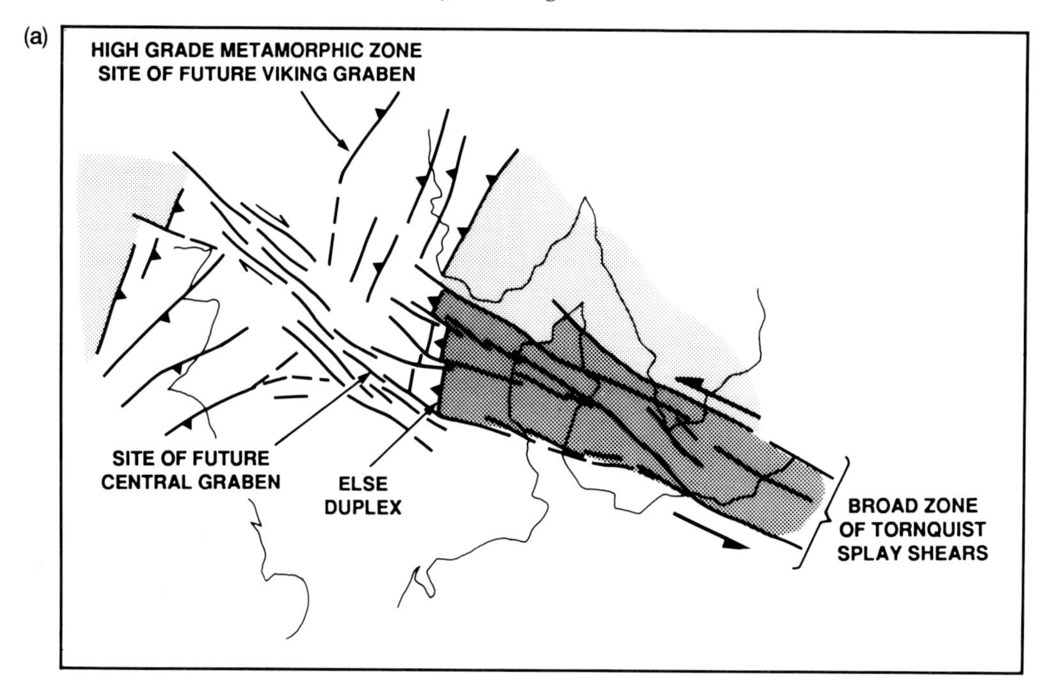

(b)

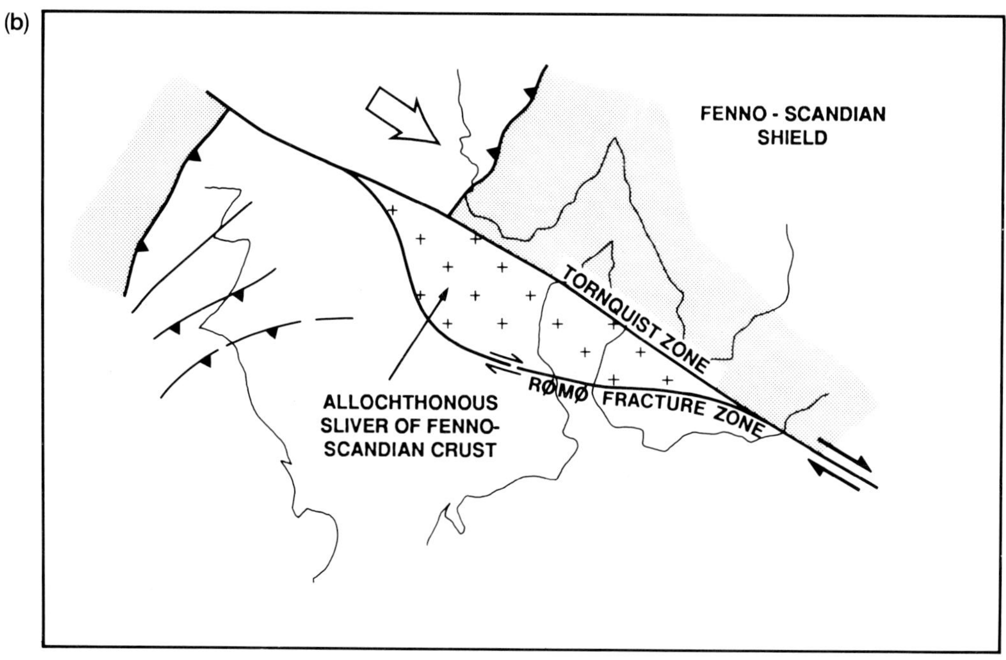

Fig. 10.12. Two interpretations of the Late Caledonian tectonic setting of the Central North Sea. In **a**, the Tornquist Zone is viewed as a broad zone of distributed shear acting as a transform for the final closure of the Iapetus Ocean, whereas in **b**, the Tornquist Zone is placed well to the north of the Ringkøbing-Fyn High, and the High is viewed as a strike-slip allochthon (after Pegrum 1984). In both interpretations the Rømø Fracture Zone is required to have functioned as a major transform fault.

viewing this Late Carboniferous–Early Permian inception of rifting in north-west Europe as being fundamentally controlled by a network of 'Tornquist-parallel' megashears, it is then possible to link the structural development as described for the RFH into this wider context.

The spatial relationship between Late Carboniferous–Early Permian rifts and major continental-scale fractures is illustrated in Fig. 10.13. It is immediately apparent from this illustration that all the rifts in north-west Europe have a predominantly N–S or NNE–SSW orientation and, furthermore, all the rifts or rift segments are linked directly (at either segment boundaries or longitudinal terminations) to major fractures whose orientations are either NW–SE or WNW–ESE (Tornquist-parallel). Indeed, the tectonic fabric is that of a rhombohedral mosaic whose corners and intersection nodes are the sites of graben development. This pattern alone is evocative of some rhombochasm-megashear mechanism to explain the initial propagation of grabens and graben systems in north-west Europe, but when the direct evidence for strike-slip motion in the NW–SE lineaments (Arthaud and Matte 1977; Ziegler 1982) is

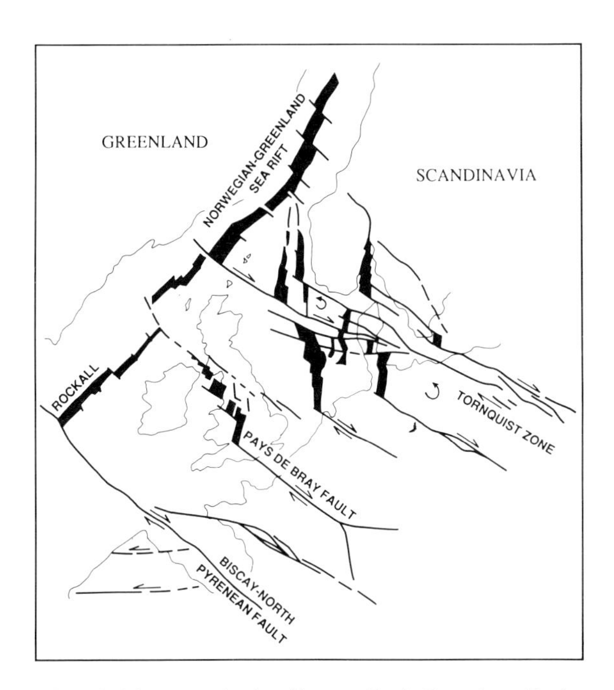

Fig. 10.13. Late Carboniferous-Early Permian rifts in north west Europe, and their relationship to 'Tornquistian' crustal-scale fractures/lineaments.

assembled, with the close spatial affinity between centres of Late Carboniferous–Early Permian magmatism and the 'megashears', the conclusion that such a mechanism was in force is strengthened even further. Indeed, given the wide dispersal and diachronous development of the grabens, their vast distance from any convergent plate boundaries, the lack of evidence for localized mantle plumes, and their uniform orientation, any alternative passive or active rift mechanisms seem highly implausible.

When this pattern of initial rift development in the North Sea is compared with the postulated Late Caledonian framework (Fig. 10.12a), a straightforward relationship emerges that is based solely on the simple principle of reactivation of a pre-existing basement fabric. It has been demonstrated by seismic mapping that the RFZ and VFZ extend close to, or possibly into, the region of the Tail End Graben, and are aligned directly with the points where the boundary fault is re-oriented so as to adopt the more north-westerly trajectory of the Central Graben. In seeking a rationale for this kinked morphology and strike reorientation (and the zig-zag pattern of the North Sea Rift as a whole), then surely we need look no further than this far western part of the Ringkøbing-Fyn High. The continuity in basement grain between the WNW–ESE-dominated fabrics of the Mesozoic Central Graben and the Palaeozoic boundary structures of the RFH is too acute to be entirely the product of coincidence. It is suggested, therefore, that the abrupt change in orientation in the North Sea Rift in the Danish Sector was a direct result of the translation of strike-slip motion along the borders of the RFH along pre-existing fracture zones, which were themselves consolidated as transform zones in the final throes of Caledonian collision.

Acknowledgements

This research was funded by Fina Exploration Ltd. The seismic data was provided by Merlin Geophysical Ltd. I would like to thank Dr H. G. Reading for supervising the research, Dr R. M. Pegrum, Dr W. H. Ziegler, Dr A. Berthelsen and Dr W. Brochwicz-Lewinski for fruitful and thought-provoking discussions, Carol Kilby for preparing the manuscript, and Tony Brown for drafting the figures. This paper was written during the tenure of a post-doctoral fellowship sponsored by Sun Oil

Britain Ltd. Constructive reviews of the manuscript by A. Berthelsen and D. Clark-Lowes are gratefully acknowledged.

References

Arthaud, F. and Matte, P. (1977). Late Palaeozoic strike-slip faulting in southern Europe and northern Africa: Result of a right-lateral shear zone between the Appalachians and the Urals. *Bull. Geol. Soc. America*, **88**, 1305–20.

Brochwicz-Lewinski, W., Pozaryski, W., and Tomczyk, H. (1984). Sinistral strike-slip movements in Central Europe in the Palaeozoic. *Inst. of Geophysics, Polish Academy of Science*, **A-13**, ***(160)***, 3–13.

Cartwright, J. A. (1987). Transverse structural zones in continental rifts—an example from the Danish Sector of the North Sea. In *Petroleum geology of north west Europe* (ed. J. Brooks and K. Glennie), 441–52. London, Graham and Trotman.

Chapin, C. E., Chamberlin, R. H., Osburn, G. R., White, D. W., and Sandford, A. R. (1978). Exploration framework of the Socorro geothermal area, New Mexico. *New Mexico Geol. Soc. Special Publication*, **7**, 114–29.

Cheadle, M. J., McGeary, S., Warner, M. R., and Matthews, D. H. (1987). Extensional structures on the western UK continental shelf: A review of evidence from deep seismic profiling. In *Continental Extension Tectonics* (ed. M. P. Coward, J. F. Dewey, and P. L. Hancock), *Geol. Soc. Special Publication*, **28**, 445–66.

Cocks, L. R. M. and Fortey, R. A. (1982). Faunal evidence for oceanic separation in the Palaeozoic of Britain. *J. Geol. Soc.*, **139**, 465–78.

Dewey, J. F. (1982). Plate tectonics and the evolution of the British Isles. *J. Geol. Soc.*, **139**, 371–412.

Dikkers, A. J. (1977). Sketch of a possible lineament pattern in north-west Europe. In *Fault tectonics in north west Europe* (ed. R. T. C. Frost and A. J. Dikkers). *Geol. Mijnbouw*, **56 (4)**, 275–85.

Dixon, J. E., Fitton, J. G., and Frost, R. T. C. (1981). The tectonic significance of post Carboniferous igneous activity in the North Sea Basin. In *Petroleum geology of the continental shelf of north west Europe* (ed. L. W. Illing and G. D. Hobson), 121–40. London, Institute of Petroleum, Heyden and Son.

Donato, J. A., Martindale, W., and Tully, M. C. (1983). Buried granites within the Mid North Sea High. *J. Geol. Soc.*, **140**, 825–37.

EUGENO-S Working Group (1988). Crustal structure and tectonic evolution of the transition between the Baltic Shield and the North German Caledonides (the Eugeno-S Project). *Tectonophysics*, **150**, 253–348.

Frost, R. T. C., Fitch, F. J., and Miller, J. A. (1981). The age and nature of the crystalline basement of the North Sea Basin. In *Petroleum geology of the continental shelf of north west Europe* (ed. L. W. Illing and G. D. Hobson), 43–57. London, Institute of Petroleum, Heyden and Son.

Latin, D. M., Dixon, J. E., and Fitton, J. G. (this volume). Rift-related magmatism in the North Sea basin.

Olsen, J. C. (1983). Structural outline of the Horn Graben. *Geol. Mijnbouw*, **62**, 47–50.

Peddy, C. P. (1984). Displacement of the Moho by the Outer Isles Thrust. *Nature*, **312**, 628–30.

Pegrum, R. M. (1984). The extension of the Tornquist Zone in the Norwegian of North Sea. *Norsk Geologisk Tidsskrift*, **64**, 39–68.

Ramberg, I. B. and Spjeldnaes, N. (1978). The tectonic history of the Oslo Region. In *Tectonics and Geophysics of continental rifts* (ed. I. B. Ramberg and E. R. Neumann), 167–94. Dordrecht, Holland, Reidel.

Rosendahl, B. R., Reynolds, D. J., Lorber, P. M., Burgess, C. F., McGill, J., Scott, D., Lambiase, J. J., and Derksen, S. J. (1986). Structural expressions of rifting: Lessons from Lake Tanganyika, Africa. In *Sedimentation in the African rifts* (ed. L. E. Frostick, R. W. Renaut, T. Reid, and J. J. Tiercelin). *Geol. Soc., Special Publication*, **25**, 29–43.

Stein, A. M. (1988). Basement controls upon basin development in the Caledonian foreland, NW Scotland. *Basin Research*, **1**,107–19.

Watson, J. (1976). Eo-Europa, the evolution of a craton. In *Europe from crust to core* (ed. D. V. Ager and M. Brooks), 59–78. London, Wiley.

Ziegler, P. A. (1982). *Geological atlas of western and central Europe*. Amsterdam, Elsevier.

11 Does the uniform stretching model work in the North Sea?

N. White

Abstract

Since the uniform stretching model for extensional sedimentary basins was first proposed by McKenzie, it has been argued that there is a significant discrepancy between the amount of extension measured in the brittle upper crust and that determined from subsidence analyses and measurements of crustal thickness. More recently there has been considerable debate as to whether the continental lithosphere extends by lithospheric simple shear or by bulk pure shear (stretching). Here, an analysis of deep and shallow seismic reflection data from part of the northern North Sea is used to address both of these issues. First, the geometry of subsidence in the North Sea is used to demonstrate that the lithospheric simple shear model is not applicable to the evolution of the North Sea. Second, within the East Shetland Terrace or Basin of the northern North Sea, it is argued that subsidence-derived stretching and extension measured from normal faulting are in good agreement and that, within error, there is no evidence for any extension discrepancy. The most significant stretching phase began at the end of the Middle Jurassic, crustal stretching factors suggesting that the poorly constrained Triassic stretching episode was considerably less important. The importance of error propagation in analyses such as the one presented here is strongly emphasized. The uniform stretching model is also shown to be of some use in predicting the magnitude and duration of footwall uplift and stratigraphic onlap. Although the conclusions presented here only strictly apply to the northern North Sea, both data quality and stratigraphic control within the East Shetland Terrace are far better than anywhere else in the North Sea. It is suggested that more complex models for the evolution of the North Sea basin cannot, in general, be justified as long as the simplest possible model adequately accounts for all major observations in areas like the East Shetland Terrace.

11.1. Introduction

It is now generally accepted that the uniform stretching model (McKenzie 1978; Jarvis and McKenzie 1980) can explain, at least in outline, the main features of many continental sedimentary basins and passive margins (Sclater *et al.* 1980; Le Pichon and Sibuet 1981; Sawyer *et al.* 1982; Royden *et al.* 1983; Ye *et al.* 1985). In the North Sea, where the model was first successfully tested, results suggest that the amount of stretching measured by crustal thinning is in good agreement with that calculated by backstripping subsidence data (Sclater and Christie 1980; Barton and Wood 1984). In both studies, Upper Jurassic stretching was shown to be more significant than the poorly constrained Triassic stretching event. However, Ziegler (1983) has pointed out that the amount of stretching measured from displacements across normal faults on a number of North Sea profiles is about a factor of two smaller than that calculated from either subsidence or crustal thinning observations. Extension discrepancies have also been reported for other extensional sedimentary basins and passive margins (Chenet *et al.* 1982; Moretti and Pinet 1987; Karner *et al.* 1987).

Attempts have been made to account for such discrepancies by a variety of arguments. Wood and

Barton (1983) argued that extension measured in the brittle upper crust may be considerably underestimated because major high-angle faults interpreted on seismic reflection data have cut through and rotated earlier faults, masking much of the extension (Proffett 1977; Jackson 1987). Substantial internal deformation within fault-bounded blocks could also account for the discrepancy (White *et al.* 1986). Alternatively, the stretching model in its simplest form (i.e. uniform pure shear and one phase of extension) could be incorrect, more complex stretching models with multiple phases of extension and/or heterogeneous stretching being required instead (Sclater *et al.* 1980; Hellinger and Sclater 1983; Rowley and Sahagian 1986; Gibbs 1987). In some cases, stretching (pure shear) models have been rejected in favour of alternative basin models, such as the lithospheric simple shear model (Wernicke 1985), which has also been applied to many different sedimentary basins and passive margins including the North Sea (Gibbs 1984; Wernicke 1985; Etheridge *et al.* 1985; Lister *et al.* 1986; Ussami *et al.* 1986; Bosworth 1987; Le Pichon and Barbier 1987; Beach *et al.* 1987; Gibbs 1987).

Application of the lithospheric simple shear model to extensional sedimentary basins has focused attention on the nature of lithospheric extension: does the lithosphere accommodate extension by displacement along a low-angle normal fault which penetrates the entire lithosphere, or does the lithosphere stretch by bulk pure shear? In the North Sea, regional deep seismic reflection profiles have been interpreted according to both of these models (compare Klemperer 1988, and Gibbs, 1987). It has been argued that each model makes very different predictions regarding fault geometries, subsidence, crustal thinning and, ultimately, source rock maturity so it is important to be able to decide which model is most appropriate.

In this review, this problem of the nature of extension is tackled first, the discussion being based largely on work already published or in press (White 1989; Klemperer and White, 1989). Then the older and more important issue of the extension discrepancy is addressed using data from the northern North Sea. It is concluded, first, that the lithospheric simple shear model in its simplest form fails to account for the distribution of subsidence in the North Sea. This model also fails to predict either the existence or the location of asthenospheric melts in the North Sea (Latin and White, 1990). Secondly, it appears that all three estimates of stretching—crustal thinning, subsidence analysis, and summed displacements across normal faults—agree within error. This result implies that the uniform stretching model adequately accounts for the major observations in the North Sea. The latest stretching event, which started in the Upper Jurassic and lasted 60 Ma, is shown to have been more important than the preceding Triassic stretching episode.

Detailed stretching calculations have only been carried out on a small portion of seismic reflection data from the North Sea Basin where both structure and stratigraphy are unambiguously clear (Figs 11.1 and 11.2). Hence any conclusions made do not necessarily hold elsewhere in the North Sea. This drawback, however, is not at issue—the burden of proof lies with those who want a more complex general basin model to demonstrate that the uniform stretching model fails when applied to the best possible data. The quality of data and stratigraphic control in the East Shetland Basin area is extremely difficult to match anywhere else in the North Sea.

A number of other important observations can be explained using a slightly modified stretching model. For example, post-rift stratigraphic onlap at basin margins (i.e. the 'Steer's Head' geometry) can be accounted for by distributing the stretching within the lithospheric mantle over a slightly broader zone than stretching within the crust (White and McKenzie 1988). A small difference in the vertical distribution of stretching does not affect the previous conclusion made concerning the extension discrepancy. On the scale of individual fault-bounded blocks, observations such as local syn-rift stratigraphic onlap and footwall uplift seem consistent with the stretching model in its simplest form.

11.2. Nature of extension: pure or simple shear?

The recent controversy concerning the nature of lithospheric extension is illustrated in Fig. 11.3 which shows two different interpretations of the same deep seismic reflection line (NSDP-84-1). According to Klemperer (1988), the depth-migrated data are consistent with the uniform stretching model, because the crust appears to have thinned symmetrically, and both syn-rift and

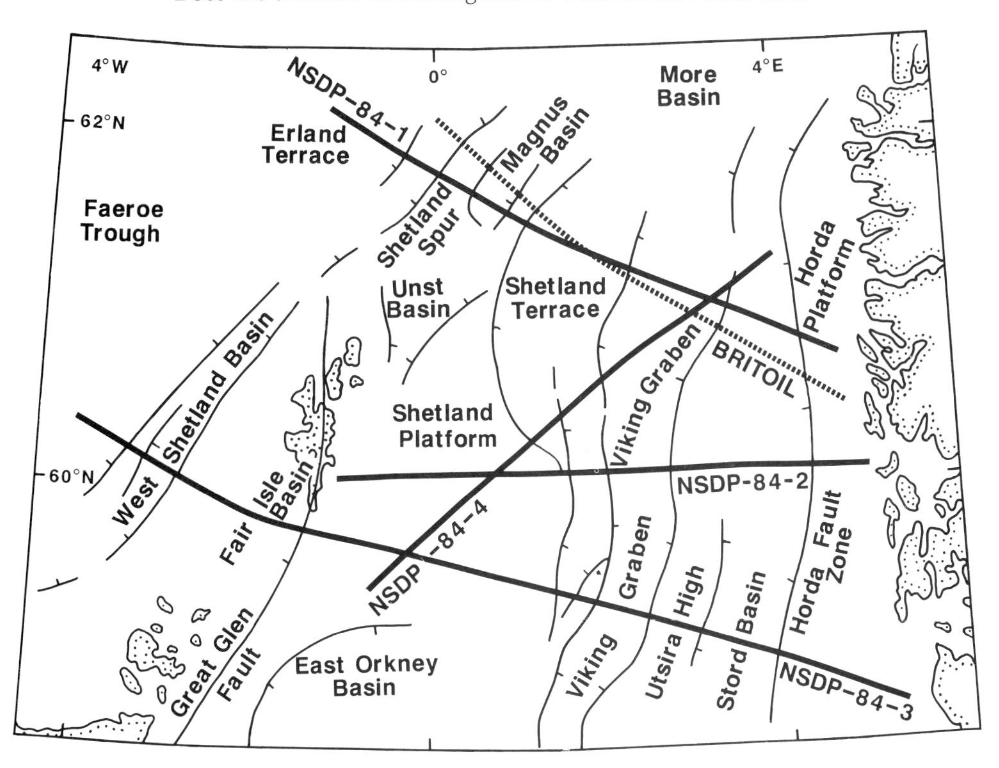

Fig. 11.1. Generalized structural map of northern North Sea showing location of NSDP-84 lines shot by Geophysical Company of Norway (GECO; solid lines) and Britoil proprietary line (dotted line). The latest episode of rifting began at the end of Middle Jurassic, lasting until Lower Cretaceous when unfaulted, post-rift sedimentation commenced.

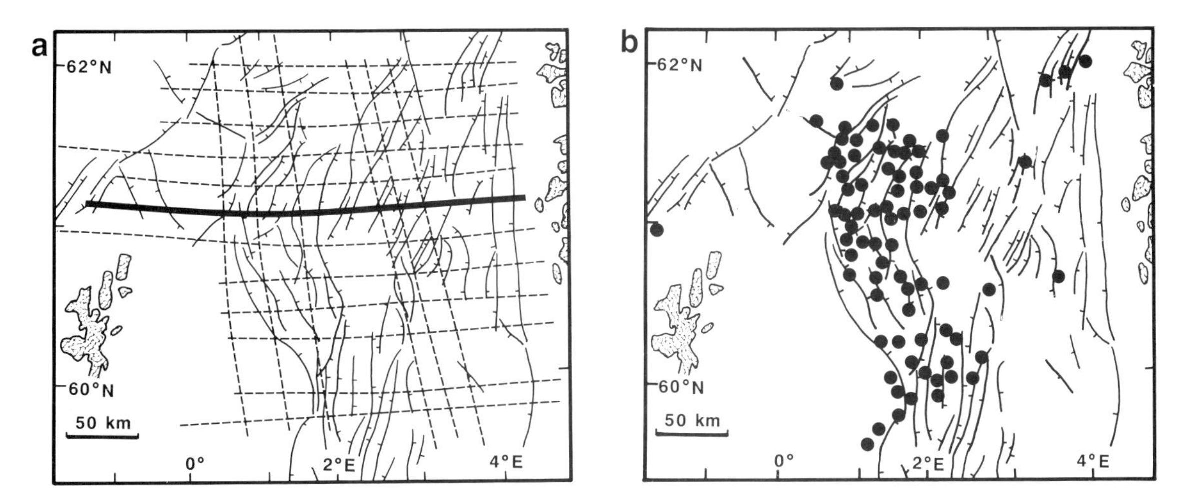

Fig. 11.2. a Location of NNST-84 seismic reflection survey, generously provided by Norwegian Petroleum Exploration Consultants (NOPEC a.s.). Solid line is NNST-84-7 (*see* Fig. 11.8.c). **b** Solid circles show location of well data used to calibrate seismic reflection data.

post-rift sediments are directly superimposed on thinned crust. Gibbs (1987), on the other hand, interprets a low-angle detachment fault on the unmigrated data which penetrates through the entire crust and into the lithospheric mantle. The geometry of the detachment fault is determined primarily from the geometry of fault-bounded blocks within the brittle upper crust (Gibbs 1983, 1984). Note that a continuous dipping relection linking upper crustal faulting with a step in the Moho is not observed on the depth-migrated profile (Fig. 11.3a). In general, the North Sea deep profiles do not directly image a low-angle detachment fault penetrating through the lower crust. Nevertheless, Beach (1986) and Gibbs argue that zones of differing reflectivity in the lower crust delineate, in places, a detachment zone into which upper crustal faults sole.

The difficulty with interpreting selective portions of lower crustal reflectivity as detachment zones is that we do not as yet fully understand the causes of such reflectivity (Cheadle *et al.* 1987; Warner, in press). If, as is strongly suspected, lower crustal reflectivity is due to intrusions of basaltic material of variable age, detailed structural interpretations of lower crustal reflectivity are difficult to justify.

Here, an alternative approach is taken. Numerical modelling of lithospheric simple shear is used to isolate well-constrained observations that can help to distinguish the two models. Then the distribution of subsidence in the North Sea is used to show that application of the lithospheric simple shear model is inappropriate.

11.2.1. *Predictions*

The lithospheric simple shear model was originally proposed because normal faulting in extensional sedimentary basins is often asymmetric over considerable distances. It is worth pointing out that asymmetry of normal faulting does not rule out the uniform stretching model. Figure 11.4 shows how asymmetric faulting can be interpreted according to a detachment-style model where the footwall to the detachment remains rigid. An alternative interpretation is also shown, where bulk pure shear is accommodated in the brittle upper crust by a combination of simple shear and fault block rotation. In this case, faulting gives way at depth to penetrative pure shear. The depth at which this transition takes place is likely to be thermally controlled, and thus bears no direct relationship to the volume of the sedimentary basins formed at the surface (Jackson *et al.* 1988).

Elsewhere, it has been argued that neither symmetry or crustal thinning nor spatial superposition of crustal thinning and subsidence can be used to discriminate between the lithospheric simple shear model and the uniform stretching model. Only the distribution of initial and thermal subsidence in a basin can be used to distinguish the two (White

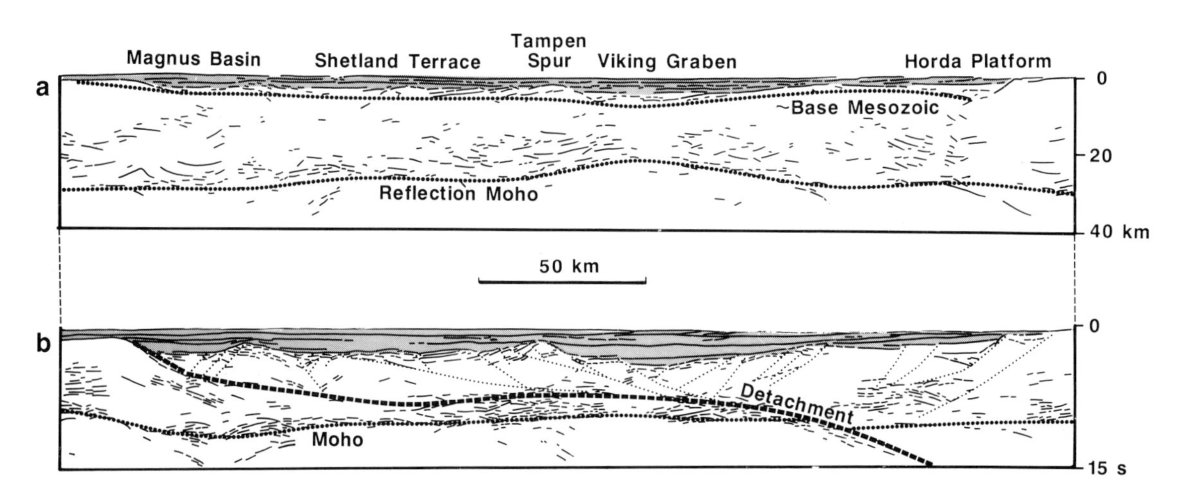

Fig. 11.3. Comparison of two different interpretations of NSDP-84-1. Shading indicates post-rift sediments (i.e. Upper Cretaceous and Tertiary). **a** Depth-migrated line drawing showing crustal thinning (after Klemperer 1988). **b** Lithospheric simple shear interpretation, showing position of inferred detachment fault (after Gibbs 1987).

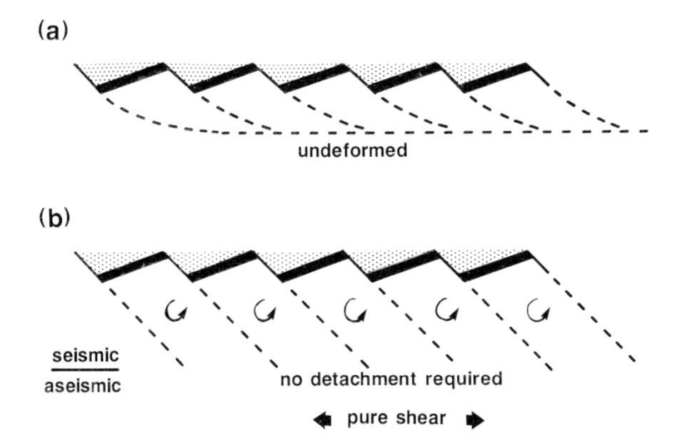

Fig. 11.4. Cartoons illustrating how one seismic profile could lead to two different interpretations. Thin solid lines = observed fault plane reflections; dashed lines = interpreted faults; thick solid lines = pre-rift sedimentary rocks; stipple = syn-rift sediment fill ('excess area' created by extension). **a** Observed rotation of fault-bounded blocks, explained by pervasive internal deformation (simple shear). Section balancing predicts shallow detachment fault, depth of which is related to amount of 'excess area'. Footwall beneath detachment assumed to stay undeformed. **b** Observed rotation explained by rigid-body rotation between inferred planar faults. No detachment required since lower boundary is thermal. 'Excess area' not related to thickness of deforming zone.

1989; Klemperer and White, 1989). The modelling on which these conclusions are based is similar to that carried out by Kusznir and colleagues (1987), Voorhoeve and Houseman (1988), Mudford (1988), Buck and colleagues (1988), and White (1988).

The lithosphere is assumed to deform by instantaneous displacement along a planar low-angle normal fault, and heat flow is assumed to be one-dimensional. Stretching is assumed to have taken place 300 Ma ago, β reaching a maximum of 2.0 in the centre of the basin. Altering these simplifying assumptions does not change the conclusions made below (i.e. changing the dip of the detachment or its shape at depth makes little difference; White 1989).

The lithospheric simple shear model predicts that crustal thinning will be symmetric about the basin, and that both initial and thermal subsidence are spatially coincident with the locus of crustal thinning (Fig. 11.5). Clearly, neither of these observations will be useful in discriminating between the two models since the uniform stretching model makes the same predictions (Fig. 11.5c). The minor amount of syn-rift uplift on the upper or distal plate (Wernicke 1985) is indistinguishable from footwall uplift associated with normal faulting (Barr 1987*a*, *b*). However, an important prediction of the lithospheric simple shear model is that the distribution of subsidence across the basin is asymmetric. Subsidence is negligible where the detachment fault comes to the pre-rift surface and increases very rapidly across the basin, reaching a maximum above the position where the detachment fault is coincident with the Moho (Fig. 11.5*a*).

Stretching models predict a symmetric distribution of subsidence (Fig. 11.5b, 11.5c). The ratio of initial to thermal subsidence is about 3:1 according to the lithospheric simple shear model (exactly the reverse of that predicted by uniform stretching). As White (1989) argued, this asymmetric pattern of subsidence for the lithospheric simple shear model is enhanced if a less idealized numerical model is used (i.e. if the two plates are not cleanly pulled away from each other, and fault-bounded blocks of upper plate material are left along the lower plate: Wernicke 1985). If stretching is distributed non-coaxially through the crust and lithospheric mantle (Fig. 11.5b), the ratio of initial to thermal subsidence is the same as for the lithospheric simple shear model (3:1).

11.2.2. *Observations*

Although crustal thinning is approximately symmetric in the North Sea, and although maximum thinning and subsidence are coincident, the predictions of the previous section imply that neither of these observations can discriminate between the two different basin models. However, the observed subsidence geometry in the North Sea can be compared with the results of numerical modelling. Depth to top Upper Cretaceous throughout the North Sea is taken to be a representative measure of the thermal subsidence which has occurred (Fig. 11.6; Barton and Wood 1984).

In the northern North Sea, Gibbs (1987) and Beach *et al.* (1986) envisage a single detachment fault dipping to the east. Although in some places the amount of subsidence increases slightly across the basin from the position where the detachment fault comes to the pre-rift surface to where the detachment fault soles along the Moho, this increase is

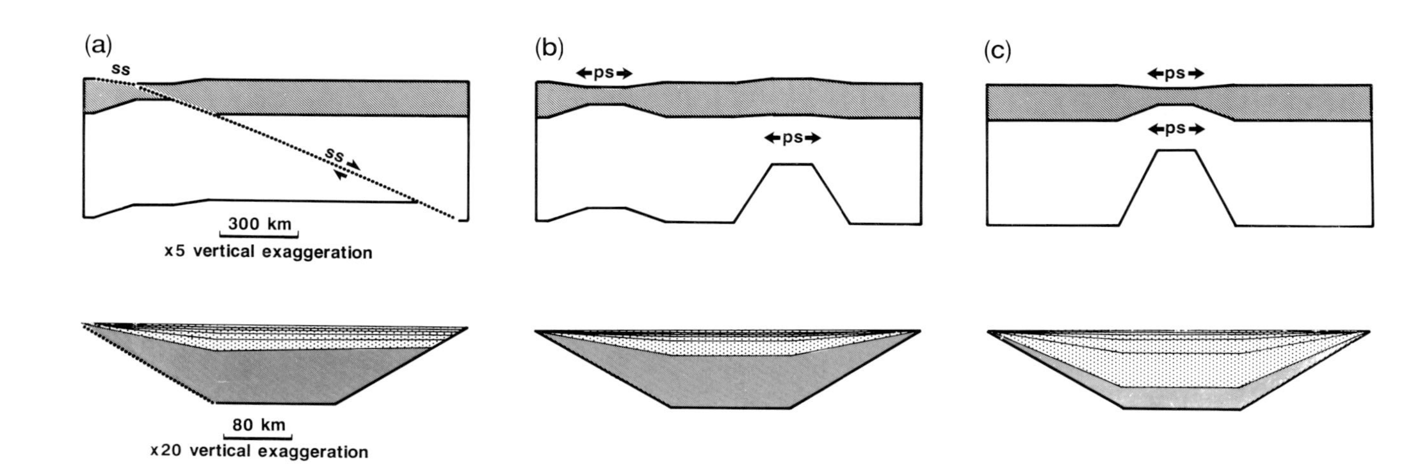

Fig. 11.5. Results of numerical modelling for **a** lithospheric simple shear, **b** non co-axial stretching, and **c** co-axial stretching. For each model are shown, (top) lithospheric configuration immediately after instantaneous stretching and (bottom, enlarged) pattern of initial subsidence (shaded) and thermal subsidence (stippled). Within thermal subsidence basin, isochrons are drawn every 60 Ma. Note much larger ratio of thermal to initial subsidence for co-axial stretching. ps = pure shear; ss = simple shear.

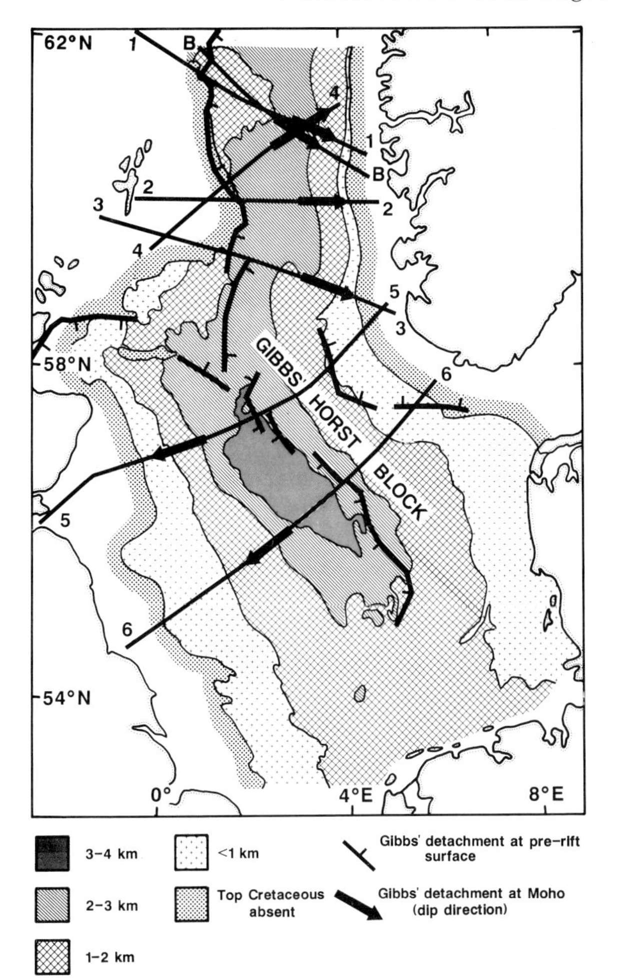

Fig. 11.6. Contour map of depth to base Cainozoic in North Sea from Day *et al.* (1981). Superimposed are subcrops of intersection of Gibbs' (1987) detachments at the pre-rift surface and at the Moho. Note interpretation of Norwegian salt platform as a horst block. Numbers indicate NSDP survey lines; BB indicates Britoil proprietary line.

certainly not by an order of magnitude. Close to NSDP-84-2 and NSDP-84-3, subsidence actually decreases across the basin, in direct contradiction to the predictions of the lithospheric simple shear model. A pair of detachment faults dipping in opposite directions is envisaged in the central and southern North Sea. Nevertheless, it is clear that subsidence still decreases away from the pre-rift expression of both detachment faults towards those positions where the detachment faults are soling along the Moho. Again, this observation is in direct contradiction with the modelling results. The horst block between two oppositely dipping detachment faults should not, according to the lithospheric simple shear model, have suffered any subsidence, but obviously it has.

Not only does the distribution of subsidence in the North Sea fail to agree with the predictions of the lithospheric simple shear model, but the ratio of initial to thermal subsidence (roughly estimated at 1:1; Barton and Wood 1984) is at variance with a predicted ratio according to the lithospheric simple shear model of 3:1. This predicted ratio was calculated for instantaneous stretching, so if finite duration stretching is allowed for (*c*.60 Ma in the North Sea) the ratio for lithospheric simple shear would be expected to have been even greater.

In summary, the lithospheric simple shear model in its simplest form fails to predict important and well-constrained observations of subsidence geometry and distribution in the North Sea. Latin and White (1990) argue that this model also fails to predict the existence or location of asthenospheric melt in the North Sea. If a lithospheric simple shear model is applicable, the 'detachment footwall' cannot have remained rigid during extension. Deformation must have occurred so that a sufficiently large thermal anomaly was introduced into those parts of the basin closest to the pre-rift expression of the detachment fault. More complex lithospheric simple shear models (Coward 1986) are, in terms of heat-flow and subsidence, indistinguishable from stretching models. Thus they cannot be used as predictive tools in young extensional sedimentary basins. In the following sections, it is argued that the structure and evolution of the North Sea can be most easily explained by the uniform stretching model.

11.3. Magnitude of extension

In the North Sea, the uniform stretching model was used first used by Sclater and Christie (1980) and Barton and Wood (1984) to show that subsidence and crustal thinning could be accounted for by a major lithospheric stretching event at the end of the Jurassic. A much less significant Triassic stretching event was also recognized. Subsequently, Ziegler (1983) pointed out that, while the amount of stretching estimated from crustal thinning was in good agreement with that determined by backstripping subsidence data, both of these estimates were a factor of two or more greater than stretching measured by summing the displacements across normal faults in the brittle upper crust. Subsequent studies in the northern North Sea and in the central North Sea confirm the existence of such a discrepancy (Badley *et al*. 1988) and try to account for it by invoking large amounts of Triassic or pre-Triassic stretching (Giltner 1987; Hellinger *et al*. 1988; Shorey and Sclater, 1988).

The extension discrepancy is often accounted for by appealing to two-layer stretching where, for example, the lower crust and lithospheric mantle are stretched by a considerably greater amount than is the brittle upper crust (Hellinger and Sclater 1983; Moretti and Pinet 1987; Karner *et al*. 1987). In some instances, *ad hoc* heating events have been proposed (Ziegler and van Hoorn, 1989). The existence of this discrepancy depends primarily on how accurately we can estimate the amount of stretching that has occurred in the brittle upper crust (β_f). Stretching factors determined from crustal thinning and subsidence analyses can be reasonably well constrained (Sclater and Christie 1980; Barton and Wood 1984). However, estimating the amount of stretching across a complicated system of normal faults is extremely difficult (see disagreement between Chenet *et al*. 1982, and Le Pichon and Sibuet 1981). Almost all of the simplifying assumptions made when measuring β_f, such as the apparent horizontal displacement (i.e. the heave) equalling the extension, or fault-bounded blocks remaining rigid, tend to underestimate extension considerably (Jackson and White 1989). In areas of very complex normal faulting, simple 'rules' for estimating stretching are likely to be even less applicable.

The purpose of this section is to re-examine critically the discrepancy issue in order to see how well it is constrained by the data. The obvious approach would be to examine a large amount of data from many different extensional sedimentary basins. This method is unsatisfactory for several reasons. Firstly, seismic reflection data quality and stratigraphic control are both, inevitably, very variable. As a result, it may often be practically impossible to make any estimates of stretching from normal faulting. A good illustration of this problem is the Plymouth Bay Basin in the English Channel. This basin is clearly extensional, since it is surrounded by extensional sedimentary basins of the same age and it is underlain by thinned crust (Cheadle *et al*. 1987). However, it appears to have no measurable extension whatsoever in its brittle upper crust. In fact, in this case it is likely that normal faulting at the bottom of the basin is obscured by basaltic flows.

Within the North Sea itself there are many areas where reasonable estimates of β_f simply cannot be made due to poor stratigraphic control and inadequate resolution—the Viking and Central Grabens *sensu stricto* fall into this category. In other areas (e.g. the southern North Sea), basin evolution is strongly affected by basin inversion and complex salt tectonics (Glennie 1986).

Here, the discrepancy issue is tackled in detail by examining a single area where data quality and stratigraphic control is excellent, where the geological history is relatively simple, and where major normal faulting can be interpreted very easily. The interpreted seismic reflection data should be so closely constrained that it can be regarded as 'data' itself.

Subsidence and normal faulting have been studied in the northern North Sea using 5000 km of regional seismic reflection data (Northern North Sea Tie, NNST-84) generously provided by Norwegian Petroleum Exploration Consultants (NOPEC) and over 150 well-logs provided by different oil companies (Fig. 11.1). In the absence of a modern refraction profile (Barton and Wood 1984), the five deep seismic reflection profiles (North Sea Deep Profiles, NSDP-84), acquired by GECO as a group shoot with British Institutions' Reflection Profiling Syndicate (BIRPS) participation, provide some constraint on crustal thickness in the northern North Sea (Klemperer 1988; Holliger and Klemperer 1989).

Previous work on basin evolution in the northern North Sea has been hampered by the lack of either detailed subsidence analyses (Beach *et al*. 1987; Badley *et al*. 1988) or calibrated seismic reflection data (Giltner 1987). The structure and evolution of the northern North Sea, especially around 61°N, is very well documented and normal faulting easily interpreted (Ziegler 1981, 1982; Glennie 1986). Complications which occur elsewhere in the North Sea, such as salt tectonics, Middle Jurassic regional uplift, volcanism, and basin inversion, do not arise.

Only the latest, Upper Jurassic, stretching phase can be modelled with any degree of certainty, although it is likely that there has been a Triassic stretching phase as well. This earlier event is not incorporated directly into the stretching calculations for two reasons. First, since stratigraphic control is poor we cannot actually measure the amount of stretching accommodated by Triassic normal faulting, or by Triassic subsidence. Secondly, any thermal anomaly associated with early Triassic rifting will have almost completely decayed away by Upper Jurassic times (Barton and Wood, 1984; Badley *et al*. 1988). Attempts to incorporate the effects of this earlier rifting episode are hampered by the lack of deeply-penetrating well data. However, the likely magnitude of Triassic rifting is estimated below, and found to be considerably smaller than the subsequent Upper Jurassic episode.

Upper Jurassic stretching is assumed to have started at 160 Ma and lasted 60 Ma with a constant strain rate (Jarvis and McKenzie 1980). The conclusions of this paper are not affected if shorter or longer rifting periods are used. Increasing or decreasing strain rates during the stretching period also do not affect the main results (Fig. 11.10). Standard values of lithospheric and crustal thicknesses and densities have been assumed (Cochran 1980; Barton and Wood 1984).

Detailed stretching calculations were carried out on the western portion of line NNST-84-7, which runs east-west just north of 61°N (i.e. across the East Shetland Terrace or Basin, Figs 11.1, 11.7, 11.8). Here, post-rift sediment thickness is relatively uniform and the tilted fault block structure is simple. Stratigraphic control is excellent since at least one well penetrates each fault-bounded block (Fig. 11.8b). Further east, in the Viking Graben itself, thicknesses of Jurassic sediments are not accurately known and the amount of extension taken up by normal faulting is extremely difficult to determine. Post-rift subsidence along the Norwegian margin has been considerably modified by the Tertiary uplift of Norway (Glennie, 1986; Badley *et al*. 1988; Biddle and Rudolph, 1988), thus yielding meaningless stretching estimates. South of 61°N, the basin narrows considerably and it is difficult to measure fault displacements, while to the north, the structure is more complicated and there is less well-control (Fig. 11.2). Consequently, the western half of line NNST-84-7 probably occurs in the best possible area of the North Sea for carrying out stretching calculations. The magnitude of subsidence in the East Shetland Terrace is also typical of that seen elsewhere in the North Sea (*c*.4 km of post-rift sediments). The only disadvantage is the lack of crustal thinning information. Fortunately, there are two deep seismic reflection profiles on either side of NNST-84-7 (NSDP-84-1 and NSDP-84-4; Fig. 11.7). Stretching factors determined from crustal thinning on NSDP-84-1 by Holliger and Klemperer (1989) have been projected onto NNST-84-7. The deep seismic reflection data were depth-converted by Holliger and Klemperer using velocities derived from well-log information.

11.3.1. *Subsidence analyses*

To calculate stretching factors from the subsidence data, observed and theoretical water-loaded subsidence curves are matched at successive points across the basin yielding β_s, the subsidence-derived

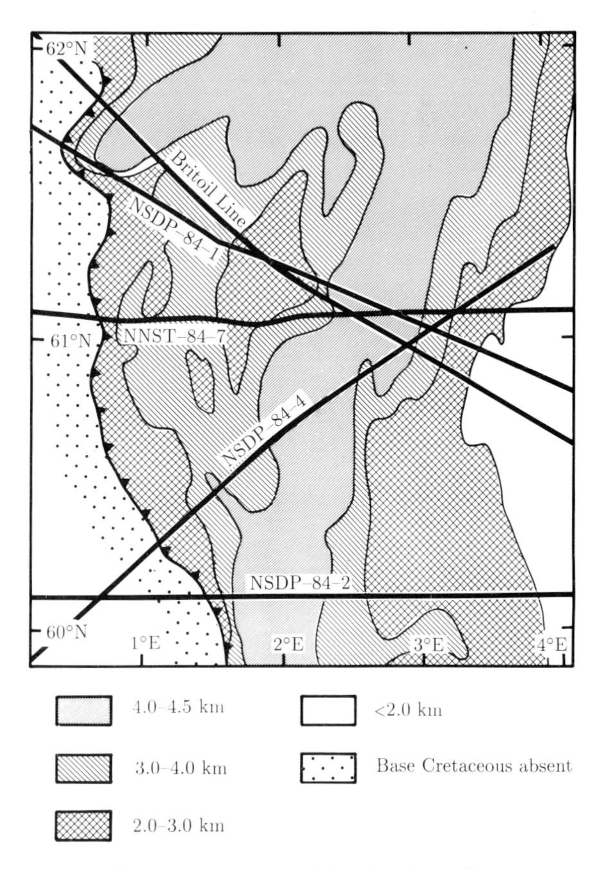

Fig. 11.7. Contour maps of depth to base Cretaceous in northern North Sea (see Figs 11.1 and 11.2.a for location of seismic reflection profiles). Subsidence data from Day *et al*. (1981).

stretching factor (Jarvis and McKenzie 1980; Barton and Wood 1984). The subsidence history at any point within a sedimentary basin is determined by removing the effects of sediment loading and fluctuating water depths at successive increments of time, using the backstripping method (Steckler and Watts 1978; Sclater and Christie 1980). The general effect of including sea-level variations (Watts and Steckler 1979; Haq *et al.* 1987) is to introduce apparent fluctuations in the subsidence rate, although the implied stretching factor will not alter (Barton and Wood 1984, fig. 22).

As in previous studies, Lower and Middle Jurassic, Triassic and older deposits were assumed to have been completely compacted prior to Upper Jurassic stretching. It is important to note that this assumption does not invalidate the backstripping process, which simply depends on what is removed from above that level designated as basement. However, ignoring compaction of Lower Jurassic, Triassic, and older sediments will have resulted in slightly more post-Jurassic subsidence than would have occurred otherwise (i.e. β_s is greater than it should be).

Lithologies were subdivided into three categories, sand, shale, and shaly sand. Sediment porosities and porosity decay lengths were taken from Sclater and Christie (1980) and Barton and Wood (1984). Palaeobathymetry was based on Wood's (1982) study of benthic foraminifera from the central North Sea, although it was assumed that water depths in the northern North Sea were generally greater (Wood, pers. comm.). Sediment loading was assumed to have occurred by local compensation since admittance studies in the North Sea have shown that flexural rigidity (i.e. elastic thickness) was negligible both during stretching and throughout post-rift cooling (Barton and Wood, 1984). As White and McKenzie (1988) pointed out, increasing flexural rigidity during post-rift cooling is not necessarily required to explain stratigraphic onlap at the basin margins (Watts *et al.* 1982).

Ideally, subsidence analyses should be carried out on detailed well data. However, wells are generally drilled on structural highs, which may have been elevated during stretching (Barr 1987*a*, *b*), so the amount of stretching can be considerably underestimated (Sawyer, 1986). Subsidence curves have therefore been determined at the centres of each block (i.e. at the average level of the topography; Figs 11.8a, 11.8b). Although there are fewer points on each curve compared with well-derived subsidence curves, the calculated stretching factors are more representative.

Maximum and minimum theoretical subsidence curves have been fitted to the observed water-loaded curves (Fig. 11.8a). The final point on each of the observed curves often departs significantly from the theoretical curve, possibly due to a rapid increase in sea-level and sediment deposition following the Pleistocene glaciation (Barton and Wood 1984). The poor fit at the end of stretching can be improved by changing the strain rate or the thickness of the lithosphere without affecting the chosen β_s values (Figs 11.9, 11.10). The total subsidence-derived extension across the East Shetland Terrace is *c.* 15 ± 2 km. Palaeowater depths were chosen to maximize extension estimates, so that measurements of β_s are probably slightly overestimated. Within the Viking Graben itself, where normal faulting cannot be adequately constrained, subsidence analysis gives $\beta_s \simeq 1.5$ for Upper Jurassic rift phase.

11.3.2. *Normal faulting*

The six major normal faults within the East Shetland Terrace currently dip at *c.*30°, the fault-bounded blocks having rotated through *c.*10° about a horizontal axis (Ziegler 1982; Glennie 1986; Badley *et al.* 1988). Uncertainties in fault dip and block rotation are at least ±5° and ±2° respectively. If these faults are assumed to be planar at depth, and if internal deformation for each block is ignored, β_f (the fault-derived stretching factor) can be calculated by dividing the present-day block width by the original block width (Morton and Black 1975; Le Pichon and Sibuet 1981; Jackson 1987). Thus the total amount of extension accommodated by all six blocks is *c.* 14 ± 3 km (Table 11.1). If block rotation is ignored and extension estimated by simply summing horizontal displacement or heave, extension will be overestimated by $(1 - \cos\omega)l$, where ω is the block rotation and l the original block width (Sclater and Célérier 1989). Within most of the blocks, there is evidence for minor antithetic faulting and downwarping of beds close to the main faults, suggesting that both internal deformation and differential compaction were important (Glennie 1986; Badley *et al.* 1988). Provided block rotation is allowed for, fault planes calculated from the geometry of beds within hangingwalls using area balance techniques (Verrall 1982; Gibbs 1983) are still approximately

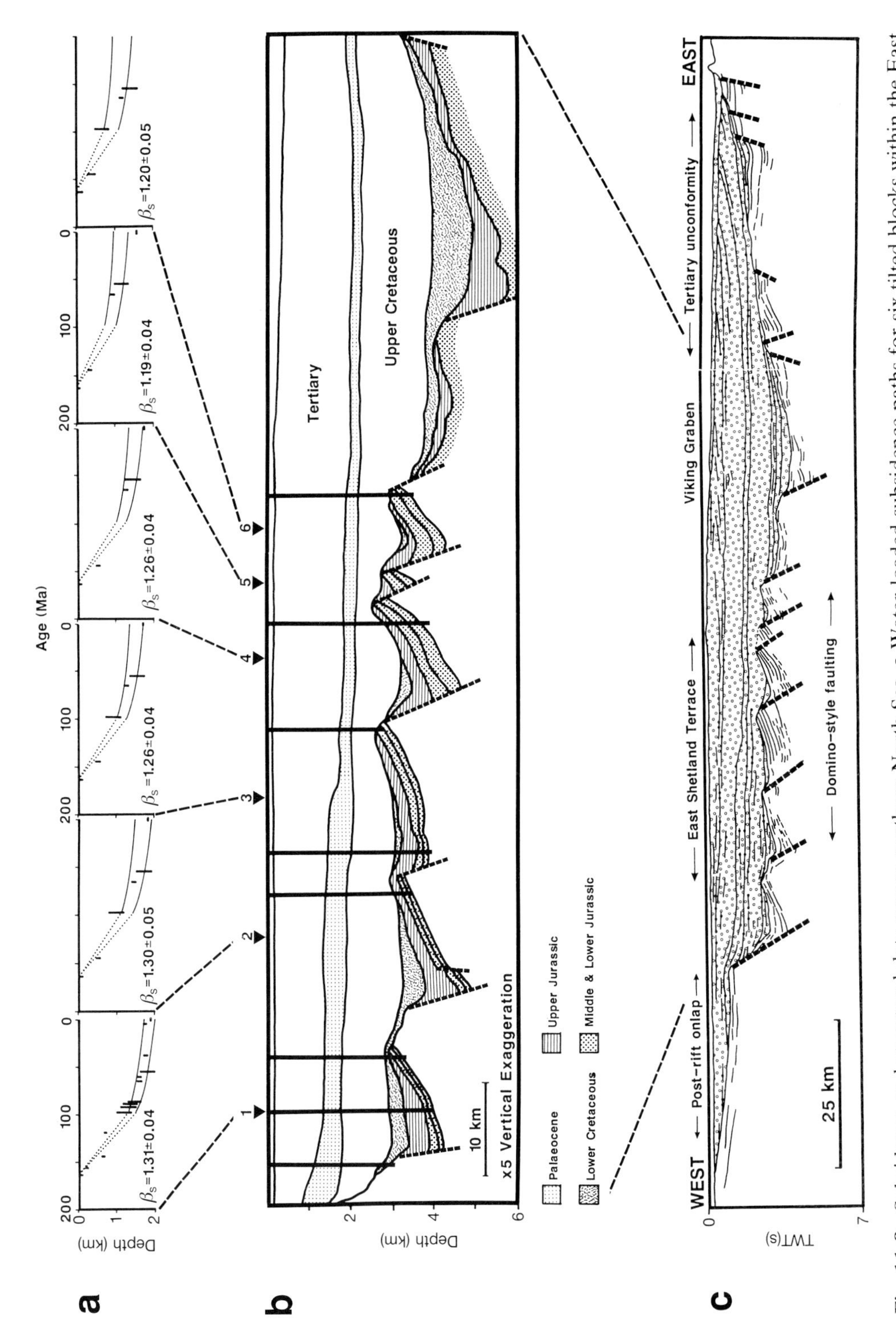

Fig. 11.8. Subsidence and structural data across northern North Sea. **a** Water-loaded subsidence paths for six tilted blocks within the East Shetland Terrace. Sediment compaction and loading effects have been removed. Bars, range of water depths at time of deposition; heavy lines, maximum and minimum predicted subsidence. β is subsidence-derived stretching factor (i.e. β_s). **b** Interpretation of western half of NNST-84-7 (i.e. East Shetland Terrace) showing location of well data (vertical lines) and subsidence analyses for six tilted blocks (numbered triangles). **c** Schematic line drawing along line NNST-84-7 (0–7 seconds two-way travel time) showing main structural features. Post-rift sediments are stippled.

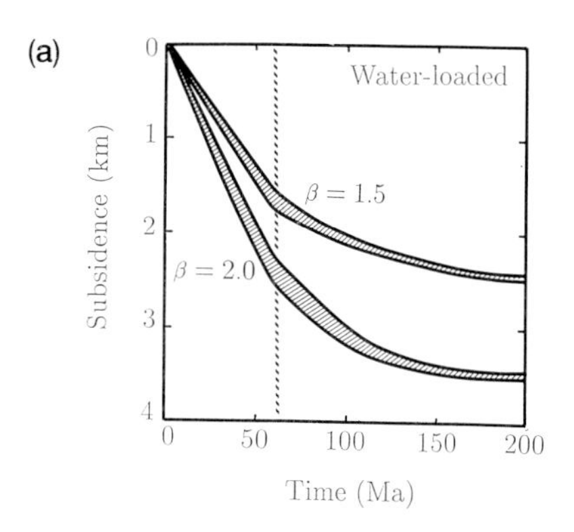

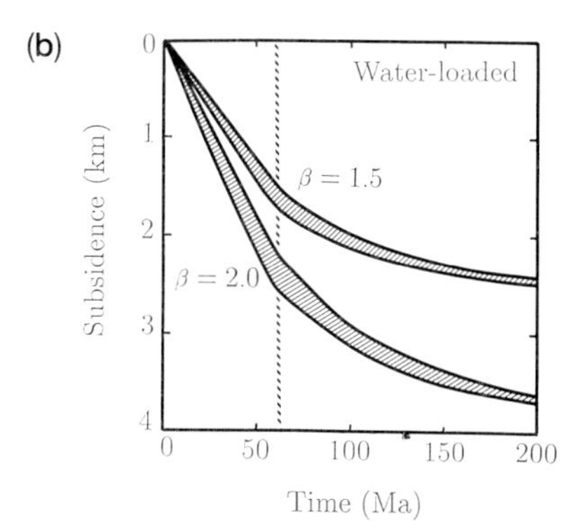

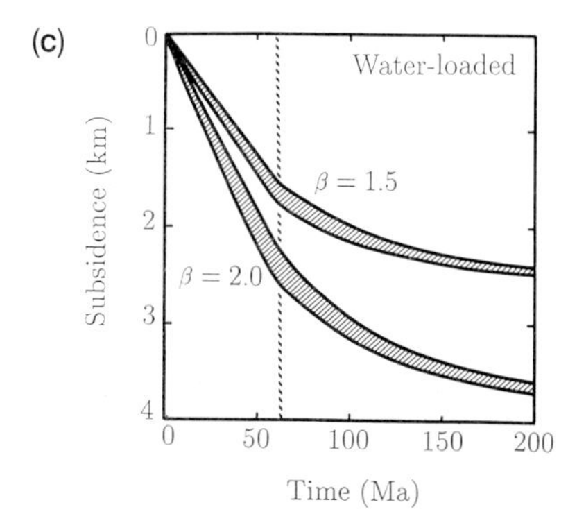

Fig. 11.9. Three sets of theoretical subsidence curves for two values of β (1.5 and 2.0) illustrating range of possible subsidence values for ± 10 per cent variation in **a** lithospheric thickness (T_l, mean value = 125 km); **b** asthenospheric temperature (T_1, mean value = 1333°C); and **c** thermal expansion coefficient (α, mean value = 3.4×10^{-5}°C^{-1}). Stretching lasts 60 Ma. Note similarity of subsidence errors in each case.

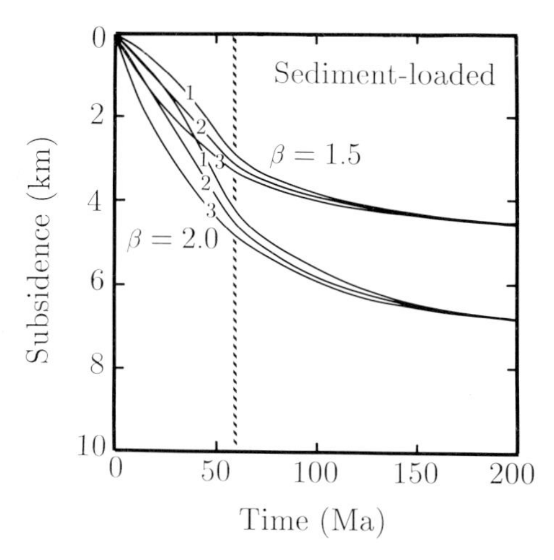

Fig. 11.10. Set of theoretical subsidence curves for two values of β (1.5 and 2.0). Three different strain rates ae used during stretching period (0–60 Ma): 1 = strain rate increases to 4 times initial value; 2 = constant strain rate; 3 = strain rate decreases to 0.25 times initial value.

planar, dips decreasing with depth by as little as 10° (White, in preparation). However, the effect of such a small decrease in dip is quite dramatic (Fig. 11.12), increasing the total amount of extension across the East Shetland Terrace to *c.* 20 km. Similar estimates of extension can be obtained from the published sections (Ziegler 1982; Glennie 1986).

11.3.3. *Crustal thinning*

Unfortunately, no seismic refraction profile comparable to that of Barton and Wood (1984) has been shot in the northern North Sea, the most recent being that of Solli (1976). Fortunately, the deep seismic reflection profiles can be used to calculate crustal stretching factors, assuming that the base of the lower crustal layering corresponds to the Moho (Mooney and Brocher 1987). According to Klemperer (1988), the crust (minus post-Permian sediments) beneath the Viking Graben is now 15 ± 3 km thick, corresponding to $\beta \simeq 2$. This estimate, combined with the fact that β_s for Upper Jurassic stretching is *c.*1.5 within the Viking Graben *sensu stricto*, suggests that the crust stretched by a factor of *c.* 1.3 in the Viking Graben during the Triassic, in agreement with results from the Central Graben

Table 11.1 Comparison of estimates of extension from normal faulting and from subsidence

Fault-bounded block	1	2	3	4	5	6
Present-day block width (km)	17.0	15.4	17.5	14.2	3.5	9.5
Present-day fault dip	50 ± 10°	35 ± 5°	35 ± 5°	30 ± 5°	30 ± 5°	30 ± 5°
Block rotation	10 ± 2°	10 ± 2°	10 ± 2°	10 ± 2°	10 ± 2°	10 ± 2°
Fault-derived stretching factor β_f	1.13 ± 0.01	1.23 ± 0.06	1.23 ± 0.06	1.29 ± 0.08	1.29 ± 0.08	1.29 ± 0.08
Fault-derived extension (km)	2.0 ± 0.02	2.9 ± 0.6	3.3 ± 0.7	3.2 ± 0.7	0.8 ± 0.2	2.1 ± 0.5
Subsidence-derived stretching factor β_s	1.31 ± 0.04	1.30 ± 0.05	1.26 ± 0.04	1.26 ± 0.04	1.19 ± 0.04	1.20 ± 0.05
Subsidence-drived extension (km)	2.9 ± 0.03	3.6 ± 0.5	3.6 ± 0.4	3.0 ± 0.3	0.6 ±	1.6 ± 0.3

(Barton and Wood, 1984). If the Upper Jurassic stretching phase is modelled by allowing for a previously thinned crust, then the Upper Jurassic β_s will be even greater than 1.5, and the Triassic β_s less than 1.3. These measurements also assume that, prior to the Triassic, crust was of 'normal' thickness and that crustal volume has not been measurably increased by the addition of basaltic melt during stretching. The scarcity of Mesozoic volcanics within the sedimentary pile of the northern North Sea justifies this assumption (Latin *et al.*, this volume). This observation agrees with recent calculations by McKenzie and Bickle (1988) which show that, for normal asthenosphere temperatures and $\beta \leqslant 2$, little melt is produced during extension.

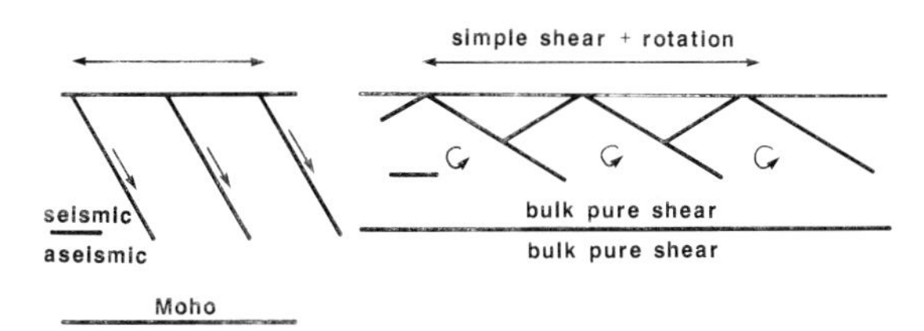

Fig. 11.11. Cartoons illustrating rotating block ('domino') style of extension. Surface length increases by factor β. Faults not shown below seismic-aseismic transition, which is drawn as being raised by extension; whether it is or not depends primarily on extensional strain rate (England and Jackson, 1987).

11.3.4. *Comparing different estimates of stretching*

Given that Ziegler's (1983) extension discrepancy is between β_f and the other two estimates, β_s and β_c (crustal stretching factor), crustal thinning data will, for the moment, be ignored while subsidence and normal faulting estimates are compared in detail. Subsequently both β_s and β_f can be compared with β_c.

Extension measurements from normal faulting and subsidence data are subject to many errors. In the former case, the dip of the faults and the rotation of the fault-bounded blocks need to be accurately known in order to apply the simplest possible 'rigid domino' model (Fig. 11.11). Given that internal deformation and modest fault dip changes with depth can considerably increase β_f (Fig. 11.12), the fault-derived extension estimates of Table 11.1 are probably minima.

As far as subsidence analyses are concerned, there are two main contributors to error. First, the back-stripping process itself, where the most significant errors are in porosities and solid densities. Secondly, theoretical subsidence curves themselves are subject to error, small errors in crust and mantle densities probably being the most significant. Aside from density errors, it is worth noting the effect of 10 per cent errors in lithospheric thickness, asthenospheric temperature, and thermal expansion coefficient. In each case, the largest error in the theoretical curves is likely to occur at the point which marks the cessation of stretching and the beginning of post-rift subsidence (Fig. 11.9). Varying the strain rate during stretching also introduces error (Fig. 11.10). As stated previously, maximum estimation of β_s were obtained by ensuring that poorly-constrained measurements (such as palaeowater depths) were overestimated rather than underestimated.

Despite deliberately maximizing values of β_s and minimizing values of β_f, the total amount of extension across the East Shetland Terrace determined from subsidence analyses is in good agreement with

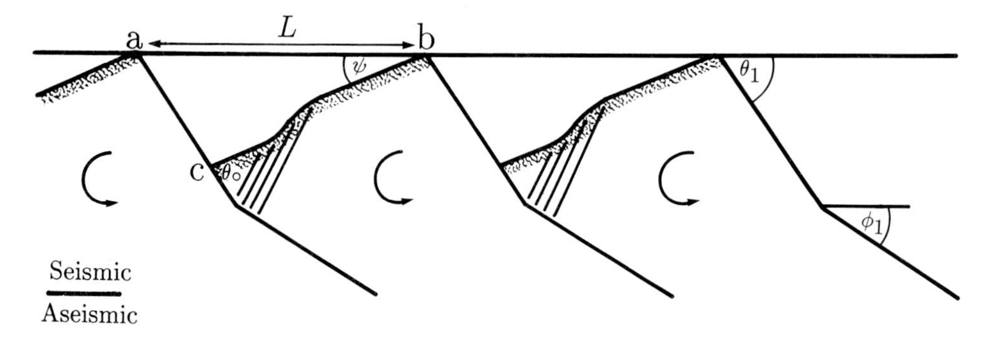

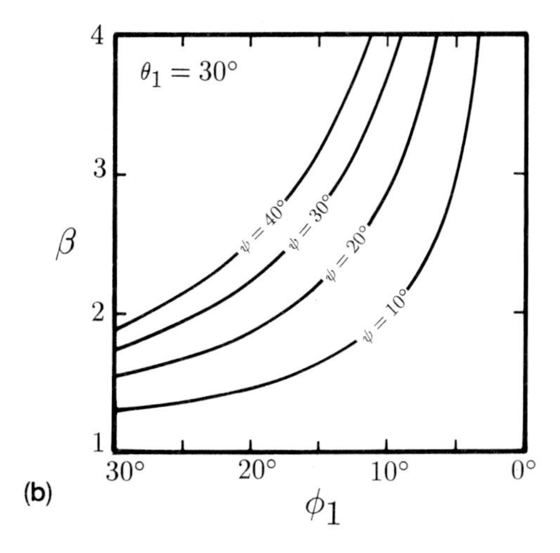

Fig. 11.12. Effect of fault curvature, **a** on the amount of extension achieved by rotating faults. ψ is amount of tilt, while θ_1 and ϕ_1 are dips of fault at surface and depth respectively; θ_0 and ϕ_0 were initial dips of fault at surface and at depth ($\phi_0 = \phi_1 + \psi$). L = |ab| = apparent distance between block crests after extension. |cb| = apparent distance between block crests prior to extension. Example in **b** shows the tilt (ψ) for a fault whose final dip at surface (θ_1) is 30° with values of ϕ_1 varying between 0° and 30°. When $\phi_1 = 30°$, fault is planar and $\beta = 1.5$ if $\psi = 20°$. If ϕ_1 decreases to 15° (i.e. fault becomes more listric) then $\beta = 2.1$.

that measured from normal faulting (15 ± 2 km, and 14 ± 3 km, respectively). Extension measurements for each fault-bounded block are shown in Fig. 11.13. Both estimates agree within error for all but the block closest to the unrifted Shetland Platform. There are two possible reasons for a discrepancy at this location. First, since it is the last rotated block, a large amount of internal deformation on either side of the fault is required to avoid forming voids at the boundary; thus β_f at this point is likely to be unrealistically small. Secondly, given the spatially rapid variation in stretching factors at this position, lateral heatflow both during and after extension may be locally important. Ignoring this effect tends to overestimate the subsidence-derived stretching factor, β_f. The decreasing trend of subsidence-derived extension from west to east across the East Shetland Terrace is discussed below.

Throughout the analysis presented in this secton, it has been assumed that the extension direction was in the plane of the section. Extension measured perpendicular to the strike of normal faulting would underestimate the true extension while that derived from subsidence would be unchanged if the overall slip-vector across the northern North Sea was oblique (e.g. NW–SE; Beach 1985; Beach *et al.* 1987). Hence conclusions made concerning the extension discrepancy are unaffected. In any case, Roberts and co-workers (this volume) argue that structural and subsidence data in the North Sea are consistent with orthogonal opening of all three arms of the North Sea rift system, with the exception of the Inner Moray Firth Basin.

The good agreement between β_f and β_s in an area where structure and stratigraphy are tightly constrained implies that, in this part of the North Sea, the amount of post-rift subsidence and the magnitude of extension measured on normal faults can be

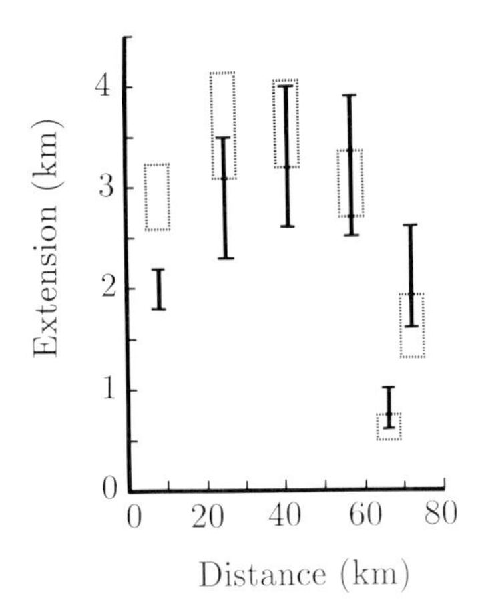

Fig. 11.13. Extension estimates for each of the six fault-bounded blocks of Fig. 11.8.b. Zero point on horizontal axis refers to edge of East Shetland Terrace as shown in Fig. 11.8.b. Solid bars = range of fault-derived extension; boxes with broken lines = range of subsidence-derived extension (*see* Table 11.1).

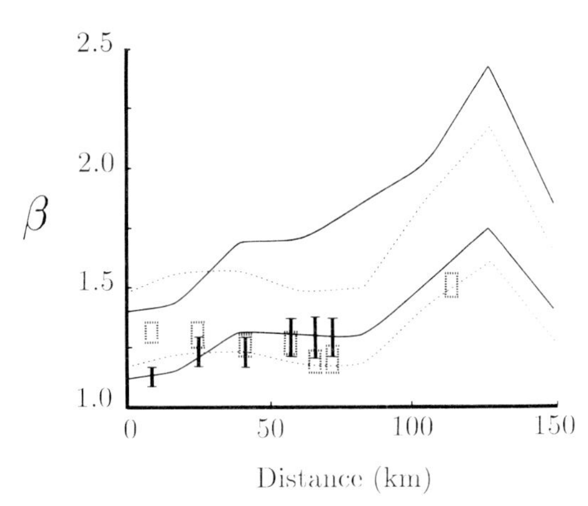

Fig. 11.14. Comparison of stretching factors calculated from crustal thinning, subsidence, and faulting. Crustal thinning information projected onto NNST-84-7 from NSDP-84-1. Pairs of solid lines = range of β factors determined from crustal thinning on deep seismic data (Holliger and Klemperer 1989); pairs of dashed lines = range of β factors determined from gravity-derived crustal thinning along deep seismic lines (Holliger and Klemperer); solid error bars = fault-derived β factors (β_f); boxes with broken lines = subsidence-derived β factors (β_s).

accounted for by the uniform stretching model. Contrary to previous studies, more complex two-layer or multi-phase stretching models, or lithospheric simple shear models are not required.

How do the above two estimates of stretching compare with β_c, the crustal stretching factor? Figure 11.14 shows all three stretching factors as a function of distance across the East Shetland Terrace and Viking Graben (crustal stretching values are projected from NSDP-84-1). Crustal stretching factors were estimated from the gravity-derived and seismic reflection profiles produced by Holliger and Klemperer (1989). The error in crustal thickness before stretching was taken as ± 1 km, and that in the estimate of present-day crustal thickness taken to be ± 1 km. Unfortunately, both errors propagate to give a large error in β_c which increases as β_c increases. The projected crustal stretching profile obtained from NSDP-84-1 is more suitably compared with NNST-84-7 than that obtained from NSDP-84-4, which crosses a much narrower part of the northern North Sea rift (Fig. 11.7).

In Fig. 11.14, β_f and β_s are slightly smaller than β_c across the East Shetland Terrace. This difference may be partly due to having projected the crustal stretching information by up to 40 km in a direction slightly oblique to the direction of strike (Fig. 11.7). If the difference is real, it can be accounted for by allowing a very minor amount of Triassic rifting. Across the East Shetland Terrace, a Triassic stretching factor of 1.0 (i.e. no extension) at the western edge increasing to 1.05 at the eastern edge will considerably improve the already reasonable match between all three stretching factors on Fig. 11.14. The slight downward trend in β_s from west to east will be flattened out since allowing for previously thinned crust causes the theoretical subsidence curves for Upper Jurassic stretching to be 'pulled up', thus increasing β_s. Over the same distance, β_c for Upper Jurassic stretching will now tend to be pushed in the opposite direction to β_s (i.e. decrease from west to east). Within the Viking Graben itself, a Triassic stretching factor of *c*.1.2 is required to pull β_s up and push β_c down so that they match. Estimated Triassic stretching for the Viking Graben is considerably less than that recently proposed by Giltner (1987), and Ziegler and van Hoorn (1989). It is, however, in good agreement with the value suggested for the Central Graben by Barton and Wood (1984).

The extremely small amount of Triassic rifting across the East Shetland Terrace (β_c *c.* 1.05 or less than 5 per cent extension) is in agreement with values of β_s which can be approximated from decompacted Triassic and Lower Jurassic sediment thicknesses (although it must be pointed out that, at present, Triassic sediment thicknesses in the northern North Sea are poorly known).

Some will argue that resolving the discrepancy in one small area does not necessarily mean that it does not exist elsewhere in the North Sea. Indeed, it could be that the East Shetland Terrace is somehow unique. Nevertheless, it is perhaps the only area in the North Sea where reasonably accurate estimates of β_f can be made. It is suggested here that basin models can only be rigorously tested on the highest quality data, and that the onus is on those who want more complicated general basin models to demonstrate that the simplest possible model cannot be applied to areas where stratigraphic control is excellent, and where the structure is relative unambiguous.

11.4. Previous studies

An important question is why previous work in the East Shetland Terrace did not also come to the conclusion that there might be no extension discrepancy. The first published work applying the uniform stretching model to the northern North Sea is that of Giltner (1987). Unfortunately, he only worked on the generalized interpretations of Ziegler (1982). These large-scale profiles are unsatisfactory both for detailed backstripping calculations and for estimating extension from normal faulting. Giltner's main conclusion was that Lower Triassic stretching was considerably more important than the better constrained Upper Jurassic event. For example, in the East Shetland Terrace he argued for a Triassic β_s of 1.25, increasing to 1.6 in the Viking Graben. His Jurassic β_s varies from 1.15 to 1.1 over the same distance. Unfortunately, Triassic thicknesses are poorly known on the flanks of the basin, and unknown in the centre. Thus Giltner accounted for the extension discrepancy by placing the most significant stretching period within the Triassic where stratigraphic control is poor, and where no estimate of extension across normal faults can be made. In the previous section, where the better constrained Upper Jurassic stretching event was analysed first, estimated Triassic and Lower Jurassic thicknesses were accounted for with much more modest amounts of stretching.

More recently, a detailed study was carried out by Badley and colleagues (1988) using seismic reflection data and well-log information along profiles very close to NNST-84-7. They also argued in favour of an extension discrepancy. In contrast to Giltner, they considered the Triassic stretching event to be less important than the Upper Jurassic event. Across the East Shetland Terrace, this team estimated that the Upper Jurassic β_s, determined from post-rift subsidence alone, was *c.* 1.38, while β_s determined from syn-rift subsidence was *c.* 1.0, and β_f was *c.* 1.1. These figures imply a significant discrepancy.

Unfortunately, Badley and colleagues' subsidence-derived stretching was calculated by estimating the amounts of syn-rift and post-rift stratigraphy in the basin, rather than by backstripping whole subsidence curves. The observed amounts of syn-rift and post-rift sediment-loaded subsidence were then each used to determine two different values of β_s with the sediment-loaded formulae of Barr (1987*a*, *b*).

There are several important problems with this approach. First, it is usually difficult to discriminate accurately between syn-rift and post-rift subsidence using interpreted seismic reflection data alone; this distinction is generally best made by backstripping the subsidence data (Barton and Wood 1984). The main problem in discriminating between syn-rift and post-rift in the North Sea arises because the amount of extension is relatively small. Hence block tilting is modest and compaction effects tend to dominate, blurring the distinction between syn-rift and post-rift.

Secondly, the approximate method used to determine β_s is critically dependent on the accuracy of one single point—the theoretical amount of subsidence at the end of rifting. As mentioned previously, this portion of the theoretical subsidence curve is more sensitive to small changes, in lithospheric thickness, asthenospheric temperature, and thermal expansion coefficient, than any other point in the theoretical curves (Fig. 11.9). Using different strain rates during stretching also has an important effect on the theoretical initial subsidence (Fig. 11.10). In the North Sea, there is no reason why strain rate should be constant, decreasing, or increasing during stretching. Note, however, that stretching factors determined from backstripped subsidence curves are not affected significantly by any of these areas.

Thirdly, basins usually have heterogeneous sediment-loads, so it is much more reliable to compare theoretical and observed water-loaded subsidence rather than the more convenient sediment-loaded curves. Errors in porosity, compaction decay lengths, and solid sediment densities all propagate and result in overall errors which are largest near the middle of the backstripped curve.

Ziegler and van Hoorn (1989) favour continuous rifting from Lower Triassic through to Upper Jurassic times, arguing that there is little evidence for two separate pulses. They also support the existence of an extension discrepancy in the northern North Sea. Their analysis of crustal thinning gives a total of 100–130 km extension, while unspecified palinspastic restorations yield only 19 km of extension across normal faults. Subsidence data were not analysed.

None of the workers mentioned above quote errors, or determine how error propagation may affect their results. Therefore it is very difficult to judge the significance of their conclusion that a large extension discrepancy exists. It is argued here that in order to assess critically the applicability of any basin model, error analysis must be undertaken.

11.5. Uplift and stratigraphic onlap

Despite the arguments in the previous section concerning the apparent lack of an extension discrepancy in the northern North Sea, there are a number of other observations in basins such as the North Sea which, at first glance, seem difficult to reconcile with the uniform stretching model. Perhaps the most widely discussed of these observations are syn-rift uplift and stratigraphic onlap, both of which may occur on a local or regional scale. On base margins, syn-rift uplift has frequently been documented (Steckler 1985). In the North Sea, such uplift has subsequently been followed by post-rift stratigraphic onlap, giving the basin a characteristic 'Steer's Head' geometry. Syn-rift uplift and onlap are also associated with individual fault-bounded blocks.

White and McKenzie (1988) argued that if stretching in the lithospheric mantle is distributed over a slightly wider zone than is stretching in the crust, a significant amount of basin margin uplift can be generated during rifting (Fig. 11.17; Leeder 1983; Rowley and Sahagian 1986). The total amount of stretching within the crust and mantle is constrained to be the same in order to avoid space problems (see White and McKenzie (1988) for details of model). When rifting ceases, the proposed distribution of stretching can result in substantial post-rift stratigraphic onlap generating the familiar 'Steer's Head' basin geometry. Neither syn-rift uplift nor post-rift onlap are generated if stretching is uniform throughout the lithosphere (Fig. 11.16). Surprisingly, however, the difference in distribution of stretching in the crust and lithospheric mantle required to generate the scale of onlap seen in basins like the North Sea may be less than 10 per cent (White and McKenzie 1988, Fig. 4a).

Basin margin uplift and stratigraphic onlap generally occur on a large scale—perhaps ranging for up to 100 km or more along strike. Uplift also occurs on a

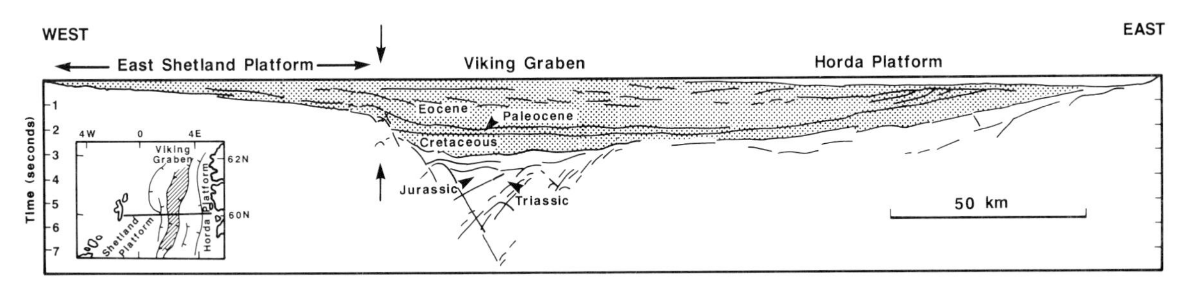

Fig. 11.15. Deep seismic profile NSDP-84-2 (top 8 seconds two-way travel time). Most recent rifting took place at end of Middle Jurassic. Shaded area = post-rift sediments ranging in age from Upper Cretaceous to present-day. In the west, stratigraphic onlap extends across East Shetland Platform. Vertical arrows indicate maximum thickness of post-rift onlap (D in Fig. 11.17c). The geometry of the eastern margin is more complex, due to late Tertiary uplift of Scandinavia.

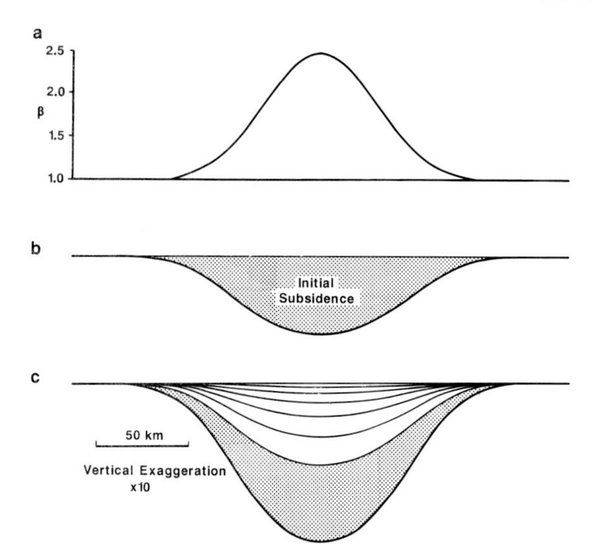

Fig. 11.16. **a** β, stretching factor in crust and upper mantle, is plotted as function of distance using a Gaussian distribution. In centre of basin, $\beta_c = \beta_m = 2.5$. **b** Initial subsidence immediately after instantaneous stretching (sediment-loaded, $\rho_s = 2.0$ Mg m^{-3}. Other parameters as in McKenzie, 1978. **c** Total subsidence 150 Ma after rifting. Thermal subsidence represented by time lines drawn every 25 Ma. Thermal subsidence dies out at same position as initial subsidence. Redrawn from White and McKenzie (1988).

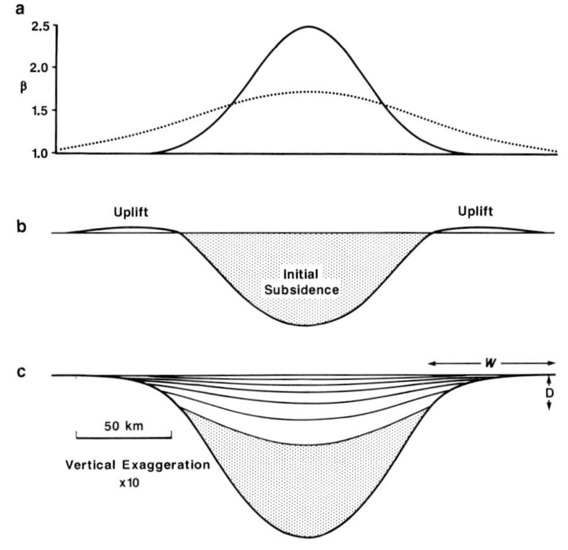

Fig. 11.17. **a** β_c, stretching factor in crust (solid line) and β_m, stretching factor in lithospheric mantle (dotted line), are plotted as function of distance. Total amount of extension is identical in each case (White and McKenzie, 1988). In centre of basin, $\beta_c = 2.5$ as before but now $\beta_m = 1.75$. All other parameters as in Fig. 11.16. **b** Initial subsidence immediately after instantaneous stretching. Note uplift along basin flanks. **c** Total subsidence 150 Ma after rifting. Thermal subsidence represented by time lines drawn every 25 Ma. Onlap produced by thermal subsidence at basin margin results in 'steer's head' geometry. W is horizontal extent of onlap and D is its maximum thickness. Note that total subsidence in Figs 11.16c and 11.17c is almost identical, the only difference being the relative proportion of initial and thermal subsidence. Redrawn from White and McKenzie (1988).

much more local scale. For example, in the northern North Sea, the footwalls of fault-bounded blocks have, in some cases, undergone substantial uplift and erosion (e.g. the Brent, Statfjord, and Snorre blocks; Yielding, 1990). Stratigraphic onlap is evident in the troughs of such blocks. Footwall uplift has been explained by sea-level fluctuations, by regional uplift and/or compression, and by flexural models. Stratigraphic onlap is often accounted for by sea-level rise (Vail and Todd 1981; Rawson and Riley 1982).

Barr (1987*a*, *b*) suggested a much more elegant way of explaining both footwall uplift and stratigraphic onlap. He showed that, when the lithosphere stretches instantaneously, the accommodation of this stretching by domino-style faulting in the brittle upper crust can result in either footwall uplift or subsidence. The amount of footwall uplift or subsidence depends mainly on the initial dip of the faults and their spacing. Subsequently, Jackson and coworkers (1988) corrected Barr's loading calculations and successfully applied the model to a tilted fault-bounded block in the Gulf of Suez where the amount of footwall uplift can be independently estimated.

To be useful in basins such as the North Sea, the effects of compaction and finite duration rifting should be included in the model (White 1988). Figure 11.18 shows two synthetic examples where stretching lasts 30 Ma. In Fig. 11.18a, the initial fault spacing is 15 km and it is clear that both footwall uplift and stratigraphic onlap are generated during rifting. Note that syn-rift stratigraphic onlap is most rapid towards the crest of each block. In Fig. 11.18b, the initial fault spacing is only 5 km and no syn-rift

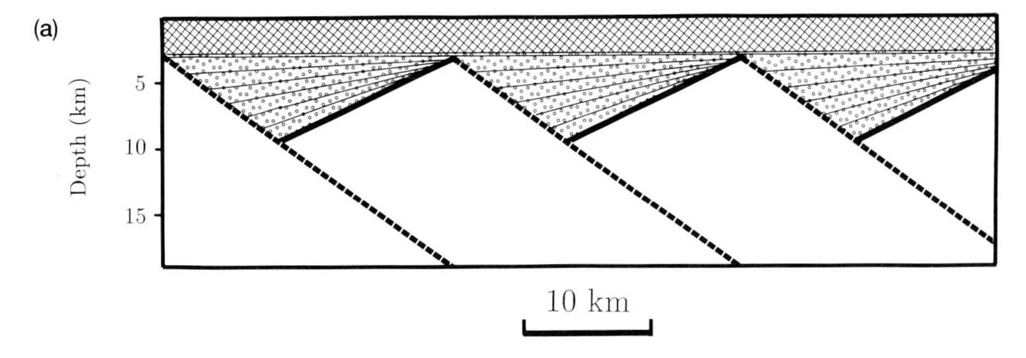

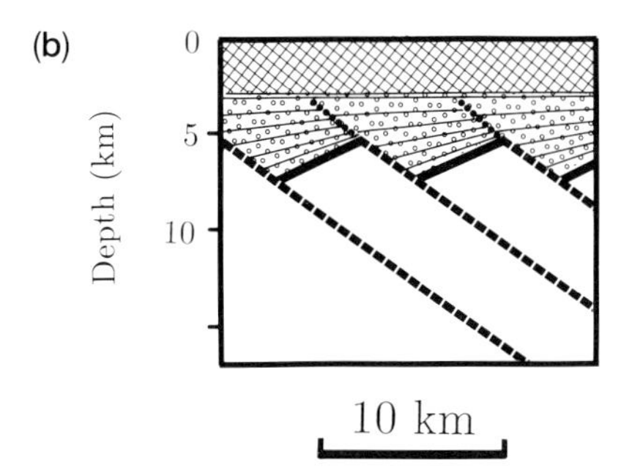

Fig. 11.18. Interaction of sediment-loaded subsidence and fault-bounded blocks for $\beta = 1.5$. Stretching begins at 150 Ma, lasting 30 Ma with constant strain rate. Initial subsidence is shown stippled; thermal subsidence is cross-hatched. **a** Initial block width = 15 km; initial fault dip = 60°; solid sediment density = 2.6 Mg m^{-3}; initial porosity = 60 per cent; compaction decay length = 2 km. **b** As for **a** with block width of 5 km.

uplift is generated. Instead, normal faults must propagate upwards through the syn-rift sediment.

11.6. Conclusions and future work

The main aim of this review is to show that the uniform stretching model in its simplest possible form can successfully account for most of the major observations in a well-understood part of the North Sea. The lithospheric simple shear model is rejected, primarily because it fails to predict the observed distribution of subsidence in the North Sea. More complex detachment models which allow the lower plate to deform are thermally indistinguishable from the uniform stretching model, while lacking its predictive power.

The most important episode of stretching began in the Upper Jurassic and lasted about 60 Ma. Although stretching undoubtedly occurred in the Triassic as well, the analysis presented here suggests that, within the northern North Sea, this earlier phase of rifting was of minor importance. In the East Shetland Terrace of the northern North Sea, where the structure is very clearly developed and where the data are tightly constrained by numerous wells, detailed calculations suggest that there is no evidence to support the widely discussed extension discrepancy. All three estimates of stretching (crustal thinning, backstripped subsidence data, and summed displacements across normal faults) agree within error.

The most important point about the extension discrepancy controversy is that its presence or absence should be demonstrable in areas with the best constrained data. The East Shetland Terrace is perhaps the only such area in the North Sea. This very selective approach is necessary mainly because of the great difficulty in making accurate measurements of the amount of extension taken up by normal faults. Application of idealized normal faulting models is always likely to underestimate extension.

Other observations in the North Sea, such as footwall uplift and local stratigraphic onlap, also appear to be consistent with the uniform stretching model. Slight modification to the model is required to account for syn-rift basin margin uplift and large-scale post-rift stratigraphic onlap. These latter observations are most simply explained by requiring stretching within the lithospheric mantle to be distributed over a slightly wider area than stretching within the crust. This vertical distribution of stretching does not affect the conclusions made concerning the extension discrepancy.

Several tectonic problems still remain to be tackled in the North Sea. The first of these is constraining the overall slip vector across the three different arms of the basin. Recently there has been considerable argument whether each rift arm opened orthogonally or whether there was a significant component of oblique slip. Roberts and colleagues (this volume) argue convincingly that subsidence and faulting observations through the North Sea are consistent with orthogonal opening of the three-armed system. However, direct structural evidence for a component of oblique slip can be elusive, and so oblique opening cannot be completely ruled out. One way to solve this problem would be to carry out a palaeomagnetic study on accurately oriented pre-rift core in a structurally simple area such as the East Shetland Terrace.

A second problem is Triassic rifting. Although it can be argued that Triassic rifting was probably considerably less significant than Upper Jurassic rifting, many workers will only be satisfied when the effects of Triassic rifting are fully incorporated into the modelling. This can only be properly done when the thickness of Triassic sediments throughout the basin is known accurately, especially in areas such as the actual Central and Viking Grabens.

Of perhaps greater importance are the dynamic reasons for a changing spatial distribution of stretching with depth through the lithosphere. As discussed above, this change is probably too small to be detected by, say, backstripped subsidence data. The only alternative way to tackle the problem would be to carry out rheological experiments using finite element methods. The results of such numerical experiments will have some bearing on the stretching-related melting problem in the North Sea (Latin *et al*., this volume).

Acknowledgements

I am grateful to Dan McKenzie and Drum Matthews for their help and encouragement, and to the British Council, Merlin Geophysical Ltd., and BIRPS for funding. Thanks also to M. Cheadle, J. Jackson, S. Klemperer, D. Latin, M. Collins, and P. Ryan. Norwegian Petroleum Exploration Consultants (NOPEC a.s.) generously allowed access to their North Sea data-set, NNST-84. Well data was provided by Chevron, Conoco, Elf, Esso, Hamilton Bros., Home Oil, Mobil, Occidental Oil Phillips, Saga, Shell, Sovereign, Statoil, Texaco, Total, and Unocal.

References

Badley, M. E., Price, J. D., Rambech Dahl, C., and Agdestein, T. (1988). The structural evolution of the Viking Graben and its bearing upon extensional modes of basin formation. *J. Geol. Soc.*, **145**, 455–72.

Barr, D. (1987*a*). Lithospheric stretching, detached normal faulting and footwall uplift. In *Continental extensional tectonics* (ed. M. P. Coward, J. F. Dewey, and P. L. Hancock). *Geol. Soc. Special Publication*, **28**, 75–94. Oxford, Blackwell Scientific Publications.

Barr, D., (1987*b*). Structural/stratigraphic models for extensional basins of half-graben type. *J. Struct. Geol.*, **9**, 4, 491–500.

Barton, P. and Wood, R. (1984). Tectonic evolution of the North Sea basin: Crustal stretching and subsidence. *Geophys. J. Roy. Astron. Soc.*, **79**, 987–1022.

Beach, A. (1985). Some comments on sedimentary basin development in the northern North Sea. *Scot. J. Geol.*, **21**, 4, 493–512.

Beach, A. (1986). A deep seismic reflection profile across the northern North Sea. *Nature*, **323**, 53–5.

Beach, A., Bird, T., and Gibbs, A. (1987). Extensional tectonics and crustal structure: deep seismic reflection data from the northern North Sea Viking Graben. In *Continental extensional tectonics* (ed. M. P. Coward, J. F. Dewey, and P. L. Hancock). *Geol. Soc. Special Publication*, **28**, 467–76. Oxford, Blackwell Scientific Publications.

Biddle, K. T. and Rudolph, K. W. (1988). Early Tertiary structural inversion in the Stord Basin, Norwegian North Sea. *J. Geol. Soc.*, **145**, 603–11.

Bosworth, W. (1987). Off-axis volcanism in the Gregory rift, East Africa: implications for models of continental rifting. *Geology*, **15**, 397–400.

Buck, W. R., Martinez, F., Steckler, M. S., and Cochran, J. R. (1988). Thermal consequences of lithospheric extension: Pure and simple. *Tectonics*, **7**, 213–34.

Cheadle, M. J., McGeary, S. E., Warner, M. R., and Matthews, D. H. (1987). Extensional structures on the UK continental shelf: a review of evidence from seismic reflection profiling. In *Continental exten-*

sional tectonics (ed. M. P. Coward, J. F. Dewey, and P. L. Hancock). *Geol. Soc. Special Publication*, **28**, 445–65. Oxford, Blackwell Scientific Publications.

Chenet, P. Y., Montadert, L., Gairaud, H., and Roberts, D. (1982). Extension ratio measurements on the Galicia, Portugal and northern Biscay continental margins: implications for evolutionary models of passive continental margins. In *Studies in continental margin geology* (ed. J. S. Watkins and C. L. Drake). *Am. Assoc. Petrol. Geologists Memoir*, **34**, 703–15.

Cochran, J. R. (1980). Some remarks on isostasy and the long term behaviour of the continental lithosphere. *Earth and Planetary Science Letters*, **46**, 266–74.

Coward, M. P. (1986). Heterogeneous stretching, simple shear and basin development. *Earth and Planetary Science Letters*, **80**, 325–36.

Coward, M. P., Dewey, J. F., and Hancock, P. L. (eds, 1987). *Continental extensional tectonics*, *Geol. Soc. Special Publication*, **28**. Oxford, Blackwell Scientific Publications.

Day, G. A., Cooper, B. A., Anderson, C., Burgers, W. F. J., Ronnevik, H. C., and Schontich, H. (1981). Regional seismic structure maps of the North Sea. In *The petroleum geology of the continental shelf of north west Europe* (ed. L. W. Illing and G. D. Hobson), 76–84. London, Institute of Petroleum, Heyden and Son.

England, P. C. and Jackson, J. A. (1987). Migration of the seismic-aseismic transition during uniform and non-uniform extension of the continental lithosphere. *Geology*, **15**, 291–4.

Etheridge, M. A., Branson, J. C., and Stuart-Smith, P. G. (1985). Extensional basin-forming structures in Bass Strait and their importance for hydrocarbon exploration. *APEA Journal*, **4**, 344–61.

Gibbs, A. D. (1983) Balanced cross-section construction from seismic sections in areas of extensional tectonics. *J. Struct. Geol.*, **5**, 153–60.

Gibbs, A. D. (1984). Structural evolution of extensional basin margins. *J. Geol. Soc.*, **141**, 609–20.

Gibbs, A. D. (1987). Deep seismic profiles in the northern North Sea. In *Petroleum geology of north west Europe* (ed. J. Brooks and K. Glennie), 1025–8. London, Graham and Trotman.

Giltner, J. P. (1987). Application of extensional models to the northern Viking Graben. *Norsk Geologisk Tidsskrift*, **67**, 339–52.

Glennie, K. W. (ed., 1986). *Introduction to the petroleum geology of the North Sea*. Oxford, Blackwell Scientific Publications.

Haq, B. U., Hardenbol, J., and Vail, P. R. (1987). Chronology of fluctuating sea levels since the Triassic. *Science*, **235**, 1156–67.

Hellinger, S. J., and Sclater, J. G. (1983). Some comments on two-layer extensional models for the evolution of sedimentary basins. *J. Geophys. Research*, **88**, 8251–69.

Hellinger, S. J., Sclater, J. G., and Giltner, J. P. (1988). Mid-Jurassic through Mid-Cretaceous extension in the Central Graben of the North Sea: estimates from subsidence. *Basin Research*, **1**, 191–200.

Holliger, K. and Klemperer, S. L. (1989). A comparison of the Moho interpreted from gravity data and from deep seismic reflection data in the northern North Sea. *Geophys. J.*, **97**, 247–58.

Jackson, J. A. (1987). Active normal faulting and crustal extension. In *Continental extensional tectonics* (ed. M. P. Coward, J. F. Dewey, and P. L. Hancock). *Geol. Soc. Special Publication*, **28**, 3–17. Oxford, Blackwell Scientific Publications.

Jackson, J. A. and White, N. (1989). Normal faulting in the upper continental crust: observations from regions of active extension. *J. Struct. Geol.*, **11**, 5, 15–36.

Jackson, J. A., White, N. J., Garfunkel, Z., and Anderson, H. (1988). Relations between normal fault geometry and vertical motions in extensional terranes: an example from the southern Gulf of Suez. *J. Struct. Geol.*, **10**, 155–70.

Jarvis, G. T. and McKenzie, D. P. (1980). Sedimentary basin formation with finite extension rates. *Earth and Planetary Science Letters*, **48**, 42–52.

Karner, G. D., Lake, S. K., and Dewey, J. F. (1987). The thermal and mechanical development of the Wessex Basin, Southern England. In *Continental extensional tectonics* (ed. M. P. Coward, J. F. Dewey, and P. L. Hancock). *Geol. Soc. Special Publication*, **28**, 517–36. Oxford, Blackwell Scientific Publications.

Klemperer, S. L. (1988). Crustal thinning and nature of extension in the northern North Sea from deep seismic reflection profiling. *Tectonics*, **7**, 803–21.

Klemperer, S. L. and White, N. (1989). Coaxial stretching or lithospheric simple shear in the North Sea? Evidence from deep seismic profiling and subsidence. In *Extensional tectonics and stratigraphy of the North Atlantic margins* (ed. A. J. Tankard and H. Balkwill). *Am. Assoc. Petrol. Geol. Memoir*, **46**, 511–22.

Kusznir, N. J., Karner, G. D., and Egan, S. (1987). Geometric, thermal and isostatic consequences of detachments in continental lithosphere: Extension and basin formation. In *Sedimentary basins and basin-forming mechanisms* (ed. C. Beaumont and A. J. Tankard). *Can. Soc. Petrol. Geology Memoir*, **12**, 185–203.

Latin, D. and White, N. (1990). Generating melt during lithospheric extension: Pure shear *vs.* simple shear. *Geology*, **18**, 327–31.

Latin, D. M., Dixon, J. E., and Fitton, J. G. (this volume). Rift-related magmatism in the North Sea basin.

Leeder, M. R. (1983). Lithospheric stretching and North Sea Jurassic clastic sourcelands. *Nature*, **305**, 510–13.

Le Pichon, X. and Barbier, J. (1987). Passive margin formation by low-angle faulting within the upper crust: the northern Bay of Biscay margin. *Tectonics*, **6**, 2, 133–50.

Le Pichon, X. and Sibuet, J. C. (1981). Passive margins: A model of formation. *J. Geophys. Research*, **86**, 3708–20.

Lister, G. S., Etheridge, M. A., and Symonds, P. A. (1986). Detachment faulting and the evolution of passive continental margins. *Geology*, **14**, 246–50.

McKenzie, D. (1978). Some remarks on the development of sedimentary basins. *Earth and Planetary science Letters*, **40**, 25–32.

McKenzie, D. and Bickle, M. J. (1988). The volume and composition of melt generated by extension of the lithosphere. *J. Petrol.* **29**, 625–79.

Mooney, W. D. and Brocher, T. M. (1987). Coincident seismic reflection/refraction studies of the continental lithosphere: a global review. *Reviews of Geophysics*, **25**, 723–42.

Moretti, I. and Pinet, B. (1987). Discrepancy between lower and upper crustal thinning. In *Sedimentary basins and basin forming mechanisms* (ed. C. Beaumont and A. J. Tankard). *Can. Soc. Petrol. Geology Memoir*, **12**, 233–9.

Mudford, B. S. (1988). A quantitative analysis of lithospheric subsidence due to thinning by simple shear. *Can. J. Earth Sciences*, **25**, 20–9.

Morton, W. H. and Black, R. (1975). Crustal attenuation in Afar. In *Afar depression of Ethiopia* (ed. A. Pilger and A. Rosler). Inter-union commission on geodynamics. Scientific Report, **14**, 55–65. Stuttgart, E. Schweizerbart'sche Verlagsbuchhandlung.

Proffett, J. M. Jr. (1977). Cenozoic geology of the Yerington district, Nevada and implications for the nature and origin of Basin and Range faulting. *Geol. Soc. Am. Bulletin*, **88**, 247–66.

Rawson, P. F. and Riley, L. A. (1982). Latest Jurassic–Early Cretaceous events and the 'Late Cimmerian Unconformity' in the North Sea area. *Am. Assoc. Petrol. Geol. Bulletin*, **66**, 2628–48.

Roberts, A. M., Yielding, G., and Badley, M. (this volume). A kinematic model for orthogonal opening of the Late Jurassic North Sea rift system, Denmark-Mid Norway.

Rowley, D. B. and Sahagian, D. (1986). Depth-dependent stretching: A different approach. *Geology*, **14**, 32–5.

Royden, L., Horvath, F., Nagymarosy, A., and Stegena, L. (1983). Evolution of the Pannonian Basin System 2. Subsidence and thermal history. *Tectonics*, **2**, 91–137.

Sawyer, D. S. (1986). Effects of basement topography on subsidence history analysis. *Earth and Planetary Science Letters*, **78**, 427–34.

Sawyer, D. S., Toksöz, M. N., Sclater, J. G., and Swift, B. A. (1982). Thermal evolution of the Baltimore Canyon Trough and Georges Bank Basin. In *Studies in continental margin geology* (ed. J. S. Watkins and C. L. Drake). *Am. Assoc. Petrol. Geologists Memoir*, **34**, 743–62.

Sclater, J. G. and Célérier, B. (1989). Errors in extension measurements from planar faults observed on seismic reflection lines. *Basin Research*, **1**, 217–21.

Sclater, J. G. and Christie, P. A. F. (1980). Continental stretching: An explanation of the post-mid-Cretaceous subsidence of the Central North Sea Basin. *J. Geophys. Research*, **85**, 3711–39.

Sclater, J. G., Royden, L., Horvath, F., Burchfiel, B. C. Semken, S., and Stegena, L. (1980). The formation of the intra-Carpathian basins as determined from subsidence data. *Earth and Planetary Science Letters*, **51**, 139–62.

Shorey, M. D. and Sclater, J. G. (1988). Mid-Jurassic through Mid-Cretaceous extension in the Central Graben of the North Sea part 2: estimates from faulting observed on a seismic reflection line. *Basin Research*, **1**, 201–15.

Solli, M. (1976). En seismisk skorpeudersokelse Norges-Shetland. Unpublished M.Sc. Dissertation, University of Bergen.

Steckler, M. S. (1985). Uplift and extension at the Gulf of Suez: Indications of induced mantle convection. *Nature*, **317**, 135–9.

Steckler, M. S. and Watts, A. B. (1978). Subsidence of the Atlantic-type continental margin off New York. *Earth and Planetary Science Letters*, **41**, 1–13.

Ussami, N., Karner, G. D., and Bott, M. H. P. (1986). Crustal detachment during South Atlantic rifting and formation of the Tucano-Gabon basin system. *Nature*, **322**, 629–32.

Vail, P. R. and Todd, R. G. (1981). Northern North Sea Jurassic unconformities, chronostratigraphy and sea level changes from seismic stratigraphy. In *Petroleum geology of the continental shelf of north west Europe* (ed. L. W. Illing and G. D. Hobson), 216–35. London, Institute of Petroleum, Heyden and Son.

Verrall, P. (1982). Structural interpretation with applications to North Sea problems. *Course Notes No.* **3**, JAPEC (UK).

Voorhoeve, H. and Houseman, G. (1988). The thermal evolution of lithosphere extending on a low-angle detachment zone. *Basin Research*, **1**, 1–9.

Warner, M. R. (in press). Seismic reflections from the lower continental crust. *J. Geophys. Research*.

Watts, A. B. and Steckler, M. S. (1979). Subsidence and eustasy at the continental margin of Eastern North America. In *Am. Geophys. Union*, *Maurice Ewing Series*, **3**, 218–34.

Watts, A. B., Karner, G. D., and Steckler, M. S. (1982). Lithospheric flexure and the evolution of sedimentary basins. *Phil. Trans. Roy. Soc. London,* **A305**, 249–81.

Wernicke, B. (1985). Uniform-sense normal simple shear of the continental lithosphere. *Can. J. Earth Sciences*, **22**, 108–25.

White, N. (1988). Extension and subsidence of the continental lithosphere. Ph.D. dissertation, University of Cambridge, UK.

White, N. (1989). Nature of lithospheric extension in the North Sea. *Geology*, **17**, 111–14.

White, N. and McKenzie, D. P. (1988). Formation of the 'steer's head' geometry of sedimentary basins by differential stretching of the crust and mantle. *Geology*, **16**, 250–3.

White, N., Jackson, J. A., and McKenzie, D. P. (1986), The relationship between the geometry of normal faults and that of the sedimentary layers in their hanging walls. *J. Struct. Geol.*, **8**, 8, 879–909.

Wood, R. J. (1982). Subsidence in the North Sea. Ph.D. dissertation, University of Cambridge, UK.

Wood, R. J. and Barton, P. (1983). Crustal thinning and subsidence in the North Sea: matters arising. *Nature*, **304**, 561.

Ye, H., Shedlock, K. M., Hellinger, S. J., and Sclater, J. G. (1985). The north China basin: An example of a Cenozoic rifted intraplate basin. *Tectonics*, **4**, 153–69.

Yielding, G. (1990). Footwall uplift associated with Late Jurassic normal faulting in the northern North Sea. *J. Geol. Soc.*, **147**, 219–22.

Ziegler, P. A. (1981). Evolution of sedimentary basins in NW Europe. In *The petroleum geology of the continental shelf of north west Europe* (ed. L. W. Illing and G. D. Hobson), 3–39. London, Institute of Petroleum, Heyden and Son.

Ziegler, P. A. (1982). *Geological atlas of western and central Europe*. The Hague, Shell Internationale Petroleum Maatschappij, B.V.

Ziegler, P. A. (1983). Crustal thinning and subsidence in the North Sea: matters arising. *Nature*, **304**, 561.

Ziegler, P. A. and van Hoorn, B. (1989). Evolution of the North Sea rift system. In *Extensional tectonics and stratigraphy of the North Atlantic margins* (ed. A. J. Tankard and H. Balkwill). *Am. Assoc. Petrol. Geologists Memoir*, **46**, 471–500.

12 Application of a flexural cantilever simple-shear/pure-shear model of continental lithosphere extension to the formation of the northern North Sea basin

G. Marsden, G. Yielding, A. M. Roberts, and N. J. Kusznir

Abstract

The Viking Graben of the northern North Sea provides an excellent opportunity to study the development of an extensional sedimentary basin. Prolific hydrocarbon discoveries have resulted in a large amount of well and seismic data, from which a series of structural cross sections have been constructed from the Shetland Platform in the west to the Norwegian coast in the east.

The development of the basin is investigated quantitatively, using a mathematical model which incorporates the rheological, thermal, and flexural isostatic consequences of lithosphere extension by simple-shear on major planar faults in the upper crust, and by distributed pure-shear in the lower crust and mantle. During faulting, the footwall and hangingwall blocks are considered as two interacting flexural cantilevers; the response of these cantilevers to the isostatic forces produced by extension generates footwall uplift and hangingwall collapse. For a set of adjacent planar faults, the lateral superposition of flexural footwall uplift and hangingwall collapse produces the familiar 'domino'-style block rotation of such multiple fault systems.

Fault displacements during a given interval of time are used as input to the numerical model in order to produce synthetic basin cross-sections and derive post-extension crustal thickness.

Both Triassic and Jurassic stages of rifting have been modelled. For a complete understanding of the post-Jurassic evolution of the Viking Graben it is essential to consider the earlier Triassic rift. Triassic rifting generates the gross structural framework on which the later Jurassic rift acts. Compaction of Triassic basin fill sequences and residual Triassic thermal re-equilibration both enhance subsidence in the Cretaceous and Tertiary. The preferred model of Triassic rifting has a total extension across the Viking Graben of 38 km.

Jurassic fault extensions, as imaged on industrial seismic reflection data, have been used to constrain modelling of syn- and post-Jurassic basin evolution. The seismically observed Jurassic extension of 22 km is able to generate the observed Upper Jurassic, Cretaceous, and Tertiary basin geometry and subsidence. The flexural cantilever model predicts isostatic uplift and erosion of Middle Jurassic footwall crests on the East Shetland Terraces.

Industrial and deep seismic data have been used to produce profiles of crustal basement thickness across the Viking Graben. The amount of crustal thinning calculated is dependent on the original (pre-Triassic stretching) crustal thickness. While the Jurassic rift model based on observed fault extension predicts the correct syn- and post-Jurassic subsidence, the model only predicts the observed crustal thinning for pre-extension crustal thicknesses less than 25 km (a value much less than the crustal thickness of 30 km seen at the basin margins). The modelling suggests that either the pre-Triassic crustal thickness was less than 25 km across the Viking Graben or, more likely, that a substantial amount of extension (possibly 40 per cent) is unimaged on seismic reflection data.

12.1. Introduction

The Viking Graben is the northernmost part of the North Sea rift system. It separates the East Shetland Platform in the west from the Norwegian Horda Platform in the east (Fig. 12.1). Hydrocarbon exploration in the Viking Graben began in the late 1960s, and received great impetus with the discovery of the giant Brent Field in 1971 (Bowen 1975). The first synthesis of the regional structure in terms of basin formation was that of Ziegler (1982). Ziegler recognized that the basin began to develop in the Early Triassic, but that the formation of the main observed fault-blocks occurred in the latest Jurassic/earliest Cretaceous. Subsequent late Cretaceous and Tertiary subsidence was largely unaccompanied by faulting. Ziegler pointed out the apparent discrepancy between a seismic refraction line (Solli 1976) which showed severe crustal thinning beneath the graben and his own extension measurements based on regional cross-sections. If the crustal thinning was caused by stretching alone (McKenzie 1978), 75–100 km of extension would be required. However, Ziegler's extension estimate (from fault heaves) was a maximum of 25 km, of which 10 km was attributed to the Early Cretaceous. To reconcile these figures, he suggested that, in addition to mechanical stretching, crustal thinning during rifting is accomplished by 'thermally-induced physico-chemical processes' termed 'sub-crustal erosion'. Ziegler (1983) reiterated his appeal to sub-crustal erosion in his discussion of Wood and Barton's (1983) stretching estimates from the Central Graben of the North Sea.

The apparent discrepancy between measured extension on faults and observed crustal thinning has been a continuing theme in subsequent papers concerned with the formation of the Viking Graben. Beach (1986) discussed a deep seismic profile across the northern Viking Graben, and qualitatively likened the Jurassic structure to Wernicke's (1985) model of basin formation by lithospheric simple shear. However, Beach and co-workers (1987, in discussion of the same seismic line) proposed that, although Triassic extension may have been on an easterly-dipping detachment, Late Jurassic extension was in a dominantly strike-slip setting. Strike-slip movement was suggested to explain why fault heaves appear to show far less extension than was implied by the crustal thinning and the basin subsidence. Total stretching over the whole profile was estimated by Beach and colleagues from the crustal thinning, to be $\beta = 1.5$, reaching a local maximum of $\beta = 2.3$ in the graben axis. Their subsidence calculations gave $\beta = 3.3$ in the graben axis. By contrast, summation of Late Jurassic fault heaves across the profile gave $\beta = 1.05$.

A further suite of deep seismic lines (NSDP) was shot in 1984 and discussed by Gibbs (1987) and Klemperer (1988). Gibbs expressed the view that the basin was dominantly extensional in origin, with all faults rooting into an easterly-dipping detachment. Again a discrepancy between fault extension ($\beta = 1.3$–1.4 across the whole profile) and subsidence calculations ($\beta = 3$ at the graben centre) was inferred. Gibbs suggested that the pre-Late Jurassic faults were not seismically imaged, but had been broken up and rotated by the younger Late Jurassic faults. Crustal underplating during extension was postulated as a qualitative explanation for the high β required by such a model of extreme extension. By contrast, Klemperer used the same seismic data set to infer that no large-scale detachments are present, and that upper crustal brittle deformation is decoupled from discrete shear zones in the mantle by ductile extension (bulk pure shear) in the lower crust. Klemperer estimated a mean extension across the basin of $\beta = 1.4$, with a maximum in the graben axis of 2.1. Slightly greater extension estimates were obtained by Zervos (1987) on the basis of gravity modelling.

Depth-dependent stretching models of the north

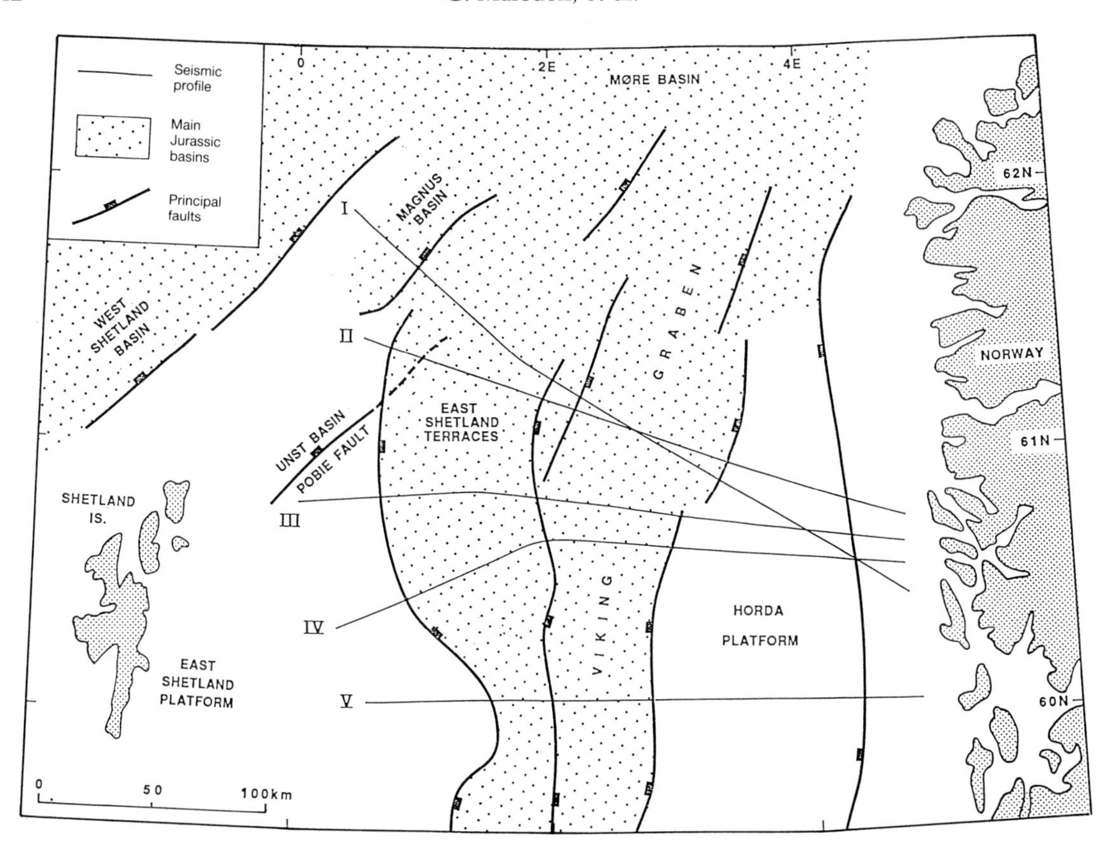

Fig. 12.1. Tectonic map of the Viking Graben area showing the location of structural cross-sections compiled in this study.

Viking Graben have been suggested by Coward (1986), and White and McKenzie (1988). Coward proposed that extensional thinning of the lower crust and the mantle was spatially concentrated in the centre of the basin. This could, potentially, produce syn-rift uplift and then enhanced thermal subsidence in the axial zone (where there is an 'excess' of post-rift sediment by comparison with the thickness predicted by a simple model of one-phase Jurassic extension). By contrast, White and McKenzie suggested that the lower lithosphere (mantle) stretching was distributed over a greater area than the crustal stretching. They used this model to explain the 'steer's head' geometry of the basin, i.e. the onlap of the thermal-subsidence sediments on to the basin margins.

Badley and co-workers (1988) and Giltner (1987) have emphasized the long subsidence history of the north Viking Graben, recognizing two separated stretching episodes (?Permo-Triassic and Late Jurassic). Badley and colleagues assumed that the thermal subsidence phase of the early rifting episode was essentially completed when the second (Late Jurassic) episode began. They then calculated β factors for the later rifting from the post-Jurassic thermal subsidence sediments, obtaining $\beta = 1.14$–1.38 at the margins and $\beta = 1.49$ in the rift axis. Pointing out that this exceeded their estimates of late Jurassic extension by faulting ($\beta = 1.1$–1.2), this group suggested that either depth-dependent stretching or 'active' thermal effects could explain the discrepancy. Giltner, however, showed that the magnitude and thermal effects of the earlier (Triassic) stretching are crucial to understanding the later subsidence history. In his modelling of subsidence along two profiles across the basin, Giltner allowed the earlier rifting to span most of the Triassic. This results in the second (Late Jurassic) stretching occurring in the middle of the first-phase thermal subsidence, rather than after it. Hence a significant portion of the Cretaceous–Tertiary subsidence was a response to the lithospheric thinning during the Triassic. Of a maximum β of 2.0 in the rift axis, Giltner estimated *c.* 1.8 occurred in the Triassic and only *c.* 1.2 in the Late Jurassic. Giltner believed that faults imaged on seismic data account adequately for the predicted Jurassic subsidence, but that faults related to Triassic extension are unobserved. Like other authors before him, he suggested that these unobserved faults may have been rotated and refaulted.

Thus, despite the abundance of seismic and well data from this prolific hydrocarbon province, a plethora of models have been created to describe its formation. Important in many of these has been the suggestion that fault-related extension cannot account for the observed crustal thinning if a simple McKenzie-type stretching model is assumed. Additional factors that have been invoked are:

(1) sub-crustal erosion (Ziegler 1982);
(2) crustal underplating (Gibbs 1987);
(3) lithospheric simple shear (Beach 1986);
(4) strike-slip movement (Beach *et al.* 1987);
(5) rotation and refaulting of Triassic structures (Gibbs 1987; Giltner 1987);
(6) depth-dependent stretching, with narrower mantle extension (Coward 1986);
(7) depth-dependent stretching, with broader mantle extension (Badley *et al.* 1988; White and McKenzie 1988);
(8) two-phase extension, with Triassic extension contributing significantly to post-Jurassic thermal subsidence (Giltner 1987);
(9) three-phase extension (Triassic, Jurassic, Tertiary) (Beach *et al.* 1987).

In the present study we attempt to address the problem of whether a discrepancy truly exists between observed crustal thinning and observed faulting using a generalized model of lithospheric extension.

12.2. Construction of the profiles

The cross-sections are based on a collation of industry seismic reflection data across the northern Viking Graben. Some profiles comprise a single regional line across the whole basin from the East Shetland Platform to the Norwegian coast, whereas others were constructed from a number of shorter near-contiguous seismic lines. The industry data are generally recorded to 5-7s TWT, and are time migrated. For deeper structure (i.e. the Moho), we have relied on modelling of the regional gravity field (Zervos 1987) and published deep seismic reflection profiles. Profiles I and II, in the north, lie close to the Britoil deep seismic line (Beach *et al.* 1987) and to the GECO deep line NSDP-1 (Klemperer 1988); Profile V, in the south is nearly coincident with NSDP-2.

Some fifty wells on or near the profiles were used to identify stratigraphic horizons on the seismic reflection profiles, and compile seismic velocity information. The following horizons were interpreted on the reflection data:

(1) sea-floor;
(2) top Palaeocene;
(3) top Cretaceous;
(4) top Lower Cretaceous (locally);
(5) base Cretaceous;
(6) intra-Upper Jurassic (locally);
(7) top Middle Jurassic;
(8) top Triassic (locally);
(9) intra-Triassic Lomvi Formation (locally);
(10) top Basement (locally).

Confidence in the reliability of the seismic picks varies greatly from the shallower to deeper levels of the stratigraphy. Tertiary and Cretaceous units are relatively well defined, but the pre-Cretaceous horizons can be much more equivocal. This is particularly so in the graben centre, where the deepest sediments are poorly imaged and often not drilled. In these cases regional trends observed in wells were used as a constraint on likely thicknesses. However the position of top basement in particular is in many areas highly speculative.

Conversion of the seismic interpretations to depth sections was performed using Badley Ashton's in-house Sattlegger Interpretive Seismic Processing system. The following depth-dependent velocity functions were derived from well data:

(1) water column $v = 1500\ ms^{-1}$
(2) post-Palaeocene $v = 1855 + 0.215z$
(where z = depth in metres)
or $v = 1735\ ms^{-1}$
on the Horda Platform

(3) Palaeocene $v = 2134 + 0.147z$
(4) Cretaceous $v = 1280 + 0.624z$
(5) Upper Jurassic $v = 2067 + 0.27z$
(6) Middle/Lower Jurassic + Triassic $v = 2072 + 0.48z$

Following Klemperer (1988), a seismic velocity of 6200 ms^{-1} was assumed for pre-Mesozoic basement.

Digitized seismic picks from the time-migrated sections were converted to depth using image-ray migration. This process takes into account the refraction of rays, which is neglected by standard time migration.

12.3. Description of structure

The gross crustal structure shown by the profiles comprises a thinning of the pre-Mesozoic basement from the adjacent platforms to the rift axis (Fig. 12.2). The overlying sediment column thickens into the rift axis, reaching a maximum thickness of perhaps 10 km (though the base of the sediments is not well-constrained in the graben centre). The Moho rises from about 30 km depth below the East Shetland Platform and the Norwegian coast to approximately 20–22 km depth below the graben axis. The region of thinnest pre-Mesozoic crust is typically about 12 km thick. If the 30 km crustal thickness seen at the basin edges is indicative of the original (pre-stretching) thickness, then a maximum stretching factor of β *c.* 2.5 is implied at the graben axis. The average β over the 250 km lengths of the profiles is about 1.4–1.5, implying a total extension of about 80 km. These figures are in broad accord with those of Zervos (1987) and Klemperer (1988), on whose results our Moho interpretations are partly based.

Profile I (Fig. 12.3) with which the modelling is principally concerned, runs NW–SE across the basin (Fig. 12.1). The north-western end shows the Magnus Basin. Strictly speaking, this is not part of the Viking Graben system but is part of a broad, *en échelon*, transfer zone between the Møre Basin to the north and the West Shetland Basin to the west (Roberts *et al.*, this volume).

Between the Magnus Basin and the Viking Graben proper are the East Shetland Terraces, an area of tilted fault blocks which are bounded mainly by east-dipping normal faults. These tilted blocks form the traps for the bulk of the hydrocarbon accumulations in the basin, such as Brent, Statfjord, Gullfaks, and Ninian. The bounding fault of Gullfaks separates the East Shetland Terraces from the Viking Graben proper, where the structure is least well known. On profiles I and II (Fig. 12.2), the eastern margin of the Viking Graben is relatively gradual, climbing up to the Horda Platform on the Norwegian side of the basin. On profiles III and IV, however, the structure is much more symmetrical, with moderate-sized faults at both edges of the graben. Profile V shows a return to an asymmetric appearance, with most of the extension apparently on a few east-dipping faults. The East Shetland Terraces are absent here, the Platform being separated from the Graben by one major fault of at least 10 km heave. This variation in structural symmetry/asymmetry across the basin challenges Beach's (1986) assertion that the Viking Graben is fundamentally an asymmetric structure. If the deep seismic line discussed by Beach had been shot less than 100 km to the south it would have revealed a largely symmetric graben.

The sedimentary pile in the north Viking Graben is divided by regional unconformities into three major, structurally-controlled sequences:

(1) Triassic–Middle Jurassic: the syn-rift and initial thermal subsidence fill of an early stretching episode;
(2) Upper Jurassic (and locally Lower Cretaceous): sediment fill associated with renewed fault activity; and
(3) Cretaceous–Tertiary thermal subsidence basin, with relatively little faulting.

12.3.1. *Triassic–Mid Jurassic*

The clearest evidence for the early stretching episode comes from seismic data across the Horda Platform (at the east end of the profiles). As discussed by Eynon (1981), Badley and colleagues (1984, 1988), and Lervik and co-workers (1989), wedge-shaped sediment packages overlying a tilted basement topography can be recognized. The major faults defining the basement blocks trend N–S and are spaced *c.* 15 km apart. Only one well (N31/6-1) has reached basement in this area, and found Lower Triassic resting on gneiss.

Further Triassic fault activity is documented from the north-west margins of the Viking Graben. The

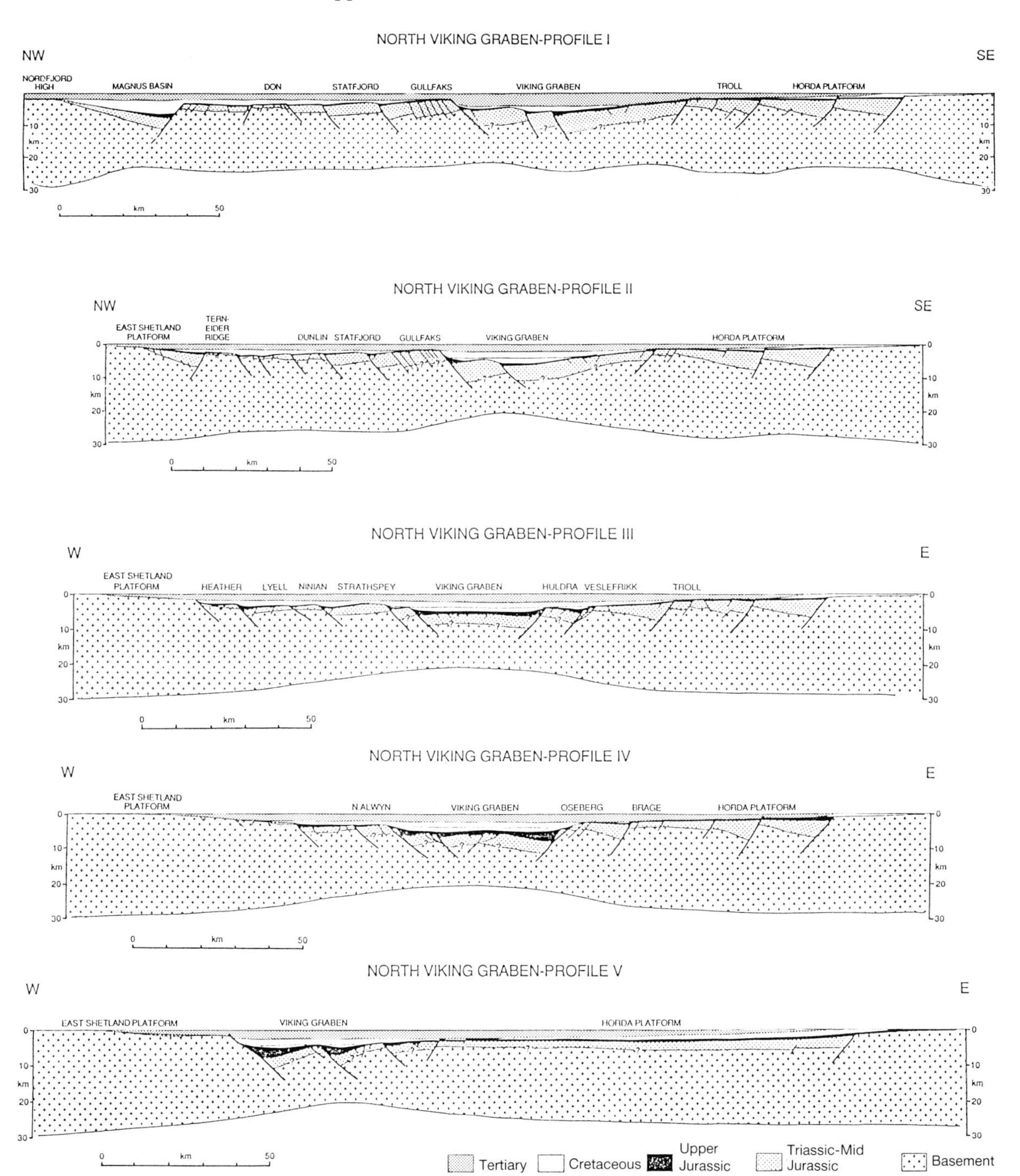

Fig. 12.2. Series of full crustal structural profiles compiled across the Viking Graben. These cross-sections represent a collation of both industry and deep seismic reflection data which has been migrated and depth-converted. Additional constraint on deeper structure is provided by gravity modelling.

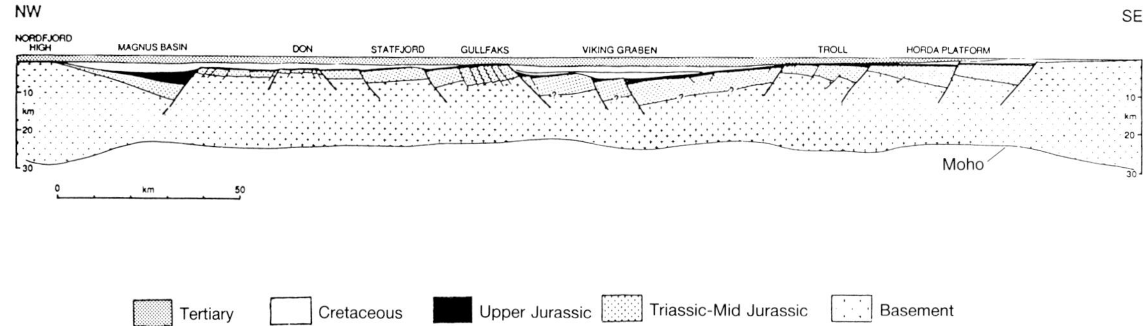

Fig. 12.3. North Viking Graben, Profile I. A crustal-scale structural cross-section through the Viking Graben and Magnus Basin. Numerical modelling described in this paper is principally concerned with this profile.

Unst Basin is a small half-graben within the East Shetland Platform (Fig. 12.1); it contains a wedge of Permo-Triassic red beds up to 3 km thick (Johns and Andrews 1985). Its controlling fault, the Pobie Fault, shows negligible post-Triassic movement. At its north-east end the Pobie Fault has been cross-cut by later Jurassic faulting, but probably continues north-eastwards as the western bounding fault of the Tern-Eider Ridge (see north-west end of profile II, and Roberts *et al*., this volume). Further to the north-west, the Magnus Basin also appears to have begun its development in the Triassic (see north-west end of profile I).

Between these marginal areas (Horda Platform and Unst Basin), the Triassic structure has been overprinted by Late Jurassic faulting and there is no direct evidence for major syn-Triassic faulting and block rotation. However, thickness data from wells for the Triassic and Lower–Middle Jurassic confirm the gross features shown on the interpreted depth profiles. Drilled Triassic thicknesses on the western margin of the basin increase from less than 100 m at the platform edge, to >*c*. 2 km (incomplete section drilled) on the Statfjord fault block (Lervik *et al*., 1989). Thickening of the Triassic eastwards towards the basin centre can be seen in a present-day up-dip direction within individual fault blocks (Speksnijder 1987). Similar patterns are seen for the overlying Lower Jurassic Dunlin Group and the Middle Jurassic Brent Group (Badley *et al*. 1988; Brown *et al*. 1987). For example, Brent Group thickness increases from <100 m at the platform margin to over 350 m near the edge of the Viking Graben proper. Combining the Lower Jurassic, Middle Jurassic, and Triassic units indicates an overall basinward thickening from about 100 m near the western margin (Triassic on basement) to >3 km on the terraces west of the graben axis. In the graben axis itself there is no direct observation of the top of the basement, but by extrapolating thickness trends we estimate that the Triassic–Mid Jurassic sequence may be as much as 5 km thick.

Badley and colleagues (1988) and Giltner (1987) interpreted the general basinwards thickening of the Triassic–Mid Jurassic sequences as indicating deposition during thermal subsidence following the early (?Early Triassic) stretching episode. The Triassic largely comprises continental red beds in monotonous sandstone-mudstone sequences (Fisher 1986; Lervik *et al*., 1989). These pass upwards into the Statfjord Formation, deposited in coastal environments (Røe and Steel 1985), followed by the marine shales of the Dunlin Group. This continental-to-marine transition implies that, during the latest Triassic and Early Jurassic, sedimentation was not keeping pace with the subsidence. The Mid-Jurassic Brent Group, however, consists of a sand-dominated, broadly deltaic sequence, and the common occurrence of coals in its upper part demonstrates a depositional surface at or near sea-level (Brown 1986). The basinward thickening of the Brent Group occurs principally across the major N–S faults (Brown *et al*. 1987), indicating the onset of the second extensional episode.

12.3.2. *Upper Jurassic (sl.)*

The second rifting episode affected the area from the Bathonian to the Ryazanian, i.e. from the latest Middle Jurassic to earliest Cretaceous. It is characterized by wedge-shaped packages of sediment that partially infill a tilted-block topography, itself created by motion on major normal faults. The upper surface of the tilted blocks is generally formed by the top of the Mid Jurassic Brent Group, locally eroded at footwall crests. This surface is onlapped by the marine shales of the Heather Formation which, in turn, is overlain by the Kimmeridge Clay Formation (=Draupne Formation). No regional uplift preceded or accompanied extension, elevation above the pre-faulting regional datum being accomplished only by local uplift of footwalls (Barr 1987). Onset of block tilting was not synchronous throughout the basin, occurring as late as Kimmeridgian on the northern Horda Platform (Badley *et al*. 1988).

The Late Jurassic (sl.) extensional faulting was responsible for creating the most readily-visible structures in the basin, i.e. the tilted fault-blocks that constitute the petroleum traps of the Brent Province. However, the actual amount of extension across these structures is relatively small, as has been noted many times in the literature (Ziegler 1983; Beach *et al*. 1987; Badley *et al*. 1988). Summation of post-Mid-Jurassic fault heaves across our depth profile I (Fig. 12.3) yields an extension of about 22 km; this estimate includes the Magnus Basin. The major normal faults themselves are generally not clearly imaged on seismic reflection data, but have to be inferred from the cut-offs of the displaced sedimentary horizons. Identification of cut-off points is in itself not an entirely objective task, particularly in the deeper parts of the section. Notwithstanding these limitations, on our depth-converted profiles the normal faults generally dip at 45–60° through the sedimentary column. There is, as yet, no clear evidence of their attitude within basement. Horizon dips within the tilted blocks are typically about 5°, and rarely more than 10°.

As discussed in the previous section, faults active in the early Triassic can be identified at the margins of the basin. Those on the Horda Platform (east end of profiles) show a relatively minor reactivation in the Late Jurassic. The Pobie Fault of the Unst Basin was inactive in the Jurassic (Johns and Andrews 1985). However, its continuation as the bounding fault of the Tern-Eider Ridge (northwest end of profile II) shows increasing reactivation northeastwards past the Tern and Eider fields (Speksnijder 1987). Thus, where the Triassic faults are known, there is evidence that at least some of them were used again in the later Jurassic extension. It remains conjectural whether the same is true within the centre of the basin where the top of the basement is not imaged. Although it is possible that early faults may have been rotated, abandoned, and cross-cut by Jurassic faults, we consider it perhaps more likely that the same faults were active in both the earlier Triassic and in the later Jurassic phases of extension.

The marine shales of the Heather Formation and the Kimmeridge Clay Formation form wedge-shaped packages a few hundred metres thick against the major normal faults. Even allowing for subsequent compaction, it is clear that these packages do not fill the half graben. The basin was therefore partially sediment-starved, i.e. sedimentation did not keep pace with the rapid hangingwall subsidence. From the amplitude of the tilted-block topography, Late Jurassic water depths must have been at least several hundred metres in the deeper parts of the half-graben, and possibly of the order of 1 km or more in the axis of the Viking Graben.

12.3.3. *Cretaceous and Tertiary*

The 'Base Cretaceous' seismic marker in the North Sea is conventionally picked at the top of the Kimmeridge Clay Formation, within the late Ryazanian (lowermost Cretaceous). On seismic reflection data the overlying Lower Cretaceous sediments appear to onlap this marker, though a highly condensed Lower Cretaceous sequence is commonly encountered in wells on fault-block crests (Rawson and Riley 1982). Throughout much of the north Viking Graben, offsets of the 'Base Cretaceous' marker are relatively small, indicating that most of the fault-related extension was completed by the end of the Ryazanian. Exceptions to this are the Horda Platform faults, and the large faults bounding the western side of the Viking Graben proper (e.g. Gullfaks fault; *see* profiles I and II).

In contrast to the Late Jurassic sediment wedges, the Lower Cretaceous as seen on seismic data appears to 'pond' in the structural lows. Thinning towards faults on the limb of a compaction syncline, above the underlying Jurassic, is the norm. Sedimentation rates during the Early Cretaceous were very low (Zervos 1986; Nelson and Lamy 1987), and continuing basin subsidence probably resulted in a progressive increase in the water depth through much of the Cretaceous. This trend was reversed in the earliest Tertiary when uplift of the Shetland/Highland block (as the Atlantic opened to the west) provided a major clastic input to the basin. During the Tertiary, sedimentation outpaced subsidence, infilling the basin to give the present relatively shallow water (typically 100–200 m). Total Tertiary-plus-Cretaceous thickness is about 3 km on the East Shetland Terraces, and about 5 km in the graben axis.

Fault activity during the later Cretaceous and Tertiary progressively diminished, becoming restricted to the major faults at the basin margins. A significant part of this late minor movement was probably caused by the continuing compaction of the Mesozoic sedimentary pile.

12.4. A flexural cantilever simple-shear/pure-shear model of continental lithosphere extension and basin formation

Deep seismic reflection profiling (Matthews and Smith 1987; Barazangi and Brown 1986) provides a powerful technique for investigating the fundamental structural architecture that controls continental extensional tectonics. The DRUM deep seismic reflection profile, acquired by BIRPS (McGeary and Warner, 1986) and shown in Fig. 12.4a, clearly shows the critical role of major crustal faults in controlling lithosphere extension and the formation of associated sedimentary basins. These faults, when depth-converted, are shown to be planar (Fig. 12.4b) and penetrate down from the surface into the lower crust. This planar geometry revealed by seismic reflection data for major basement extensional faults is also supported by earthquake seismology (Jackson 1987).

To date no unequivocal example of a major dip-slip fault or shear zone passing continuously from the surface down into the upper mantle, as suggested by the model proposed by Wernicke (1985), has been observed on any deep seismic section (Kusznir and Matthews 1988). The major planar faults imaged on deep seismic data appear to be restricted to the cool, brittle, topmost part of the lithosphere corresponding to the seismogenic layer. Beneath the seismogenic layer (which typically has a thickness between 10 and 15 km) deformation takes place by a plastic rather than by a brittle mechanism (Jackson and McKenzie 1983; Kusznir and Park 1987). Within this region of plastic deformation in the lower crust and mantle, lithosphere extension will be achieved by pure shear (i.e. distributed stretching; McKenzie 1978), rather than by the simple shear (i.e. faulting) of the upper lithosphere. A cartoon summarizing the contrasting modes of deformation by simple shear on planar faults in the upper and middle crust and by pure shear below is shown in Fig. 12.5. The extension by faulting in the upper lithosphere must be balanced by an equal amount of extension by pure shear in the middle and lower lithosphere.

Extensional sedimentary basins form as a consequence of the geometric, thermal, and flexural isostatic response of the lithosphere to extension. Sedimentary basin geometry, crustal structure and subsidence history are controlled by the interaction of these three effects.

1. Geometric response. Lithosphere extension by coupled simple shear/pure shear gives rise to crustal thinning. Simple shear (faulting) thins the upper crust by creating a surficial hole, while pure shear thins the lower crust.
2. Thermal response. Both simple shear and pure shear extension of the lithosphere perturb the lithosphere temperature field. Faulting (simple shear) juxtaposes hot footwall against cool hangingwall, while the pure shear elevates the asthenosphere/lithosphere boundary and increases the geothermal gradient.
3. Flexural isostatic response. Both the thinning of the crust, and the perturbation and subsequent re-equilibration of the lithosphere temperature field modify the lithosphere density field, generating isostatic loads and subsidence/uplift. These isostatic loads are distributed flexurally (Watts *et al*. 1982), as testified by the existence of footwall uplift and fault block rotations. The thermal and crustal thinning isostatic loads are supplemented by sediment fill loading within the basin and erosional unloading of the uplifted flanks. The thermal, sediment fill, and erosion loads vary with time.

While the vertical shear construction (Verrall 1981) is commonly used to model the collapse of hangingwall onto footwall in sediments during extension, this construction is inappropriate for modelling the behaviour of crystalline basement within the hangingwall of the major planar faults, which control continental extension at the lithosphere scale. The vertical shear construction assumes that the hangingwall block is infinitely

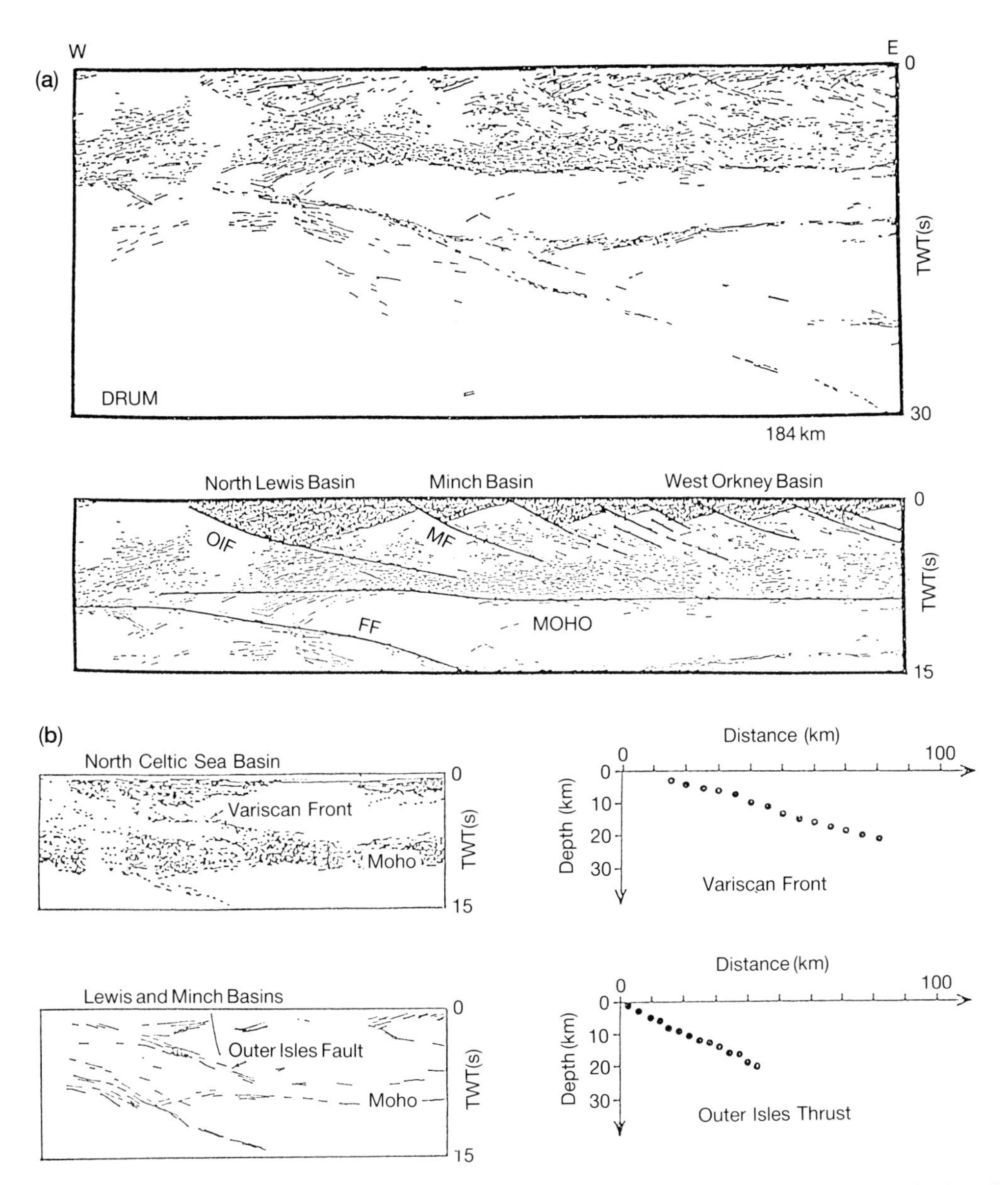

Fig. 12.4. a Line drawing and interpretation of the DRUM (McGeary and Warner 1986) deep seismic reflection profile, shot to the north of Scotland, showing the control of major basement faults on lithosphere extension and the formation of extensional sedimentary basins. OIF = Outer Isles Fault; MF = Minches Fault; FF = Flannan Fault. **b** Line drawings of deep seismic reflection profiles over the North Celtic Sea, and Lewis and Minch basins. Both basins are underlain by dipping reflectors interpreted as major basement faults. When depth-converted using stacking velocities, these major basement faults are shown to be planar from the surface down to 20 km depth.

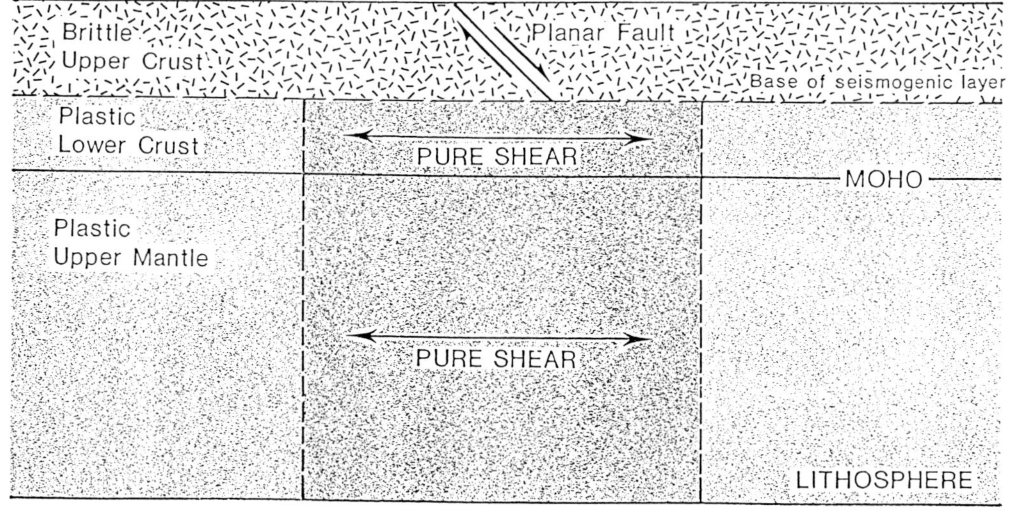

Fig. 12.5. Within the cool, brittle upper crust, lithosphere extension takes place by planar faulting (simple shear). Beneath the brittle/ductile transition, within the plastic lower crust and mantle, lithosphere extension takes place by distributed stretching (pure shear).

weak, and collapses onto a rigid footwall. The assumption that hangingwall crystalline basement is infinitely weak is obviously unrealistic. Rather it is to be expected that basement footwall and hangingwall blocks have similar mechanical properties at the lithosphere scale.

In response to this expectation, a mathematical model has been developed in which both footwall and hangingwall blocks of the major planar faults penetrating the seismogenic layer behave as two mutually self-supporting flexural cantilevers. The planar faulting within the upper lithosphere is assumed to give way beneath 20 km depth to distributed plastic (pure shear) deformation within the lower crust and mantle. All loads are assumed to be flexurally distributed. The mathematical formulation of the geometric, thermal, and flexural isostatic aspects of the flexural-cantilever coupled simple-shear/pure-shear model of continental lithosphere extension is described in greater detail by Kusznir, Marsden, and Egan (in press).

In Fig. 12.6, syn-rift and post-rift crustal structure and basin geometry as predicted by the flexural cantilever model are shown, following lithosphere extension by 5 km on a planar fault dipping at 60°. Footwall uplift and hangingwall subsidence occur through flexure in response to the isostatic forces generated by extension.

In Fig. 12.7a the consequences of lithosphere extension on a set of four planar faults, each with an extension of 5 km, is shown at the syn-rift stage. For the internal fault blocks the constructive and destructive interference of the flexural bending for footwall and hangingwall blocks produces the familiar 'domino-style' block rotations of multiple extensional fault systems. While the internal blocks are rotated they suffer little internal deformation. The tops of the fault blocks are elevated above sea-level.

Figure 12.7b shows the same model but following thermal subsidence. The uplifted fault block tops shown in Fig. 12.7a may be eroded at the syn-rift stage. The resulting basin geometry and crustal structure after thermal subsidence is shown in Fig. 12.7c. A single post-rift thermal subsidence basin is formed above a series of discrete syn-rift sub-basins.

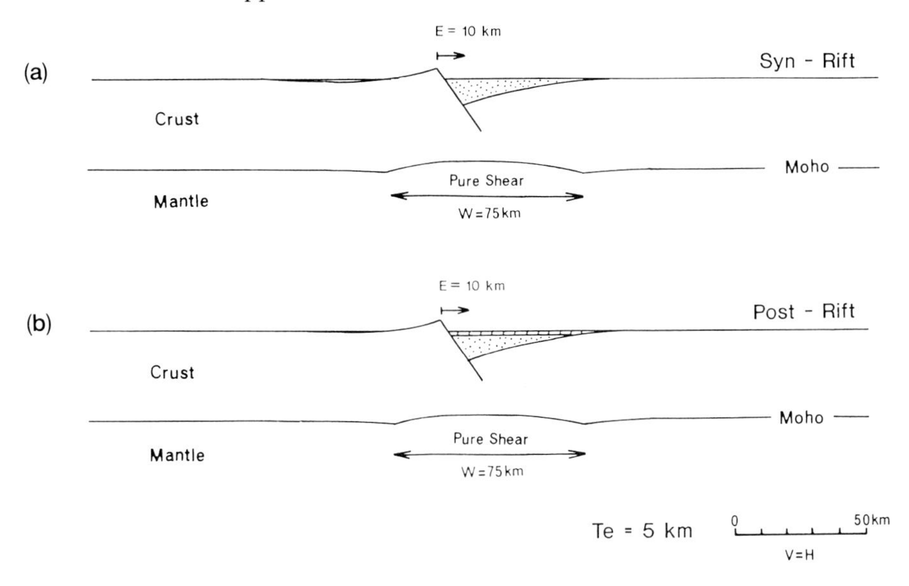

Fig. 12.6. Crustal structure and basin geometry calculated by the flexural cantilever model after 10 km of extension (E) on a single planar fault at both **a** syn-rift and **b** post-rift (150 Ma) stages of basin evolution. Effective elastic thickness (Te) is 5 km. Pure shear width (W) = 75 km. Syn-rift basin fill in dotted ornament, post-rift fill in diagonal ornament.

12.5. The application of the flexural cantilever model to the formation of the Viking Graben

As previously discussed, the tectonic evolution of the Viking Graben can be broadly divided into four components:

(1) Triassic rifting;
(2) post-Triassic rift thermal subsidence;
(3) Late Jurassic rifting;
(4) post-Jurassic rift thermal subsidence.

Any realistic quantitative model of the formation of the Viking Graben must incorporate each of these stages of the basin history.

12.5.1. *Modelling the Triassic rift event and its subsequent thermal subsidence*

It is critical that Triassic rifting and subsequent thermal subsidence are adequately modelled for a number of reasons.

1. To generate the crustal structure that existed prior to Jurassic rifting, in order to establish the gross structural framework upon which all later events are superimposed.
2. To compute the inherited thermal anomaly that existed prior to the Jurassic rift event. Any thermal perturbations of the lithosphere at this time will have the effect of enhancing the Cretaceous–Tertiary thermal subsidence.
3. To predict the thickness and geometry of Triassic syn-rift and post-rift basin fill. The compaction of these sediments, when subjected to further sedimentary loading following Jurassic rifting, is essential to a complete understanding of post-Jurassic basin evolution.

The major problem in modelling Triassic rifting is that of constraint on the Triassic extension and the period of thermal contraction and subsidence which followed it. Notwithstanding these limitations, the Triassic sediment thickness interpreted from seismic data, and by extrapolating thickness trends from well data, has been used as a starting point in modelling (Fig. 12.8).

This profile represents a compacted Triassic–Mid Jurassic sequence which has been dissected and extended by late Jurassic faulting. Decompaction of

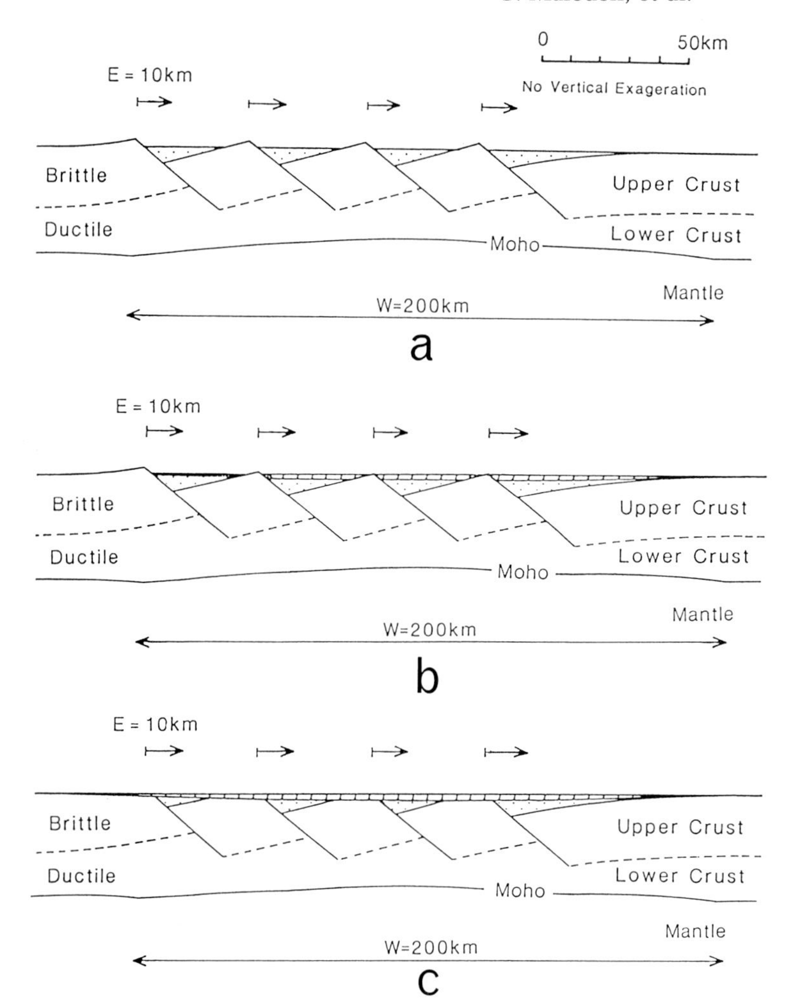

Fig. 12.7. Crustal structure and sedimentary basin geometry following lithosphere extension by 10 km on each of four planar faults. **a**, syn-rift; **b**, post-rift at 100 Ma after extension; **c**, post-rift at 100 Ma after syn-rift erosion of uplifted fault block crests. E = extension, W = pure shear width.

this profile, and restoration to account for the seismically-observed Jurassic extension of *c.* 22 km, yields the profile shown in Fig. 12.8b. This represents the basin geometry existing just prior to Jurassic rifting. It shows Triassic–Mid-Jurassic sediments overlying a tilted block basement topography and thickening from <100 metres at the basin margins to *c.* 7 km at the graben axis.

Decompaction was carried out assuming an exponential relationship between porosity ϕ and depth z such that:

$$\phi = \phi_0 e^{-cz}$$

where ϕ_0 = surface porosity, and c = exponential decay constant. Compaction parameters used were ϕ_0 = 56 per cent and c = 0.39 km^{-1}, representative of a shaley sandstone lithology (Sclater and Christie 1980).

The flexural cantilever model has been applied to this profile in an attempt to reproduce the observed sediment thicknesses. Active rifting may have spanned the whole of the Triassic (250–213 Ma) but, for the purposes of modelling, rifting is taken to be an instantaneous event occurring in the mid-Triassic (230 Ma). Post Triassic rift thermal subsidence is considered to have lasted for 60 Ma before being interrupted by renewed extension in the late Middle Jurassic (*c.* 170 Ma).

The effective elastic thickness, fault extension, and fault position for the Triassic rift models are

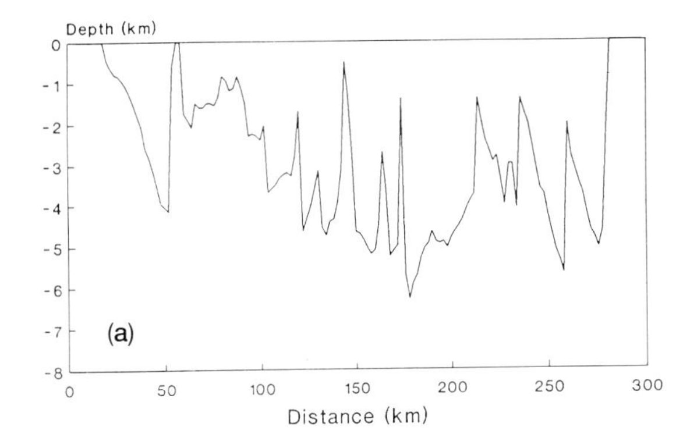

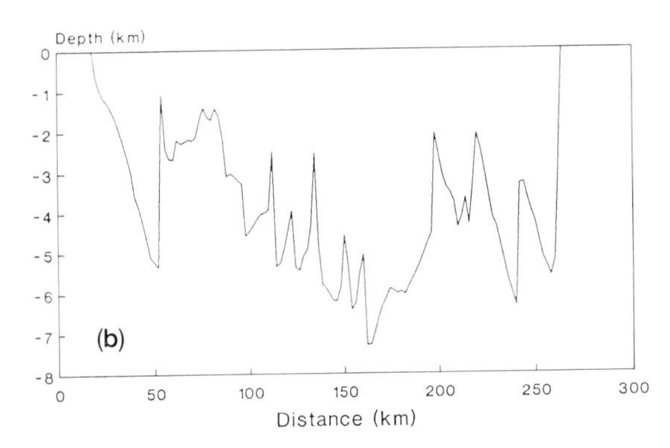

Fig. 12.8. **a** Present-day thickness profile of Triassic–Middle Jurassic sediments along Profile I as inferred from seismic data and extrapolation of thickness trends from well data. **b** Same profile after decompaction and restoration to account for seismically-observed Jurassic fault extension.

considered to be adjustable parameters. They have been determined by iterative adjustment to obtain a best fit between the computed model and the observed/inferred basin geometry and depth. Erosion of surface uplifts is incorporated in the model. Figure 12.9 is the preferred model for Triassic rifting, and shows the computed crustal structure and basin geometry at mid-Jurassic times. The total Triassic extension within the preferred model is *c.* 38 km, with an effective elastic thickness used of *c.* 6 km. The pre-Mesozoic basement at the graben axis has been thinned to *c.* 21 km with a maximum sediment accumulation of *c.* 7 km.

The Viking Graben model at this time is a relatively symmetrical structure with fault block terraces developed at both margins. Basin geometry is that of a series of discrete syn-rift half-graben sediment wedges overlain by a regional Triassic–Mid Jurassic thermal subsidence sequence up to *c.* 3 km thickness. Comparison of the computed sediment thickness profile with that observed (Fig. 12.10) reveals a good match across most of the profile, particularly in the Horda Platform and axial regions. A substantial discrepancy of *c.* 2 km occurs on the East Shetland Platform.

The lithosphere temperature field, perturbed by Triassic rifting, is not fully relaxed at the onset of renewed extension in the Middle Jurassic.

12.5.2. *Modelling the Late Jurassic rift event and its associated thermal subsidence*

Present-day observed crustal profiles allow the identification of the major extensional faults on which displacement occurred during Late Jurassic rifting. These are taken to be the structures which penetrate (or are thought to penetrate) crystalline basement. Fault locations are noted or, in the case of fault-block-crest erosion, fault locations prior to denudation are estimated. The heave of each of these structures is measured from the cut-off positions of a marker horizon at the top of the Middle Jurassic Brent Group, which generally forms the upper surface of the tilted fault blocks. This horizon has been chosen as it records only the post Mid-Jurassic fault displacements.

From the positions and heaves of each of the Jurassic rift faults a simplistic restoration of their starting positions prior to translation has been made. The present-day fault dips have been measured, and corrections made for their rotation during extension. The results of the restoration is a structural framework which can be superimposed on the computed model of post-Triassic crustal configuration.

The flexural cantilever model has been applied to Jurassic rifting assuming extension to be instantaneous and occurring at *c.* 170 Ma in the late Middle Jurassic.

Total fault extension (*c.* 22 km) and fault position for the Jurassic rift event are fully constrained by observation. The perturbed lithosphere temperature field predicted by the Triassic rift model has been carried forward. Rift-related subsidence is known to have been rapid, and far outpaced sedimentation, leading to substantial water depths (Ziegler 1982). This is simulated in the model by the assumption of a syn-rift bathymetry across the whole profile of

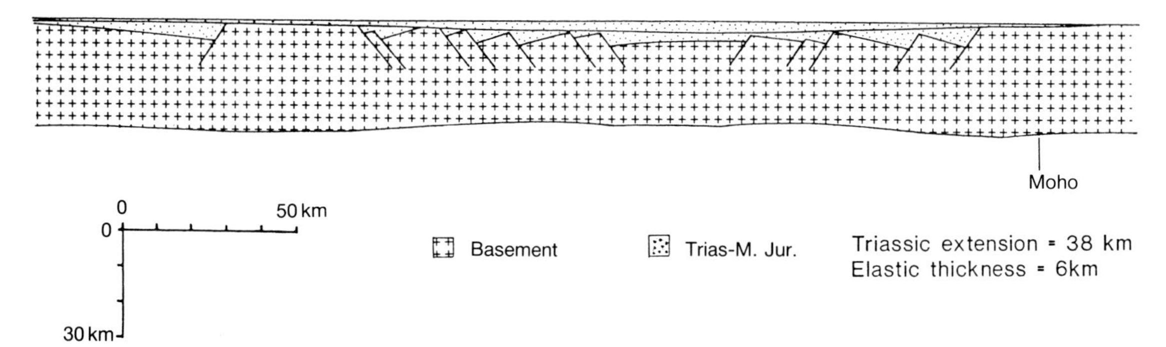

Fig. 12.9. Sedimentary basin geometry and crustal structure predicted for the Viking Graben, Profile I, at the end of active Triassic rifting, and following *c.* 60 Ma of thermal subsidence. Triassic extension = 38 km, and flexural rigidity corresponds to an effective elastic thickness of 6 km. The distribution and geometry of both rift and post-rift sediment sequence is shown.

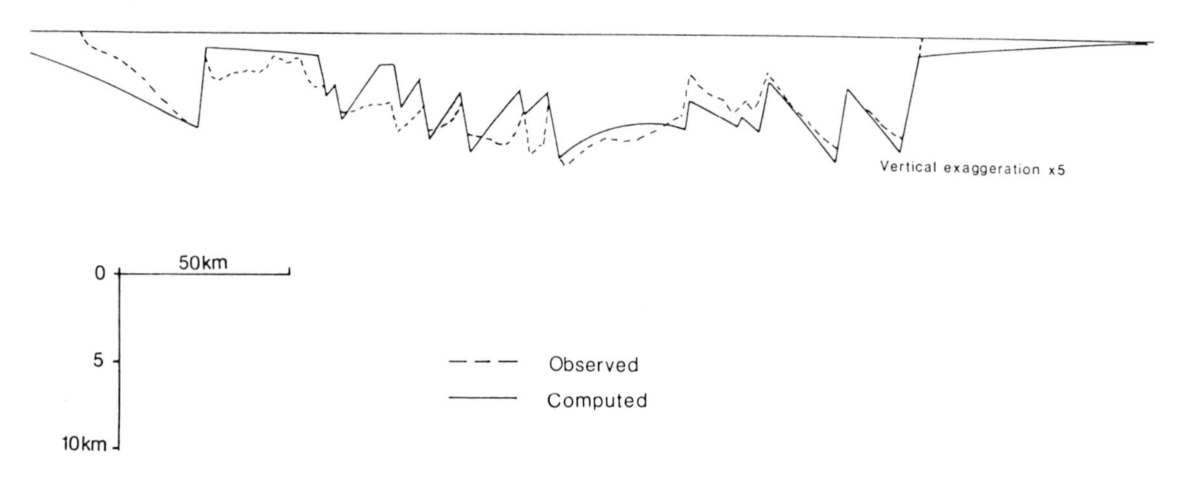

Fig. 12.10. Comparison of the Triassic–Middle Jurassic sediment thickness along Profile I predicted by modelling, and that derived from seismic observation. Vertical exaggeration is ×5.

800 metres. Compaction of earlier sediments is computed according to the exponential equation previously used in decompaction. An effective elastic thickness of 3 km is found to give the best overall match between observed and predicted basin and structural geometries. This is a value in accordance with previous North Sea estimates (Barton and Wood 1984).

Fig. 12.11 shows the computed crustal structure and basin geometry for the preferred model at the syn-Jurassic stage of graben evolution. The thickness of crystalline basement at the graben axis has been reduced to *c.* 15 km. The graben has developed into a more asymmetrical structure, with extension on the easterly-dipping faults bounding the western margin of the basin predominating. Some of the Triassic faults in the model have been abandoned in favour of new Jurassic faults, but many of the Triassic structures have been reactivated (Fig. 12.9). Small, thin, sediment-starved syn-rift sub-basins have developed in the hangingwalls of each of the major faults, the exception being the Magnus Basin where over 3 km of syn-rift sediments have developed in a large half-graben in the hangingwall of the major fault bounding the western end of the East Shetland Terraces.

A small amount of local elevation above sea-level datum is predicted for fault block crests due to

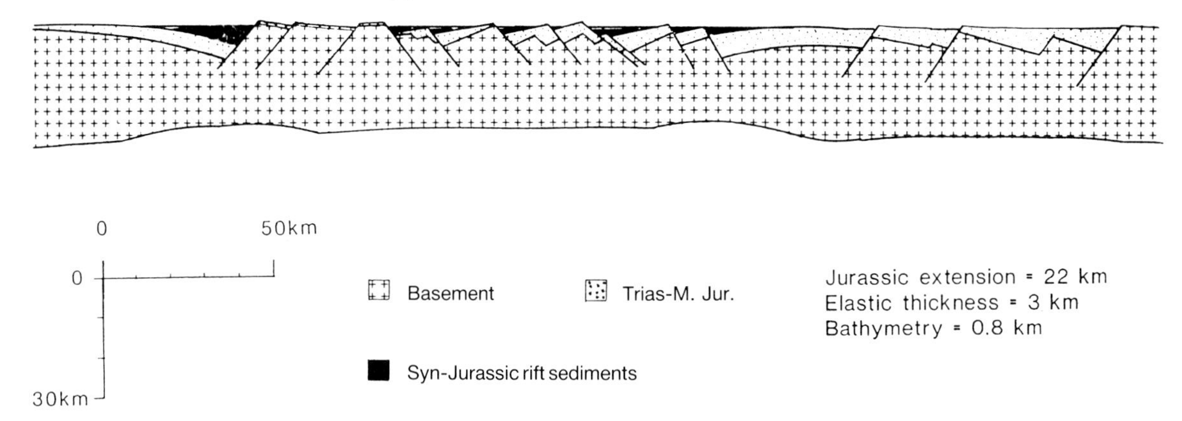

Fig. 12.11. Basin geometry and crustal structure predicted for the Viking Graben, Profile I, at the syn-Jurassic rift stage of evolution. Seismically-observed Jurassic extension = 22 km, and flexural rigidity corresponds to an effective elastic thickness of 3 km. Syn-rift bathymetry = 800 metres.

isostatic footwall uplift. This is seen most markedly on the faults with large displacements where tectonic unloading of the footwall is greater, e.g. Magnus fault, Gullfaks fault.

This syn-Jurassic rift model has been allowed to relax thermally to the present day. Sedimentation since rifting has been assumed to catch up with and then keep pace with post-rift subsidence, such that present day bathymetry is negligible. Sediment filling and loading of the basin throughout the thermal subsidence stage are computed at small time steps, and at each time step all earlier sediments are compacted using an inverse algorithm to that used in decompaction. Sediment filling of the basin is to sea-level. All surface uplifts have been flexurally eroded. The syn-rift effective elastic thickness of 3 km has been used throughout thermal subsidence.

Figure 12.12 shows the preferred present-day model of crustal structure and basin geometry and, at an enlarged vertical scale, Fig. 12.13 compares computed and observed post Mid-Jurassic stratigraphies. Direct comparisons can be drawn between model and observation.

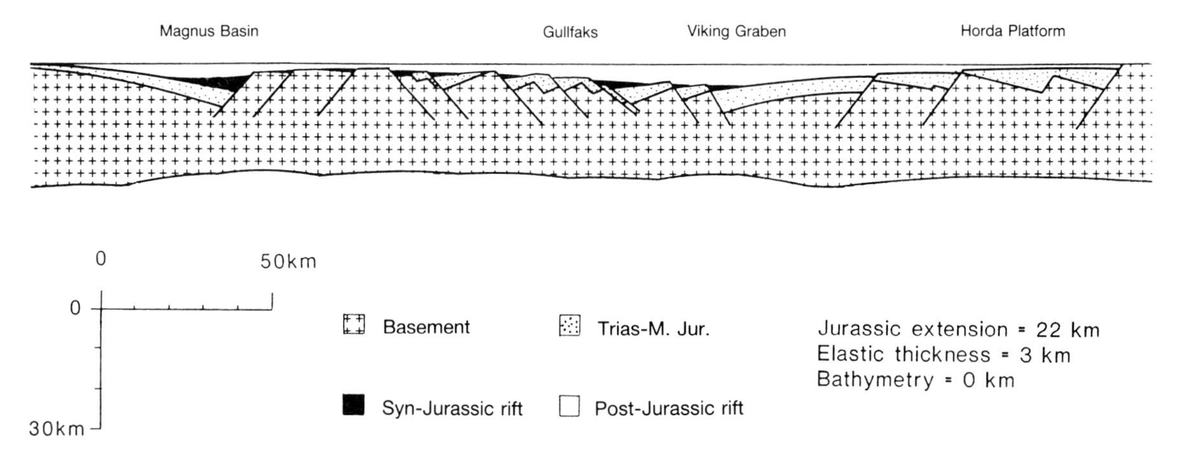

Fig. 12.12. Present-day basin geometry and crustal structure predicted by modelling for the Viking Graben, Profile I. A flexural rigidity corresponding to an effective elastic thickness of 3 km has been used throughout post-Jurassic rift subsidence. Present-day bathymetry is assumed to be negligible.

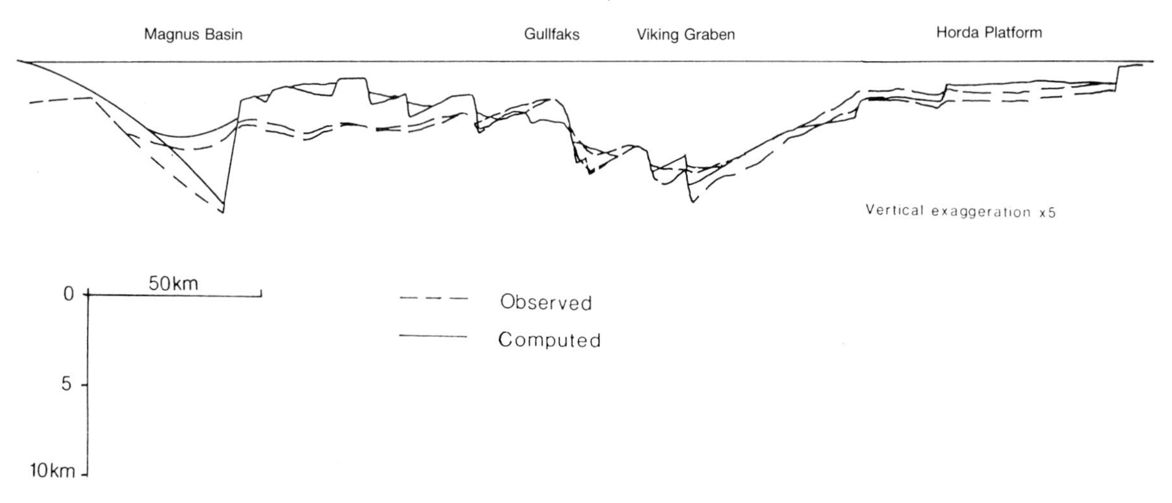

Fig. 12.13. Comparison of the syn- and post-Jurassic rift basin geometry predicted by modelling and that observed from seismic reflection data. Vertical exaggeration is ×5.

1. The model predicts extended crustal thickness in the order of *c.* 14–15 km beneath the graben axis and the Magnus Basin and *c.* 22 km beneath the platform areas.
2. The overall basin architecture and structural geometry agree well with that observed.
3. The model successfully predicts the total basin depth: *c.* 11 km in the Magnus Basin and *c.* 12 km in the Viking Graben.
4. The model reproduces both the observed thickness and distribution of Cretaceous/Tertiary thermal subsidence sedimentation. An exception to this is on the western margin of the East Shetland Terraces, where subsidence is underestimated by *c.* 1.5 km. Subsidence was also substantially underestimated here in the Triassic rift model. This may be due to an underestimate of extension from seismic data due to footwall uplift and erosion and the consequential loss of stratigraphic markers from which extension can be estimated. In addition, an unmodelled major fault to the west of the Magnus Basin would cause additional subsidence in this area.
5. Syn-rift sediment wedge thickness and geometry are similar to observation in the graben axis and Magnus Basin areas. Syn-rift subsidence is underestimated in the region of Troll and on the Horda Platform.
6. The position of the basin Cretaceous post-rift unconformity is reasonably replicated.
7. The amount of fault block erosion predicted is slightly overestimated. One reason for this may be that, in the observed profiles, the fault block crests are seen to be affected by numerous small faults which would have the effect of reducing the block crest elevation. As only the major structures have been used for modelling purposes the predicted block crest elevation will be too high, thus leading to an overestimate of material removed by erosion.
8. The predicted Moho is deeper and exhibits less topography than that interpreted from deep seismic reflection data and gravity inversion (compare Fig. 12.3 with Fig. 12.12).

12.6. Thinning of crustal basement and β estimates

The extensional stretching factor β may be estimated in several ways:

(1) thinning of crustal basement assuming conservation of crustal volume during extension;
(2) basin subsidence; and
(3) summation of seismically-observed fault extensions.

Since β factors determined by different methods have their own definitions, caution should be exercised when comparing β factors derived in different ways.

Crustal basement thickness profiles can be used to determine the crustal stretching β if the assumption is made of a uniform crustal thickness prior to extension. The β estimates will depend on the original (pre-extension) crustal thickness value. In Fig. 12.14, β profiles based on seismically-observed crustal basement thickness (Fig. 12.3) are plotted for original crustal thicknesses of 25 km and 30 km. The β profiles show a bimodal distribution, with peaks across the Viking Graben and Magnus Basin. The maximum observed crustal stretching β values for the Viking Graben are 2.51 and 2.1 for original crustal thicknesses of 30 km and 25 km respectively. These β estimates are obviously highly dependent on where top basement and Moho have been picked on the seismic data.

β profiles predicted by basin modelling constrained by seismically-observed Jurassic fault extensions are shown in Fig. 12.15. β profiles have been determined in two ways, from modelled crustal basement thickness, and from modelling of syn-rift and thermal subsidence. β profiles are shown individually for Permo-Triassic and Jurassic rifting, and for the combined total extension of both rifts. The maximum β values based on modelled crustal thinning are greater than those estimated from subsidence modelling. The maximum β value derived from subsidence modelling is greater for the Triassic rift than for the Jurassic rift. Configuration of the model does not allow the determination of crustal thinning β profiles for Jurassic rifting. Maximum model β values for the combined Triassic and Jurassic rifts are 1.97 and 1.54 for crustal thinning and thermal subsidence respectively. All β profiles show peaks over the Viking Graben and Magnus Basin.

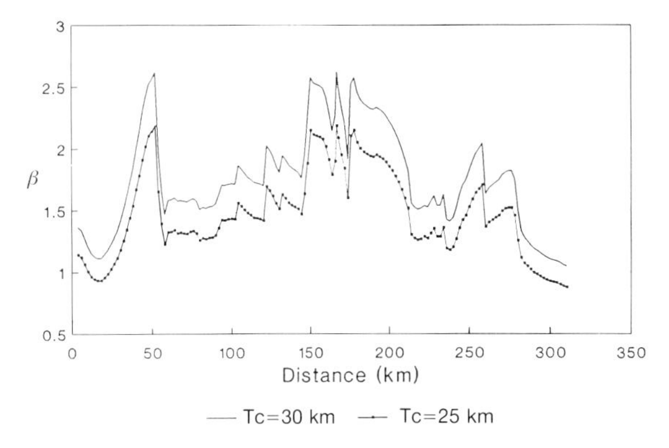

Fig. 12.14. β profiles computed from observed crustal basement thickness data across the Viking Graben and Magnus Basin (Profile I), assuming pre-extension crustal thicknesses (Tc) of 25 km and 30 km.

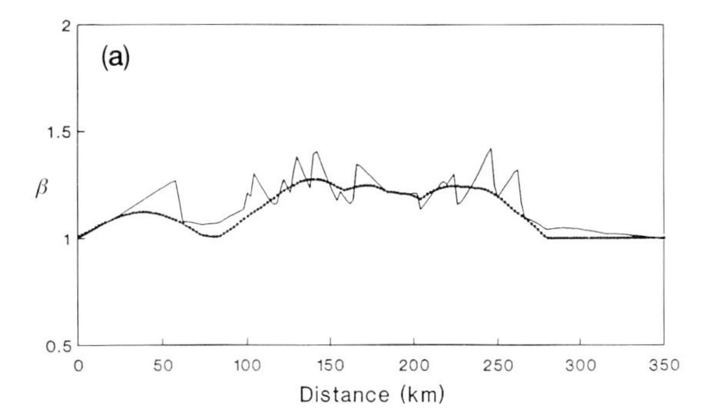

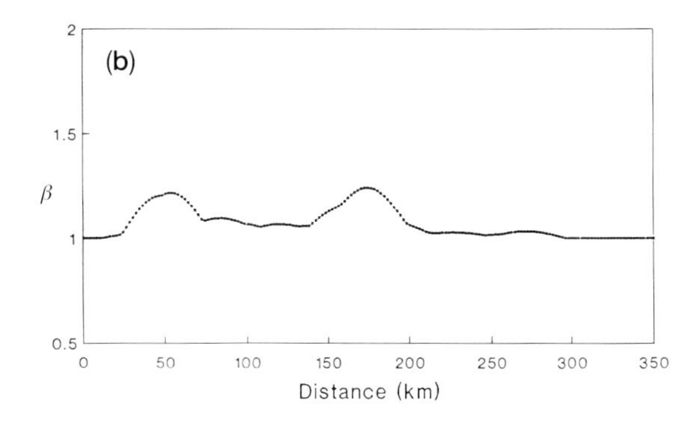

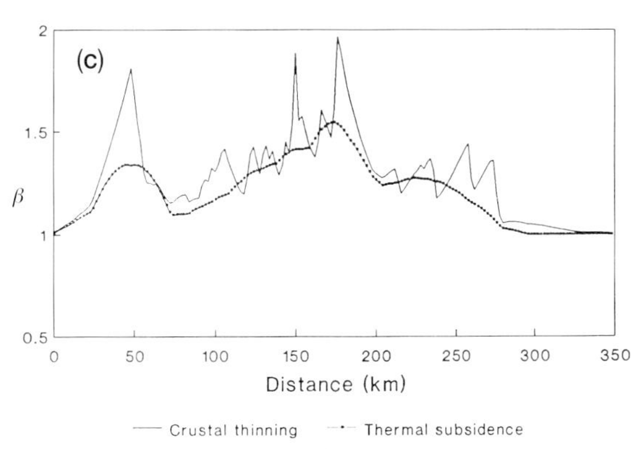

Fig. 12.15. β profiles obtained from basin modelling across the Viking Graben and Magnus Basin. Profiles are shown for **a** Triassic, **b** Jurassic, and **c** combined Triassic/Jurassic rifting. Note that the crustal-thinning β factor could not be determined for Jurassic rifting, due to the configuration of the model.

Table 12.1 Summary of β estimates

β type	Authors	β at rift axis
Crustal thinning β	Ziegler 1982	2.5–3.0
	Beach *et al.* 1987	2.3
	Klemperer 1988	2.1
	This study (observed assuming T_c = 30 km)	2.51
	This study (observed assuming T_c = 25 km)	2.1
	This study (model prediction)	1.97
β derived from thermal subsidence	Beach *et al.* 1987	3.3
	Giltner 1987	1.8–2.0
	Badley *et al.* 1988	1.49
Modelling of syn-rift and thermal subsidence	This study (Permo-Triassic rifting)	1.27
	This study (Total rifting)	1.54
		β along whole profile
Jurassic fault heave summation	Badley *et al.* 1988	1.20
	This study	1.24

β stretching factors for the Viking Graben have been estimated by several workers using a range of different methods. The maximum rift-axis β values derived from previous work are summarized in Table 12.1, which also shows values from this study. Our observed maximum β values derived from crustal thinning (2.10–2.51) and whole-profile β from Jurassic fault heave summation (1.24) are in general agreement with previous estimates.

Crustal thinning β profiles across the Viking Graben, derived from seismically-observed crustal basement thickness and from combined Triassic and Jurassic modelling, are compared in Fig. 12.16. The model-derived crustal thinning β agrees more closely with the seismically-determined β profile if an original pre-extension crustal thickness of 25 km is assumed (Fig. 12.16a), maximum β values being 1.97 and 2.1 for model and observation respectively. If, however, the 30 km crustal thickness observed beneath the basin margins is taken to be indicative of the original crustal thickness, then a substantial discrepancy exists between the crustal-thinning β profiles determined seismically and by subsidence modelling (Fig. 12.16b), maximum values being 2.51 and 1.97.

Various possibilities exist to resolve this paradox.

1. The Triassic and Jurassic extension of the Viking Graben is superimposed on an earlier Devonian rift. The possibility exists that Devonian extension had locally thinned the crust of the Viking Graben region to a thickness of 25 km.
2. The 22 km Jurassic extension determined from seismic reflection is a substantial underestimate. The crustal thinning β profile discrepancy (Fig. 12.16b) of 2.51 versus 1.97 would suggest that the seismically-imaged extension represents only 60 per cent of the total Jurassic extension.
3. Triassic extension, obtained from modelling Triassic sediment thickness, is an underestimate.
4. The observed crustal thinning is an overestimate, as a result of incorrect seismic interpretation of top basement and Moho.

Of these possibilities, 2 and 4 are favoured.

12.7. Conclusions

A total Mesozoic extension of 60 km is estimated for the Viking Graben. This constitutes a predicted Triassic extension from modelling of 38 km and a seismically-observed Jurassic extension of 22 km. This amount of extension has been shown, using the flexural cantilever model, to produce the observed basin geometry and subsidence. Triassic and Jurassic rifting within the models require an effective elastic thickness for flexural isostatic response of 6 km

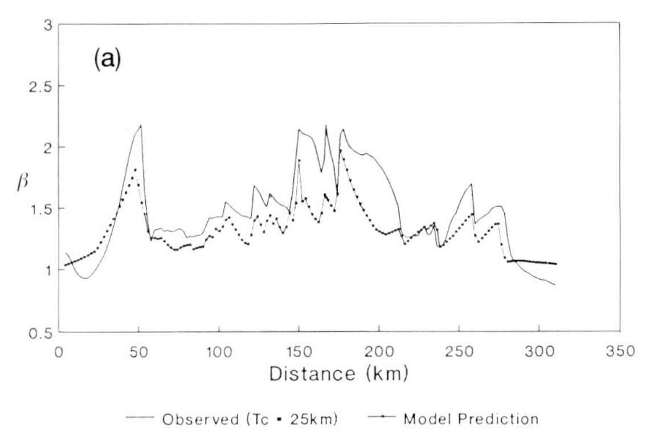

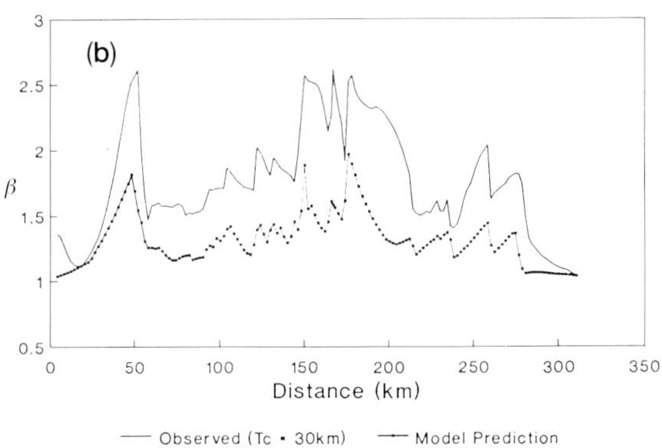

Fig. 12.16. A comparison of observed and modelled crustal thinning β profiles across the Viking Graben and Magnus Basin for pre-extension crustal thicknesses (Tc) of **a** 25 km and **b** 30 km.

and 3 km respectively. These values give the best agreement between model and observation for each stage of basin evolution, and are in general agreement with previous estimates for the North Sea basin (Barton and Wood 1984).

We have attempted to model, as accurately as possible, the observed basin and fault-block geometries. It is realized, however, that because of the numerous variables and uncertainties involved, the modelled solution is non-unique. Given a different set of input parameters, for example additional extension, alternative solutions could be determined.

The flexural cantilever model differs greatly from pure shear models of lithospheric extension (McKenzie 1978; Royden and Keen 1980) in that the rift created by the flexural cantilever model is more localized, being generated by simple shear (faulting) rather than pure shear; it is also deeper, due to the assumption of a finite flexural strength for the lithosphere. The flexural cantilever model, unlike the pure shear models, predicts footwall uplift due to isostatic unloading during rifting. The flexural cantilever model assumes that thermal subsidence is flexurally, and therefore regionally, distributed. These differences between the flexural cantilever and pure shear models account in part for why the flexural cantilever is better able to predict observed structural and basin geometries.

The Cretaceous and Tertiary subsidence within the Viking Graben is not only a consequence of thermal subsidence generated by Jurassic rifting and associated basin fill compaction, but also depends on the earlier Triassic rift event. In order to model the Cretaceous and Tertiary subsidence it is necessary to model Triassic rifting also. The Triassic rift generates the structural framework on which the later Jurassic rift acts. In addition, residual thermal perturbations from the Triassic rift serve to enhance Cretaceous and Tertiary subsidence, as does the compaction of the Triassic rift basin fill when subjected to further sedimentary loading during the Late Jurassic, Cretaceous, and Tertiary.

The modelling shows that, in order to generate the observed, relatively-thin Jurassic syn-rift sequence and thick Cretaceous and Tertiary post-rift sequence, it is necessary to assume that the Jurassic rift was starved of sediment. The effect of substantial bathymetry during rifting is to starve the basin of syn-rift sediments but to enhance post-rift sedimentation.

The crustal thinning β profiles for the combined Triassic and Jurassic rifts, based on industrial and deep seismic reflection data, and the crustal thinning β predicted by the model agree more closely if the original pre-Triassic crustal thickness was 25 km. Such a pre-Triassic thinning may have occurred in the Devonian; however, this hypothesis does not appear to be testable. If the pre-Triassic crustal thickness was 30 km, as suggested by the crustal thickness of relatively unextended basin margins, then, in order to generate the observed crustal stretching β, it is estimated that a further 40 per cent extension is required above that observed seismically. An underestimate of observed fault extension from seismic reflection data across the Viking Graben has been proposed by Watterson and Walsh (pers. comm.) on the basis of fault population statistics and the lower limit of seismic resolution (Childs *et al*., in press).

As mentioned in the previous section, care should be exercised when comparing or discussing β values derived from crustal thinning and subsidence studies. For the combined Triassic/Jurassic rifts the model-predicted β values are substantially less when derived from subsidence ($\beta_{max} = 1.54$) than when derived from crustal thinning ($\beta_{max} = 1.97$). The two different β values are obtained from the same model; the difference arises simply in the way β is defined. The subsidence-derived β equates best with the pure-shear stretching β originally defined by McKenzie (1978). This pure shear β gives the clearest indication of the elevation of the geothermal gradient and the lithosphere/asthenosphere boundary during extension.

The maximum pure-shear β values determined from modelling basin geometry and subsidence in this study are 1.27 and 1.24 for Triassic and Jurassic rifting respectively. These amounts of pure shear extension, for the individual Triassic and Jurassic rifts, would not be expected to generate asthenospheric melting and magmatic underplating of the crust beneath the Viking Graben for normal values of asthenosphere temperature (McKenzie and Bickle 1988). The separation of 60 Ma between Triassic and Jurassic rifting would have allowed the two rifts to behave independently in terms of melt generation.

Acknowledgements

We thank our colleagues at Badley Ashton Associates and within the Department of Earth Sciences, University of Liverpool for discussion and ideas. In addition, we thank Peter Ziegler and Derek Blundell for their encouragement. Gary Marsden was supported in this research by a NERC studentship.

References

Badley, M. E., Egeberg, T., and Nipen, O. (1984). Development of rift basins illustrated by the structural evolution of the Oseberg feature, Block 30/6, offshore Norway. *J. Geol. Soc.*, **41**, 639–49.

Badley, M. E., Price, J. D., Rambech Dahl, C., and Agdestein, T. (1988). The structural evolution of the northern Viking Graben and its bearing upon extensional modes of basin formation. *J. Geol. Soc.*, **145**, 455–72.

Barazangi, M. and Brown, L. (1986). *Reflection seismology: the continental crust*. Washington D.C., *Am. Geophys. Union*, *Geodynamics Series*, **13** and **14**.

Barr, D. (1987). Lithospheric stretching, detached normal faulting and footwall uplift. In *Continental extensional tectonics* (ed. M. P. Coward, J. F. Dewey, and P. L. Hancock). *Geol. Soc. Special Publication*, **28**, 75–94.

Barton, P. and Wood, R. (1984). Tectonic evolution of the North Sea basin: crustal stretching and subsidence. *Geophys. J. Roy. Astron. Soc.*, **79**, 987–1022.

Beach, A. (1986). A deep seismic reflection profile across the northern North Sea. *Nature*, **323**, 53–5.

Beach, A., Bird, T., and Gibbs, A. (1987). Extensional tectonics and crustal structure: deep seismic reflection data from the northern North Sea Viking Graben. In *Continental extensional tectonics* (ed. M. P. Coward, J. F. Dewey, and P. L. Hancock). *Geol. Soc. Special Publication*, **28**, 467–76.

Bowen, J. M. (1975). The Brent oil-field. In *Petroleum and the continental shelf of north west Europe* (ed. A. W. Woodland), 353–60.

Brown, S. (1986). Jurassic. In *Introduction to the petroleum geology of the North Sea* (ed. K. W. Glennie), 133–59. Oxford, Blackwell Scientific Publications.

Brown, S., Richards, P. C., and Thompson, A. R. (1987). Patterns in the deposition of the Brent Group (Middle Jurassic) UK North Sea. In *Petroleum geology of north west Europe* (ed. J. Brooks and K. Glennie), 899–913. London, Graham and Trotman.

Childs, C., Walsh, J. J., and Watterson, J. (1990). A method for estimation of the density of fault displacements below the limits of seismic resolution in reservoir formations. In: *North Sea Oil and Gas reservoirs – II*, 309–18. Graham and Trotman.

Coward, M. P. (1986). Heterogeneous stretching, simple shear and basin development. *Earth and Planetary Science Letters*, **80**, 325–36.

Eynon, G. (1981). Basin development and sedimentation in the Middle Jurassic of the northern North Sea. In *Petroleum geology of the continental shelf of north west Europe* (ed. L. W. Illing and G. D. Hobson), 196–204. London, Institute of Petroleum, Heyden and Son.

Fisher, M. J. (1986). Triassic. In *Introduction to the Petroleum geology of the North Sea* (ed. K. W. Glennie), 113–32. Oxford, Blackwell Scientific Publications.

Gibbs, A. (1987). Deep seismic profiles in the northern North Sea. In *Petroleum geology of north west Europe* (ed. J. Brooks and K. Glennie), 1025–8. London, Graham and Trotman.

Giltner, J. P. (1987). Application of extensional models

to the Northern Viking Graben. *Norsk Geologisk Tidsskrift*, **67**, 339–52.

Jackson, J. A. (1987). Active normal faulting and crustal extension. In *Continental extensional tectonics* (ed. M. P. Coward, J. F. Dewey, and P. L. Hancock). *Geol. Soc. Special Publication*, **28**, 3–17.

Jackson, J. A. and McKenzie, D. P. (1983). The geometrical evolution of normal fault systems: *J. Struct. Geol.*, **5**, 471–82.

Johns, C. and Andrews, I. J. (1985). The petroleum geology of the Unst Basin, North Sea. *Marine and Petroleum Geology*, **2**, 361–72.

Klemperer S. (1988). Crustal thinning and nature of extension in the northern North Sea from deep seismic reflection profiling. *Tectonics*, **7**, 803–32.

Kusznir, N. J. and Matthews, D. H. (1988). Deep seismic reflections and the deformational mechanisms of the continental lithosphere. *J. Petrology, Special Lithosphere Issue*, 66–87.

Kusznir, N. J. and Park, R. G. (1987). The extensional strength of the continental lithosphere: its dependence of geothermal gradient, crustal composition and thickness. In *Continental extensional tectonics* (ed. M. P. Coward, J. F. Dewey, and P. L. Hancock). *Geol. Soc. Special Publication*, **28**, 35–52.

Kusznir, N. J., Marsden, G., and Egan, S. S. (in press). A flexural-cantilever simple-shear/pure-shear model of continental lithosphere extension: applications to the Jeanne d'Arc Basin, Grand Banks and Viking Graben, North Sea. In: *The Geometry of Normal Faults* (ed. A. H. Roberts, G. Yielding, and B. Freeman), *Geol. Soc. Special Publication.*

Lervik, K. S., Spencer, A. M., and Warrington, G. (1989). Outline of the Triassic stratigraphy and structure in the central and northern North Sea. Submitted to *Correlation in petroleum geology* (ed. J. D. Collinson). Norwegian Petroleum Society, 173–89.

McGeary, S. and Warner, M. R. (1986). Seismic profiling the continental lithosphere. *Nature*, **317**, 795–7.

McKenzie, D. (1978). Some remarks on the development of sedimentary basins. *Earth and Planetary Science Letters*, **40**, 25–32.

McKenzie, D. and Bickle, M. J. (1988). The volume and composition of melt generated by extension of the lithosphere. *J. Petrology*, **29**, 625–79.

Matthews, D. H. and Smith, C. (ed.; 1987). Deep seismic reflection profiling of the continental lithosphere. *Geophys. J. Roy. Astron. Soc.*, **89**, 1–447.

Nelson, P. H. H. and Lamy, J. M. (1987). The Møre/West Shetland area: a review. In *Petroleum geology of north west Europe* (ed. J. Brooks and K. Glennie), 775–84. London, Graham and Trotman.

Rawson, P. F. and Riley, L. A. (1982). Latest Jurassic-Early Cretaceous events and the 'Late Cimmerian unconformity' in North Sea area. *Am. Assoc. Petrol. Geologists Bulletin*, **66**, 2628–48.

Røe, S.-L. and Steel, R. (1985). Sedimentation, sea-level rise and tectonics at the Triassic-Jurassic boundary (Statfjord Formation), Tampen Spur, northern North Sea. *J. Petrol. Geology*, **8(2)**, 163–86.

Royden, L. and Keen, C. E. (1980). Rifting processes and thermal evolution of the continental margin of eastern Canada determined from subsidence curves. *Earth and Planetary Science Letters*, **51**, 343–61.

Sclater, J. G. and Christie, P. A. F. (1980). Continental stretching: an explanation of the post-mid-Cretaceous subsidence of the central North Sea basin. *J. Geophys. Research*, **85**, 3711–39.

Solli, M. (1976). En Seismik skorpeunderskelse Norge-Shetland. Unpublished Ph.D. thesis, University of Bergen.

Speksnijder, A. (1987). The structural configuration of Cormorant Block IV in context of the northern Viking Graben structural framework. *Geol. en Mijnbouw*, **65**, 359–79.

Verrall, P. (1981). Structural interpretation with applications to North Sea problems. *JAPEC*, Course Notes, **3**.

Watts, A. B., Karner, G. D., and Steckler, M. S. (1982). Lithosphere flexure and the evolution of sedimentary basins. In *The evolution of sedimentary basins* (ed. P. Kent, M. H. P. Bott, D. P. McKenzie, and C. A. Williams). *Phil. Trans. Roy. Soc., London*, **A305**, 249–81.

Wernicke, B. (1985). Uniform-sense normal simple shear of the continental lithosphere. *Can. J. Earth Science*, **22**, 108–25.

White, N. J. and McKenzie, D. (1988). Formation of the 'steer's head' geometry of sedimentary basins by differential stretching of the crust and mantle. *Geology*, **16**, 250–3.

Wood, R. and Barton, P. (1983). Crustal thinning and subsidence in the North Sea. *Nature*, **302**, 134–6.

Zervos, F. A. (1986). Geophysical investigations of sedimentary basin development: Viking Graben, North Sea. Unpublished Ph.D. thesis, University of Edinburgh.

Zervos, F. (1987). A compilation and regional interpretation of the northern North Sea gravity map. In *Continental extensional tectonics* (ed. M. P. Coward, J. F. Dewey, and P. L. Hancock). *Geol. Soc. Special Publication*, **28**, 477–93.

Ziegler, P. A. (1982). Faulting and graben formation in western and central Europe. *Phil. Trans. Roy. Soc., London*, **A305**, 113–43.

Ziegler, P. A., reply by Wood, R. and Barton, P. (1983). Discussion on: Crustal thinning and subsidence in the North Sea. *Nature*, **304**, 561.

13 Hydrocarbon plays and rifting in the northern North Sea

A. M. Spencer and R. M. Pegrum

Abstract

Exploration in the northern North Sea since 1966 has resulted in some 270 discoveries, with originally recoverable resources of *c*. 8.5×10^9 Sm^3 oil equivalent. These hydrocarbons are associated with a Late Jurassic to Early Cretaceous rift system buried beneath a Cretaceous and Tertiary cover. Late Jurassic, organic-rich, syn-rift mudstones provide the main source rocks but reservoir rocks, principally sandstones, occur in every system from Devonian to Oligocene. Rifting was important in providing structural traps, and post-rift cooling caused the subsidence necessary for hydrocarbon generation.

13.1. Introduction

The pattern of the Late Jurassic to Early Cretaceous rift system of the northern North Sea is revealed by an isopach map of the Upper Jurassic strata (Fig. 13.1), which shows three converging rifts, the Viking, Moray Firth, and Central Grabens. They are asymmetric-faulted troughs, and are flanked by eroded highs, which are often the footwall blocks to the major boundary faults. In this summary article we wish to show the clear relationship between this rift system and the hydrocarbon finds. Rifting also affected the northern North Sea in Triassic times, but had a quite different pattern (Lervik, *et al*. 1989); this earlier rifting has had little influence on the hydrocarbon finds and is not discussed here. Further details on the hydrocarbon plays and finds described here can be found in Pegrum and Spencer (1990).

In the Central Graben rifting began in Callovian times and continued through late Jurassic times, waning during the early Cretaceous. The rifting was transtensional, extensional collapse being accompanied by oblique-slip and strike-slip offsets, which allowed structural inversions in late Jurassic and Cretaceous times. The presence of Zechstein evaporites resulted in widespread detachment of the Triassic and Jurassic 'cover' from the sub-Zechstein 'basement'. The 'cover' deformed by gravity mechanisms, listric normal faults detaching downwards onto the salt, whereas the 'basement' has rotated fault blocks revealing the crustal extension.

In the north-west, near the Middle Jurassic Forties volcanic centre, the Central Graben merges with the Moray Firth rift system. There the different sedimentary fill and the absence of thick Zechstein evaporites resulted in a different tectonic style. Rifting also began later and continued longer into Early Cretaceous times. Complex fault patterns with local compression and inversion indicate that dip-slip movements were accompanied by oblique-slip or strike-slip movements. The Moray Firth rift terminates westward against NNE-trending faults of the Great Glen strike-slip system.

The Viking Graben is structurally simpler, with a northerly trend comprising three NNE-trending *en échelon* elements (Fig. 13.1), termed North, Central and South. Each is a half graben bounded in the west by major normal faults. Syn-rift sequences are 1–3+ km thick on the east side of these faults, whilst on the west, major footwall uplift and erosion occurred. The main rifting phase was late Jurassic. Faulting may have migrated outwards from the rift axis with time, the earliest occurring in late middle Jurassic times and the latest during Early Cretaceous times (Badley *et al*. 1988).

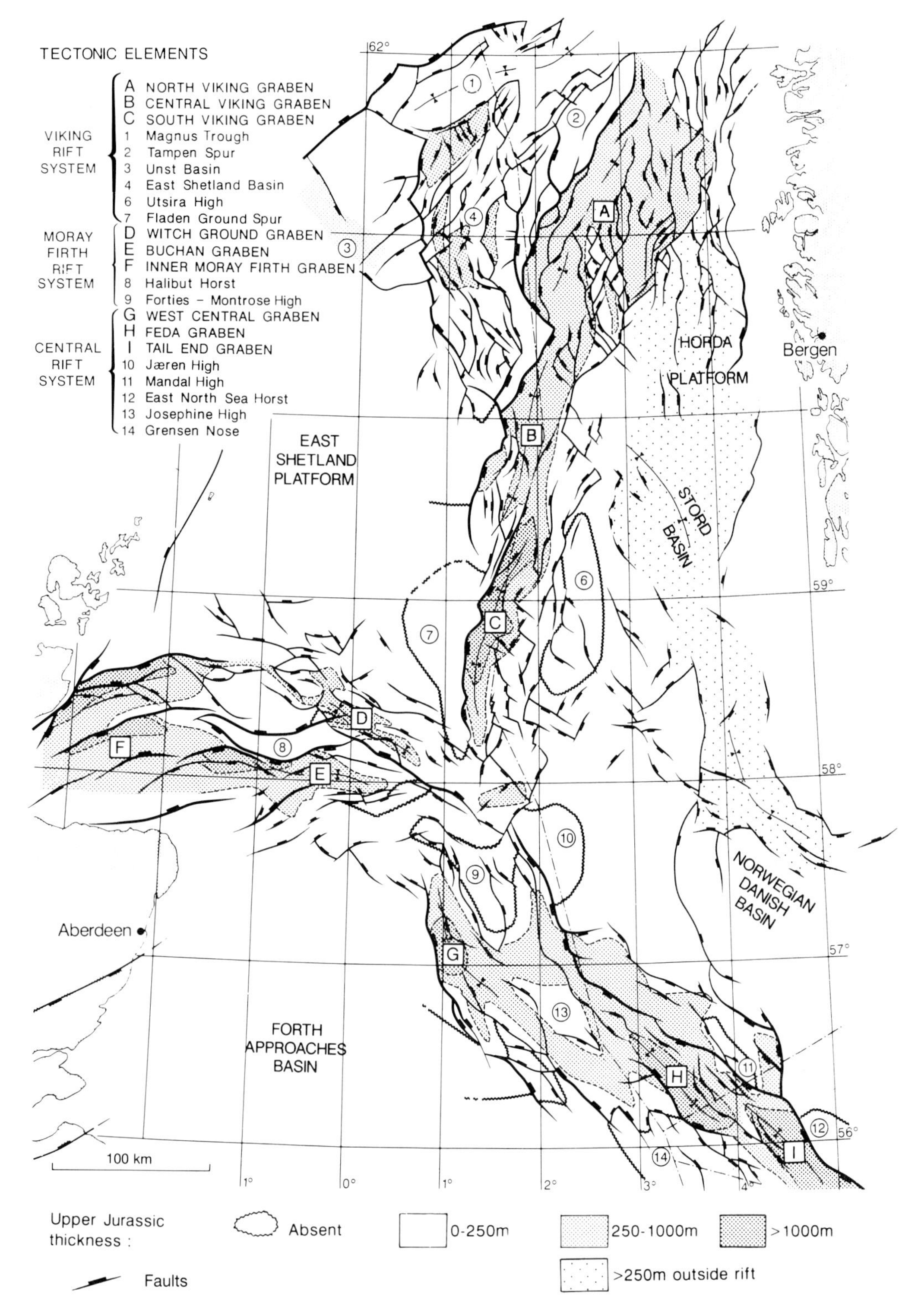

Fig. 13.1. Tectonic map of the late Jurassic to early Cretaceous rift system.

13.2. Source rocks and maturity

Late Jurassic to Early Cretaceous rifting resulted in widespread subsidence (Fig. 13.1), and marine Upper Jurassic to lowest Cretaceous strata are present throughout the basin. The thickest sequences exceed 3 km, whilst incomplete sequences, sometimes only a few metres thick, cover highs and fault blocks. These strata are predominantly mudstones and range from Callovian to Ryazanian in age (Doré *et al.* 1985). At many levels there are black shales with high radioactivity ('hot' shales). Total organic carbon contents are high, ranging up to 15 per cent (Stow and Atkin 1987). Many studies have indicated that the 'hot' shales are the principal hydrocarbon source rocks of the northern North Sea (Van den Bark and Thomas 1980; Schou *et al.* 1985; Mackenzie *et al.* 1987).

The 'hot' shales achieved maturity during burial from Cretaceous to Recent times. The Cretaceous and Tertiary depocentres overlie the rift system, and the continuous regional subsidence is generally assumed to have resulted from post-rift thermal contraction. The area within which the 'hot' shales have generated oil is shown on Figs 13.2, 13.3, and 13.4; it closely follows the locations of the rift zones. Oil generation began over wide areas during Eocene times, and gas generation was achieved in Neogene times. Due to the continuous subsidence, the widest area of generation and the maximum rank of generation occur at the present day. This simple maturity history does not apply in one area, the Inner Moray Firth, where thick Upper Jurassic strata only locally reached the oil-window before maturation was terminated by major Palaeogene uplift. The oil discovered there may have been sourced from Devonian lacustrine rocks.

13.3. Hydrocarbon plays

The hydrocarbon geology can be divided into hydrocarbon plays with different reservoir ages: pre-Jurassic, Lower-Middle Jurassic, Upper Jurassic, Lower Cretaceous, Chalk, and Palaeogene. The tectonic history enables these divisions to be grouped into 'pre-rift, 'syn-rift', and 'post-rift' plays. These two classifications can be applied throughout the northern North Sea, the only uncertainty being in equating the syn-rift group with the Upper Jurassic and Lower Cretaceous divisions, for the onset and close of rifting vary from area to area.

Large areas of the northern North Sea sedimentary basin fall outside the rift system. These include the East Shetland Platform, the Forth Approaches Basin, Horda Platform, Stord Basin, and the Norwegian–Danish Basin. In some of these, thick Upper Jurassic mudstone source rocks are present, with reservoir sandstones and chalks. These areas are largely lacking in hydrocarbons, however, for the potential source rocks have never been buried deeply enough to generate hydrocarbons. The cycle of rift collapse, infill, post-rift cooling, and subsidence is therefore crucial to the hydrocarbon plays.

13.3.1. *Pre-rift plays*

Pre-Jurassic play

The pre-Jurassic play includes fields with reservoirs ranging in age from Devonian (Buchan), to Rotliegendes (Argyll), Zechstein (Auk) and Triassic (Snorre). In all the fields the reservoir is unconformably overlain by Upper Jurassic or Cretaceous strata; they are located in the deeply-eroded highs which formed during the Late Jurassic rifting episode (Fig. 13.2). All are comlicated, eroded fault traps whose structure formed during the Late Jurassic rifting. The fields are close to areas with mature source rocks and have short migration routes.

Lower-Middle Jurassic play

This play is of outstanding importance in the Viking Graben, containing many of the largest fields. The oldest reservoirs are the non-marine to marginal marine sandstones of the Rhaetian to Sinemurian Statfjord Formation. The most important reservoirs are the sandstones of the Brent Group, deposits of a delta system which reached its maximum extent in late Bajocian times and retreated southwards during Bathonian and Callovian times (Graue *et al.* 1987). Middle Jurassic sandstones are important reservoirs as far south as the Beryl Field, but in the south Viking Graben, in the Outer Moray Firth, and in the Central Graben, the rock sequence is different. Middle Jurassic extrusive basalts are locally more than 1 km thick there and are interbedded with fluvial strata. Lower Jurassic rocks are generally absent, and Upper Jurassic or Cretaceous rocks often rest unconformably on eroded Permo-Triassic rocks. The incomplete and local distribution of

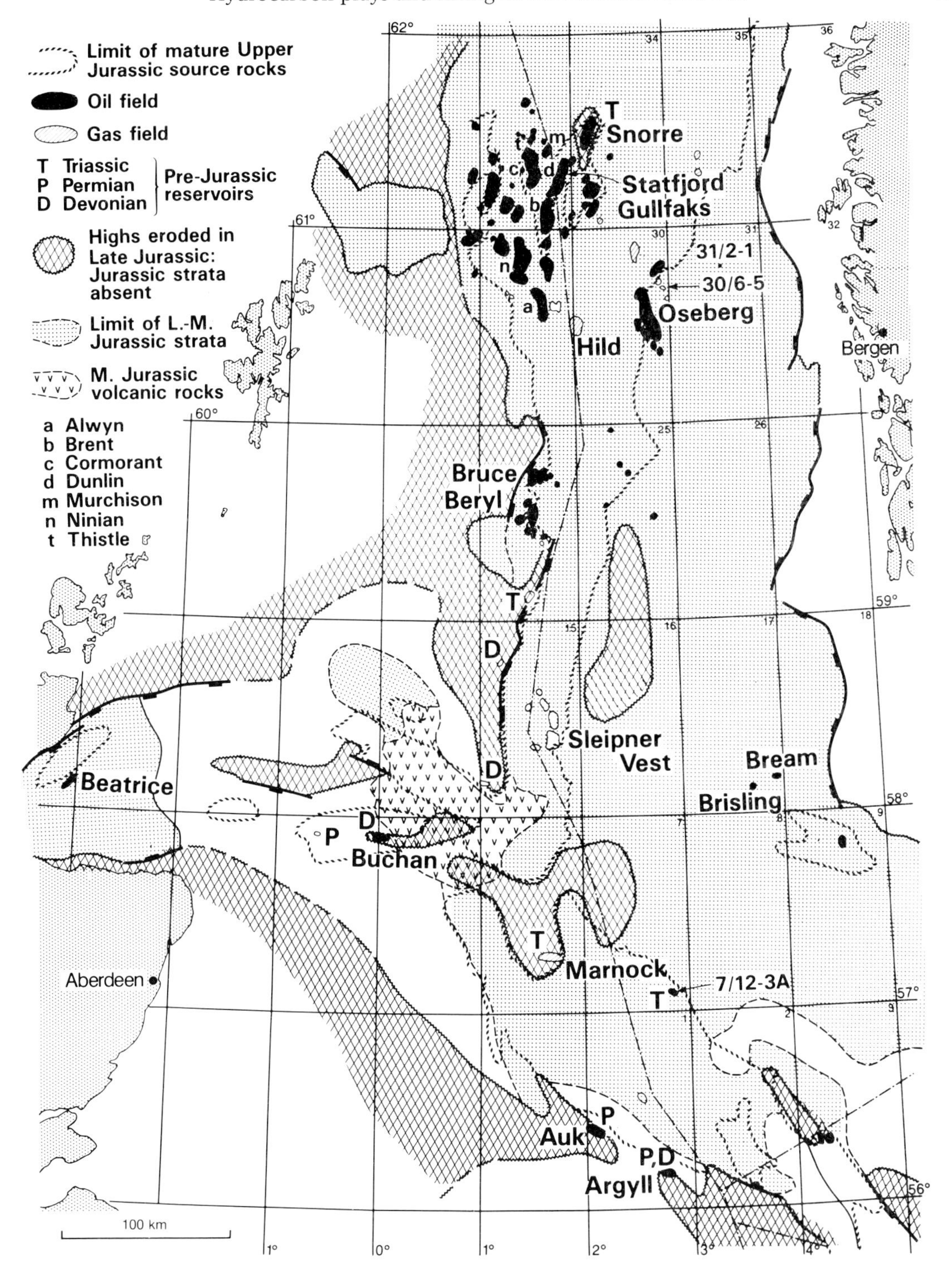

Fig. 13.2. Map of pre-rift plays. Fields with Devonian, Permian, and Triassic reservoirs are located on the highs eroded in late Jurassic times. The remaining fields have Lower and/or Middle Jurassic reservoirs.

Lower and Middle Jurassic strata indicates that the Outer Moray Firth and Central Graben areas were part of an uplifted region prior to Late Jurassic rifting.

The majority of the hydrocarbon traps of the play are tilted fault blocks. On the western flank of the Viking Graben, the rotated fault blocks generally dip to the west; on the eastern flank they dip to the east. Major erosion of the Lower-Middle Jurassic reservoirs, as a result of footwall uplift during Late Jurassic rifting, has occurred on many of the fault blocks (Spencer and Larsen, in press). In addition, some of the fault blocks have 'gravity glide' listric faults detaching at shallow depths in shale intervals (e.g. Statfjord Field). The traps are sealed vertically by unconformable Upper Jurassic or Cretaceous shales. The Upper Jurassic mudstones are the source rocks, so migration routes into the traps were short. Subsidence in Cretaceous and Tertiary times buried the fault blocks without disrupting them, and allowed the source rocks to mature and generate hydrocarbons.

13.3.2. *Syn-rift plays*

Upper Jurassic play

The syn-rift clastic marine strata which accumulated during the rifting movements show great lateral variations in thickness and lithology. As well as containing thick mudstones with source rocks, they include important sandstone reservoirs (Fig. 13.3). In the Outer Moray Firth the sequence begins with pre-rift deltaic strata, overlain by marine sandstones (Piper Formation, Oxfordian-Kimmeridgian). Above come turbiditic sandstones (Claymore Formation, Kimmeridgian to Volgian), which mark the onset of extensional collapse (Harker *et al.* 1987; Boote and Gustav 1987). In the Central Graben, shallow marine sandstones accumulated inside the active graben margins (e.g. Ula Formation, Oxfordian to Kimmeridgian; Spencer *et al.* 1986). In the southern Viking Graben, syn-rift submarine fan and turbiditic conglomerates and sandstones 1–3 km thick accumulated along the giant Brae-trend fault system (Harms *et al.* 1981; Stow *et al.* 1982). Shallow marine sheet sandstones in the Troll area form a sedimentary wedge of Bathonian to Kimmeridgian age, produced by uplift of the Norwegian mainland; they were faulted in Early Cretaceous times. Kimmeridgian sandstones in the Magnus Field were deposited as submarine fans prior to Cretaceous faulting and tilting (De'Ath and Schuyleman 1981).

The traps of the Upper Jurassic fields are very varied. The Troll Field is shallow (only 1000 m beneath the sea floor) but is enormous (770 km^2) and comprises several gently east-dipping fault blocks. The Brae 'trend' continues for 100 km along the down-faulted side of the Fladen Ground Spur and includes more than ten finds. They are classic syn-rift hydrocarbon accumulations, trapped in the hangingwall of a major fault, movement on which was responsible for the supply of the reservoir clastics. In the Outer Moray Firth the Piper Field trap comprises three parallel, tilted fault-blocks; gentle movements started during the deposition of the late Jurassic reservoir sands but the trap formed principally in Early Cretaceous times. The Claymore Field is more complex with oil in Upper Jurassic sandstones, in onlapping Lower Cretaceous sandstones, and even in Permian and Carboniferous rocks (Maher and Harker 1987). In the Central Graben the largest oil field, Fulmar, has a domal trap, perhaps produced by salt withdrawal during Late Jurassic times (Johnson *et al.* 1986). It lies within 4 km of the main graben boundary fault, and so half-graben detachment faulting may also have occurred, as suggested along strike at Clyde Field (Gibbs 1984). On the north-east flank of the Central Graben the Ula Field lies in a fairway containing ten oil finds with traps ranging from domes to hangingwall closures involving stratigraphic truncations.

Lower Cretaceous play

Rifting continued into early Cretaceous times in some areas but movements on many of the major faults had now ceased, so Lower Cretaceous marine mudstones drape and infill the post-rift relief. Cretaceous strata on the highs commonly rest unconformably on older rocks (the 'base Cretaceous unconformity'), but in basin centres there is a continuous sequence up into Cretaceous strata (Rawson and Riley 1982). In the Moray Firth, rifting continued through Early Cretaceous times and sandstone-rich sequences accumulated. Due to Tertiary uplift these rocks are exposed at the sea floor in the Inner Moray Firth, so the only hydrocarbon finds are in the Outer Moray Firth where the reservoirs are sealed and in contact with mature source rocks. In the north-east of the North Sea, minor gas finds occur in stratigraphically trapped, westerly-thickening, submarine fan sandstones.

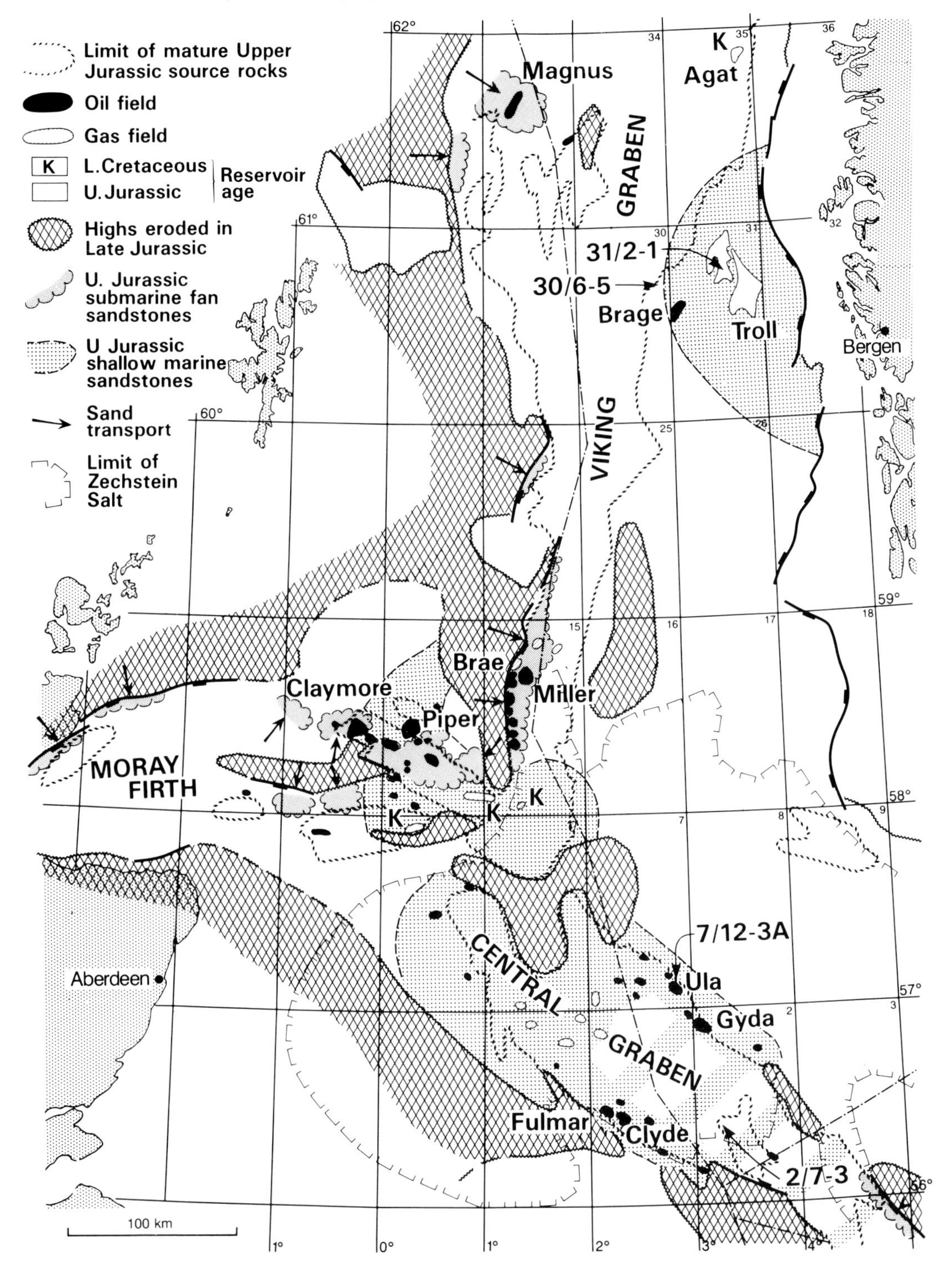

Fig. 13.3. Syn-rift play map. Fields with Lower Cretaceous reservoirs occur only in the Outer Moray Firth and at Agat. All other fields have Upper Jurassic reservoirs.

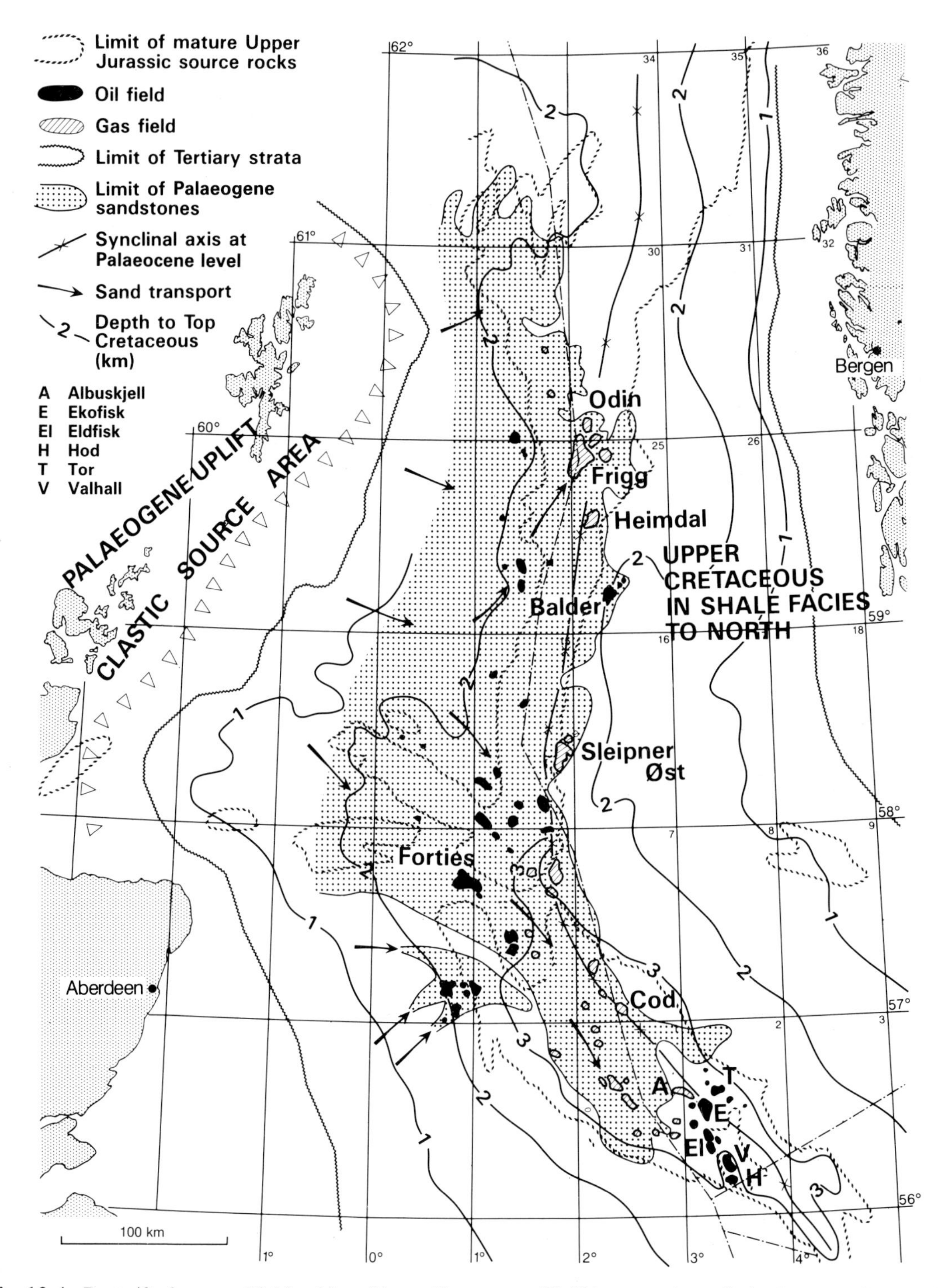

Fig. 13.4. Post-rift play map. Fields with an Upper Cretaceous (Chalk) reservoir are limited to a small area where the chalk is thick, deeply buried, overpressured, underlain by mature source rocks and not overlain by Paleogene sandstones. Note that many of the Paleogene fields are not directly underlain by mature source rocks.

13.3.3. *Post-rift plays*

Upper Cretaceous play

Rifting had essentially ceased by the end of Early Cretaceous times. Later subsidence was greatest along the axes of the earlier rifts and there Upper Cretaceous strata locally exceed 1000 m. In the Central Graben inversion movements were widespread, indicating continued but localized fault activity.

Upper Cretaceous to Danian strata are mostly in a chalk facies as far north as 57°N, but only in a small area in the Central Graben does the chalk contain hydrocarbons. There it is an unexpectedly good reservoir, due to the early migration of oil and the development of high overpressures (Hardman 1982). Depositional controls are also important; the most porous intervals are in debris-flow chalks, which accumulated near rising salt and inversion structures and along the oversteepened flanks of the trough (Kennedy 1987). All the chalk fields have traps with four-way dip closure, but stratigraphic variations control the distribution of the productive reservoir zones. Salt-supported chalk fields include Albuskjell, Ekofisk, Edda, Eldfisk, Tor, and Tommeliten (D'Heur 1987; Pekot and Gersib 1987). Valhall and Hod fields occupy local culminations on the Lindesnes inversion ridge.

The Chalk play is delimited by the pinch-out line of the Palaeogene sandstones (Fig. 13.4) which provide an escape route for hydrocarbons and prevent the development of overpressure (Cayley 1987) and the retention of porosity in the subjacent chalk.

Paleogene play

Basin-wide subsidence continued into Tertiary times, the depocentre approximately coinciding with the buried rift zone. The Tertiary sequence is up to 3 km thick and is dominated by mudstones, but deltaic, submarine fan, and turbiditic sandstones are present. Palaeocene and Lower Eocene sandstones are important reservoirs. They were input as a result of uplift and erosion of the north of Scotland area, associated with the initial opening of the North Atlantic Ocean. Towards the east these sandstones thin and pass into marine mudstones.

The early Tertiary basin axis exerted an important control on the submarine fan sands, for the turbidity currents were unable to flow far to the east of the axis.

Several trap types occur. Despite a lack of regional tectonism, many of the traps are structural in that they have four-way dip closure. The Forties and Montrose fields are drape structures over buried, older highs. Other fields lie near the eastern pinch-out of the sandstones (Balder, Sleipner Øst); domal traps occur above Zechstein salt diapirs (Cod).

All the Palaeogene fields have been ultimately sourced from Jurassic source rocks. Fields located near the graben axes (e.g. Frigg, Heimdal, Cod) overlie deeply-buried Upper Jurassic source rocks, and vertical migration paths are probable (Goff 1983). In other fields (e.g. Forties, Balder, Sleipner Øst) mature Upper Jurassic source rocks are absent vertically beneath the accumulations; lateral hydrocarbon migration within the Palaeogene sandstones has taken place.

Acknowledgements

The authors acknowledge with thanks the permission of Statoil to publish this article.

References

Badley, M. E., Price, J. D., Rambech Dahl, C., and Agdestein, T. (1988). The structural evolution of the northern Viking Graben and its bearing upon extensional modes of basin formation. *J. Geol. Soc.*, **145**, 455–72.

Boote, D. R. D. and Gustav, S. H. (1987). Evolving depositional systems within an active rift, Witch Ground Graben, North Sea. In *Petroleum geology of north west Europe* (ed. J. Brooks and K. Glennie), 819–34. London, Graham and Trotman.

Cayley, G. T. (1987). Hydrocarbon migration in the central North Sea. In *Petroleum geology of north west Europe* (ed. J. Brooks and K. Glennie), 1029–38. London, Graham and Trotman.

De'Ath, N. G. and Schuyleman, S. F. (1981). The geology of the Magnus oilfield. In *Petroleum geology of the continental shelf of north-west Europe* (ed. L. V. Illing and G. D. Hobson), 342–51. London, Institute of Petroleum, Heyden and Son.

D'Heur, M. (1987). Albuskjell. In *Geology of the Norwegian oil and gas fields* (ed A. M. Spencer *et al.*), 51–62. London, Graham and Trotman.

Doré, A. G., Vollset, J., and Hamar, G. P. (1985). Correlation of the offshore sequences referred to the Kimmeridge Clay Formation—relevance to the Norwegian sector. In *Petroleum geochemistry in exploration of the Norwegian shelf* (ed. B. M. Thomas), 27–37. London, Graham and Trotman.

Gibbs, A. D. (1984). Clyde field growth fault secondary detachment above basement faults in the North Sea. *Am. Assoc. Petrol. Geologists Bulletin*, **68**, 1029–39.

Goff, J. C. (1983). Hydrocarbon generation and migration from Jurassic source rocks in the Shetland Basin and Viking Graben of the northern North Sea. *J. Geol. Soc.*, **140**, 445–74.

Graue, E., Helland-Hansen, W., Johnson, J., Lømo, L., Nøttvedt, A., Rønning, K., Ryseth, A., and Steel, J. (1987). Advance and retreat of Brent delta system, Norwegian North Sea. In *Petroleum geology of north west Europe* (ed. J. Brooks and K. Glennie), 915–37. London, Graham and Trotman.

Hardman, R. F. P. (1982). Chalk reservoirs of the North Sea. *Bulletin, Geol. Soc. of Denmark*, **30**, 119–37.

Harker, S. D., Gustav, S. H., and Riley, L. A. (1987). Triassic to Cenomanian stratigraphy of the Witch Ground Graben. In *Petroleum geology of north west Europe* (ed. J. Brooks and K. Glennie), 809–18. London, Graham and Trotman.

Harms, J. C., Tackenberg, P., Pollock, R. E., and Pickles, E. (1981). The Brae oilfield area. In *Petroleum geology of the continental shelf of north west Europe* (ed. L. W. Illing and G. D. Hobson), 352–7. London, Institute of Petroleum, Heyden and Son.

Johnson, H. D., Mackay, T. A., and Stewart, D. J. (1986). The Fulmar oil-field (central North Sea): geological aspects of its discovery, appraisal and development. *Marine and Petroleum Geology*, **3**, 99–125.

Kennedy, W. J. (1987). Sedimentology of late Cretaceous–Paleocene chalk reservoirs, North Sea Central Graben. In *Petroleum geology of north west Europe* (ed. J. Brooks and K. Glennie), 469–82. London, Graham and Trotman.

Lervick, K. S., Spencer, A. M., and Warrington, G. (1989). Outline of Triassic stratigraphy and structure in the central and northern North Sea. In *Correlation in petroleum geology* (ed. J. D. Collinson), pp. 173–89. London, Graham and Trotman.

Mackenzie, A. S., Leythauser, D., Muller, P., Radke, M., and Shaefer, R. G. (1987). The expulsion of petroleum from Kimmeridge clay source rocks in the area of the Brae oilfield, UK continental shelf. In *Petroleum geology of north west Europe* (ed. J. Brooks and K. Glennie), 865–78. London, Graham and Trotman.

Maher, C. E. and Harker, S. D. (1987). Claymore oil field. In *Petroleum geology of north west Europe* (ed. J. Brooks and K. Glennie), 835–46. London, Graham and Trotman.

Pegrum, R. M. and Spencer, A. M. (1990). Hydrocarbon plays in the northern North Sea. In *Classic petroleum provinces* (ed. J. Brooks). *Geol. Soc. Special Publication* **50**, 441–70.

Pekot, L. J. and Gersib, G. A. (1987). Ekofisk. In *Geology of the Norwegian oil and gas fields* (ed. A. M. Spencer *et al.*), 73–87. London, Graham and Trotman.

Rawson, P. F. and Riley, L. A. (1982). Latest Jurassic–early Cretaceous events and the 'Late Cimmerian unconformity' in the North Sea area. *Am. Assoc. Petrol. Geol. Bulletin*, **66**, 2628–48.

Schou, L., Eggen, S., and Schoell, M. (1985). Oil-oil and oil-source rock correlation, northern North Sea. In *Petroleum geochemistry in exploration of the Norwegian shelf* (ed. B. M. Thomas *et al.*), 101–17, London, Graham and Trotman.

Spencer, A. M. and Larsen, V. B. (in press). Fault traps in the northern North Sea. In *Tectonic events responsible for Britain's oil and gas* (ed. R. F. P. Hardman). *Geol. Soc. Special Publication*.

Spencer, A. M., Home, P. C., and Wiik, V. (1986). Habitat of hydrocarbons in the Jurassic Ula trend, Central Graben, Norway. In *Habitat of hydrocarbons on the Norwegian continental shelf* (ed. A. M. Spencer *et al.*), 111–27, London, Graham and Trotman.

Stow, D. A. V. and Atkin, B. P. (1987). Sediment facies and geochemistry of Upper Jurassic mudrocks in the Central North Sea area. In *Petroleum geology of north west Europe* (ed. J. Brooks and K. Glennie), 797–808. London, Graham and Trotman.

Stow, D. A. V., Bishop, C. D., and Mills, S. T. (1982). Sedimentology of the Brae oilfield, North Sea: fan models and controls. *J. Petrol. Geol.*, **5**, 129–48.

Van den Bark, E. and Thomas, O. D. (1980). Ekofisk: first of the giant oilfields in Western Europe. *Am. Assoc. Petrol. Geol., Memoir*, **30**, 195–224.

Index